Origins of Life on the Earth and in the Cosmos

SECOND EDITION

Origins of Life on the Earth and in the Cosmos

SECOND EDITION

Geoffrey Zubay

Professor of Prebiotic Chemistry
Fairchild Center for the Biological Sciences
Columbia University
New York, NY 10027

ACADEMIC PRESS

A Harcourt Science and Technology Company

San Diego   San Francisco   New York   Boston   London   Sydney   Tokyo

Cover photo credit: Images courtesy of NASA.

This book is printed on acid-free paper. ⊗

Copyright © 2000, 1996 by ACADEMIC PRESS

All Rights Reserved.
No part of this publication may be reproduced or transmitted in any form or by any means, electronic or mechanical, including photocopy, recording, or any information storage and retrieval system, without permission in writing from the publisher.

Requests for permission to make copies of any part of the work should be mailed to: Permissions Department, Harcourt, Inc., 6277 Sea Harbor Drive, Orlando Florida 32887-6777

Academic Press
A Harcourt Science and Technology Company
525 B Street, Suite 1900, San Diego, California 92101-4495, U.S.A.
http://www.apnet.com

Academic Press
Harcourt Place, 32 Jamestown Road, London NW1 7BY, UK
http://www.hbuk.co.uk/ap/

Harcourt/Academic Press
A Harcourt Science and Technology Company
200 Wheeler Road, Burlington, Massachusetts 01803
http://www.harcourt-ap.com

Library of Congress Catalog Card Number: 99-64634

International Standard Book Number: 0-12-781910-X

PRINTED IN THE UNITED STATES OF AMERICA
99 00 01 02 03 04 EB 9 8 7 6 5 4 3 2 1

CONTENTS

PART II   **Logic of Living Systems**

6. Cells, Organelles, and Biomolecules

7. Metabolic Strategies and Pathway Design

PART III   **Biochemical and Prebiotic Pathways: A Comparison**

17. Early Developments in Polypeptide Synthesis

18. Lipid Metabolism and Prebiotic Synthesis of Lipids

19. Properties of Membranes and Their Evolution

PREFACE

The goal of this book is to describe key events in the evolution of living systems with an emphasis on the early events that led to the origin of life and other precellular events. The level of presentation has been designed to make the text most suitable for college students who are looking beyond the basic curriculum to a subject that is in the forefront of research and development. It is difficult to specify the prerequisites for such a course because it skims over many disciplines: geology, astrophysics, chemistry, molecular biology, and evolution. I would suggest exposure to a 1-year college chemistry course because the emphasis in the main part of this text is on the chemistry of the origin of life. One or two terms of biology also would be helpful but are not considered absolutely essential. The coverage given to the other scientific disciplines is done at a level where it should be understood without formal training in these subjects.

HISTORY OF THE SUBJECT

In 1905, the astronomer Simon Newcomb proposed that because Earth was a representative planet orbiting a representative star, life must be abundant throughout the universe. Around the same time, in 1903, Svante Arrhenius proposed that life on Earth was seeded by spores originating from another planet outside our solar system. Theories of panspermia, that life on Earth did not originate here but was transported here from another planet, have been elaborated on by others to include planned voyages by advanced civilizations.

Wherever the first living system did arise, all fossil and biochemical information points to a gradual evolution of complex forms of life from simpler single cell organisms (prokaryotes) over a period of a few billion years. Although these findings do not preclude the possibility of panspermia, they are consistent with the view that life arose on Earth. Indeed, it is difficult to imagine a planet that would have been more ideal for the origin of life than Earth.

In the late 1920s, the English biologist J. B. S. Haldane and the Russian chemist A. I. Oparin independently suggested that life may have originated from abiological materials on this planet. It was not until the 1950s, however, that serious experiments were performed to test the idea that biological molecules could be reconstructed from abiological materials. In 1950, Melvin Calvin at Berkeley, and shortly afterward Harold Urey at the University of Chicago, initiated such "prebiotic" experiments. In both cases the experiments consisted of mixing together simple compounds to make organic molecules of biological significance. For instance, in Urey's laboratory Stanley Miller mixed methane, ammonia, hydrogen, and water together; after passing an electric discharge through the mixture for a considerable length of time, he was able to detect certain simple amino acids. Although these early experiments were crude and possibly not performed under realistic prebiotic conditions, they were significant because they represented a beginning of the experimental approach to the study of the origin of life.

There has been a slow growth of laboratories engaged in the origin of life research. By the late 1960s there were about 20 laboratories involved in such experiments and today there are probably around 100.

There could be no better time to take this subject seriously. Currently we have a detailed description of over half of the reactions that take place in the simplest living cells and we have a deep enough understanding of organic chemistry to arrive at realistic ideas of how life is likely to have originated. I suspect that we are only a few years away from very plausible model systems for the origin of life.

ORGANIZATION OF THE TEXT

This text consists of 24 chapters divided into four parts. In Part I, which consists of five chapters, the events that occurred between the origin of the universe (15 billion years ago) and the formation of Earth (about 4.6 billion years ago) are described. Earth is pictured as a planet that was ideally suited for the origin of life.

Part II contains four chapters that deal with the basic strategies of living systems that are well known to us because of the scientific accomplishments of the twentieth century.

In Part III, which contains 11 chapters, specific aspects of biochemistry are compared with the types of chemistry that occurred around the time of the origin of life. This comparison is interesting both for its similarities and for its differences. We can point only to a few very close parallels between the biochemical world and the prebiotic world but the number of such parallels appears to be on the increase as we learn more.

Part IV consists of four chapters that deal with general and specific questions on biological evolution. The evolution of organisms is considered in Chapter 21, while the evolution of specific processes within living systems is dealt with in the final three chapters.

Geoffrey Zubay

PART I

Creation of an Environment Suitable for the Origin of Life

The first part of this book contains brief chapters and one not-so-brief chapter in which the primary concerns are the events that took place between the time of the origin of the Universe (15 billion years ago) and the time just prior to the origin of life on Earth (3.8 billion years ago.) The Universe began with the explosion of a dense ball of energy that rapidly became converted into matter that has been moving away from the center of the explosion ever since. Chapter 1 presents some of the evidence for this view as of 2 years ago. I suspect that this view will prevail. However, we must take note of some of the controversy that has come to our attention in the past 2 years. First, there was a view expressed that there were some stars that were older than the Universe.

Without going into the details, that view has been discounted. The current view is that the outward expansion of the Universe is actually accelerating. This view replaces the previous notion that the Universe expansion rate is decreasing. These matters are clearly unsettled and the sophisticated arguments that are being debated go beyond the scope of this text. Chapter 2 deals with the formation of the elements. We see that all elements having a mass greater than helium must have been formed in the centers of stars and were not released until these stars exploded. From the remnants of these explosions, second generation stars such as the Sun were formed. Chapter 3 deals with some very elementary notions about chemistry. Chapter 4 discusses the timing of planetary formation and explains the element abundances of the planets. The emphasis is on why element abundances on the planets differ from what they are in the Universe as a whole. Finally, Chapter 5 focuses on planet Earth—the segregation of its mass into core, mantle, and crust and the evolution of its water supply and atmosphere. By comparison with the other bodies in our solar system, Earth emerges as a very special planet. It was special when it was formed and it has become even more special as its surface has been extensively influenced by the activities of living organisms.

CHAPTER 1

Origin of the Universe

Whereas the focus of this text is the origin of life and the evolution of the biosphere we call Earth, it is appropriate to discuss some of the events that preceded. We begin with a brief discussion of occurrences from the time of the origin of the Universe to the establishment of a habitable planet.

NEWTON'S UNIVERSE WAS INFINITE AND STATIC

In Newton's time the Universe was pictured as an infinite sea of stars in fixed positions. The only movements that astronomers were aware of were those of the planets about the Sun and satellites about the planets. Rejection of this static view of the Universe required sophisticated astronomical measurements that could not be made until the 20th century.

Newton argued that the stars were scattered across an infinite expanse of space in more or less fixed positions. In proposing this model for the Universe, Newton's attention was focused on the balance of gravitational forces. If the Universe were only finite or if the stars were clustered in only one part of the Universe, the gravitational forces should cause these stars to be drawn together into one huge mass. Because Newton was not aware of any movement between the stars, it seemed most likely that the gravitational forces must be in balance.

Concern over this model was first expressed by Johannes Kepler in the 1600s and subsequently by Heinrich Olbers in the 1800s. If the Universe were truly infinite and contained stars more or less uniformity distributed throughout space, then we would expect it to be filled with stars and light in every direction. As a result the sky should be bright at all times and there would be no darkness at nighttime. This dilemma, known as Olbers' paradox, was not resolved until the 20th century when an entirely new dynamic model was proposed for the Universe.

HUBBLE'S UNIVERSE WAS FINITE AND EXPANDING

In the early part of the 20th century, Einstein's theory of relativity changed our thinking about space and time but even Einstein did not reject the concept of an infinite static Universe despite the inconsistency of this model with his own theory. In his later years Einstein said that this was the biggest mistake he had ever made. The point is that no matter how attractive a theory may seem, it is difficult to make much progress without experimental observations. The experiments that were to provide us with the currently accepted model for the Universe were performed by Edwin Hubble in the 1920s. In Hubble's model the Universe began with all the mass and the energy concentrated at a point; an explosion known as the Big Bang followed. Current indications are that this explosion occurred about 20 billion years ago and that all matter powered by the force of this explosion is still being propelled outward in all directions from the center of this explosion.

THE DOPPLER EFFECT SHOWS THAT ALMOST ALL GALAXIES ARE MOVING AWAY FROM US

Hubble's model for the Universe was the outcome of research that permitted him to estimate the movement of the stars relative to Earth. The main way such studies are made is by analyzing the light they emit. One of the important facts we can determine from such an analysis is the speed at which a stellar body is moving relative to our observation point on Earth. To appreciate how this is done, we must understand some of the basic properties of light.

Light is an electromagnetic field that oscillates in space and time. It interacts with matter in packets called *photons*, each of which contains a fixed amount of energy, which is a function of its frequency of oscillation. The relationship between the energy of a photon ϵ and the frequency ν of its oscillating field is given by

$$\epsilon = h\nu, \tag{1}$$

where h is known as Planck's constant. The frequency ν is the number of oscillations

---

BOX 1A   Explaining Exponentials

In astronomy we often deal with very small or very large numbers. It is convenient to express these numbers as the product of a simple number and 10 raised to a power indicated by a superscript. For example, we have stated that 1 nm is equal to 10^{-9} m. This is the same as saying 1×0.000000001 m. In this case the superscript has a minus sign so that the number is very small. An example of a very large number is speed of light, which we have indicated is 3×10^{10} cm/s in a vacuum. This is equivalent to 30,000,000,000 cm/s.

---

per second at a given point in space. The wavelength λ of the oscillations is conveniently expressed in nanometers (1 nm = 10^{-9} m). It depends on both ν and the velocity v with which light travels through space:

$$\lambda = \frac{\nu}{v}. \tag{2}$$

Light travels with a velocity c of 3×10^{10} cm/s in a perfect vacuum (Box 1A). This speed is reduced when it is passes through space that is occupied by matter. Blue light has a wavelength in the region of 450 nm ($\nu = 6.7 \times 10^{14}$/s) and an energy per photon of about 2.8 electron volts (eV). Red light has a wavelength of about 650 nm. Radiation with wavelengths much below 400 nm or above 750 nm is invisible to the human eye, and some prefer not to call this light. However, all radiation obeys essentially the same laws.

The light we observe from a distant star or galaxy is a composite of different frequencies. This fact can be demonstrated with a prism that permits resolution of light of different wavelengths. For example, when visible white light is passed through a prism, it is bent according to its frequency; blue light is bent less than red light and so on (Fig. 1).

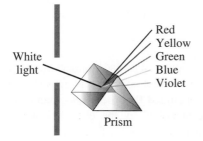

FIGURE 1   When white light is passed through a prism, it breaks up into a characteristic pattern of light with many colors.

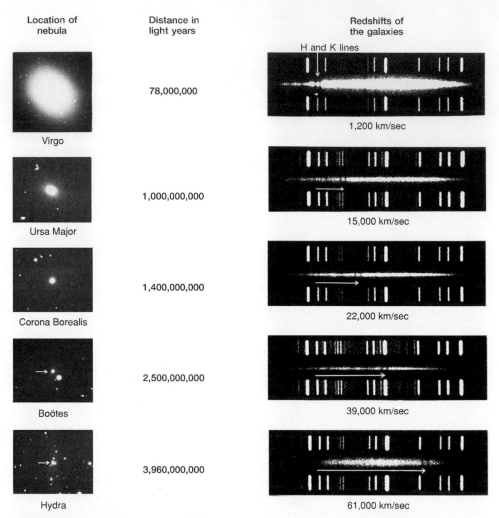

FIGURE 2 Galaxies and their light spectra. On the left are shown photos of five galaxies taken with the Hale Observatory telescope. Because these objects are probably similar in size, Virgo must be located much closer to Earth than to Hydra. Also shown, on the right, are light spectra from these galaxies. The white arrows show the displacement of an easily identified pair of dark lines from its position in a light spectrum for the Sun (or for a laboratory arc). The recession velocities corresponding to these arrow lengths are given. As can be seen, the more distant the object, the greater its recession velocity.

In observing light from any particular star, astronomers noted that superimposed on an almost continuous spectrum of light of different frequencies was a series of sharp dark bands (Fig. 2). They concluded that these dark bands must correspond to the wavelengths of light that are absorbed by the elements surrounding that body. Each element has a particular set of frequencies that it absorbs preferentially. This selective absorption is due to the interaction of the photons with the electrons of the element. When a photon is absorbed by an electron, the electron is catapulted from an orbit of relatively low energy to an orbit with a higher energy level. Because each element contains electrons at different energy levels, the frequency of absorbed light is characteristic of the element. As a result, the frequency of the absorbed light identifies the element and the intensity of the absorption indicates the amount of the element.

For a great many stars the spectra of absorbed bands are quite similar in their general patterns, an indication that most of the elements associated with different stars are quite similar. Quantitative differences in the extent of absorption at different frequencies permit the assessment of the relative concentrations of different elements for a particular star. When spectra are examined in detail, thousands of these bands become apparent. Thus a spectral analysis of absorption bands gives a detailed accounting of the relative amounts of the different elements in the solar gases that surround different stellar bodies.

For a long time astronomers used these absorption patterns to measure the relative abundance of the elements making up the atmospheres of neighboring stars. As more powerful telescopes became available, astronomers were able to extend their spectral analyses to more distant objects. To their surprise they found that, although the general pattern of absorption bands remained quite similar, for these more distance objects there was a shift in the location of the dark lines on the otherwise continuous spectrum. For example, a line that appeared in the blue part of the spectrum from the Sun or a nearby star was found in the green part of a spectrum of a very distant star, a line that appeared in the yellow part of the Sun's spectrum was present in the orange part of another more distant galaxy's spectrum, and so on. For any particular galaxy the pattern of dark lines always shifted to longer wavelengths. The extent of the shift was a characteristic of the galaxy. This effect was very puzzling to astronomers when it was first discovered, and a great deal of effort was made to correlate the extent of the shift with other characteristics of the galaxies. The most striking correlation was found with respect to the distance of the galaxy. The greater the spectral shift of the pattern of darkened lines, the farther away the galaxy appeared to be (see Fig. 2).

To understand the significance of the spectral shifts it was necessary to have a theory explaining their cause. This theory was provided by a 19th-century Viennese scientist named Doppler. The Doppler effect related the apparent frequency of a wave motion to the relative velocity between source and observer.

We find examples of the Doppler effect in our daily lives. For example, the engine of a high-speed racing car makes a high-pitched sound on approaching a stationary observer and shifts to a low-pitched sound once it has passed. Sound is a wave motion that travels through air at a velocity of 740 miles per hour. If the racing car is moving at the rate of 148 miles per hour the frequency of sound impulses reaching a stationary observer's ears will be 20% higher as the car approaches and 20% lower after it has passed. This frequency change accounts for the considerable variation in the pitch heard by a stationary observer. By contrast, the racing car driver hears a sound with a constant, intermediate pitch. This is because the onboard listener is traveling at the same rate as the vehicle that is producing the sound. Thus the sound heard is a function of the relative velocity between the source of the sound and the listener.

Could the Doppler effect explain the shift in frequency of the absorption bands in the observed stellar spectra from different galaxies? If so, the calculated speeds would have to be much higher than that of a racing car. Because light travels at the rate of 670 million miles an hour, a shift in the spectrum of the light reaching us from a distant galaxy corresponding to a 10% reduction in frequency would mean that the galaxy must be speeding away from us at a recessional velocity one-tenth the speed of light, or 67 million miles an hour (3×10^{10} cm/s). The general picture of the recessional velocities that has been obtained by comparing the shifts in spectra from observed galaxies is that all galaxies are speeding away from us and that the farther away they are, the faster is the rate at which their distance is increasing. This pattern can be explained if all matter is moving out from a point source at approximately the same speed from an explosion that occurred a long time ago.

RATE OF SEPARATION AND DISTANCE DATA SUGGEST THAT THE UNIVERSE ORIGINATED ABOUT 20 BILLION YEARS AGO

The relationship between distance in galaxies and their redshifts has led to one of the most important astronomical discoveries of the 20th century. The finding that virtually all objects are receding from us can most likely be explained by the fact that we live in an expanding Universe. To take full advantage of this relationship, it is necessary that more precise measurements of recessional velocity and distance be made.

The recessional velocity is relatively easy to measure precisely. The redshift z is defined as

$$z = \frac{\lambda - \lambda_0}{\lambda_0}, \tag{3}$$

where λ_0 is unshifted wavelength and λ is the observed wavelength. The recessional velocity v may be calculated from z because $z = v/c$, where c is the velocity of light. For z value of 0.05 we calculate that $v = 0.05\ c$. For a large value of z we must use the relativistic equation

$$\frac{v}{c} = \frac{(z + 1)^2 - 1}{(z + 1)^2 + 1}. \tag{4}$$

In contrast to recessional velocities, distances are most difficult to estimate. In fact, there is no way that they can be measured with certainty for stars and galaxies that are very far away. The problem of measuring the distance of an object without actually measuring it directly has been solved by surveyors using the method of triangulation. This is illustrated in Fig. 3. Imagine that the goal is to determine the exact distance of

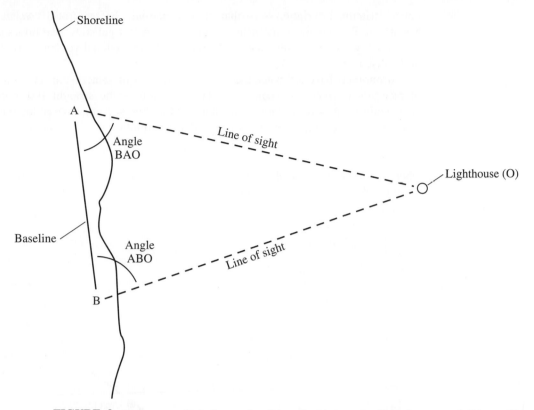

FIGURE 3   Surveyors method of measuring distance by trigonometry. From the measured distance AB and the angles BAO and ABO, the distance AO and BO may be calculated by trigonometry.

a lighthouse from either point A or point B in the shoreline. The surveyor would measure the distance between A and B directly and then determine the angle between the line of light of the lighthouse (AO or BO) and the line AB. From this information the distances AO and BO can be calculated by trigonometry. The known distance of Earth on either side of the Sun can be used in the same way to determine the distances to nearby stars. However, nearby stars are not of great interest to us because they have very small recessional velocities. For very distant objects in the Universe, the angle subtended by the object of interest to the two reference points is so close to zero that it does not give us a perceptible triangle. For the purpose of estimating distances to far away stars or galaxies, brightness is the main criterion used. For stars that are close enough to resolve as individual stars, one compares the brightest stars in the galaxies. For more distant stellar objects where it is impossible to resolve individual stars, the brightness of the galaxies themselves is compared; the hope is that there is not too much variation in brightness so that the distance will be reciprocally related to the brightness. By making many measurements on different galaxies, one hopes to obtain a reasonably accurate assessment of the ratio of the recessional velocities as a function of the distance.

Astronomers have estimated the distance of dozens of galactic clusters as a function of recessional velocity. When these data are plotted, the straight line relationship displayed in Fig. 4 is produced. The slope of this line is a constant called the Hubble constant. Hubble's law is most easily stated as a formula,

$$v_r = H_0 r, \tag{5}$$

where v_r is the recessional velocity, r is the distance, and H_0 is the hubble constant. From the data plotted on this graph we find that

$$H_0 = 15 \text{ km/s/Mly}, \tag{6}$$

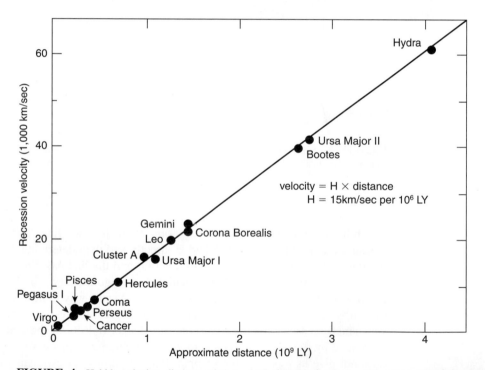

FIGURE 4 Hubble velocity–distance relation of 15 clusters of galaxies. The alignment of the data along a nearly straight line implies that recession velocity varies regularly with distance. For instance, galaxies at 1 billion light years (LY) have a recession velocity of 15,000 km/s, those at 2 billion LY have a recession velocity of 30,000 km/s. and so forth. (From Kutter, G. S. *The Universe and Life*, 1987, Sudbury, MA: Jones & Bartlett publishers. www.jbpub.com. Reprinted with permission.)

where Mly stands for 10^6 light years. In other words, for each million light years to a galaxy, the galaxy's recessional velocity increases by 15 km/s. For example, a galaxy located 100 million light years from Earth should be rushing away from us at a speed of 1500 km/s. Most astronomers prefer to speak in units of millions of parsecs, termed megaparsecs (Mpc), instead of millions of light years (Mly). By using that unit,

$$H_0 = 50 \text{ km/s/Mpc}. \tag{7}$$

Units of the Hubble constant sometimes are written with exponents instead of slashes:

$$H_0 = 15 \text{ km s}^{-1} \text{ Mly}^{-1} = 50 \text{ km s}^{-1} \text{ Mpc}^{-1}. \tag{8}$$

As one might suspect, the exact value of the Hubble constant is a topic of heated debate among astronomers today, simply because of the uncertainty in determining distances.

To calculate the time elapsed since the Big Bang, imagine watching a movie of any two galaxies separated by a distance r receding from each other with a velocity v. Now run the film backward, and observe the two galaxies approaching each other as time runs in reverse. We can calculate the time to T_0 it will take for the galaxies to collide by using the simple equation

$$T_0 = \frac{r}{v}. \tag{9}$$

Employing the Hubble law, $v = H_o r$, to replace the velocity, v in this equation, we get

$$T_0 = \frac{1}{H^0} = \frac{1}{50 \text{ km/s/Mpc}} = 20 \text{ billion years}. \tag{10}$$

Because the separation r has canceled, T_0 is the same for all galaxies. This is the time in the past when all galaxies were crushed together, the time the Big Bang occurred.

The true age of the Universe may be somewhat greater than this because the speed at which galaxies have been separating probably has been gradually increasing since the time of the Big Bang.

QUASARS HAVE ANOMALOUSLY HIGH REDSHIFTS

Whereas all matter is being propelled by the Big Bang that occurred about 20 billion years ago, there is also movement that results from other mostly gravitational factors. A most obvious example is the orbital motion of the planets around the Sun. The sun and all the planets are subject to a translational motion that resulted from the Big Bang. However, relative to one another, the planets have movement of a different sort.

A more puzzling type of movement is observed by unusually luminous stellar bodies that are located at great distances from us but show recessional velocities of a much greater magnitude than is consistent with their brightness. The redshifts recorded for so-called *quasars* frequently have z values greater than 1, indicating relativistic speeds. For example, the quasar known as OH 471 has a redshift of $z = 1.4$, which corresponds to a recessional velocity of greater than 90% of the speed of light. The highest recorded redshift for a quasar is $z = 4.7$. Based on their brightness, quasars are believed to be much closer to us than would be suggested by Hubble's law. Current opinion favors the notion that quasars are being accelerated by very dense bodies in their vicinity known as *black holes*. For the present all that we need to know about black holes is that they are objects whose gravity is so strong that the escape velocity

---

BOX 1B   Properties of Blackbody Radiation

The amount of energy radiated by an object depends on its temperature. The hotter the object, the more energy it emits in the form of electromagnetic radiation. The dominant wavelength of the emitted radiation also depends on the temperature of the object. A hot object emits most of its energy at short wavelengths, whereas a cool object emits most of its energy at long wavelengths.

While a white body reflects a great deal of the incident radiation that comes on it, a blackbody absorbs most of the incident radiation. An idealized blackbody absorbs all the radiation falling on it and emits a continuous spectrum of radiation at equilibrim. Because a blackbody reflects no electromagnetic radiation, the radiation that it emits is entirely the result of its temperature. The temperature of a blackbody and the dominant wavelength (λ_{max}) of the energy it emits are inversely related by the equation

$$\lambda_{max} = \frac{2.9 \times 10^{-3}}{T},$$

where λ_{max} is measured in meters, and T is measured in degrees Kelvin. This relation is called Wien's law.

To a first approximation the Sun and most stars are good examples of blackbodies because they absorb almost all the radiation falling on them. Consequently, we may estimate the Sun's surface temperature from Wien's law. The maximum intensity of sunlight is at a wavelength of about 500 nm = 10^{-7} m. From Wien's law, we find the Sun's surface temperature to be $T_{\odot} = \dfrac{2.9 \times 10^{-3}}{5 \times 10^{-7}} = 5800$ K. A subscript with a circle and a center dot refers to the Sun.

BOX 1C   Black Holes

A black hole is a region of space in which matter is so concentrated and the pull of gravity is so powerful that nothing, not even light, can emerge from it. Black holes represent the ultimate triumph of gravity over all other forces.

By definition, a black hole cannot be seen. Its presence must be detected through indirect evidence: the vast whirlpools of matter being sucked in by consuming gravity at ever increasing velocities.

Scientists have uncovered evidence that supermassive black holes probably lurk at the core of nearly all galaxies. They also have detected for the first time confirming evidence for the existence of the boundary of no return that surrounds a black hole—an event horizon across which matter and energy pass in one direction only, falling in but never coming back out.

exceeds the speed of light (Box 1B). Hence they emit no visible radiation and can be detected only by their gravitational effects. One may wonder why we do not see quasars with blueshifts. Probably the light of a quasar oriented in this way would be absorbed by the black hole (Box 1C) before it reached us.

ISOTROPIC BACKGROUND RADIATION IS BELIEVED TO BE A REMNANT OF THE BIG BANG

In the early 1960s, Arno Penzias from Princeton University and Robert Wilson from Bell Telephone Labs were experimenting with a new microwave horn antenna designed to relay telephone calls. Initially they were annoyed by the fact that no matter in what direction they pointed their horn they picked up a weak background radiation with a wavelength of about 1 mm. Eventually they realized that this radiation was coming from outer space with approximately equal intensity in all directions. For a blackbody a wavelength of 1 mm corresponds to a temperature of about 2.7°C (see Box 1B). The detection of this radiation was considered so important that Penzias and Wilson were eventually awarded the Nobel Prize for its detection. This cosmic background radiation is believed to be a vestige of very high energy photons that existed shortly after the big bang. As the Universe expanded these very short wavelength photons become stretched. This stretching process is referred to as a cosmological redshift to distinguish it from the Doppler redshift that is caused by an object's motion through space. A most remarkable aspect of cosmic background radiation is that it is almost perfectly isotropic, which reflects its ancient origin.

CURRENT EVIDENCE SUGGESTS THAT THE RATE OF EXPANSION OF THE UNIVERSE IS INCREASING

The present rate of expansion of the Universe is given by Hubble's constant. To measure the change in expansion rate we must be able to look into the past or into the future. We will settle for looking into the past by focusing our observations on objects that are very far away. There are two reasons we are interested in this question. First, it affects our estimate of the age of the universe. If the expansion rate is decreasing, the Universe is probably younger than calculated from the Hubble constant. On the other hand, if the expansion rate is increasing, then the Universe actually would be older than calculated by the Hubble constant. The second and more profound reason we are interested in the rate of expansion is that it has an effect on the future of the Universe. A change in expansion rate is the best indicator of whether the Universe will keep expanding or the expansion will stop or even reverse so that a contraction process will ensue.

Cosmologists tell us that the ultimate fate of the Universe is a matter of its average density. The estimated critical density required to just halt expansion of the Universe is 5×10^{-10} g/cm^3 which is equivalent to about three hydrogen atoms per cubic meter. The estimated density of the universe is still considerably below this. However, it keeps rising as new stellar objects continue to be discovered. These new objects

---

BOX 1D   The Apparent Magnitude of a Star Is a Function of Its Absolute Magnitude and Its Distance

By convention the *absolute magnitude* of a star is the magnitude it would have if it were located a distance of exactly 10 parsecs (pc) from Earth. Absolute magnitude is a very useful quantity, because it gives a measure of the intrinsic brightness of a star. *Apparent magnitude* is a measure of the light energy arriving at Earth. Apparent magnitude tells us how bright a star appears in the sky. The farther away a source of light, the dimmer it appears.

Astronomers have derived an equation that relates a star's apparent magnitude (m), its absolute magnitude (M), and its distance (d, measured in parsecs) from Earth:

$$m - M = 5 \log d - 5.$$

From this equation it should be apparent that if the quantities of two of the variables m, M, and d are known, the third one may be calculated. This equation has been most useful for estimating the distance of far away supernovas.

include dark objects such as burned out stars that no longer emit visible radiation and black holes that are very dense objects detected only by their gravitational effects.

While efforts to obtain a more accurate measure of the density of the universe continue, more direct evidence indicates that the expansion rate of the universe is increasing. The evidence comes from close scrutiny of a class of stars that explode violently. Such an event is referred to as a *supernova*. All supernovas begin with a sudden rise of about a millionfold in brightness. For this reason supernovas can be observed over enormous distances. It is believed that most supernovas have the same intrinsic brightness at their peaks regardless of their distance from Earth. This means that the intrinsic brightness of a supernova can be approximated by the constant that can be determined by measuring the brightness of close-by supernovas where the distance can be accurately estimated. By having fixed on a value for M, the apparent brightness m of a very distant supernova may be used to obtain a value for the distance (Box 1D). Comparison of this with the measured recessional velocity for several dozen distant supernovas gives values that indicate the rate of expansion of the universe is increasing.

SUMMARY

In this chapter we have considered the evidence supporting the hypothesis that the Universe began with a Big Bang that resulted in a rapidly and continuously expanding system.

1. From the shift in the spectra of light reaching us from distant galaxies, it has been determined that the distance between Earth and all galaxies is increasing. The more distant the galaxy, the greater the velocity of separation.
2. From the distance of different galaxies and the speed with which they are moving relative to one another, it has been estimated that all matter and energy originated from a single location in the Big Bang about 20 billion years ago.
3. The isotropic background radiation that is observable in all directions is believed to reflect radiation that was produced immediately after the Big Bang.
4. The Universe may keep expanding or the expansion may give way to arrest or even contraction. Current indications from measured recessional velocities of very distant supernovas favor the notion of indefinite expansion.

Problems

1. On average galaxies at a distance of 100 million light years are moving away from us with a velocity of 1500 km/s. Can this information be used to estimate the age of the Universe? (Answer: 20 billion years.) Show how you get this answer and indicate what assumptions you used.
2. One way to explain the Doppler shift is by claiming that Earth is at the center of the Universe. Why is this very unlikely?
3. How many seconds are in a light year?

4. What is the difference between a cosmological redshift and a Doppler redshift?
5. If you were an observer on a quasar, would you expect to find a direct proportionality between the recessional velocities of galaxies and their distance from you?
6. What recessional velocity is suggested by a redshift z of 3? How far away would you expect a galaxy to be that gave rise to this redshift?
7. In Fig. 2 calculate the recessional velocity for Hydra.

References

General

Engelbrektkson, *Astronomy through Space and Time*. Dubuque, IA:Wm. C. Brown, 1994. This presents the basic information and contains details of how further information has been gained in recent years with the help of NASA space probes. It is useful backup text for much of what is covered in Chapters 1, 2 and 4 here.

Glanz, J. New Light on Fate of the Universe. *Science* 278:799, 1997; Glanz, J. Astronomers See a Cosmic Antigravity Force at Work, *Science* 279:1298, 1998. (Exploding stars billions of light years away suggest that the universe may expand forever.)

Kaufman, W. J. *The Universe,* Freeman Press.

(This is a somewhat more rigorous treatment than Engelbrektkson but is very readable.)

Watson, A. The universe Shows Its Age, *Science* 279:981, 1998. (A cosmic embarrassment is fading. By some new measures, the oldest stars no longer appear to be older than the universe as a whole.)

For the More Adventurous

Grandlay, J. E. Black Holes Take Centre Stage, *Nature* (London) 371:561–562, 1994.

Hawking, S. W. *A Brief History of Time.* New York: Bantam Books, 1998. (This is exciting but with many difficult passages.)

Linde, A. The Self-Reproducing Inflationary Universe, *Sci. Am.* November: p 48–57, 1994. (Recent versions of the inflationary scenario describe the universe as a self-generating fractal that sprouts other inflationary universe.)

Maddox, J. Beyond Einstein's Theory of Gravitation, Nature (London) 374: 759, 1995.

Ouyed, R., Pudritz, R. E., and Stone, J. M. Episodic Jets from Black Holes and Protostars, *Nature* (London), 385:409–414, 1997.

Overbye, D. Weighing the Universe, *Science* 272:1426–1428, 1996. (Astronomers are making inventories of the unseen mass in the universe to learn its composition and fate.)

Perlmutter, S. et al. Discovery of a supernova explosion at Half the Age on the Universe, *Nature* (London), 391:51–56, 1998. (The ultimate face of the universe, infinite expansion or a big crunch, can be determined by using the redshifts and distances of very distant supernovas to monitor changes in the expansion rate.)

Urey, H. *The Planets: Their Origin and Development.* New Haven: Yale University Press, 1952. (This is of historical interest only.)

Wilson, R. W. The Cosmic Microwave Background Radiation—Wilson's Nobel Address, *Science* 205:866–874, 1979.

CHAPTER 2

Formation of the Elements

Although there are more than 100 known elements, the vast majority of matter in the Universe is composed of just two, hydrogen and helium (Fig. 1). In this chapter we explain how elements were formed and why their abundances are heavily skewed toward the lighter elements. Prior to the Big Bang it seems likely that all substance was confined to a point source containing an incredible amount of energy. Within seconds after the Big Bang the conversion of energy to mass began (Fig. 2) but it was many years before the amount of mass in the Universe was equal to the amount of energy (approximately one-half million). The neutrons that were first formed were unstable on their own, decaying into protons and electrons. By 12 min, the *half-life* of a free neutron, most matter was equally divided between neutrons and protons and electrons. In the dense mass of rapidly expanding matter, frequent collisions between neutrons and protons led to the formation of considerable helium, which was quite stable. Very few larger nuclei were produced at this time so that by the end of day one, most matter in the Universe consisted of a mixture of hydrogen, which contains only a single proton, and helium, which contains two protons and two neutrons.

    The hydrogen and helium atoms produced in the early stages of the expanding Universe eventually coalesced into loosely knit nebulas that condensed further to form stars; ultimately, clusters of stars formed galaxies. Big stars and small stars go through

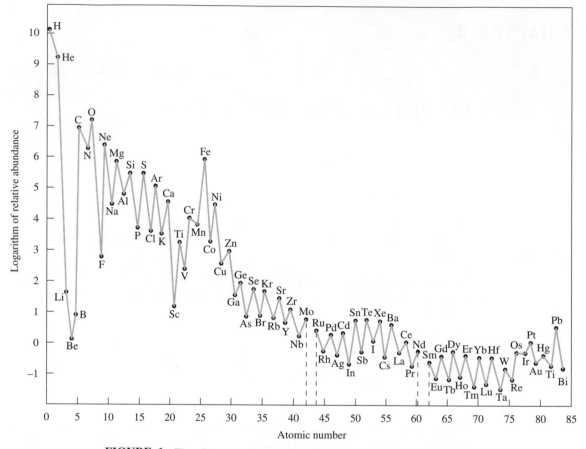

FIGURE 1   The relative abundances of the elements in the Sun, as determined from the solar spectrum. Abundances are plotted relative to 10^6 silicon atoms.

very different stages of development. Interwoven with this evolution of the stars is the further conversion of hydrogen into helium and of helium into the remaining elements.

CHEMICAL COMPOSITION OF THE SUN APPROXIMATES CHEMICAL COMPOSITION OF THE UNIVERSE

Before we can discuss the composition of the Universe we must explain how estimates on composition were obtained. Because stars formed from the gravitational collapse of clouds of gases, the chemical composition of a star should be representative of the cloud from which it was formed. Thus if we could determine the chemical

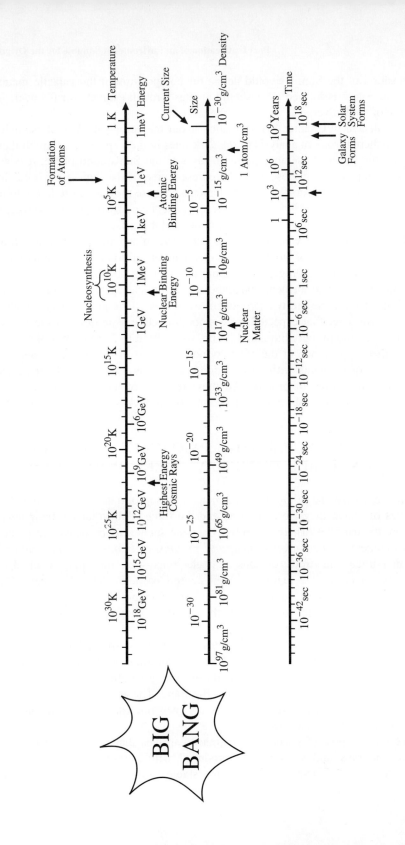

FIGURE 2 History of the Universe.

composition of the Sun, we could define the composition of the galactic matter from which the Sun was formed. This should give us a rough estimate for the average composition of the Universe.

As indicated in Chapter 1, information about the composition of stars such as our Sun is gathered from an analysis of the dark lines in their spectra. Although these dark lines are the result of the absorption of gases in the Sun's atmosphere, it is assumed that the composition of these gases is representative of the Sun as a whole. To some extent this estimate favors the lighter elements because they are more likely to be closer to the Sun's surface. However, this effect is attenuated by the strong convection currents that exist between the interior of the Sun and its surface.

The abundances of elements in the Sun determined in this way are plotted in relation to the number of atoms of each element per million atoms of silicon in Fig. 1. This plot shows that hydrogen and helium are the most abundant elements. There is a precipitous decline in element abundances with increasing atomic number. Superimposed on this decline are two prominent exceptions. One is the much greater abundance of iron over that expected for a smooth decline. The other is the much lower amounts for lithium, beryllium, and boron. In addition to these exceptional features, a saw-toothed appearance to the curve is notable. This is due to the generally lower abundances of elements with an odd number of protons over neighboring elements with an even number of protons. All these features of the abundance curve can be explained by the mode of origin of the elements and their stability once formed.

FIVE STABLE SUBATOMIC PARTICLES AND MANY MORE UNSTABLE ONES HAVE BEEN IDENTIFIED

Before we discuss the origin of the elements it is appropriate to discuss the subatomic particles of which they are composed and the forces that influence their interaction. First are the particles. There are only five stable subatomic particles that make up the known content of the Universe: *neutrons, protons, electrons, neutrinos,* and *photons.* The first three—neutrons, protons, and electrons—are the building blocks of the elements. Neutrinos are extremely light neutral particles created in certain nuclear reactions. Despite their lightness they are so abundant that they may constitute a significant fraction of the total mass of the Universe. Photons are quanta of energy that are believed to have no rest mass.

Two other stable elementary particles have been postulated but never detected. They are the *graviton* and the *gluon.* These particles are believed to be created whenever mass is violently accelerated, as during the gravitational collapse of a star, the falling of matter into a black hole, or the Big Bang. The graviton accounts for the force of gravity, and the exchange of gluons between nuclear particles accounts for the stability of the atomic nucleus.

There are a host of unstable elementary particles that were first detected among the cosmic rays and subsequently produced with particle accelerators (Table 1). They are very short-lived and decay into stable elementary particles in fractions of

TABLE 1
Some Elementary Particles Found in the Universe

	Particle	Symbol	Rest energy (Mev)[a]	Electric Charge	Threshold[b] Temperature
Quarks	Up	u	900	$\frac{2}{3}$e	348
	Down	d	300	$-\frac{1}{3}$e	348
Leptons	Neutrino	$\nu, \bar{\nu}^c$	0.00001	0	0.0001
	Electron	e^-, e^+	0.5110	$-1(+1)$	5.930
	Muon	μ^-, μ^+	105.66	$-1(+1)$	1,226.2
	Mesons	π°	134.96	(0)	1,556.2
		π^+, π^-	139.57	$+1(-1)$	1,619.7
Baryons	Proton	$p, \bar{p}^d$	938.26	$+1(-1)$	10,888
	Neutron	$n, \bar{n}^e$	939.55	0	10,903
Bosons	Graviton	—	$\leq 10^{-36}$	0	—
	Gluon	—	≤ 100	0	—
	Photon	γ	0	0	—

[a] Mev is equivalent to 1.8×10^{-30} kg.

[b] The threshold temperature of a particle is its rest energy divided by the Boltzmann constant; it is the temperature above which the particle can be freely created from thermal radiation.

[c] A bar above a symbol indicates an antiparticle.

[d] A proton contains 2 up quarks and 1 down quark.

[e] A neutron contains 2 down quarks and 1 up quark.

seconds following their production. These include *quarks* from which protons and neutrons are composed.

FOUR TYPES OF FORCES ACCOUNT FOR ALL INTERACTIONS IN THE UNIVERSE

Four types of forces are alleged to account for all interactions in the Universe (Table 2). These are *gravitational forces, electromagnetic forces,* and the *strong* and *weak nuclear forces.*

Gravitational forces account for the universal attraction that material objects have for one another. Although the relative strength of gravitational forces is comparatively weak at short distances, the influence of gravitational forces is additive and for massive objects it becomes a dominant force with profound influences within the Universe.

Electromagnetic forces account for all known chemical reactions and most reactions between light and matter.

The role of the two nuclear forces are quite different. The strong nuclear force is responsible for the large amounts of energy released in nuclear reactions, such as the energy of radioactive decay from the explosive energy of an atomic bomb. Of greatest importance the strong nuclear force accounts for the nuclear reactions that take place

TABLE 2
The Four Forces

Force	Relative strength at 10^{-15} m	Particles exchanged	Particles affected	Range	Example
Strong nuclear	1	Gluons	Quarks	10^{-15} m	Holds nuclei together
Electromagnetic[a]	10^{-2}	Photons	Charged particles	Infinite[b]	Holds atoms together
Weak nuclear	10^{-4}	Gluons	Quarks, electrons, neutrinos	10^{-16} m	Released on β decay
Gravitational[c]	10^{-38}	Gravitons	All particles	Infinite	Holds the solar system together

[a] $F_{\text{electric}} \propto \dfrac{Q_1 Q_2}{R^2}$, where Q is the charge and R is the distance of separation of the interacting charged particles Q_1 and Q_2.

[b] Infinite in principle but not in fact because of the cancellation effect of opposite charges.

[c] $F_{\text{gravitational}} \propto \dfrac{M_1 M_2}{R^2}$, where M_1 and M_2 are the masses of the interacting bodies and R is there center-to-center distance of separation.

in the center of stars. The exchange of gluons between nuclear particles is believed to account for the attractive nature of the strong nuclear force.

The weak nuclear force by itself cannot hold anything together. The weak nuclear force is at work in certain kinds of radioactive decay, such as the transformation of a neutron into a proton with the release of an electron and an antineutrino.

PRIOR TO STAR FORMATION THE ONLY ELEMENTS FORMED IN SIGNIFICANT AMOUNTS WERE HYDROGEN AND HELIUM

The first phase of element creation began with the Big Bang. Cosmologists have speculated that prior to the Big Bang the Universe was massless with all the energy concentrated at a point. The Big Bang unleashed this energy that resulted in a rapid conversion of energy into mass and an expansion process that continues to the present day. When energy is converted into mass, it follows the relationship proposed by Einstein:

$$E = mc^2,$$

where E is the energy, m is the mass, and c is the velocity of light. The energy of the photons initially was so high that all elementary particles could be freely created from the conversion of thermal radiation. As the Universe expanded the temperature rapidly decreased and so did the capacity for thermal radiation to be converted into elementary particles. The description given here represents a possible scenario for the first phase of element production (See Fig. 2).

At 10^{-35} s following the Big Bang the approximate temperature was 10^{27} K. At this temperature considerable energy was spontaneously converted into quarks and leptons as well as antiquarks and antileptons. This was a period of rapid expansion.

By 10^{-12} s the temperature had dropped to 10^{15} K. Quarks and leptons were no longer created spontaneously. Collisions between particles and antiparticles resulted in their reversible conversion into photons. Leptons and antileptons separated into electrons and positrons, and neutrinos and antineutrinos.

At 10^{-6} s the temperature had dropped to 10^{13} K. At this temperature quarks and antiquarks combined into protons and antiprotons as well as neutrons and antineutrons. Annihilation reactions involving protons and neutrons no longer occurred reversibly. The Universe contained protons, neutrons, electrons and positrons (in nearly equal numbers), neutrinos and antineutrinos, photons, and gravitons.

At 1 s the temperature had dropped to a mere 10^{10} K. At this temperature neutrinos and antineutrinos stopped interacting with one another.

By 15 s the temperature had dropped to 3×10^9 K. The existing photons no longer had sufficient energy to produce electron–positron pairs. Electrons and positrons continued to annihilate one another leaving a slight excess of electrons.

From 1 min to about 5 min the temperature gradually dropped from about 1.3×10^9 to 600×10^6 K. During this period of primordial nucleosynthesis ^{1}H$^+$ (76%), ^{4}He^{2+} (24%), and traces of ^{2}H$^+$ and ^{3}He^{2+} were formed by the following reactions:

$$p^+ + n \rightarrow {}^2H^+ + \gamma$$
$$p^+ + p^+ \rightarrow {}^2H^+ + e^- + \nu$$
$$p^+ + {}^2H^+ \rightarrow \gamma + {}^3He^{2+}$$
$${}^3He^{2+} + {}^3He^{2+} \rightarrow {}^4He^{2+} + 2p^+ + \gamma.$$

With the exception of the first reaction the same reactions occur in the interior of the Sun. At 600×10^4 K nucleosynthesis came to an end.

Finally, after half a million years had passed and the temperature had dropped to 3000 K, it became possible for nuclei and electrons to form stable complexes. The original photons evolved very little after this. They passively partook in the cosmic expansion with their wavelengths expanding as the cosmos expanded and they remain to this day as the 3 K cosmic background radiation discussed in Chapter 1.

ELEMENTS BETWEEN HELIUM AND IRON WERE PRODUCED IN THE CENTERS OF STARS

The hydrogen and helium produced in the immediate aftermath of the Big Bang formed dust clouds that condensed into stars. Most of the remaining elements are believed to have been formed in the interiors of the stars. During the transformation of diffuse clouds of helium and hydrogen gas into compact stars, an enormous amount of gravitational energy was converted into heat. So much heat was produced that the

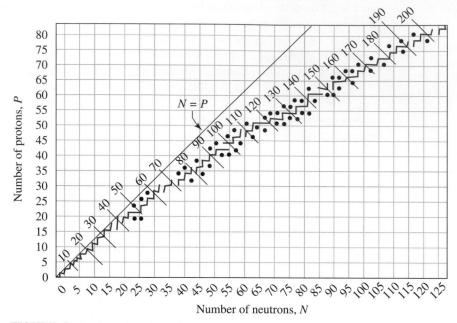

FIGURE 3   Stable combinations of neutrons and protons give rise to a belt of stability. For elements of low atomic number, the ratio of protons to neutrons in stable isotopes is about 1:1. For elements of high atomic number, it approaches 1:1.5.

core of a newly formed star became hot enough to start a "nuclear fire." For the nuclei in a star to react, they must make direct contact. This requires tremendous collision velocities to overcome the repulsion between positively charged nuclei. For example, for two protons to collide they must have velocities equivalent to a temperature of about 60 million degrees centigrade. Only at the centers of stars are the concentrations of nuclear particles and the temperatures adequate to produce such nuclear reactions.

The abundances of nuclei that form in the stars are a function of which types of nuclei can be formed and also of the stability of the nuclei that are formed. Only certain combinations of neutrons and protons form stable units (Fig. 3). The stable nucleus with the most neutrons and protons is bismuth, ^{209}Bi, which contains 209 nuclear particles. All nuclei larger than this are unstable. From Fig. 3 we can see that stable nuclei lie in a narrow band of isotopes running from ^{1}H to ^{209}Bi. We also can see that the ratio of protons to neutrons for stable nuclei is near unity for elements with a low atomic number and gradually rises to 1.5 by the time bismuth is reached. The unstable nuclei that are transiently formed decompose into stable nuclei by radioactive disintegration. The time required for disintegration of an unstable substance is expressed in terms of its *half-life*, the amount of time it takes for exactly half of the substance to disintegrate (Box 2A). A favored type of disintegration

---

BOX 2A   Radioactive Decay and Half-Lives

The simplest type of reaction we can imagine is that for an irreversible unimo-
lecular process such as the decay of an unstable isotope N that decays with a
fixed rate constant,

$$\frac{-d[N]}{dt} = k[N], \tag{1A}$$

where k is the rate constant, $[N]$ is the concentration of radioactive isotope, and
$(-d[N])/dt$ is the rate of disappearance of N. This rate is proportional to the first
power of the concentration of isotope, and the reaction is accordingly described
as a first-order reaction. By methods of calculus it can be shown that the
preceding equation is equivalent to

$$N = N_0 e^{-kt}, \tag{2A}$$

where N_0 is the number of undecayed nuclei at time $t = 0$.

The half-life $t_{1/2}$ of a radioactive sample is defined as the time at which the
number of radioactive nuclei has decreased to one-half the number at $t = 0$. At
this time

$$e^{-kt_{1/2}} = \frac{1}{2}. \tag{3A}$$

By taking natural logarithms of both sides and solving for $t_{1/2}$ we find

$$kt_{1/2} = \ln 2 \tag{4A}$$

$$t_{1/2} - \frac{\ln 2}{k} = 0.693 k^{-1}. \tag{5A}$$

Half of the original radioactive isotope in the sample decays in a time interval
$t_{1/2}$, half of the remaining isotope decays in a second time interval $t_{1/2}$, and so on.

---

involves β-particle decay, in which a neutron within the atomic nucleus decomposes
into a proton and an electron (Fig. 4).

As we have seen the helium atom is formed by a complex series of collisions
involving four hydrogen atoms. The overall reaction results in the release of an enormous amount of heat that can be quantitatively estimated from the difference in mass
between the four ^{1}H atoms and the single ^{4}He atom:

Mass of 4 ^{1}H atoms	6.696×10^{-24} g
Mass of 1 ^{4}He atom	-6.648×10^{-24} g
	0.048×10^{-24} g.

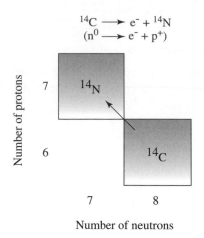

$$^{14}C \longrightarrow e^- + {}^{14}N$$
$$(n^0 \longrightarrow e^- + p^+)$$

FIGURE 4   The β decay process for the decay of ^{14}C to ^{14}N. Symbols: n = neutron (no net charge); e^- = electron (negative charge); p^+ = proton (positive charge).

From Einstein's equation that relates mass and energy, $E = mc^2$, it may be calculated that 0.048×10^{-24} g is equivalent to an energy of 1×10^{-12} cal [4.184×10^{-12} joules (J)].

The amount of heat obtained from the conversion of hydrogen to helium is so enormous that once a star's nuclear fire is ignited, its further collapse is stemmed by the back pressure created from the escaping heat. As a rule, the star stabilizes in size and burns smoothly until most of the nuclear fuel is used up. This takes a long time. For example, the Sun, which has been burning for 4.6 billion years, probably will not run out of hydrogen fuel for another 5 billion years.

Most visible stars emit light created by the burning of hydrogen to form helium. Slowly but surely most of the remaining hydrogen in the Universe is being converted into helium. The hydrogen that fuels most stars eventually gets used up. In a large star the supply of hydrogen is exhausted much more rapidly than in small stars like our Sun. Once this has happened, the nuclear fire cools and the star begins to collapse again. The heat released by this renewed collapse causes the core temperature to rise to new heights until the ignition temperature for helium is reached. Because a helium nucleus has twice the positive charge of a hydrogen nucleus, a much higher temperature is required for effective collisions (about 200 million Kelvin).

When two helium nuclei collide, they form a beryllium (^8Be) nucleus. The beryllium nucleus does not survive for long because it is very unstable. If another ^4He collides with the ^8Be nucleus before it decomposes, a carbon nucleus ^{12}C is formed. Similarly, an oxygen nucleus, ^{16}O, is formed when a carbon nucleus reacts with an additional ^4He nucleus. Carbon and oxygen are the main nuclei that result from helium burning.

In a large star this cycle of fuel depletion, renewed collapse, core temperature rise, and ignition of a less flammable nuclear fuel may be repeated several times (Fig. 5). A cycle involving carbon burning is followed by a cycle of neon burning, a cycle of

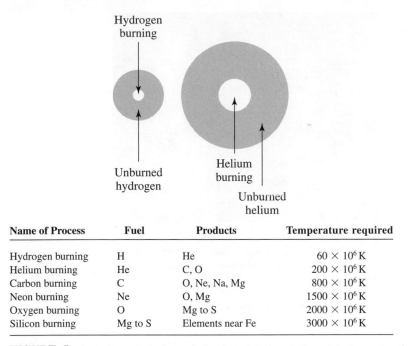

Hydrogen
burning

Unburned
hydrogen

Helium
burning

Unburned
helium

Name of Process	Fuel	Products	Temperature required
Hydrogen burning	H	He	60×10^6 K
Helium burning	He	C, O	200×10^6 K
Carbon burning	C	O, Ne, Na, Mg	800×10^6 K
Neon burning	Ne	O, Mg	1500×10^6 K
Oxygen burning	O	Mg to S	2000×10^6 K
Silicon burning	Mg to S	Elements near Fe	3000×10^6 K

FIGURE 5   A star burns its hydrogen before it starts to burn helium. A hotter nuclear fire is required to burn helium. As nuclei get bigger, higher and higher ignition temperatures are required.

oxygen burning, a cycle of silicon burning, and so on. Element formation by this mechanism stops at the element iron, ^{56}Fe. This is because the masses of nuclei heavier than iron are greater than the masses of nuclei that are merged to formed them. Thus further element formation does not release energy; instead it requires the input of energy. This "energy wall" at ^{56}Fe explains why iron is roughly 1000 times more abundant than would be predicted from a smooth decline in element abundances (see Fig. 1), but it does not provide us with any clues as to how elements of greater masses are formed.

FORMATION OF ELEMENTS HEAVIER THAN IRON STARTS WITH NEUTRON CAPTURE

When the core of our Sun runs out of hydrogen about 5 billion years from now, it will resume its collapse. However, because the Sun is just barely massive enough to generate the temperature necessary to start a helium fire, it will probably collapse permanently and quietly after it has burned a small amount of its helium. A small star that has undergone this quiet collapse and cooling off is called a *white dwarf.*

FIGURE 6   Supernova remnant. Optical image of the Crab nebula M1 (NGC 1952), a supernova remnant (SNR) in the constellation of Taurus 6300 light years (LY) away from Earth. The supernova explosion was recorded in 1054 A.D. by Chinese astronomers. M1 is the brightest SNR and is now almost 6 LY in diameter. Near its center, lies a rapidly spinning neutron star (not identifiable here), The Crab pulsar, which represents the core of the exploded star. The Crab pulsar emits pulses of radiation with a period of 0.033 s. (*Source*: Royal Observatory, Edinburgh/Science Photo Library.)

While small stars like the Sun undergo quiet deaths, big stars frequently undergo violent deaths in which their contents are blown apart. Such explosions result in a tremendous burst of light energy and are called *supernovas* (Fig. 6).

Supernovas have provided astronomers with a rare opportunity to observe the contents of the inside of stars. Recall that most forms of spectral analyses are limited to telling us about the elements present on or above the star's surface. However, new elements that are relatively short-lived might never make it to the surface, so a supernova provides a unique opportunity to see such elements. Technetium is an element fitting this description. It has two isotopes with moderately long half-lives: ^{92}Tc (2.6×10^6 years) and ^{98}Tc (4.2×10^6 years). Although these isotopes persist for millions of years after production, they are not seen in ordinary stars. However, the spectral bands for technetium are seen in the nebulas produced by a supernova blast. This type of observation provides us with the most direct support for the hypothesis that elements are being formed in the core of stars.

Supernovas are most important because they provide an effective means for building elements heavier than iron. During the explosion that occurs in a supernova, a host of nuclear reactions occur that create free neutrons. Because a neutron has no net charge, it is not repelled by any nucleus it happens to encounter. As a consequence, neutrons freely make effective contact with any nucleus regardless of how slowly they are moving. In the close-packed conditions that exist inside stars, the neutrons created during nuclear reactions encounter a nucleus long before they have time to undergo spontaneous decay to a proton and an electron. Many of these encounters are with iron nuclei. When an iron nucleus absorbs a neutron, it becomes heavier. In the supernova explosion these neutron hits are very frequent. An iron nucleus that is hit by one neutron is very likely to be hit by another and another. As a result, the iron nucleus gets heavier and heavier until finally it cannot absorb any more neutrons. At this point the iron nucleus with its extra neutrons seeks a more stable configuration by undergoing radioactive decay, converting some of its neutrons into protons and electrons. The conversion of one neutron to a proton and an electron creates a cobalt atom. The cobalt atom can absorb neutrons one after another until it too becomes saturated. Subsequently, it undergoes a similar neutron decay process and becomes a nickel atom. The buildup to larger nuclei stops only when nuclei get so big that they fragment into much larger decay products, a process known as nuclear fission.

The entire process of repeated neutron bombardment takes place in a very short time following a supernova explosion. This is because these neutrons are rapidly produced and rapidly dispersed. The unstable neutron-rich isotopes so produced emit one electron after another by β decay until they have achieved a stable neutron-to-proton ratio. For most nuclei this process is quickly completed, but in some cases the nuclei have very long half-lives, and the process can continue for thousands or even millions of years.

The rapid neutron bombardment process does not account for all the stable nuclei that are observed in the mass range from ^{56}Fe to ^{238}U. Another neutron capture process takes place during the less traumatic smooth nuclear burn period that characterizes most of a star's history. In this situation the time between neutron hits is much greater, giving the nuclei that have already been hit a much greater opportunity to decay before being hit again. As a result, unstable nuclei that might be hit by additional neutrons are more likely to decay before this can happen. The path followed by this low flux neutron bombardment process runs along the belt of stability illustrated in Fig. 3.

ISOTOPES WITH EVEN NUMBERS OF PROTONS ARE FAVORED IN ELEMENT FORMATION

It remains for us to explain the saw-toothed appearance of the element abundance plot (see Fig. 1). This is due to the greater stability of neutron–proton combinations that contain an even number of protons and an even number of neutrons. The preference for even–even combinations destines odd-numbered elements to have only one isotope and gives the even-numbered ones several. As a result, the even-numbered elements have a greater abundance than their immediate neighbors have.

SUMMARY

In this chapter we have considered the events that took place from the time of the Big Bang that gave rise to the more than 100 elements that comprise a significant fraction of the material substances of the Universe.

1. The chemical substances of the Universe are estimated from the chemical substances present on the surface of the Sun. This analysis indicates a general decline in the relative abundances of elements with increasing atomic number. The element iron is much more abundant than would be expected for a smooth decline, while the elements lithium, beryllium, and boron have abundances much lower than would be expected. These relative abundances can be understood as a function of the mechanisms of formation and the stability of the nuclei so formed.

2. Five stable subatomic particles have been identified: neutrons, protons, electrons, neutrinos, and photons. Two other stable elementary particles have been postulated but never detected: the graviton and the gluon. In addition to these, many more unstable particles have been detected. Most of these are components of the stable subatomic particles.

3. All interactions in the Universe can be explained by four types of forces. Gravitational forces account for the universal attraction between all material objects. Electromagnetic forces account for all chemical reactions. Strong and weak nuclear forces account for all reactions that take place within and between different nuclei.

4. To explain element abundances we must start with what happened immediately after the Big Bang when an explosion triggered the conversion of a point source of energy into matter. Within seconds after the explosion, combinations of neutrons and protons were able to form. However, the only elements produced in significant amounts during this very early phase of Universe evolution were hydrogen and helium. As the Universe expanded and cooled, the hydrogen and helium gases eventually agglomerated into clouds that condensed further to form galaxies of stars.

5. Further reactions involving hydrogen and helium nuclei required a concentration of mass and energy that only could be achieved within the cores of stars. Here the elements with masses up to that of iron were produced by nuclear bombardment. In the process, a great deal of energy was released that served to keep the process of element formation going.

6. Element production above iron is an energy-requiring process and had to follow a different route. These heavier elements were probably produced by a process that starts with neutron capture, followed by decay of some of the neutrons that have been captured. Some neutron captures occur when a star blows up. Others occur by slow neutron production throughout most of the life of the star.

Problems

1. The estimate for the amount of iron in the Sun comes from spectral analysis. Do you think such an estimate is likely to be accurate? If it is not, why not? How might you get a better estimate for the Sun, and for the Universe?
2. Why is iron so much more abundant than the elements immediately preceding it and immediately following it?
3. Could Mg arise from Al by β decay?
4. Describe a likely route to the nucleosynthesis of N.
5. Why are carbon and oxygen the main nuclei to result from He burning?

6. Are the original photons formed in the Big Bang confined to the boundaries of the expanding Universe? Explain.

7. How much time does it take for 75% of a radioisotope to decompose if its half-life is 1000 s?

References

See Chapter 1. for general references.

Liss, T. M. and Tipton, P. I. The Discovery of the Top Quark, *Sci. Am.* September: 54–59, 1997. (Violent collision between a proton and an antiproton creates a top quark and an antitop. These decay to other particles, typically producing a number of jets and possibly an electron or a positron.)

CHAPTER 3

Beginnings of Chemistry

Immediately following the Big Bang the temperatures were incredibly high. Within minutes after the rapid expansion began, the energy-dominated Universe became transformed into a mass-dominated Universe but it still took about a half a million years before the expanding Universe had cooled to the point where electrons could assume stable orbitals around individual hydrogen or helium nuclei. Even then the chemistry that could occur was extremely limited because of the absence of significant amounts of other elements. This had to await first, the formation of stars; second, the formation of additional elements at the centers of the "first-generation" stars; then explosions or supernovas leading to the release of these additional elements so that "second-generation" stars like our Sun and its solar system could form; and finally, temperatures compatible with the formation of stable chemical compounds. All this must have taken a billion years or longer.

In this brief chapter we limit ourselves to a discussion of some of the elementary aspects of chemical reactions. Of the four types of forces that direct all reactions in the Universe (see Chapter 2), only electromagnetic forces are involved in chemical reactions.

ATOMS ARE COMPOSED OF PROTONS, NEUTRONS, AND ELECTRONS

All atoms contain a centrally located nucleus with protons and neutrons (except for the hydrogen nucleus, which contains a single proton). Surrounding the nucleus in a series of approximately concentric shells are the electrons. Electrons in the innermost shell are the hardest to remove from the atom, while electrons in the outermost shell are the easiest to remove. The outermost shell is called the *valence shell* because in chemical reactions between atoms, electrons are usually added to or removed from this shell.

Electrons rotate around the nucleus at such high speeds that it is impossible to know their exact location at any given time. For simplicity, electrons are often depicted as small spheres occupying circular orbits much like planets orbiting the Sun (Fig. 1). A more realistic depiction is described by the electron orbital model, which indicates the volume of space in which a particular electron is likely to be found most of the time (Fig. 2). The shell closest to the nucleus consists of a single orbital containing up to two electrons. The second and third shells each consist of four orbitals, and each of these orbitals contains two electrons.

Subatomic particles are distinguished by mass and electrical charge. A proton has a mass of one unit and a positive electrical charge of one unit. A neutron also has a mass of one unit, but it has no net charge. An electron has a mass only about 1/1800 of that of a nuclear particle, but it has a negative charge equivalent in magnitude to the positive charge possessed by a proton. The net charge and mass of an atom are equal to the sum of the charges and masses, respectively, of its constituent particles. In calculating mass we can usually ignore the mass of the electrons. Thus an atom's mass is very nearly equal to the total number of protons and neutrons that it contains.

PERIODIC TABLE IS ARRANGED TO EMPHASIZE ELECTRON STRUCTURE

Atoms of a single type, each with the same number of protons, constitute an element. In the Periodic table (Fig. 3) the elements are arranged in a way that is most useful to chemists. Elements are ordered from left to right by *atomic number* (number of protons). Elements in the same vertical column have similar chemical characteristics because they have the same number of electrons in their valence shell. Thus elements in the leftmost column have only one electron in their valence shells, while elements in the adjacent column have two electrons in this shell and so on. Finally, in the rightmost column the valence shell is fully occupied, which accounts for the inactivity of He and the other inert gases. Atoms that have the same number of protons but different numbers of neutrons are referred to as isotopes of the same element. Carbon's most prevalent isotope has six neutrons and an atomic mass of 12. Other, much less abundant isotopes of carbon have atomic masses of 13 and 14. For the purpose of our discussion

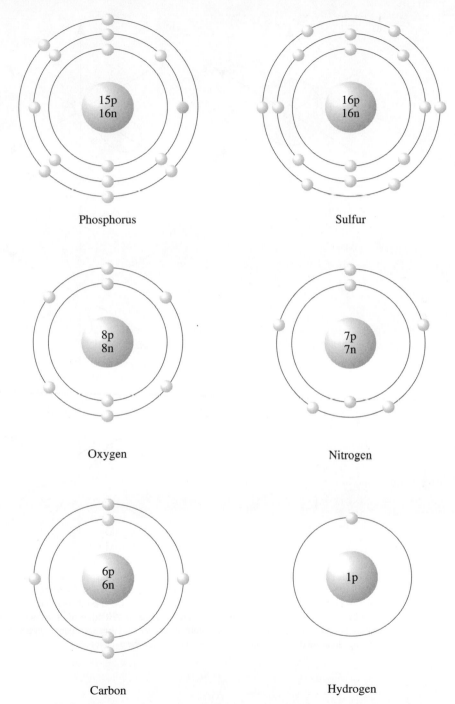

Phosphorus

Sulfur

Oxygen

Nitrogen

Carbon

Hydrogen

FIGURE 1   The atoms of life. These Bohr models show the structure of the six most common atoms found in living systems.

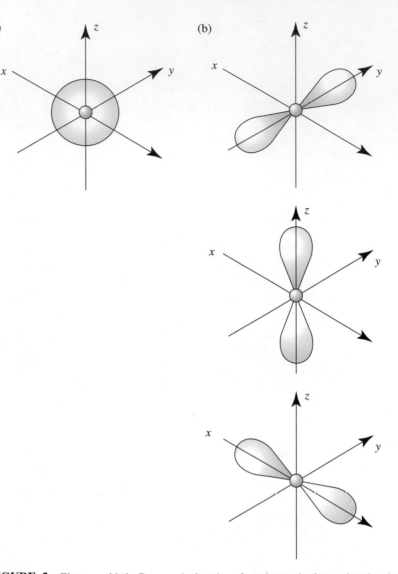

FIGURE 2   Electron orbitals. Because the location of an electron is always changing, it is more accurately represented by an orbital, which is the volume of space in which an electron is likely to be found 90% of the time. (a) The first energy level consists of one spherical orbital containing up to two electrons. The second energy level has four orbitals, each describing the distribution of up to two electrons. One of these orbitals is spherical; the other three are dumbbell-shaped and arranged perpendicular to one another, as the axis lines indicate (b). The nucleus is at the center where the axes intersect.

1	2	3	4	5	6	7	8	9	10	11	12	13	14	15	16	17	18
1 H 1.008																	2 He 4.003
3 Li 6.941	4 Be 9.012											5 B 10.81	6 C 12.01	7 N 14.01	8 O 16.00	9 F 19.00	10 Ne 20.18
11 Na 23.00	12 Mg 24.31											13 Al 26.98	14 Si 28.09	15 P 30.97	16 S 32.06	17 Cl 35.45	18 Ar 39.95
19 K 39.10	20 Ca 40.08	21 Sc 44.96	22 Ti 47.90	23 V 50.94	24 Cr 52.00	25 Mn 54.94	26 Fe 55.85	27 Co 58.93	28 Ni 58.71	29 Cu 63.55	30 Zn 65.38	31 Ga 69.72	32 Ge 72.59	33 As 74.92	34 Se 78.96	35 Br 79.90	36 Kr 83.80
37 Rb 85.47	38 Sr 87.62	39 Y 88.91	40 Zr 91.22	41 Nb 92.91	42 Mo 95.94	43 Tc 96.91	44 Ru 101.07	45 Rh 102.91	46 Pd 106.4	47 Ag 107.87	48 Cd 112.41	49 In 114.82	50 Sn 118.69	51 Sb 121.75	52 Te 127.60	53 I 126.90	54 Xe 131.30
55 Cs 132.91	56 Ba 137.34	57 La 138.91	72 Hf 178.49	73 Ta 180.95	74 W 183.85	75 Re 186.2	76 Os 190.2	77 Ir 192.22	78 Pt 195.09	79 Au 196.97	80 Hg 200.59	81 Tl 204.37	82 Pb 207.2	83 Bi 208.98	84 Po (210)	85 At (210)	86 Rn (222)
87 Fr (223)	88 Ra 226.03	89 Ac (227)	104 Rf (257)	105 Ha (260)													

FIGURE 3   The Periodic table. The Periodic table was originally devised in the 19th century to illustrate the groupings in which the chemical and physical properties of the elements recur when the elements are arranged in order of increasing atomic weight. Modern versions of the table arrange the elements slightly differently—by atomic number, instead of atomic weight—and they incorporate elements not known until this century.

In the version shown here, the number above the element symbol is the atomic number. The number below it is the average mass number, calculated from the masses of the various isotopes of the element. Elements in the lanthanide series (atomic numbers 58–71) and the actinide series (atomic numbers 70–103) are omitted because they play no role in living systems. Of the remaining elements, 24 play some role in them. Those elements used in major amounts are indicated by the darkest shading; those used in lesser amounts, by medium shading; and those used in trace amounts, by light shading.

we can focus on the most commonly occurring isotopes for each of the major elements found in biomolecules.

ATOMS CAN COMBINE TO FORM MOLECULES

Except for the inert gases (He and the other elements in the He column of the Periodic table), atoms are prone to interaction with other atoms to form molecules. Hydrogen, oxygen, and nitrogen each readily forms simple diatomic (two-atom) molecules. Invariably molecules have properties that are very different from those of the constituent elements. For example, a molecule of sodium chloride contains one atom of sodium (Na) and one atom of chlorine (Cl). Sodium is a highly reactive silvery metal, while chlorine is a corrosive yellow gas. When equal numbers of sodium and chlorine atoms interact, there is a vigorous chemical reaction and the white crystalline solid sodium chloride (NaCl) is formed.

Molecules are described by writing the symbols of the constituent elements and indicating the numbers of atoms of each element in the molecule as subscripts. For

TABLE 1
Chemical Bonds

Type	Chemical basis	Strength	Example
Ionic	Attraction between oppositely charged ions	Strong	Sodium chloride
Covalent	Sharing of electron pairs between atoms	Strong	Carbon–carbon bonds
Hydrogen	Attraction of a hydrogen atom with a partial positive charge to negatively charged atoms in neighboring molecules	Weak	Cohesiveness of water

example, the sugar molecule glucose is represented as $C_6H_{12}O_6$, which indicates that it contains 6 atoms of carbon and oxygen and 12 atoms of hydrogen.

Atoms react with one another by gaining, losing, or sharing electrons to produce molecules. Such an interaction is called a chemical reaction. For example, if hydrochloric acid (HCl) is added to sodium hydroxide (NaOH), sodium chloride (NaCl) and water (H_2O) are formed. A chemical reaction entails the making and breaking of attractive linkages, called bonds, between atoms (Table 1). The type of chemical bond that forms between atoms depends on the number of electrons the atoms have in their outermost valence shells.

IONIC BONDS FORM BETWEEN OPPOSITELY CHARGED ATOMS

Atoms of some of the elements most prevalent in living organisms (carbon, nitrogen, oxygen, phosphorus, and sulfur) are most stable chemically when they have eight electrons in their valence shells. The chemical tendency toward filled valence shells is called *the octet rule*. One way to satisfy the tendency toward filled valence shells is for an atom with one, two, or three electrons in its valence shell to lose them to an atom with, correspondingly, seven, six, or five electrons in its valence shell. This is a kind of chemical give-and-take. A sodium atom, for example, has a single electron in its outermost shell. Therefore sodium has a strong tendency to lose this single electron, because then it will have precisely eight electrons in its outermost shell. By contrast, chlorine has a strong tendency to gain a single electron to fill its valence shell, which contains only seven electrons.

The result of gaining or losing electrons, while the proton number remains unchanged, is an atom with a net positive or negative charge, respectively. Such an atom is called an *ion*. The attraction between oppositely charged ions results in an *ionic bond* (see Table 1). In this sodium chloride example, the oppositely charged ions, Na^+ and Cl^-, attract each other in an ordered manner, and a crystal of salt results (Fig. 4).

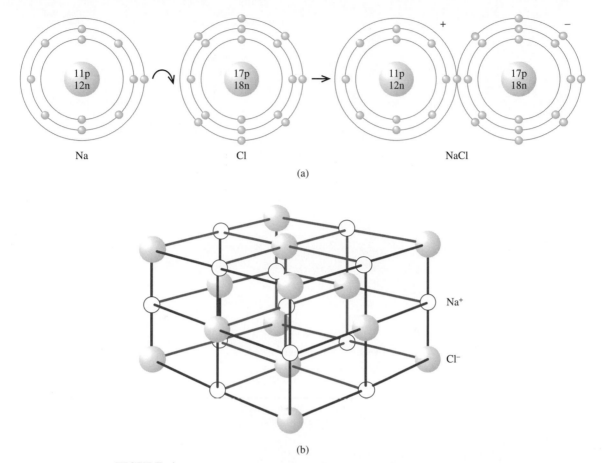

(a)

(b)

FIGURE 4   An ionically bonded crystal of sodium and chloride ions. (a) A sodium atom (Na) can donate the one electron in its valence shell to a chlorine atom (Cl), which has seven electrons in its outermost shell. The resulting ions (Na^+ and Cl^-) bond to form the compound sodium chloride (NaCl), better known as table salt. The octet rule has been satisfied. (b) The ions that constitute NaCl are arranged in a repeating pattern.

COVALENT BONDS FORM BETWEEN ATOMS THAT SHARE ELECTRON PAIRS

Atoms that have three, four, or five electrons in their valence shells are more likely to share electrons in a *covalent bond* than to swap them in the electron give-and-take of an ionic bond. Carbon, with four electrons in its outer shell, is a good example. It can attain the stable eight-electron configuration in its outer shell by sharing electrons with four hydrogen atoms, each of which has one electron in its only shell. The resulting compound is methane, CH_4 (Fig. 5).

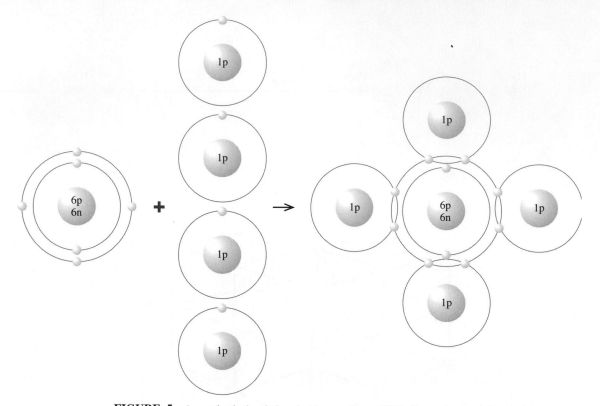

FIGURE 5 A covalently bonded molecule—methane (CH₄). One carbon and four hydrogen atoms complete their outemost shells by sharing electrons.

Two or three electron pairs also can be shared in covalent bonds, which are termed double and triple bonds, respectively. Carbon atoms can form single bonds with all the hydrogen atoms in methane. They form a double bond with the oxygen in formaldehyde, and they form a triple bond with the hydrogen cyanide (Fig. 6).

$$\begin{array}{ccc} & \text{H} & \\ & \overset{\cdot\cdot}{\underset{\cdot\cdot}{\text{C}}} & \\ \text{H} : & \text{C} & : \text{H} \\ & \overset{\cdot\cdot}{\underset{\cdot\cdot}{\text{H}}} & \end{array} \qquad \begin{array}{c} \text{H} \\ \overset{\cdot}{} \\ \overset{\cdot}{}\text{C} :: \text{O} \\ \text{H}\overset{\cdot}{} \end{array} \qquad \text{H} : \text{C} ::: \text{N}$$

$$\begin{array}{c} \text{H} \\ | \\ \text{H}-\text{C}-\text{H} \\ | \\ \text{H} \end{array} \qquad \begin{array}{c} \text{H} \\ \diagdown \\ \text{C} = \text{O} \\ \diagup \\ \text{H} \end{array} \qquad \text{H}-\text{C} \equiv \text{N}$$

FIGURE 6 Covalent linkages formed by carbon with hydrogen, oxygen, and nitrogen. Each shared electron pair is represented by two dots or a single straight line.

THE STRUCTURE OF WATER AND THE
INTERACTION OF WATER WITH OTHER
WATER MOLECULES

THE INTERACTION OF WATER
WITH OTHER MOLECULES

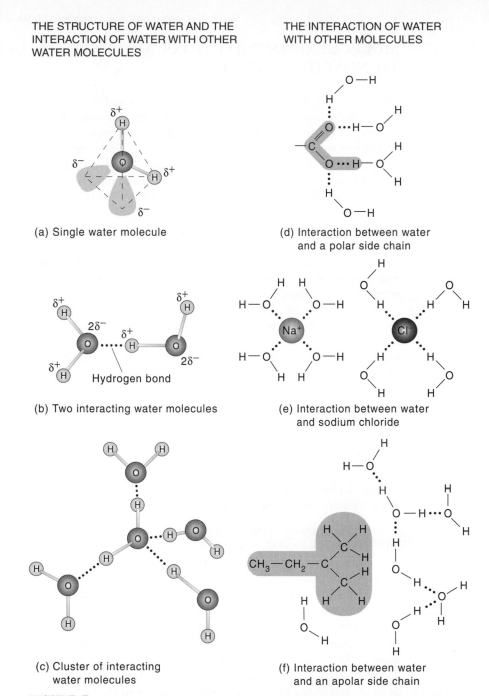

(a) Single water molecule

(b) Two interacting water molecules

Hydrogen bond

(c) Cluster of interacting
water molecules

(d) Interaction between water
and a polar side chain

(e) Interaction between water
and sodium chloride

(f) Interaction between water
and an apolar side chain

FIGURE 7   The structure of the water molecule and the secondary structures formed by multiple water molecules. (a) Water is a highly polar molecule. The hydrogen carries a partial positive charge (δ^+), the oxygen, a partial negative charge (δ^-). (b) Two water molecules interact to form a hydrogen bond between the oxygen of one molecule and one hydrogen of another molecule. (c) In bulk water a network of hydrogen bonds is formed. (d) Water interacts with polar molecules by forming hydrogen bonds. (e) Ions also interact with water. The cations (positively charged ions) interact with the oxygen atoms, and the anions (negatively charged ions) interact with the hydrogen atoms. (f) Apolar molecules such as hydrocarbons do not interact electrostatically with water molecules.

MOLECULAR INTERACTIONS ARE LARGELY DUE TO NONCOVALENT FORCES

The strong interactions that lead to ionic bonds and covalent bonds determine the primary structures of molecules. When molecules interact with each other, very often the interaction occurs not by forming more ionic or more covalent bonds, but instead by forming weaker bonds resulting from *polar* and *apolar* interactions.

Polar interactions are easier to understand. Polar molecules are produced when the electrons in a covalent bond are not equally shared. In a water molecule (H_2O), for example, the single oxygen atom has a stronger hold on the shared electrons than do the two hydrogen atoms. As a result, the shared electrons are closer most of the time to the oxygen atom. Because of the unequal (polar) sharing, the hydrogen atoms have a partial positive charge, while the oxygen atoms have a partial negative charge. The partial charges lead to a strong attraction between nonbonded hydrogen and oxygen atoms in liquid water. The result is what is referred to as a *hydrogen bond* (see Table 1 and Fig. 7).

Molecules that are apolar also are attracted to each other, in this case by so-called *van der Waals forces*. Although both polar and apolar bonds are much weaker than ionic and covalent bonds, they frequently outnumber the stronger bonds in the liquid state, when molecules are nearly close-packed. For this reason the weaker forces usually exert a decisive influence on the structure of water and of many biopolymers that exist in an aqueous environment (see Chapter 6).

SUMMARY

In this chapter we have considered the basic aspects of chemistry that involve the interaction of electrons between different atoms.

1. In their neutral form at temperatures below 3000 K, atoms contain an equal number of protons and electrons.

2. The electrons surround the nucleus in a series of approximately concentric shells. It takes more energy to remove an electron from an inner shell than from an outer shell. The outermost shell is called the valence shell because in chemical reactions electrons are usually added or removed from this shell.

3. Atoms of the elements most prevalent in living organisms (carbon, nitrogen, oxygen, phosphorus, and sulfur) are most chemically stable when they have eight electrons in their valence shells. One way to satisfy the tendency toward filled valence shells is for an atom with one, two, or three electrons in its valence shell to lose them to an atom with correspondingly, seven, six, or five electrons in its valence shell. This transfer of electrons creates negatively charged and positively charged ions. The attraction between oppositely charged ions results in an ionic bond.

4. Atoms that have three, four, or five electrons in their valence shells are more likely to share electrons in a covalent bond.

5. Weak electromagnetic forces, which may be polar or apolar, lead to interactions between nonbonded atoms or molecules.

Problems

1. Does the electronic structure for phosphate (PO_4^{3-}) satisfy the octet rule for all the atoms involved?

2. The electronic structure of the hydrogen molecule (H_2) clearly does not satisfy the octet rule. Explain why.

3. In methane (CH_4), carbon is found in its most reduced form while in carbon dioxide (CO_2) it is found in its most oxidized form. In fact we often speak of carbon as having a formal valence of -4 in methane and a formal valence of $+4$ in carbon dioxide. Explain.

4. Figure 3 indicates that most of the elements used in biosystems have a low mass. Why do you suppose this is the case?

5. A free radical is an atom or a molecule that contains an unpaired electron. Molecular oxygen (O_2) is sometimes described as a diradical. Can you suggest how molecular oxygen could form a diradical considering the number of electrons it contains?

6. Can you describe the electronic structure for ozone (O_3)?

Reference

Any up-to-date general chemistry text will do if your basic chemistry is rusty.

CHAPTER 4

Element Abundances of the Planets

The extent to which a planet is endowed with specific elements important to its surface environment is closely tied to the details of the formation process. Clues as to the time and manner in which our planetary system formed come from three sources: orbits and densities of the planets, chemical analyses of rocks from Earth, and similar analyses of meteorites. In this chapter we describe element abundances of the different planets, their time of formation, the key bits of evidence that led to these descriptions.

PLANETS MUST HAVE FORMED FROM THE SAME NEBULA AS THE SUN

The solar system contains nine planets that orbit the Sun. A number of the physical characteristics of these planets are given in Table 1. All these properties were determined by direct astronomical observations. As can be seen, there is a tendency for the smaller planets to be located closer to the Sun, the one exception being Pluto. The *sidereal period* is the time it takes to make one orbit around the Sun. Once again, we see the biggest contrast between the four inner planets and the four larger, outer planets, which have considerably longer sidereal periods. This difference is related to both the orbital speed and the longer pathway that is taken by the outer planets. Whereas all the orbits are close to circular, they do have some elliptical aspect, which is described

TABLE 1
Some Physical Properties of the Planets

Planet	Diameter (km)	Sidereal period (years)	Rotation period (days)	Maximum distance from the Sun (10^6 km)	Orbital eccentricity	Mean orbital speed (km/s)
Mercury	4,878	0.2408	58.6	57.9	0.206	47.9
Venus	12,104	0.6152	243.0[a]	108.2	0.007	35.0
Earth	12,756	1.0000	0.997	149.6	0.017	29.8
Mars	6,794	1.8809	1.026	227.9	0.093	24.1
Jupiter	142,800	11.86	0.41	778	0.048	13.1
Saturn	120,000	29.41	0.43	1,426	0.056	9.6
Uranus	51,120	84.94	0.65[a]	2,871	0.046	6.8
Neptune	49,528	164.79	0.67	4,497	0.010	5.4
Pluto	2,290	248.6	6.4	5,914	0.248	4.7

[a]Rotation of Venus and Uranus is retrograde so that the Sun rises in the west and sets in the east for these two planets.

by a major axis a and a minor axis b. The orbital eccentricity is given by $(1 - b^2/a^2)^{1/2}$, so the closer the orbit is to a perfect circle, the smaller this quantity. The distance from the Sun for the different planets varies by two orders of magnitude. The shorter this distance is, the higher the orbital speed necessary to maintain a stable orbit. Kepler's third law states that the sidereal period P is proportional to the semimajor axis of the elliptical orbit by the equation

$$P^2 = ka^3, \tag{1}$$

where k is a constant.

The distance from the Sun also strongly influences the amount of solar radiation received by the planet, making Mercury unbearably hot, and Uranus, Neptune, and Pluto, cold. In addition to orbiting around the Sun, each planet rotates on its own axis. There is a tendency for most planets to make a full rotation in 1 day or less, the two exceptions being Mercury and Venus. The unusually slow rotation of Venus stands in sharp contrast to many other properties of Venus, which closely resemble those of Earth.

A most revealing find is that all the planetary orbits lie in the same plane, and all the planets except for Venus and Uranus rotate in the same direction, which is the same as the Sun's spin. The orbital characteristics are best explained by the hypothesis that the stellar cloud from which the Sun formed was spinning and that the Sun and the planets were formed from the same cloud around the same time.

PLANETS DIFFER IN MASS, DENSITY, AND COMPOSITION

As an aid to understanding the overall composition of a planet or other solar body, it would be advantageous to have some way of determining its overall density. The size can be estimated from external dimensions. If we could determine the mass we also could determine the average density, because density is the ratio of mass to volume. The easiest way to estimate the mass is from the gravitational influence a planet exerts on the orbits of its satellites. Because all planets except Mercury and Venus have satellites, this method is generally applicable. A satellite that is orbiting a planet is constantly being accelerated as it changes its direction of movement. Its centripetal acceleration is equal to its velocity squared divided by its orbital radius, $a = v^2/R$. From Newton's second law of motion the force associated with a satellite of mass M_1 is related to its centripetal acceleration by the equation

$$F = M_1 a = M_1 v^2/R \cdot \qquad (2)$$

From Newton's law of gravitation this force also is given by the equation

$$F = G M_1 M_2/R, \qquad (3)$$

where G is the universal constant of gravitation, M_2 is the mass of the planet, and M_1 is the satellite mass. Equating these two forces and transposing gives us a simple relationship for the mass of the planet in terms of measurable quantities v and R

$$M_2 = v^2 R/G \cdot \qquad (4)$$

In metric units G has the value 6.672×10^{-11} Newton (N) $\cdot$ m^2/kg^2 where 1 N is equal to 10^5 dyn.

From Eq. 4 it should be noted that the rotational velocity of the satellite is independent of the satellite mass. This is true as long as the satellite mass is much less than the mass of the planet. A similar situation exists for the sidereal periods of the planets that are a function of their distance from the Sun but independent of their mass (see Table 1).

Another way of determining the planet mass is by the deflection in the path of a comet or a spacecraft that passes near the planet. The planet's gravitational pull, which is directly related to its mass, causes a deflection in the path of the comet or spacecraft from which the mass of the planet may be calculated by the laws of Newtonian mechanics.

It is not surprising to see that the *Jovian planets* (Jupiter, Saturn, Uranus, and Neptune) are much more massive than the *terrestrial planets* (Mercury, Venus, Earth, and Mars), because we already have seen that the planets in the former group are much larger. What is most informative is that the average density of the terrestrial planets is much higher (Table 2). This means that the average material of which the inner terrestrial planets are composed is denser. To what can we attribute this greater density if all the planets formed by accretion from the same nebula? The most obvious answer is that the terrestrial planets were not able to retain many of their less dense elements because of their lower gravitational pull, and also perhaps because of their higher average surface temperatures, which is due to their proximity to the Sun. The tendency of a planet to retain a volatile is best described by the speed that must be attained to

TABLE 2
Planet Mass and Density[a]

Planet	Mass (kg)	Average density (g/cm^3)
Mercury	3.3×10^{23}	5.4
Venus	4.9×10^{24}	5.2
Earth	6.0×10^{24}	5.5
Mars	6.4×10^{23}	3.9
Jupiter	1.9×10^{27}	1.3
Saturn	5.7×10^{26}	0.7
Uranus	8.7×10^{25}	1.3
Neptune	1.0×10^{26}	1.7
Pluto	1.3×10^{22}	2.0
Moon	7.3×10^{22}	3.3

[a]The quantities mass and weight are often confused. The mass of an object is a measure of the total amount of material of the object; it is usually expressed in gram (g) or kilogram (kg) units. Weight is the force with which an object presses down on the ground due to the pull of gravity; it is usually expressed in pounds, newtons, or dynes. Newton's second law can be used to relate mass and weight. For example, the acceleration caused by Earth's gravity is 32 ft/s^2, or 9.8 m/s^2. From the second law, the force with which an average male presses down on the ground is

$$79 \text{ kg} \times 9.8 \text{ m/s}^2 = 774 \text{ N} = 174 \text{ lb.}$$

This answer is only correct when one is standing on Earth. A person would weigh less on the Moon and considerably more on Jupiter even though their mass would be the same.

escape the gravitational pull of the planet. This *escape velocity* V_{esc} is given by the equation

$$V_{esc} = \left(\frac{2\,GM}{r} \right)^{1/2}, \tag{5}$$

where M is the mass of the planet, r is its radius, and G is the universal gravitational constant. The escape velocity for different planets is recorded in Table 3.

TABLE 3
Escape Velocities

Planet	In miles per hour	In km/s
Mercury	9,600	4.3
Venus	23,000	10.3
Earth	25,300	11.2
Mars	11,000	5.0
Jupiter	133,000	59.5
Saturn	79,600	35.6
Uranus	47,400	21.2
Neptune	52,800	23.6

For a given planet, the tendency for a gaseous molecule to escape is a function of the mass m of the molecule and the ambient temperature of the planet. This is so because the average velocity of a molecule is a function of these two parameters:

$$v = \left(\frac{3kT}{m} \right)^{1/2},$$
(6)

where T is the absolute temperature and k is the Boltzmann constant.

This equation actually gives the velocity of an average molecule of mass m at any given temperature. Because there is a distribution of velocities around the average, we would like to know how this average velocity relates to the likelihood that over an extensive time period the molecule in question is likely to be retained. To a first approximation, a planet retains a gaseous molecule if the escape velocity is six times greater than the average speed of the gaseous molecule (Box 4A).

On Earth the average speed of oxygen molecules is 0.48 km/s at 295 K, while the average speed of hydrogen molecules is 1.9 km/s. Because the escape velocity on Earth is 11.2 km/s, this means that Earth is likely to lose its hydrogen, while it is likely to retain its oxygen. In fact, the half-life of hydrogen in Earth's atmosphere is believed to be about a million years, which means that over geologic time any hydrogen in the atmosphere is likely to get lost.

When Earth was being formed, the surface temperature must have been much higher. A surface temperature 10 times the present surface temperature is conceivable. This would raise the average molecular velocity more than threefold, so that even a molecule the size of molecular oxygen would not be safe from escape. In fact, there are many other forms in which the element oxygen could find itself, for example, in a molecule of water, of carbon monoxide, or of nitrous oxide. Oxygen molecules also could dissociate into oxygen atoms by radiation exposure. When considering the likelihood of escape for a particular element from the atmosphere, all the different forms in which the element might exist must be considered.

CHEMICAL CLUES CONCERNING EARTH'S COMPOSITION COME FROM DENSITY CONSIDERATIONS AND ANALYSIS OF METEORITES

Surface rocks in Earth's crust have densities that vary from 2.6 to 3.0 g/cm^3. Nevertheless, the average density of Earth is 5.5 g/cm^3. Even if one takes into account the compression of the solid mass due to gravity, this difference is a clear indication that the composition of Earth's crust is not representative of the composition of Earth as a whole. Thus the bulk of the planet must include considerably denser substances, which exist at depths below those to which we can drill. We must

BOX 4A   Two Views of Gravity

Despite the success of Newton's law of gravitation in explaining planetary movements and retention of elements, Newton's law was challenged in 1916 by Albert Einstein in his general theory of relativity. According to this theory the mass of an object alters the properties of space and time around the object. Einstein eliminated the idea of a force of gravity. In many cases the general relativity theory gives almost exactly the same answers as Newtonian theory. A notable exception is the apparent gravitational deflection of light that passes close of the Sun. When light from a distant star observed on Earth passes close by the Sun, it bends (Fig. 4A). Newton's law of gravity cannot explain this bending if photons have no rest mass. However, this effect can be explained by the general theory of relativity as the result the mass of the Sun has on the curvature of space in its vicinity. There are some indications that suggest photons have a rest mass. This would effect these arguments.

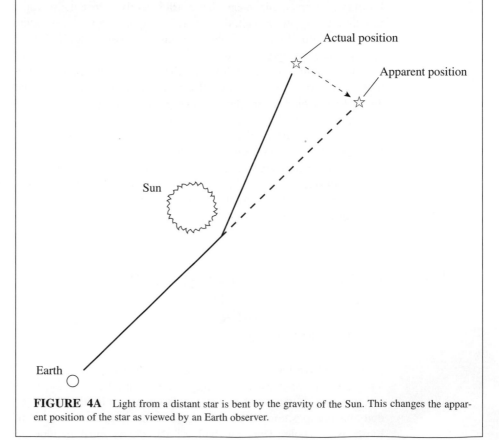

FIGURE 4A   Light from a distant star is bent by the gravity of the Sun. This changes the apparent position of the star as viewed by an Earth observer.

therefore use an indirect means to gather information about the overall composition of Earth. One way to do this is from the chemical composition of meteorites. Because meteorites are believed to have formed from the same solar nebula as Earth, they should reflect the overall composition of Earth except for the volatiles, which they would have lost because of the lack of significant gravitational forces in small bodies.

Most meteorites that have landed on Earth are probably fragments of small rock objects called asteroids that orbit the Sun. The majority of meteorites are made of stony material; Some of these, called *chondrites*, contain small spherules known as *chondrules*. Chondrulelike substances never have been found in Earth rocks. It is conjectured that they were formed from molten materials around the time of the origin of the Solar System.

For the purpose of using meteorites to estimate the overall composition of Earth we turn to a special class of chondritic meteorites known as *carbonaceous chondrites*. In carbonaceous chondrites, chondrules coexist with minerals that are unstable above 100°C, so we may conclude that these chondrites, except for the chondrules they contain, must have formed at low temperatures. Hence the carbonaceous chondrites should give the best estimate for the composition of the solar nebula, because they are unlikely to have lost substances that may have volatilized at the higher temperatures where the chondrules themselves formed.

An analysis for different metals present in carbonaceous chondrites shows that their relative amounts are strikingly similar to those found for the solar spectrum, except for the highly volatile elements. The three most prominent chondritic metals are magnesium, silicon, and iron. In considerably smaller amounts we find aluminum, calcium, nickel, and sodium. In still smaller amounts we find chromium, potassium, manganese, phosphorus, titanium, and cobalt. All other metals are present only in trace quantities. A detailed analysis for different chemical forms indicates that the four most abundant elements in chondrites are magnesium, silicon, iron, and oxygen; and that the bulk of the magnesium is present in silicates or oxides, while the bulk of the silicon is present in silicates or silicon dioxide. The oxygen that is ignored in a pure metal analysis is present in both silicates and oxides.

Iron appears to be present both as metal and as iron complexes with silicon, oxygen, and sulfur. The presence of considerable iron in the metallic form is consistent with the high density of Earth. Metallic iron has a density of 7.5 g/cm^3. Because there is essentially no metallic iron in Earth's crust, this form of iron must be buried within Earth's interior. Because metallic iron is denser than any of the silicates or oxides, it seems likely that it could have sunk to the core of Earth, especially if Earth was hot enough to become molten at some time in its past. This possibility is discussed in the next chapter. A glance at the densities of the other terrestrial planets (see Table 2) indicates that except for Mars they have the same high densities. Thus Mercury, Venus, and Earth probably have similar ratios of oxides and silicates to metallic iron. Judging from densities alone, Mars—for unknown reasons—does not appear to possess much metallic iron.

TERRESTRIAL ABUNDANCES OF THE ELEMENTS ARE A FUNCTION OF ELEMENT ABUNDANCES IN THE UNIVERSE, CHEMICAL REACTIONS, AND LOSS OF VOLATILES

The abundances of elements on Earth, as estimated from their abundances in chondrites, are strongly correlated with the relative abundances of the elements in the Universe as measured by their abundances in the solar spectrum. These abundances have been influenced, however, by the tendency of the elements to exist in nonvolatile forms. All elements other than the inert gases have some tendency to take part in chemical reactions.

Table 4 compares the abundances of 10 elements in the solar spectrum and in the chondritic meteorites, relative to 10^6 silicon atoms. These comparisons show that the vast majority of hydrogen, helium, and neon atoms have been lost in the meteorites. This is not surprising because, although hydrogen exists in a number of forms, all these are quite volatile (Table 5). Helium and neon are both inert gases of low atomic number and hence of low mass. Because neither of them makes chemical combinations, helium and neon are bound to escape in a situation where the gravity is low, regardless of the temperature. A small amount of carbon is retained, probably because of the formation of relatively nonvolatile carbonate salts. Nitrogen is retained even more effectively than carbon, possibly because nitrogen can form salts when it is present as NH_4^+ and NO_3^- or NO_2^-. Oxygen retention in the meteorites is excellent. This is undoubtedly due to the affinity of oxygen for metals and silicon in silicates and oxides. The retention of magnesium and silicon also is excellent. Again, this is undoubtedly due to these elements forming highly nonvolatile silicates and oxides.

TABLE 4
Atomic Abundances of the 10 Most Common Elements in the Universe[a]

Atomic No.	Element	Relative abundance in solar spectrum[b]	Relative abundance in chondrites[b]
1	Hydrogen (H)	2×10^{10}	—
2	Helium (He)	1.2×10^9	Trace
6	Carbon (C)	8.0×10^6	2×10^3
7	Nitrogen (N)	1.8×10^6	5×10^4
8	Oxygen (O)	1.4×10^7	3.7×10^6
10	Neon (Ne)	8×10^5	Trace
12	Magnesium (Mg)	8×10^5	9.4×10^5
14	Silicon (SI)	10^6	10^6
16	Sulfur (S)	4×10^5	1.1×10^5
26	Iron (Fe)	6×10^5	6.9×10^5

[a]Adapted from Kaufmann, W. J., III. *The Universe*, by W. J. Kaufmann III, © 1985, 1988, 1991, 1994, 1999 by W. H. Freeman and Company. Used with permission. (New York: W. H. Freeman, 1994) p. 134.

[b]Relative to 10^6 silicon atoms.

TABLE 5
Some Common Forms of the 10 Major Elements
Found on the Primitive Earth

Element	Common forms
H	H_2, H_2O, CH_4, NH_3
He	He
C	CH_4, CO, CO_2, CO_3^{2-}
N	N_2, NH_3, NO_2^-, NO_3^-
O	O_2, H_2O, O^{2-} in silicates and oxides
Ne	Ne
Mg	Mg^{2+} in silicates and oxides
Si	Si^{4+} in silicates and silicon dioxide
S	S, H_2S, and S^{2-} in sulfides
Fe	Fe metal and Fe^{2+} and Fe^{3+} in silicates, oxides, and sulfides

The retention of sulfur is fair, probably because of the formation of relatively nonvolatile iron sulfides. Finally, the retention of iron is excellent, because it is highly nonvolatile in all its chemical forms.

One could go on to consider other element abundances, but the principles involved in element retention by chondrites and terrestrial planets should be clear from this short list. For a given element, its abundance is mainly a function of two factors: (1) relative abundances of that element in the Universe as a whole and (2) ability of the element to exist in nonvolatile chemical forms.

JOVIAN PLANETS ARE MOST LIKELY TO HAVE RETAINED THEIR VOLATILES

The much greater sizes of Jovian planets (see Table 2) give them considerably higher escape velocities. Because these planets also have considerably lower surface temperatures, they are likely to have retained a good many of their volatiles, including hydrogen. This supposition is strongly supported by the substantially lower densities of the Jovian planets (see Table 2). We limit ourselves to discussing some of the properties of Jupiter.

Jupiter, the largest of all the planets, is believed to have retained virtually all its hydrogen, and therefore, overall, it probably has an element composition very similar to that of the Sun. Indeed, it has been reckoned that if Jupiter had been a bit larger, it might have turned into a star with a self-sustaining nuclear fire. However, because it did not, Jupiter must have a more stratified element distribution than the Sun has.

The gross structure of Jupiter consists of an atmosphere rich in molecular hydrogen, with much smaller amounts of methane (CH_4) and ammonia (NH_3). Below this atmosphere there exists an enormous body of liquid hydrogen, which is so highly compressed that it behaves like a metal. At the center, Jupiter is believed to have a highly compressed rocky core about 13 times the mass of Earth. The temperature at the center of Jupiter has been estimated at about 25,000 K. This drops to 165 K at the surface.

Jupiter and Saturn each has more than a dozen satellites. Four of Jupiter's satellites—Io, Europa, Ganymede, and Callisto—rival Mercury in size and could be classified as terrestrial planets. All four of these satellites orbit Jupiter in the plane of the planet's equator. The largest of Saturn's satellites, Titan, has been much studied because of its thick opaque atmosphere, which is rich in methane, nitrogen, hydrogen, and a complex mixture of hydrocarbons. Most of the nitrogen on Titan is in the form of molecular nitrogen instead of ammonia, because of the great sensitivity of ammonia to ultraviolet (UV) radiation received from the Sun.

ISOTOPE DATING OF CERTAIN METEORITES INDICATES THAT EARTH FORMED ABOUT 4.6 BILLION YEARS AGO

Geologists have spent a great deal of time studying the age of rocks that are believed to have originated on Earth. In general, rocks found on dry land are far older than those found on the sea floor. The oldest known rocks that have been dated are believed to be about 3.8 billion years old. We cannot expect that further searches will uncover rocks much older than this, but neither can we take 3.8 billion years as an accurate estimate for the age of Earth. The reason for this is that Earth's crust is subject to constant upheaval. In time, material that is presently on the surface probably will be subducted into the depths of Earth, and new material, which has been subject to melting in Earth's interior, will have taken its place. In view of this constant change, we cannot get a reliable estimate for the age of Earth from the age of rocks that currently occupy its surface (Box 4B).

Faced with this problem, geologists turned once again to investigating the properties of the chondrules that are found in chondritic meteorites. Because chondrules are

---

BOX 4B   The Lunar Surface Is Older Than Earth's Surface

The lunar surface has been subject to far less turnover than Earth's surface and therefore an age determination of lunar surface rocks might be expected to give a more accurate estimate of the age of the Solar System than the analysis of surface rocks on Earth. The entire lunar surface is covered with a layer of fine powder and rock fragments produced by continuous bombardment. Typical rock specimens brought back by Apollo astronauts are between 4.0 and 4.6 billion years old.

believed to have formed from molten materials around the time of the origin of the Solar System, a determination of their age should give us a good estimate for the age of the Solar System and therefore for the age of Earth and other planets, which are believed to have formed at about the same time.

How does one go about determining the age of ancient objects like this? The standard approach is to use a method called isotope dating. In the early 1950s, shortly after radioisotopes had been discovered, W. F. Libby used isotope dating to fix the age of mummies and/or ancient fossils believed to be those of our primate ancestors. The isotope on which Libby focused his attention was ^{14}C, which has a half-life of 6400 years. The half-life is the time required for the disintegration of exactly half of the unstable substance (see Box 2A). For a first-order reaction the half-life is equal to $0.693 \, K^{-1}$ where K is the first-order decay constant for a radioisotope. Libby reasoned that when the mummies and fossils were formed, they incorporated carbon at a ^{14}C to ^{12}C ratio that probably had not changed for a long time. However, after incorporation, the ^{14}C to ^{12}C ratio would continue to drop because of the disintegration of the ^{14}C. The estimates that Libby obtained by this approach still are considered to be the most reliable, given the likely age of the specimens he was studying.

If one is looking at rock specimens, for which the expected ages may be much longer, it is necessary to use isotopes with far longer half-lives. The working rule is that the most accurate estimates by isotope dating are obtained when the half-life of the isotope under scrutiny is comparable with the age of the substance being investigated. In the case of rocks that might go back to the time of the origin of Earth, the isotopes used should have half-lives on the order of billions of years.

The most precise estimates for the age of chondrules have come from measurements of the abundances of the radioactive isotope for rubidium (^{87}Rb), which has a half-life of 49 billion years and disintegrates into strontium (^{87}Sr) by β decay, the process in which a neutron is converted into a proton and an electron (see Chapter 2). By determining the ratio of these two isotopes in chondrules and by correcting for the ^{87}Sr that was believed to have arisen from other sources, the ages of many chondrules have been estimated. For a series of chondrules from different meteorites, the estimates varied over a very narrow range from 4.52 to 4.63 billion years. The near constancy of the ages of chondrules from widely different sources gave reassurance that all chondrules probably formed around the same time and are probably a reliable source for estimating the age of the Solar System and the planets within it.

SUMMARY

The chemical composition of Earth and its fellow planets is a function of the formation process. Clues as to how the planet system originated come from three sources: (1) planetary orbits and densities; (2) direct chemical analyses of rocks from Earth; and (3) chemical composition of meteorites.

1. All planets orbit the Sun in the same direction and all the orbits lie in the same plane. This is consistent with the hypothesis that the Sun and planets formed from the same spinning cloud.

2. From a determination of the abundances of long-lived radioisotopes and their daughter products in meteorites, it has been calculated that planetary formation occurred 4.56 billion years ago.

3. The mass of each planet is determined from the gravitational influence it exerts on the orbits of its moons and on the space probes sent out from Earth.

4. From the observed size of the planets and their measurable masses the densities can be computed. Density determinations of the planets provide clues concerning the chemical composition of the planets. The densities for the four inner planets are much higher than those of the four main outer planets. Clearly, the four major outer planets (Jovian planets) must be made of elements with lower densities than the four inner (terrestrial) planets have.

5. The best evidence concerning the composition of terrestrial planets comes from meteorites. These analyses indicate that four elements — oxygen, silicon, magnesium, and iron — account for most of the mass of Earth and other terrestrial planets.

6. The solar abundances of various elements, together with their chemical affinities, are the two factors that explain which elements should dominate the composition of a planet.

7. Unlike the terrestrial planets, the Jovian planets — Jupiter, Saturn, Neptune, and Uranus — were able to retain a major portion of their volatiles because of greater gravitational pulling. Indeed, the densities of Jupiter and Saturn are so low that they must consist mainly of hydrogen and helium gas.

Problems

1. If the orbits of Jupiter and Earth were exchanged tomorrow, how do you expect that this would change their properties in the future?

2. If Mar's orbit and Earth's orbit had been exchanged at the time of formation of the Solar System, how would this have affected their development to the present time?

3. Derive Kepler's third law (Eq. 1) from Newton's second law of motion (Eq. 2) and his universal law of gravitation (Eq. 3).

4. What is the difference in escape velocity for hydrogen in molecular hydrogen and hydrogen in methane?

5. The Sun's density is 1.4 g/cm^3. What does this tell you about its composition?

6. What is meant by the statement that the Sun is a second-generation star and what is the evidence for this?

7. A carbon specimen found in a cave believed to have been inhabited by cavemen contained one-eighth as much ^{14}C as an equal amount of carbon in living matter. What is the approximate age of the specimen? Use a half-life for ^{14}C of 5730 years.

References

See general references given in Chapter 1.

Helmley, R. Sound and Fury in Jupiter, *Science* 269: 1233, 1995. (Jupiter is a ball of mostly hydrogen and helium. The high pressure and temperature cause hydrogen to form a plasma with poorly understood properties.)

McKinnon, W. B. Galileo at Jupiter—meetings with remarkable moons, *Nature (London)* 390:23, 1997. (The four large moons of Jupiter form the most coherently organized planetary system known.)

CHAPTER 5

Geologic, Hydrologic, and Atmospheric Evolution of Earth

Of the four terrestrial planets, only our Earth has evolved into a habitable planet. Mercury is barren, with very high surface temperatures. Venus is quite hot and is blanketed by a dense atmosphere of carbon dioxide; it has volcanic activity but its high carbon dioxide content and lack of water make it unsuitable for the origin or the maintenance of living forms. Mars, again, is barren, showing signs of having contained considerable surface water at one time. Presently, it shows very little surface activity. Like Venus, Mars has an atmosphere that is primarily composed of carbon dioxide, but at much lower concentrations. By contrast, Earth is covered with living forms, but even if we were to ignore these it would be obvious that Earth is "geologically and hydrologically alive." Two-thirds of its surface is covered with liquid water. The surface is segmented into several massive plates, which slowly move in different directions. Wherever two plates meet, there is earthquake and volcanic activity.

To what do we attribute the strikingly different attributes of Earth compared with those of the other terrestrial planets, which probably started with similar compositions? In this chapter we explore answers to this questions. In the process we discuss the present features of Earth and how they evolved, with special attention given to what Earth was like around the time of the origin of life. We start with a description of the nonvolatiles that constitute 99% of Earth's mass; next we discuss water; and finally, the atmospheric features.

EARTH HAS A LAYERED STRUCTURE

If we were to take a chondritic meteorite and heat it in a crucible until it melted, it would separate into a lower layer of metallic iron and an upper layer of metal silicates and oxides. The ratio of the two layers would vary from meteorite to meteorite. This separation into two layers results because of the considerably higher density of metallic iron (7.5 g/cm^3) compared with the density of metal silicates and oxides (3.5 g/cm^3). We might expect that the same type of separation occurred on Earth at some time after it was formed. During the process of formation, Earth must have become extremely hot as a result of gravitational heat produced by the accretion process. It is easy to imagine that the process elevated the temperature of Earth's mass to the melting point.

Consistent with this picture, Earth is believed to have a layered structure with a core principally composed of metallic iron (Fig. 1). The very center of the core is believed to be solid, because of the extreme pressures induced by gravitational forces, while the rest of the core is believed to be liquid metal of approximately the same composition. Proceeding outward from the liquid core, we find a solid layer called the *mantle*, which principally is composed of metal silicates and oxides. Because silicates and oxides have a considerably higher melting temperature than metallic iron has, there is no contradiction in having a mixture of solid silicates and oxides in direct contact with a liquid iron core. Closer to the surface, in the topmost region of the mantle, is a thin layer called the *asthenosphere*, which is quite soft because, at the reduced pressures this close to the surface, it is very near the melting point for mantle material. The uppermost layer of Earth, the *lithosphere*, is only about 100 km thick. This is the layer that is often called the crust. It is considerably more rigid than the asthenosphere because of the greatly reduced temperature this near the surface. On the surface itself we find two distinctly different layers. The first, found on dry land masses, principally is composed of *granites*, while the second, found on ocean floors, is composed of denser *basalts*. We have more to say about basalts and granites in a later section.

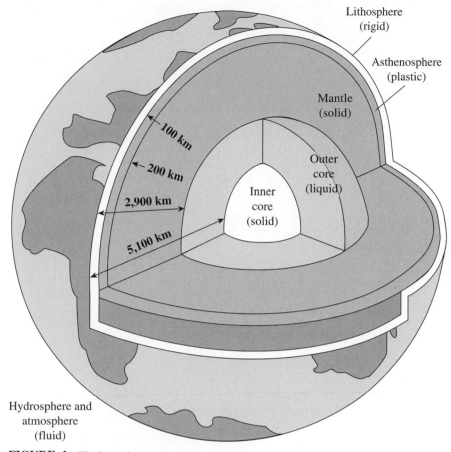

FIGURE 1 The layered structure of inner Earth.

EVIDENCE FOR THE LAYERED STRUCTURE COMES FROM STUDIES OF SEISMIC WAVES

It is approximately 6300 km from the surface to the center of Earth. The deepest drillings go no more than about 10 km into Earth, and erosions expose rocks that are little more than 20 km below the surface. This means that our understanding about the vast inner structure of Earth is all indirect.

Most information concerning Earth's interior has come from seismological investigations of the transmission of vibrations originating from earthquakes. These waves are of three different types: (1) *primary waves* (*P waves*), which are longitudinal waves of compression and rarefaction, where the wave moves in the direction of

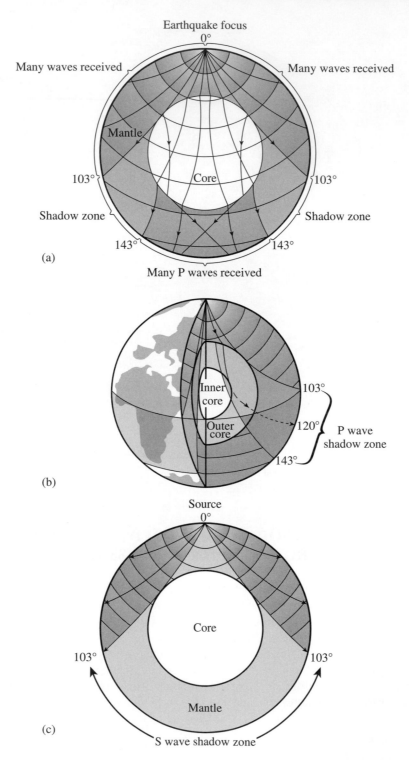

(a)

(b)

(c)

propagation of the waves motion; (2) *secondary waves* (*S waves*), which move more slowly in the direction of wave travel (the particles responsible for S waves move back and forth at right angles to the direction of wave propagation); and (3) waves that travel on or near the surface, thus telling us relatively little about the interior structure of Earth.

Seismologists have set up stations around the globe so that they can monitor the vibrations resulting from the seismic waves that originate from earthquakes. Because P waves move more rapidly than S waves, they are the first waves to be recorded by seismographs after an earthquake. P waves can travel through both solids and liquids, but S waves can travel only through solids. If S waves reach a solid–liquid border, they are deflected. The result is a very different distribution for the two types of waves (Fig. 2). The analysis of the combined patterns for the two types of waves demonstrates a solid–liquid discontinuity at a depth of about 2800 km. An additional solid core within the liquid core has been detected by the fact that even the P waves appear to be deflected at this second liquid–solid boundary.

From these seismological investigations and density considerations a description of the layered structure of Earth has emerged. Earth is composed of a massive solid layer called the *mantle*, which contains metal silicates and oxides surrounding a predominantly liquid iron core. There are additional indications that support this overall picture. For example, pieces of mantle that have thrust up through discontinuities in Earth's surface contain silicates and oxides but no iron metal. On the other hand, support for a liquid iron center comes from observations of Earth's magnetic field, which can be explained by eddy currents generated in a liquid iron core. The rotation of Earth on its own axis results in a distortion of Earth's shape from a perfect sphere. The poles are slightly flattened and the equator has a distinct bulge. The magnitude of this distortion could be explained if Earth's mass is concentrated at its center, which is exactly what would be expected from a dense iron core.

The mass of Earth's core has been estimated at about 1.9×10^{27}g, while the mass of the mantle is estimated at about 4.1×10^{27} g. From lavas containing mantle materials it is estimated that the mantle is about 8% by weight iron, principally in the form of oxides. The metallic content of the core is estimated to be about 85% iron by weight. The remaining core is believed to contain those substances that are soluble in liquid iron. From these estimates we conclude that there are about 0.2×10^{27} g of iron in the mantle and 1.6×10^{27} g of iron in the core. Because the total mass of Earth is 6.0×10^{27} g, Earth appears to contain 30% iron by weight, most of which is concentrated in the dense metallic core.

Of the other three terrestrial planets only Mars appears to lack a significant liquid iron core. This is not surprising because the much lower density of Mars indicates that it contains considerably less metallic iron.

FIGURE 2   Seismic waves delineate the layered structure of Earth. (a) Penetration of mantle and liquid core by P waves. (b) Deflection of P waves by inner core. (c) Deflection of S waves by liquid core.

CORE FORMATION OCCURRED DURING THE FIRST 100 MILLION YEARS OF EARTH'S HISTORY

Most probably Earth was formed by the gradual accumulation of substances present in the solar nebula. As Earth grew, it must have become increasingly hot because of the release of gravitational energy. This heating effect was undoubtedly augmented to some extent by heat released from the decomposition of radioisotopes. However, gravitational energy alone could have provided more than enough heat to melt the entire planet. Thus it has been calculated that gravitational energy released from planetary growth would have raised the temperature of Earth by 29,000°C, if all the energy was converted into heat. In a molten state we would expect a separation to occur with the denser metallic substances sinking to the core; it seems likely that this separation took place early in Earth's history, when the heat effect of accretion was maximal.

The timing of phase separation is supported by isotopic evidence that exploits the selective solubility of liquid iron for different metals. For example, lead is very soluble in liquid iron, while uranium is not. As a result, core formation would have produced an abrupt decrease in the ratio of lead to uranium in Earth's mantle. This in turn would have influenced the ratio of ^{206}Pb and ^{207}Pb to ^{204}Pb in the mantle. This is because the lead isotopes ^{206}Pb and ^{207}Pb are produced by the decay of isotopes of uranium, while ^{204}Pb is not. The earlier the time of core formation, the greater should be the ratio of the lead isotopes produced by decay to the long-lived uranium isotopes and to ^{204}Pb. From the relative abundances of these isotopes in ancient Earth rocks it has been estimated that the separation of lead and uranium must have occurred about 100 million years after the formation of Earth.

CONTINENTAL MOVEMENTS REFLECT GEOLOGIC ACTIVITY WITHIN THE MANTLE

There is ample evidence that, over time, the continents have migrated to different positions across the face of Earth. Current thinking is that 200 million years ago all the land masses were closely connected, and that only since then have they separated into the Eastern and Western Hemispheres and other smaller land masses. There are many indications that this has happened. For example, the shapes of eastern South America and western Africa complement each other (Fig. 3), and the similarities in the fossils they contain suggest that these two continents were at one time connected.

The idea of whole continents moving seems extraordinary but the evidence for it is overwhelming. How do we explain it? For this we have to go back to a consideration of the different layers of Earth. Recall that the uppermost layer of the mantle, known as the asthenosphere, is quite soft. The asthenosphere is more fluid than the lithosphere above it or the mantle below it because of a balance between the effects of temperature and pressure on the solid-to-liquid phase transition. The temperatures are sufficiently high and the pressures are sufficiently low to create a semiliquid condition uniquely in the asthenosphere. The lithosphere, which is located immediately above the asthenos-

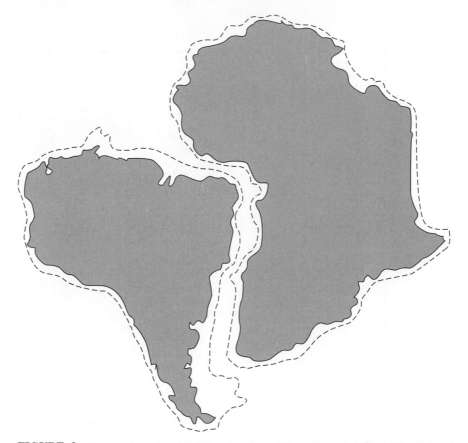

FIGURE 3   The complementary fit of the outer edges of the continental shelves of South America and Africa suggests that these two continents were at one time connected and were subsequently separated by continental drift.

phere, is quite brittle because of the greatly reduced temperature resulting from its proximity to the surface. In effect, the solid lithosphere floats on the slippery semiliquid surface of the asthenosphere. The brittle lithosphere is broken into large fragments. Currently, the lithosphere is fragmented into 13 large plates, called *tectonic plates*, which move as units (Fig. 4).

Given this description, we would not be surprised to see some plate movement as a function of time, and given their large size, we would not be surprised to see great upheavals when two plates happen to collide with one another. The first question we must ask is why these giant solid masses keep moving. The belief is that the movement of tectonic plates is powered by *convection currents* in the asthenosphere, which arise because of the heat differential between the upper and the lower mantle. These convection currents push new molten rock (magma) toward the surface along rifts located beneath the oceans, at locations where the plates are slightly separated (Fig. 5).

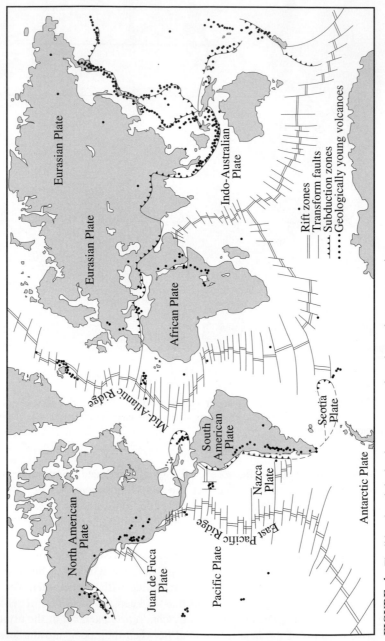

FIGURE 4 The lithosphere is broken into numerous plates that tend to move as units.

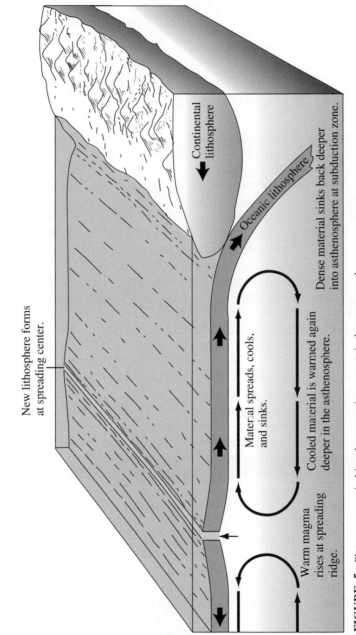

New lithosphere forms
at spreading center.

Continental
lithosphere

Oceanic lithosphere

Dense material sinks back deeper
into asthenosphere at subduction zone.

Material spreads, cools,
and sinks.

Cooled material is warmed again
deeper in the asthenosphere.

Warm magma
rises at spreading
ridge.

FIGURE 5 Plate movement is driven by convection currents in the upper mantle.

The upward pressure along the rifts forces the plates to spread farther apart, pushing them toward plates associated with solid land masses as pictured in Fig. 5. To understand what happens at a *subduction zone*, where an oceanic plate collides with a land plate, we must say something about the composition of these two different types of plates.

TECTONIC PLATES ARE COMPOSED OF OLD GRANITES AND YOUNG BASALTS

Basalts and granites are examples of *igneous* rocks, that is, rocks formed under conditions of intense heat, so that they became solid only on cooling from a liquid state. The ocean-covered plates primarily are composed of basalt, a complex mixture of minerals of which the major components are pyroxene (Fe, mg) SiO_3 olivine (Fe, mg)$_2$ SiO_4, and a calcium–sodium feldspar (an aluminum silicate of calcium and sodium). By contrast, the major rock species found on the land masses is granite, which is composed of quartz (SiO_2) and a potassium feldspar (an aluminum silicate of potassium). Granites have a significantly lower density (about 2.7 g/cm^3) than basalts have (about 3.2 g/cm^3). This is one of the reasons that the relatively thick granitic plates are above water. If they were denser, they would tend to "sink" into the soft asthenosphere. As an expanding basaltic plate makes contact with a granitic plate, something has to give. Because the granitic crust is less dense, it tends to be pushed upward into a mountain range, while the more dense basaltic crust tends to turn downward into the asthenosphere. The latter is said to be *subducted*.

Given this description of basalts and granites, we might expect the basaltic layers to be young compared with the granitic layers. When we speak of the age of igneous rocks like basalts and granites, we are referring to the time since they were last crystallized from the liquid state. When basalt is subducted, it becomes liquified in the asthenosophere, and new, reformed basalt appears at the midoceanic rifts (see Fig. 4). Basalts rarely exceed 100 million years in age, so that the majority of them were formed in the most recent 2% of Earth's history. It seems likely that the turnover of basalts was even more rapid in the past, but even at the current rate there would have been 50 complete turnovers of the basaltic layers in Earth's history. By contrast, granites have ages as high as 3.8 billion years, so some granites have hardly turned over at all.

There is a marked compositional difference between a basalt and the mantle from which it originates (Table 1). Some of this is due to a difference in the liquefaction temperatures in the mixture of minerals from which the mantle is composed. The minerals that become liquefied in the asthenosphere are those with lower melting temperatures. This explains why there is a much lower magnesium content in basalt than in the mantle, and why there is a considerably higher content of both sodium and potassium in basalt (see Table 1). There is another stage in the mineral fractionation process that takes place at the end of its journey, in the subduction region. During subduction, partial melting occurs and some of the melted rock spews forth from

TABLE 1

Gross Chemical Abundance of Different Elements in Mantle, Chondritic Meteorites, Basalt, and Granite[a]

Element	Mantle	Chondritic meteorites	Basalt	Granite
O	43.5	32.3	44.5	46.9
Fe	6.5	28.8	9.6	2.9
Si	21.1	16.3	23.6	32.2
Mg	22.5	12.3	2.5	0.7
Al	1.9	1.4	7.9	7.7
Ca	2.2	1.3	7.2	1.9
Na	0.5	0.6	1.9	2.9
K	0.02	0.1	0.1	3.2

[a]Abundances are expressed in weight percent.

volcanoes to become part of the granitic surface. This process probably accounts for some of the compositional differences between basalt and granite. As Table 1 shows, there is even less magnesium, and even more of both sodium and potassium, in granite than in basalt.

This leaves the elevated content of calcium in the basalt to be explained. The calcium source may be the very thick layers of calcium carbonate, also called calcite ($CaCO_3$), in the sediments that cover ocean floors. At the hot subduction zones, some of this material may become mixed in with basaltic minerals. It seems highly likely, however, that considerable calcite decomposes into lime (CaO) and carbon dioxide at the subduction zones; this decomposition thereby releases a good deal of the trapped carbonate as bicarbonate and carbon dioxide, which gets into the atmosphere. The same process would result in the release of some of the calcium to maintain the calcium concentration of the oceans.

VOLCANOES AND QUAKES REFLECT THE EXISTENCE OF CONVECTION CELLS IN THE UPPER MANTLE

Since early in Earth's history, volcanoes have been of great importance in supplying gases to the atmosphere. The main gases currently supplied are water and carbon dioxide, but on primitive Earth volcanoes also were the main suppliers of certain gases that were almost certainly essential for the origin of life. A casual relationship between volcanoes and convection cells arising from convection currents in the upper mantle is indicated by the fact that both earthquakes and volcanoes are strongly concentrated in zones where tectonic plates meet. The main exception to this pattern is a seemingly random distribution of some volcanoes in the Pacific. The latter, however, occur in regions that are described as mantle hot spots, where the basaltic layers are especially thin.

LIQUID WATER COVERS TWO-THIRDS OF EARTH'S SURFACE

Water is the most abundant component of living cells, so we are very concerned about the distribution of water, how it got to the surface, and what has kept it there. Other terrestrial planets have very little surface water. The supposition is that at an earlier stage in their history they had considerable surface water, but have since lost it.

Earth is only about 0.5% water by weight, and yet two-thirds of the lithosphere is covered by water. This is because of the low density of water, which causes it to rise to the surface. Almost 90% of Earth's water is found as a liquid in the oceans, lakes, rivers, and streams. About 2% of all surface water is frozen. Only about 10^{-5}% exists as a gas in the atmosphere. It is estimated that about 10% of Earth's water exists in the solid form as hydrates of silicates and oxides in the lithosphere and the mantle. Because of the high affinity of silicates and oxides for water, a good deal of water is regularly subducted with the basalts, only to make its way back to the surface as basalts rise again at the oceanic rifts.

Originally water was probably formed in the solar nebula by the interaction of hydrogen and oxygen. However, then, like most other volatiles, the water accreted by Earth was largely lost during the early phases of planet formation. Perhaps the only water left was that trapped as hydrates in the metal silicates and oxides. Even under today's conditions, water can get lost in a variety of ways. It can react with substances containing ferrous compounds on the surface or in the mantle to form hydrogen and ferric compounds. (Ferrous compounds are those containing iron in the $+2$ valence state, whereas ferric compounds contain iron in the $+3$ valence state.) Alternatively, water can be decomposed by low-wavelength ultraviolet (UV) radiation emanating from the Sun. Photosynthesis also decomposes a good deal of water to molecular oxygen, while using the hydrogen so produced to manufacture new compounds. Water also can escape to outer space in the high reaches of the atmosphere. With these facts in mind, let us look in greater detail at the history of water loss versus water retention by Earth.

On primitive Earth the loss of water was probably far greater than it is now, for numerous reasons. In the very early history of Earth, the great heat must have allowed for many water molecules to reach escape velocity and thereby get lost to outer space. After Earth cooled down, this form of water loss was greatly reduced.

In the first 2 billion years of Earth's history, much water was decomposed by ferrous iron compounds in the crust and the mantle. The hydrogen thereby released was very likely to have escaped because of its low mass, whereas the oxygen produced in this way was trapped in ferric iron compounds and probably did not reach the atmosphere in any significant amounts.

On primitive Earth, there was no protective ozone layer (see the later section on atmospheric composition). Therefore, low-wavelength UV radiation from the Sun reached Earth's surface in much greater amounts and probably decomposed substantial quantities of water to hydrogen and oxygen. Once again the hydrogen was very likely to have escaped, while the oxygen was probably trapped by reaction with other compounds. Radiation-directed decomposition of water still takes place in the upper reaches of the atmosphere, but so little water makes its way to such a height that the amount now lost by radiation-directed decomposition is miniscule.

Currently, very little water is decomposed in the mantle, as evidenced by the small amounts of molecular hydrogen released in volcanic gases. This is probably because the Fe^{+3} to Fe^{+2} ratio in the mantle has increased to a level where the decomposition water by this route is no longer a thermodynamically favored reaction. Finally, due to the high concentration of atmospheric oxygen, any hydrogen that is released from volcanoes is much more likely to become trapped by reaction with atmospheric oxygen to reform water than to escape to outer space.

Given that Earth has been able to retain so much water, why is there so little on the other terrestrial planets? In the case of Mercury the weak gravitational pull that results from its low mass, and the elevated surface temperatures together are sufficient to account for the early and permanent loss of water. Venus has the same mass as Earth, but it is closer to the Sun and it has a very slow rate of rotation. We believe that Venus must have lost a great deal of water because, in the small amount of water still left on Venus; the deuterium to hydrogen ratio is 150-fold what it is on Earth. Deuterium is a heavy isotope of hydrogen. Its presence on Venus in such a high ratio abundance suggests the fractional loss of a large amount of water that would preferentially carry the lighter isotope. Mars presents a more puzzling picture, because it probably possessed substantial surface water at one time, as evidenced by its fluvial features. Perhaps Mars has retained a good deal in its interior that is locked in because Mars lacks tectonic activity and therefore volcanoes.

Because molecular oxygen has played such an important role in preserving the water on Earth's surface, the loss of water from the surface of Venus and Mars may have been due in large part to the lack of molecular oxygen in their atmospheres. Atmospheric oxygen not only traps the hydrogen resulting from water decomposition, but also shields surface water from destruction by low-wavelength radiation through the formation of an ozone layer. Of course, a substantial concentration of molecular oxygen is probably closely tied to the development of efficient photosystems that are associated with many life forms on this planet. Perhaps, on planets where the temperatures are such that the water is not permanently frozen, photosystems that produce molecular oxygen are absolutely essential for the long-term preservation of a water supply.

MAGNETIC FIELDS PROTECT SOME PLANETS FROM THE SOLAR WINDS

Solar winds that emanate from the outer layers of the Sun's atmosphere in all directions can be very destructive to planetary atmospheres. These winds comprise a constant flow of charged particles (mostly protons and electrons) that unimpeded can strip the atoms in a planet's atmosphere of one or more of its electrons. Planets that generate a magnetic field repel and deflect the impinging particles in the solar wind, thereby creating an elongated "cavity" in the solar wind called a *magnetosphere*. Of the four inner terrestrial planets, Earth and Mercury have strong magnetic fields that offer them adequate protection to the deleterious effects of solar winds. These magnetic fields are generated by the eddy currents in the liquid iron cores of these two planets. Venus and Mars do not produce significant magnetic fields. In the case of Mars this is due to the absence of a liquid iron core while in the case of Venus, which has a significant liquid iron core, this is due to the very slow axial rotation of the planet. It is not possible to give precise estimates for the damage to the atmospheres of Venus and Mars that has resulted from the lack of a magnetosphere, but it seems likely that in both cases it has contributed to destructive atmospheres and the lack of water on both of these planets.

SURFACE TEMPERATURE HAS BEEN DELICATELY BALANCED FOR ALMOST 4 BILLION YEARS

When we consider that the Universe has temperatures ranging from near absolute zero in outer space to millions of degrees in the centers of stars, we can appreciate the remarkable temperature control that has existed on the surface of Earth. We know that Earth's surface temperature has been moderate for a very long time. This is because sedimentary rocks, which arise from sediments that have become compacted into layers to form solid rock bodies, form only when liquid water is present. The isotopic dating of sedimentary rock indicates that these rocks have been forming more or less continuously for the past 3.8 billion years. Because liquid water exists only over a small range of temperatures, this means that this narrow range of temperatures must have been maintained over substantial portions of Earth's suface during the same time period.

EARTH'S INTERMEDIATE DISTANCE FROM THE SUN HAS HELPED TO MODERATE EARTH'S SURFACE TEMPERATURE

A planet's surface temperature is a function of the heat it receives and the heat it releases. When initially formed, all planets in our Solar System were extremely hot as a result of the heat generated by the accretion process. This internal heat has been supplemented by the heat that results from disintegration of radioisotopes and from chemical reactions. Internal sources of heat are much more important for some planets than for others. For example, it is estimated that Jupiter and Saturn receive more than twice as much heat from internal sources as they receive from the Sun. For the inner terrestrial planets, however, the heat received from the Sun is the major source of heat.

At the same time, the internal heat plays a very important role. In the case of Earth, convection cells in the mantle are powered by the heat differential between the molten core and the solid–liquid surface. Without an internal source of heat, convection cells would die, no movement of plates would occur, and most volcanic emissions would cease. Such a change would most likely have a major effect on the mineral balance in the crust and the gas balance in the atmosphere.

The heat received from the Sun is a function of the Sun's luminosity and the planet's distance from the Sun. The Sun's luminosity has been continuously changing. It is believed to have been at one time only 70% of what it is now, and for this reason we only can wonder how Earth was able to maintain surface temperatures high enough to sustain life over the past few billion years. In the future, by contrast, the Sun's luminosity is expected to increase very substantially, so much so that in the next 2 or 3 billion years Earth's surface temperature will probably rise to the point where the planet will no longer be able to sustain life.

For the past 4 billion years, however, it appears that Earth has been ideally stationed with respect to its distance from the Sun. In this regard we only have to remind ourselves that the heat obtained from a point source is inversely proportional to the square of the distance from the point source. If we look at the other terrestrial planets, we find that there is a sharp falloff in surface temperature that correlates with the distance from the Sun. Mercury is the hottest; Venus is the next hottest. Both of these planets have surface temperatures that are too hot to sustain life. On the other hand, Mars, which is farther from the Sun than Earth, is too cold to sustain life on its surface. Although there are many moderating factors that affect surface temperature, the distance from the Sun may be the most important one in determining which planets are going to be suitable for the origin and maintenance of life.

The amount of solar radiation that a planet receives depends not only on the distance from the Sun and the Sun's luminosity, but also on how much of that radiation is actually absorbed by the planet. This factor can vary considerably. Physicists refer to a blackbody as one that absorbs all the impinging radiation. In the case of Earth, about 30% of the solar radiation is reflected instead of being absorbed. Ice and snow are much more reflective than a dense forest is, so what is on the exposed surface has a big effect on the amount of radiation absorbed. If Earth were totally covered by ice and snow, it might reflect so much of the incident radiation that it would never be able to warm up again.

WATER AND CARBON DIOXIDE HAVE HELPED TO MODERATE EARTH'S SURFACE TEMPERATURE

To maintain a relatively constant surface temperature, a planet must lose approximately the same amount of heat as it absorbs. Because planets exist in a near vacuum, the only way they can exchange heat is by radiation. The temperature of Earth's surface has hovered over a narrow range for a very long time, so we know that heat losses and heat gains have been reasonably in balance. Indeed, it seems likely that there are compensatory mechanisms at work to control the surface temperature. Because Earth is not a luminous body like the Sun, we know that the heat it loses must have resulted from an average radiation of considerably longer wavelength than that of visible light. A major mechanism for regulating Earth's surface temperature exploits this wavelength difference.

Before discussing this mechanism, let us consider how the presence of water has helped in "buffering" the surface temperature. In this regard, water has two important properties. First, it has a very high capacity. A great deal of heat input is necessary to convert ice to water (70 cal/g or 293 Joules/g), and even more heat is necessary to convert water to a gas (540 cal/g or 2707 Joules/g). Likewise, a great deal of heat is released in making the reverse conversions. The high heat capacity for these phase transitions of ice to water and water to gas, and the heat required to either heat or cool liquid water have helped Earth to resist temperature changes. Prolonged periods of excessive heat or cold would be required to produce irreversible changes. The second important property of water is the decrease in density that water undergoes on freezing. Very few liquids have this property. Because of its lower density, when ice forms on the surface of a body of water, it floats instead of sinking and makes an insulating layer that retards the freezing of the liquid water beneath it. If water behaved like most other liquids, becoming denser on freezing, the oceans might have become permanetly frozen long ago.

Although the presence of water may slow down the effects of receiving too much or too little heat, it does nothing to reverse a long-term trend in either direction, and may in fact make matters worse. If the temperature drops sufficiently, large surface areas of water freeze, and the greater reflectivity of ice causes Earth to cool down even further. Conversely, if Earth heats up, then the atmospheric content of water increases. The result is further heating, because gaseous water inhibits the rate of heat loss from Earth by what is called the greenhouse effect. This greenhouse effect of gaseous water is partially offset by the back reflection of solar radiation by clouds.

What is needed is some way of counteracting the effects of too high or too low a temperature. In other words, if Earth starts to become warmer, is there a system that can be brought into play to cool it down? Alternatively, if the surface temperature drops, is there some mechanism for raising it?

To understand the answer, it is necessary first to under stand how the *greenhouse effect* works. Let us consider a glass-enclosed greenhouse that is exposed to sunlight. The rays of sunlight heat the inside of the greenhouse because glass is transparent to radiation having the wavelengths of visible light. The inside of the greenhouse gets warmer, but the warmth so received is not able to get out of the greenhouse because the longer wavelength infrared radiation (heat) that is given off cannot penetrate the glass. The heat is trapped and as a result the greenhouse can get quite warm. Like glass, certain atmospheric gases, called greenhouse gases, can absorb infrared radiation and thus prevent the escape of heat from Earth.

If we survey the main gases in Earth's atmosphere (Table 2), we find only two gaseous molecules that are effective at absorbing infrared radiation, water, and carbon dioxide. We already have seen that water has properties that keep it from being an

TABLE 2
Atmospheres of the Terrestrial Planets[a]

	Venus	Earth	Mars
Surface pressure	100	1	0.006
Composition (in relative molecular abundance)			
CO_2	>0.98	3×10^{-4}	0.96
N_2	<0.02	0.78	0.025
O_2	<10^{-6}	0.21	2.5×10^{-3}
H_2O	10^{-6}	0.01–0.04	10^{-3}

[a]Data from Walker, J. *Evolution of the Atmosphere,* New York: Macmillan, 1977, p. 220. Reprinted by permission of Prentice-Hall, Inc., Upper Saddle River, NJ.

effective temperature regulator. What we need is a gaseous molecule that is more abundant when the temperature drops and less abundant when it rises. Carbon dioxide is just such a gas.

There are many reactions that involve atmospheric carbon dioxide as part of a cycle, and we do not intend to cover all of them. Carbon dioxide is fixed by plants in organic material that is eaten by animals, which then return carbon dioxide back to the atmosphere by respiration. The largest reservoir of carbon dioxide, however, is that found in calcium carbonate deposits on the ocean floor. The formation of such deposits does not require living organisms, although they have contributed to the process because many sea organisms have shells composed of almost pure calcium carbonate. Atmospheric carbon dioxide is incorporated into organic and inorganic deposits at a higher rate when the temperature increases, and therefore an increase in temperature tends to lower the amount of carbon dioxide in the atmosphere. By contrast, a lower temperature slows the rate of incorporation of carbon dioxide into organic and inorganic deposits and therefore will lead to an increase in atmospheric carbon dioxide. Because carbon dioxide is a greenhouse gas—one that traps heat on the planet—the higher the atmospheric carbon dioxide content, the greater is the trapping of heat. We can see that the climate effect of carbon dioxide concentration in the atmosphere goes in the right direction to make carbon dioxide an effective temperature regulator.

Whereas carbon dioxide might be the answer to regulation of the surface temperature on Earth today, we have no indication of how much carbon dioxide was available in Earth's early history, so we do not know if this mechanism for control of Earth's surface temperature could have operated in those times.

The catastrophic effects of having too much carbon dioxide are evident on Venus, which has an atmosphere 100 times as dense as that of Earth, composed almost solely of carbon dioxide (see Table 2). If one were to take all the carbon on Earth and convert it into carbon dioxide, this is about the amount of carbon dioxide that we would expect in our atmosphere. As we have seen earlier, Venus has lost almost all its water. The severe greenhouse effect that exists on Venus is due to this loss, because most mechanisms for carbon dioxide removal from the atmosphere require water. On Mars, another planet without surface water, most of the atmosphere is composed of carbon dioxide also, but the amount of carbon dioxide is several orders of magnitude lower. The difference between the two planets, as far as the atmospheric carbon dioxide content is concerned, may be explained by the fact that Venus has had considerable convection within its mantle and volcanic activity as well. The convection cells that are powered by the heat differential between the core and the asthenosphere probably provided the main mechanism for transporting and releasing carbon dioxide and water into the atmosphere. The lack of such convection cells on Mars probably means that a great deal of its carbon dioxide and water are trapped as carbonates and hydrates, respectively, in its mantle.

EARTH'S ATMOSPHERE IS SUBDIVIDED INTO FOUR REGIONS

The atmosphre varies in composition and temperature as a function of elevation; it extends to several thousand kilometers above the lithosphere. The total mass of the atmosphere (5.1×10^{21}g) is only one-millionth the mass of Earth itself (5.98×10^{27}g). Almost 99% of the current atmosphere is composed of two molecules, nitrogen (78%) and oxygen (21%) (see Table 2). The remaining atmosphere is accounted for by argon (0.9%), water (up to 4%), carbon dioxide (0.034%), and trace amounts of other gases. The most important gas present in trace amounts is *ozone* (O_3), which constitutes only about 10^5% of the atmosphere. Ozone is present primarily in the stratosphere (see following discussion). There, it absorbs most of the Sun's potentially destructive UV rays, those of wavelengths below 300 nm. Without this thin layer of ozone it would be virtually impossible for living forms to survive on the exposed surface of Earth because of the damage that UV radiation does to the chromosomes in living systems.

The atmospheric density decreases exponentially with distance above the lithosphere. Atmospheric physicists define four regions in the atmosphere (Fig. 6).

1. The *troposphere* comprises the first 10 to 15 km adjacent to Earth's surface and contains 80 to 85% of the total atmospheric mass. Almost all the water vapor is present in the troposphere. Climate control is mainly a result of interactions between the surface and the troposphere. In the troposphere the general trend is toward decreasing temperature with increasing height.

2. The *stratosphere* extends from the troposphere up to elevations of about 50 km. Most of the ozone is confined to this segment of the atmosphere. A positive temperature gradient in the stratosphere (see Fig. 6) is generated by absorption of solar UV radiation by the ozone.

3. Beyond the stratosphere we find the *mesosphere* (50 to 85 km). Here the correlation of temperature with height is again negative, as in the troposphere.

4. In the *thermosphere* (80 to 500 km) the temperature once more rises with height. The chemical composition of the atmosphere is fairly uniform up to 100 km except for

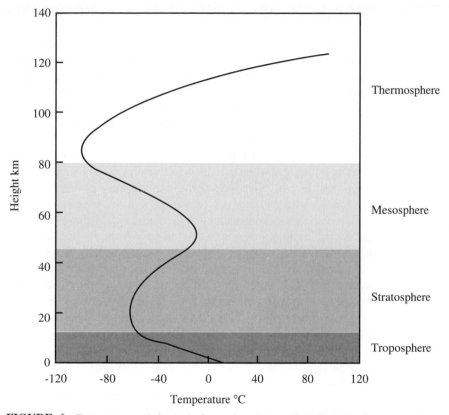

FIGURE 6   Temperature variation in the four regions that constitute the atmosphere.

the water vapor and the ozone. Above 100 km the chemical composition is strongly influenced by photochemical reaction induced by solar and cosmic radiation.

The composition of the atmosphere has varied considerably over the course of Earth's history. Geologic and biological factors have a strong influence on atmospheric composition today, as they have had in the past. We already have discussed some of the factors that influence the carbon dioxide content. Large amounts of oxygen are present in the atmosphere as a by-product of photosynthesis by plants and microorganisms. Oxygen was only a minor component of the atmosphere until about 2.2 billion years ago because photosynthetic organisms began to produce oxygen from the oxidation of water (see Chapters 11, 21, and 23). Concurrent with the rise in oxygen there was a sharp fall in the atmospheric content of molecular hydrogen, caused by the intense reaction between oxygen and hydrogen in the formaion of water. The concentrations of nitrogen, carbon dioxide, and other gases over the course of geologic time are a matter of some speculation.

PREBIOTIC ATMOSPHERE WAS A REDUCING ONE

Of great concern to us is the composition of Earth's atmosphere during the time around the origin of life. This is because many of the reactions resulting in chemical products essential to the origin of life are believed to have been formed in the primitive atmosphere, utilizing gases present as starting materials (see Chapters 11, 13, and 14).

In Chapter 3 we took a brief look at how atoms of different substances can combine to produce new substances. Such reactions often involve a transfer of electrons between the interacting substances. The loss of electrons by an atom or group of atoms is called *oxidation*. (Despite the name, oxygen does not need be involved in the process, although it often is.) The converse process, the gain of electrons by an atom or group of atoms, is called *reduction*. An atmosphere rich in substances that can give up electrons to other substances is referred to as a *reducing atmosphere*.

Given the gross composition of the Universe, in the 1950s Harold Urey proposed that Earth's early atmosphere was a strongly reducing one, containing large quantities of methane (CH_4), ammonia (NH_3), and hydrogen (H_2). However, what he did not take into account was the likelihood that the intense heat produced during the formation process probably would have driven off most of its volatiles (see Chapter 4). Thus most hydrogen and many other small molecules present in the gaseous state would have escaped to outer space. Only small amounts of these gases that happened to become trapped in the solid mass of Earth were retained.

These trapped gases are most likely to have reached Earth's surface through volcanic emissions. At the present time, such emissions contain most carbon in the form of carbon dioxide and most nitrogen in the form of nitrogen gas; mere traces of hydrogen gas or more reduced forms of carbon or nitrogen are found. It would have been impossible to synthesize the essential starting materials for the origin of life from such a gas mixture.

Faced with this dilemma, geochemists and organic chemists have reconsidered the early chemistry of Earth in some detail, trying to trace the likely changes that have taken place in this chemistry in the last 4.6 billion years. The valence state of iron in the mantle is considered to be a very important factor in determining the composition of Earth's early atmosphere. As we have seen earlier, the solid portion of the newly formed Earth was composed mainly of only four elements: iron, silicon, oxygen, and magnesium. The predominant valence states of silicon, oxygen, and magnesium were fixed at $+4$, -2, and $+2$, respectively. Only iron is likely to have existed in more than one valence state in significant quantities. It has been estimated that about 85% of the iron was in the 0 valence state and that the remaining iron was partitioned between the $+2$ and $+3$ valence states in the form of iron silicates, oxides, and sulfides. When Earth was very young, it was a homogeneous molten mass. Because most of the iron was in the 0 valence state, any iron in the $+3$ valence state would probably have been reduced to the $+2$ valence state. Isotope studies we referred to indicate that within 100 million years after Earth's formation the metallic iron separated into a central core. The other forms of iron were left behind in the mantle, with virtually all the iron there in the $+2$ valence state.

BOX 5A   Hydrogen May Have Been Formed in Prebiotic Times with the Help of Dissolved Ferrous Salts

In prebiotic times the absence of oxygen had two major effects on the environment. The absence of oxygen implies the lack of an ozone layer, which would have allowed low-wavelength UV radiation to reach the surface. In the absence of oxygen it is also quite likely that the oceans were laden with ferrous salts. Exposure of the oceans to UV radiation is quite likely to have encouraged the decomposition of water and the release of hydrogen. In this reaction ferrous ion would be oxidized to the ferric state, which would have precipitated either as the hydroxide or as other ferric salts. This possibility has been noted by others. For example, Mazerall and colleagues (see references) have found that the production of hydrogen by this route can be facilitated either thermally or photochemically. In the prebiotic world this reaction could have been a major source of hydrogen.

Do we have any other examples of this occurring in our Solar System? Possibly on Mars this reaction was allowed to go unimpeded until all the water was used. The rust-colored appearance of the Martian surface and the mysterious disappearance of considerable water from the Martian surface is consistent with this possibility.

The trapped gases would have had to travel a good distance through the mantle to reach Earth's surface. As long as all the iron in the mantle was in the $+2$ valence state, it is unlikely that any oxidation of these trapped gases would have occurred on their journey to Earth's surface. Hence volcanic emissions in Earth's early history most likely contained more highly reduced gases than today's emissions contain. Currently it is believed that most of the carbon in volcanic emissions would have been in the form of carbon dioxide, carbon monoxide, small amounts of methane, and significant amounts of hydrogen gas. In addition to the hydrogen trapped in the original accretion process, much of the accompanying hydrogen was probably formed in the mantle from the reaction of ferrous oxide with water to form ferric oxide and hydrogen. Very little hydrogen can be formed in this way today, because of the Fe^{3+} to Fe^{2+} ratio that currently exists (Box 5A).

There is still some debate over whether the nitrogen in early volcanic emissions was in the 0 valence state as nitrogen gas, in the -3 valence state as ammonia, or in some other state. However, even if the nitrogen existed in the relatively inert form of nitrogen gas, the larger quantities of carbon monoxide and hydrogen would have created a situation favorable for the formation of starting molecules for the origin of life (see Part III).

Clearly, the valence state of iron in the mantle must have shifted to account for the fact that current volcanic emissions contain only traces of carbon monoxide and hydrogen. This shift has been attributed to the selective loss of hydrogen that worked its way to the surface. Because of its small mass, hydrogen has a much higher tendency to escape Earth's gravitational pull than any other gaseous molecules have. The steady loss of hydrogen must have resulted in a gradual loss of chemical-reducing power from Earth, as indicated by a corresponding increase in the Fe^{3+} to Fe^{2+} ratio in the mantle. Loss of hydrogen from Earth's atmosphere was ultimately brought under control by the appearance of appreciable concentrations of molecular oxygen in the atmosphere, an event that took place about 2 billion years ago. This oxygen has served as a trap for the hydrogen by reacting with it to form water. The shift in the Fe^{3+} to Fe^{2+} ratio is believed to have taken a great deal of time (about 2 billion years), and the ratio is believed to have stabilized as the loss of the hydrogen from the atmosphere subsided.

The key question for the origin of life is were there reduced gases in the atmosphere around the time of the origin of life? To answer this question we must turn to the geologists for information as to when life originated. Current information indicates that Earth had cooled down sufficiently within 600 million years from the time of its formation so that organic compounds could have been synthesized without the danger of rapid thermal decomposition. The first fossils indicating the existence of living cells are about 3.5 billions years old. There are some indications that the first life forms could have existed as early as 3.8 billion years ago. Hence the origin of life and the reactions that led to it must have occurred at some time within the 500-million-year period from 4.0 to 3.5 billion years ago. This is believed to have been well within the period when the Fe^{3+} to Fe^{2+} ratio was quite low. Thus, Earth is very likely to have had a reducing atmosphere, with some methane, carbon monoxide and hydrogen, during this crucial period.

SUMMARY

In this chapter we have considered the changes that have taken place on Earth since its formation that have given rise to its current geologic, hydrologic, and atmospheric features.

1. From the vibrations made by earthquakes, seismologists have been able to determine the internal structure of our planet. They have found that there is a discontinuity at a depth of 2800 km. Material above this discontinuity has a density expected for magnesium silicates, while the properties of the core are consistent with its being formed primarily from molten iron.

2. Measurements on lead and uranium isotopes in meteorites and ancient Earth rocks suggest that the separation of Earth into mantle and core occurred during the first 0.1 billion years of its history.

3. The composition of the crust is quite different from that of the mantle. On dry land the crust is composed mostly of granites, while on the ocean floor it is composed mostly of basalts. The differences in age and chemical composition of the basalts and granites are due to the way in which they were formed.

4. The elevation difference between continental crust and ocean crust is a consequence of the density difference of the two crusts. The granitic continental crust is thicker and dense than the basaltic ocean crust.

5. The crusts are brittle layers that exist on top of a semiliquid layer. They are broken into plates that slide as independent units.

6. When a basaltic plate collides with a granitic plate, it tends to push the granitic plate upward, while the basaltic plate becomes subducted and mostly returns to the mantle.

7. The zones where different plates meet are also zones of earthquakes and volcanic activity.

8. Earth is only 0.5% water, but because water is less dense than other terrestrial materials are, it has floated to the surface, two thirds of which is covered by water.

9. A magnetic field that surrounds Earth protects its atmosphere from the damaging ionizing effects of the "solar winds."

10. During the process of accretion most volatile substances were lost from Earth and other terrestrial planets. New gases accumulated by outgassing of the fully formed planet.

11. The surface temperature of Earth is a function of several factors: (1) the amount of light energy it receives from the Sun; (2) the reflectivity of its surface; and (3) the action of greenhouse gases, which reflect back some of the energy that would otherwise be lost to outer space.

12. The composition of Earth's atmosphere has varied considerably over the past 4 billion years. The oxygen that presently constitutes 21% of the atmosphere originated mostly from the production of oxygen by photosynthetic organisms. Small amounts of oxygen have been converted to ozone in the upper atmosphere.

13. The atmosphere around the time of the origin of life was probably a reducing atmosphere, containing much more methane, carbon monoxide and hydrogen.

Problems

1. Of the terrestrial planets only Mars lacks volcanic activity. There is a very good reason for this. Please explain.

2. When a basaltic plate collides with a granitic plate, the basaltic plate subducts. What do you suppose happens when two granitic plates collide?

3. If we can drill only down 10 km into Earth's interior, how do we know that the mantle contains a much higher concentration of Mg than the lithosphere contains?

4. If life had not developed on Earth, in what ways might Earth be different now?

5. If Earth had an atmosphere rich in oxygen 4 billion years ago, would this have interfered with the origin of life?

6. Why do we refer to Mars as being geologically dead?

7. If Earth's temperature had dropped long enough to freeze all the water, is it possible that it ever would have regained a temperate climate?

8. What roles did iron play in influencing the composition of the atmosphere?

9. If life had developed on Earth 3.8 million years ago but phtosynthesis had failed to develop, in what ways might Earth be different at the present time?

10. Of the terrestrial planets only Earth and Mercury generate significant magnetic fields. Explain.

References

Angel, R. The Earth's Mantle Remodelled. *Nature (London)* 385: 490–492, 1997. (The composition of the upper mantle seems close to resolution.)

Azar, D. and Rodhe, H. Targets for Stabilization of Atmospheric CO_2, *Science* 276: 1818, 1997. (There is concern by some that the carbon dioxide level in the atmosphere is steadily rising and that this may produce a global warming effect.)

Holland, H. D. Evidence for Life on Earth More Than 3850 Million Years Ago, *Science* 275: 38–40, 1997.

Holland, K. H. and Ahrens, T. J. Melting of $(Mg, Fe)_2$ SiO_4 at the Core–Mantle Boundary of the Earth, *Science* 275: 1623–1630, 1997.

Jeanloz, R. and Lay, T. The Core–Mantle Boundary, *Sci. Ame.* May: 48–56, 1993. (This interactive zone may be the most dynamic part of the planet, directly affecting Earth's rotation and magnetic field.)

Kass, D. M. and Yung, Y. L. The Loss of Atmosphere from Mars, *Science* 274: 1932–1933, 1996.

Kasting, J. F. Warming Early Earth and Mars, *Science* 276: 1213–1214, 1997. (The Sun's lower luminosity during this period, and the lower levels of CO_2 during this period lead to the proposal that reducing gases like methane may have contributed to the greenhouse effect.)

Kerr, R. A. Deep-Slabs Stir the Mantle, *Science* 275: 813–815, 1997. (New images of Earth's interior may end a long-running debate by showing that cast-off slabs of surface rock sink to the very outskirts of the core, mixing the mantle from top to bottom.)

Mauzerall, D., Borowska, Z., and Zielinski, I. Photochemical and Thermal Reactions of Ferrous Hydroxide, *Origins Life Evol. Biosphere* 23: 105–114, 1993. (In oceans containing ferrous hydroxide water can react to form hydrogen.)

Molina, M. Rescuing the Ozone Layer, *Sci. Am.* November: 40, 1997. (The ozone layer in the stratosphere is thinning out; this could have a profound effect on surface life.)

Rye, R., Kuo, P. H., and Holland, H. E. *Nature (London)* 378: 603–605, 1995. (Atmospheric carbon dioxide concentrations before 2.2 billion years ago are discussed. The estimated carbon dioxide concentration of the atmosphere during this period of lower solar luminosity would not have been sufficient to maintain moderate surface temperatures.)

Song, X. and Richards, P. G. Seismological Evidence for Differential Rotation of the Earth's Inner Core, *Nature (London)* 382: 221–224, 1996. (The inferred rotation rate is on the order of 1 per year faster for the inner core than the daily rotation rate of the mantle and crust.)

Squyres, S. W. and Kasting, J. F. Early Mars: How Warm and How Wet? *Science* 265: 744–749, 1994.

Storey, B. C. The Role of Mantle Plumes in Continental Breakup: Case Histories from Gondwanaland, *Nature (London)*377: 301–308, 1995. (The breakup of the Gondwanland supercontinent, which started about 180 million years ago, provides an excellent case history against which to test models.)

Taylor, S. R. and McLennan, S. M. The high-standing continents owe their existence to the earth's long history of plate-tectonic activity, *Sci. Am.* January: 76–81, 1996.

Vidale, J. E. and Lay, T. Phase Boundaries and Mantle Convection, *Science* 261: 1401–1403, 1993.

Walker, J. C. G. *Earth History*, Boston: Jones Bartlett, 1986. (This is easy reading but dated.)

Wood, B. J. Hydrogen: An Important Constituent of the Core? *Science* 278: 1727, 1997. (Earth's core may contain a significant amount of hydrogen.)

Hamblin, W. K. *The Earth's Dynamic Systems,* 5th ed. New York: Macmillan 1985. (This is an excellent backup text for the geologic and meterological issues addressed in this chapter.)

PART II

Logic of Living Systems

6. Cells, Organelles, and Biomolecules
7. Metabolic Strategies and Pathway Design
8. Biochemical Catalysis
9. Storage, Replication, and Utilization of Biochemical Information

In Part I we traced the steps from the origin of the Universe to the construction of planet Earth. We saw that 3.9 billion years ago Earth had inviting geologic and chemical features that should have been favorable for the origin of life: an abundance of surface water at moderate temperatures; and an atmosphere with simple compounds containing carbon, nitrogen, oxygen, and hydrogen, the four elements that form the major components of bioorganic molecules. If we were to follow this story chronologically, we would next consider the chemical reactions that led to the first living forms. However, it would not be good pedagogy to do this because the reader must first be informed of the key characteristics of living systems. In Part II we pinpoint the fundamental features common to all living systems. Then in Part III we are in an advantagous position to consider how these systems evolved from the prebiotic world.

Part II begins with Chapter 6, which is an overview of the structures and functions of chemicals found in living systems. Then in Chapter 7 the organization of biochemical pathways and metabolic strategies inherent in pathway design are described. In Chapter 8 we consider mechanisms of biochemical catalysis. Finally, in Chapter 9 we deal with the synthesis of nucleic acids and proteins and outline some discoveries that have been made about the genetic code.

CHAPTER 6

Cells, Organelles, and Biomolecules

Living organisms are highly organized chemical systems. This is the beginning of four chapters in which we describe the fundamental characteristics of living systems. In this chapter we focus on the magnificent molecular structures that are common to most living cells.

THE CELL IS THE FUNDAMENTAL UNIT OF LIFE

Microscopic examination of any living organism reveals that it is composed of membrane-enclosed structures called cells. The enclosing membrane is called the *cell membrane* or the plasma membrane. Cells vary greatly in size and shape. In Fig. 1 we show prototypical animal and plant cells and a variety of shapes and sizes of some well-known bacteria. The cells in bacteria are fully functional as single cells, but sometimes they are connected into long chains that give the deceptive appearance of a multicellular organism. A true multicellular organism contains different types of cells with specialized structures and functions.

    The plasma membrane is a delicate, semipermeable, sheetlike structure enclosing the entire cell. By forming an enclosure it prevents gross loss of the intracellular contents; its semipermeable character permits the selective absorption of nutrients and the selective removal of metabolic waste products. In many plant and bacterial cells, a *cell wall* encompasses the plasma membrane. The cell wall is mechanically stronger

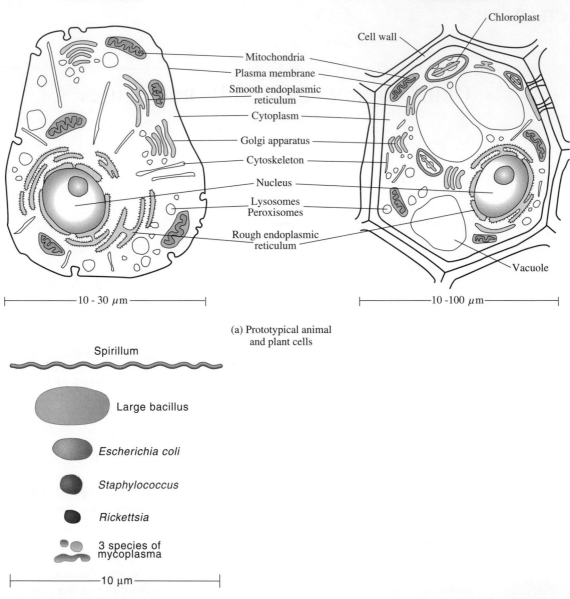

ANIMAL CELL

PLANT CELL

Chloroplast

Cell wall

Mitochondria
Plasma membrane
Smooth endoplasmic
reticulum
Cytoplasm
Golgi apparatus
Cytoskeleton
Nucleus
Lysosomes
Peroxisomes
Rough endoplasmic
reticulum

Vacuole

10 - 30 μm

10 -100 μm

(a) Prototypical animal
and plant cells

Spirillum

Large bacillus

Escherichia coli

Staphylococcus

Rickettsia

3 species of
mycoplasma

10 μm

(b) Bacterial cells

FIGURE 1 Cells are the fundamental units in all living systems, and they vary tremendously in size and shape. All cells are functionally separated from their environment by the plasma membrane that encloses the cytoplasm. (a) Generalized representations of the internal structures of animal and plant cells (eukaryotic cells). Plant cells have two structures not found in animal cells: cellulose cell wall, exterior to the plasma membrane, and chloroplasts. (b) The many different types of bacteria (prokaryotes) are all smaller than most plant and animal cells. Bacteria, like plant cells, have an exterior cell wall, but it differs greatly in chemical composition and structure from the cell wall in plants. Like all other cells, bacteria have a plasma membrane that functionally separates them from their environment. Some bacteria also have a second membrane, the outer membrane, exterior to the cell wall.

than the plasma membrane because it is made of a covalently cross-linked, three-dimensional network that keeps the cell from being deformed or disrupted under stress.

The content enclosed by the plasma membrane is called the *cytoplasm*. The purely liquid portion of the cytoplasm is called the *cytosol*. Within the cytoplasm there are a number of macromolecules and even larger structures, many of which can be seen by high-power light microscopy or electron microscopy. Many structures enclosed by membranes are called *organelles*. Organelles commonly found in plant and animal cells include the *nucleus*, the *mitochondria*, the *endoplasmic reticulum*, the *Golgi apparatus*, the *lysosomes*, and the *peroxisomes* (see Fig. 1). *Chloroplasts* are an important class of organelles found exclusively in plant cells. Each type of organelle represents a special biochemical factory in which certain biochemical products are synthesized or otherwise metabolized. In addition to organelles, animal and plant cells contain filamentous structures collectively termed *cytoskeleton*, which is important for maintaining the three-dimensional integrity of the cell.

The organelles and cytoskeleton that are found in plant, fungi, and animal cells are not present in bacteria. The biochemical functions associated with organelles are frequently present in bacteria, but they are organized in a different way.

The terms *prokaryote* and *eukaryote* refer to the most basic division between cell types. The fundamental difference between them is that the cells of eukaryotes contain a nucleus, whereas those of prokaryotes do not. The cells of prokaryotes usually lack most of the other membrane-bounded organelles as well. Plants, fungi, and animals are examples of eukaryotes, and bacteria are examples of prokaryotes.

Cells are organized in a variety of ways in different living forms. Prokaryotes of a given type produce cells that are very similar in appearance. The most common prokaryotes are bacteria, which exist as single cells. Each bacterial cell replicates by a process of binary fission, in which two identical daughter cells arise from an identical parent cell. If we ascend the evolutionary tree, we find simple eukaryotes that also exist as single nonassociating cells. Farther along the tree we find eukaryotes of increasing complexity, containing many cells. In multicellular organisms the same organism possesses cells of many different types, with specialized structures and functions. In humans, which contain about 10^{14} cells, we find a high degree of complexity. Humans contain more than a hundred different cell types. There are specialized cells associated with skin, connective tissue, nervous tissue, muscle, blood, sensory functions, and reproductive organs (Fig. 2). In such a complex organism, the capacity of different cells for replication is limited. When a skin cell or a muscle cell replicates, it makes more cells of the same tissue type. The only cells capable of reproducing an entire organism are the germ cells, that is, the sperm and the egg.

BIOMOLECULES ARE COMPOSED OF A SMALL NUMBER OF LIGHT ELEMENTS

Despite the enormous complexity of the cell and the wide range of elements available in Earth's crust, the cell is composed mostly of four elements: hydrogen, carbon, nitro-

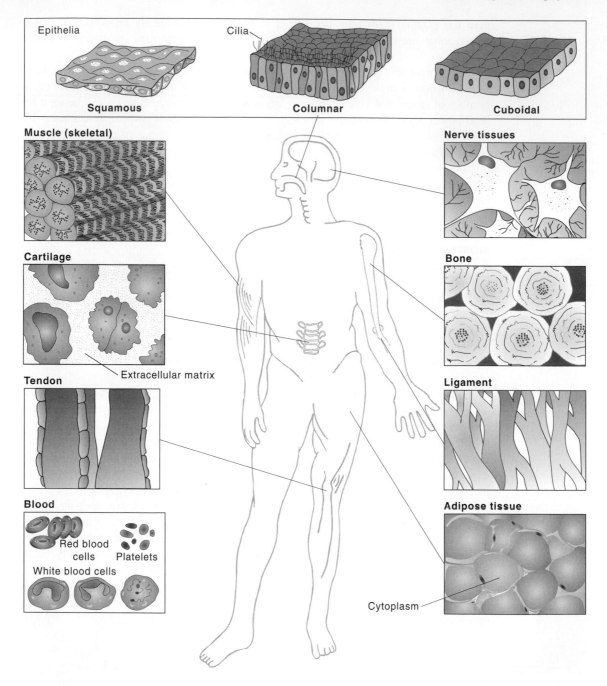

TABLE 1

Distribution of the 24 Elements Used in Biological Systems[a]

Element	Atomic no.	Earth's crust	Ocean	Human body
Hydrogen (H)	1	2,882	66,200	60,562
Carbon (C)	6	56	1.4	10,680
Nitrogen (N)	7	7	<1	2,440
Oxygen (O)	8	60,425	33,100	25,670
Fluorine (F)	9	77	<1	<1
Sodium (Na)	11	2,554	290	75
Magnesium (Mg)	12	1,784	34	11
Silicon (Si)	14	20,475	<1	<1
Phosphorus (P)	15	79	<1	130
Sulfur (S)	16	33	17	130
Chlorine (Cl)	17	11	340	33
Potassium (K)	19	1,374	6	37
Calcium (Ca)	20	1,878	6	230
Vanadium (V)	23	4	<1	<1
Chromium (Cr)	24	8	<1	<1
Manganese (Mn)	25	37	<1	<1
Iron (Fe)	26	1,858	<1	<1
Cobalt (Co)	27	1	<1	<1
Nickel (Ni)	28	3	<1	<1
Copper (Cu)	29	1	<1	<1
Zinc (Zu)	30	2	<1	<1
Selenium (Se)	34	<1	<1	<1
Molybdenum (Mo)	42	<1	<1	<1
Iodine (I)	53	<1	<1	<1

[a]Amounts are given in atoms per 100,000.

gen, and oxygen (Table 1). Two others, phosphorus and sulfur, although present in smaller amounts, play key roles in cellular structure and metabolism. These six elements are found in the organic compounds of the cell. The remaining elements listed in Table 1 are inorganic constituents, found primarily in the fluids that bathe the cell. They are important in a variety of ways.

FIGURE 2 Specialized cell types found in the human. Although all cells in a multicellular organism have common constituents and functions, specialized cell types have unique chemical compositions, structures, and biochemical reactions that establish and maintain their specialized functions. Such cells arise during embryonic development by the complex processes of cell proliferation and cell differentiation. Except for the germ cells (sex cells), all cell types contain the same genetic information, which is faithfully replicated and partitioned to daughter cells. Cell differentiation is the process whereby some of this genetic information is activated in some cells, resulting in the synthesis of certain proteins and not of other proteins. Thus specialized cells come to have different complements of enzymes and metabolic capacities.

TABLE 2

Types of Covalent Linkages Most Commonly Found in Biomolecules

	H	C	O	N	P	S
H						
C	—C—H	—C—C— C=C				
O	—O—H	—C—O— C=O				
N	N—H	—C—N= C=N—	—			
P	—	—	P—O P=O	—		
S	—S—H	—C—S	S—O— S=O	—	—	—S—S—

BIOCHEMICAL REACTIONS ARE A SUBSET OF ORDINARY CHEMICAL REACTIONS

Even though the total number of biochemical reactions is very large, it is still much smaller than the potential number of reactions that can occur in chemical systems in general. This simplification results partly from the fact that only a limited number of elements account for the vast majority of substances found in living cells.

The types of covalent linkages most commonly found in biomolecules also are quite limited (Table 2). Only 16 different types of linkages account for the vast majority of linkages found in biomolecules. All the elements form single or double bonds, except for hydrogen, which can form only single bonds; moreover, all the elements exist primarily in a single valence state (Table 3). Despite this overall simplicity, many other valence states can be found in unusual cases, and some of these are very important. For example, the biochemistry of nitrogen involves consideration of all the valence states of nitrogen from $+5$ to 0 to -3. A major source of nitrogen available to biosystems is gaseous nitrogen found in the atmosphere (valence state 0). Despite its

TABLE 3

Most Common Valences Displayed by Atoms in Covalent Linkage

Element	Valence
H	$+1$
C	-4 to $+4$
O	-2
P	$+5$
N	-3
S	$+6, -2, -1$

omnipresence there are only a very limited number of organisms that can convert gaseous nitrogen into a biochemically useful form.

CELLS AND THEIR ORGANELLES ARE COMPOSED OF SMALL MOLECULES AND MACROMOLECULES

To appreciate the functions are reactions that take place in the cell we must think in terms of molecules. The water molecule is the major component of the cell, accounting for about 70% of its weight (Table 4). Consequently, most of the other components exist in an aqueous environment. Because water is a highly polar solvent, other polar molecules tend to be *hydrophilic*; that is, they have a high affinity for water and thus are very soluble. By contrast, apolar substances tend to be *hydrophobic*; that is, they have a low affinity for water and thus are poorly soluble in water.

Except for water, most of the molecules found in the cell are *macromolecules* (very large molecules). These macromolecules can be classified into four different categories: *lipids*, *carbohydrates*, *proteins*, and *nucleic acids*. Each type of macromolecule possesses distinct chemical properties that suit it for the functions it serves in the cell.

Lipids are primarily hydrocarbon structures (Fig. 3). They tend to be poorly soluble in water, and are therefore particularly well suited to serving as a major component of the various membrane structures found in cells. Lipids also serve as a convenient, compact way to store chemical energy.

Carbohydrates, like lipids, are primarily hydrocarbon structures, but they also contain many polar hydroxyl groups (—OH), which makes them very soluble in water. Large carbohydrate molecules called polysaccharides consist of many small,

TABLE 4
The Approximate Composition of a Bacterial Cell

	Percentage of total cell weight	Number of types of each molecule
Water	70	1
Inorganic ions	1	20
Sugars and precursors	3	200
Amino acids and precursors	0.4	100
Nucleotides and precursors	0.4	200
Lipids and precursors	2	50
Other small molecules	0.2	~200
Macromolecules (proteins, nucleic acids, and polysaccharides)	22	~5000

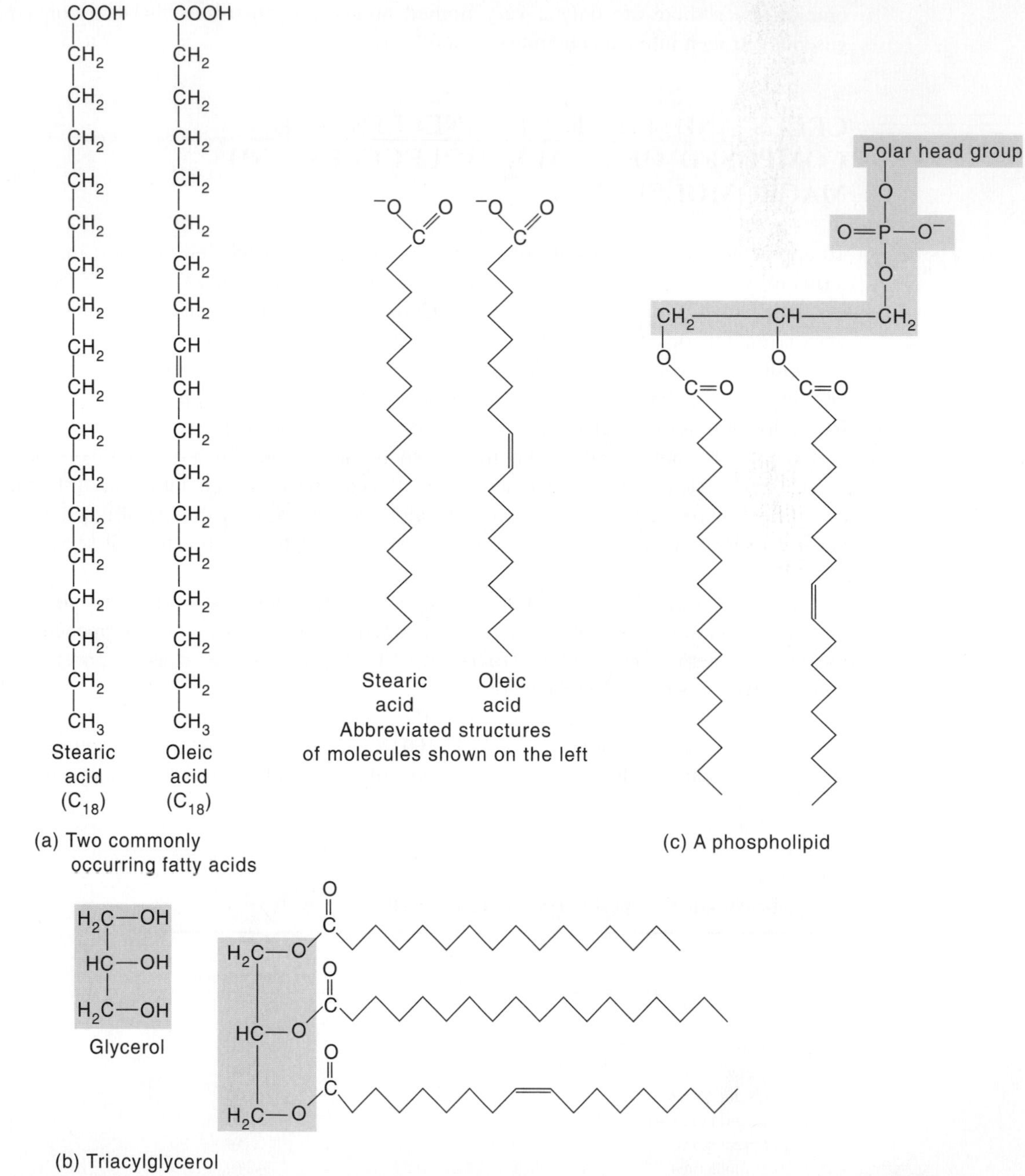

(a) Two commonly
    occurring fatty acids

Stearic acid (C$_{18}$), Oleic acid (C$_{18}$)

Stearic acid, Oleic acid — Abbreviated structures of molecules shown on the left

(c) A phospholipid

Polar head group

(b) Triacylglycerol

Glycerol

FIGURE 3   The structures of common lipids. (a) The structures of saturated and unsaturated fatty acids, represented here by stearic acid and oleic acid. (b) Three fatty acids covalently linked to glycerol by ester bonds form a triacylglycerol. (c) The general structure for a phospholipid consists of two fatty acids esterified to glycerol, which is linked through phosphate to a polar head group. The polar head group may be any one of several different compounds, for example, chloline, serine, or ethanolamine.

ringlike sugar molecules, the sugar monomers, attached to one another (by bonds called glycosidic bonds) in a linear or branched array to form the sugar polymer (Fig. 4). In the cell, such polysaccharides often form storage granules that may be readily broken down into their component sugars. With further chemical breakdown these sugars release chemical energy and also may provide the carbon skeletons for the synthesis of a variety of other molecules. Important structural functions also are served by polysaccharides. Linear polysaccharides form a major component of plant cell walls. Bacterial cell walls are constructed from cross-linked polysaccharides.

Proteins are the most complex macromolecules found in the cell. They are composed of linear polymers called *polypeptides*, which contain amino acid monomers connected by peptide bonds (Fig. 5). Each amino acid contains a central carbon atom attached to four substituents: (1) a carboxyl group ($-COOH$), (2) an amino group ($-NH^+_3$), (3) a hydrogen atom, and (4) an R group (Box 6A). The R group gives each amino acid its unique characteristics. There are 20 different amino acids (Fig. 6). Some R groups are charged, some are neutral but still polar, and some are apolar.

The properties of a given protein depend on the sequences of amino acids found in its polypeptide chains. After synthesis, the linear polypeptide molecules fold in a highly specific way to form a unique three-dimensional structure. Many proteins are composed of two or more polypeptides held together by noncovalent forces. In such a structure the individual polypeptide chains are referred to as subunits. Certain proteins function in structural roles. Some structural proteins interact with lipids in membrane structures. Others aggregate to form part of the cytoskeleton that gives the cell its shape. Still others are the chief components of muscle or connective tissue. The substances known as *enzymes* form a major class of proteins; these proteins function as catalysts that direct and accelerate biochemical reactions. Each enzyme functions in a highly specific manner, usually catalyzing only one type of reaction. Even the simplest cells probably contain more than a thousand different types of enzymes.

Nucleic acids are the largest macromolecules in the cell. These exist as very long linear polymers, called *polynucleotides*, which are composed of *nucleotide* monomers. A nucleotide contains (1) a five-carbon sugar molecule, (2) one or more phosphate groups, and (3) a nitrogenous base. The nitrogenous base gives the nucleotide a distinct character (Fig. 7). Of special interest are the nucleic acids known as *deoxyribonucleic acid* (*DNA*). DNA contains the genetic information that is inherited by each daughter cell during growth when cells divide. The DNA usually exists as nucleoprotein (DNA-protein) complexes called *chromosomes*. Prior to cell division, chromosomal DNA replicates and segregates so that an identical complement of DNA goes to each of two newly formed daughter cells.

Eukaryotic cells are more complex than those of prokaryotes. As we have seen, they contain a nucleus, whereas prokaryotic cells do not. Eukaryotic cells also, as a rule, contain more DNA; and this DNA is partitioned into two, three, or many chromosomes, versus only one per cell in prokaryotes. In both prokaryotes and eukaryotes, all cells of the same organism are identical in their chromosome content. In eukaryotes most of the chromosomes are localized in the nucleus. Thus the DNA is isolated from the main body of the cytoplasm. Some organelles, however (notably the mitochondria

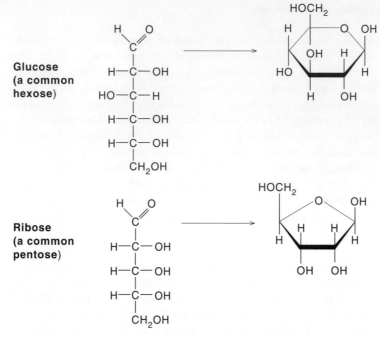

Glucose
(a common
hexose)

Ribose
(a common
pentose)

(a) Two common monosaccharides that circularize in aqueous solution

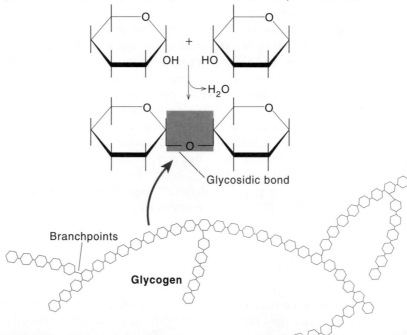

Glycosidic bond

Branchpoints

Glycogen

(b) Polysaccharides composed of covalently linked monosaccharides

FIGURE 4 Monomers and polymers of carbohydrates. (a) The most common carbohydrates are the simple six-carbon (hexose) and five-carbon (pentose) sugars. In aqueous solution, these sugar monomers circularize to form ring structures. (b) Polysaccharides are usually composed of hexose monosaccharides covalently linked together by glycosidic bonds to form long straight-chain or branched-chain structures.

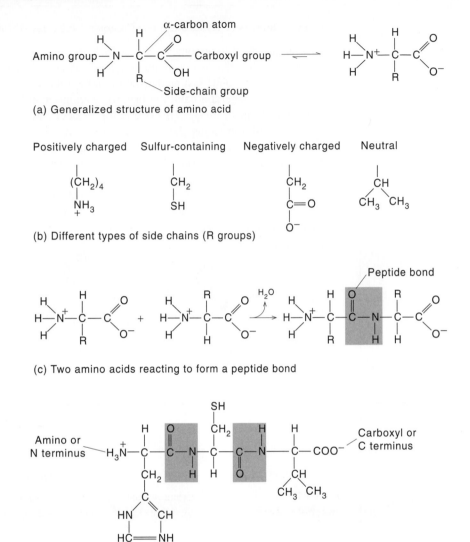

(a) Generalized structure of amino acid

(b) Different types of side chains (R groups)

(c) Two amino acids reacting to form a peptide bond

(d) Many amino acids reacting to form a polypeptide chain

FIGURE 5   Amino acids and the structure of the polypeptide chain. Polypeptides are composed of L-amino acids covalently linked together in a sequential manner to form linear chains. (a) The generalized structure of the amino acid. The zwitterion form, in which the amino group and the carboxyl group are ionized, is strongly favored. (b) Structures of some of the R groups found for different amino acids. (c) Two amino acids become covalently linked by a peptide bond, and water is lost. (d) Repeated peptide bond formation generates a polypeptide chain, which is the major component of all proteins.

If we examine the biomonomers from which the related macromolecules are composed, we find that many of them are asymmetrical or chiral. The amino acids are a case in point. Just as the right hand is related to the left hand by its mirror image, in general a naturally occurring amino acid likewise is related to its nonidentical mirror image. This is true for 19 out of the 20 amino acids most commonly found in proteins, the one exception being glycine.

Chirality of amino acids stems from the chiral or asymmetrical center, the tetravalent α-carbon atom. The α-carbon is a chiral center only if it is connected to four different substituents. Thus the α-carbon of glycine is not a chiral center because it is connected only to three different substituents, one carboxyl, one amino, and two hydrogens. However, the α-carbon atom of alanine is a chiral center because it is connected to one carboxyl, one amino, one methyl, and one hydrogen (Fig. 6A).

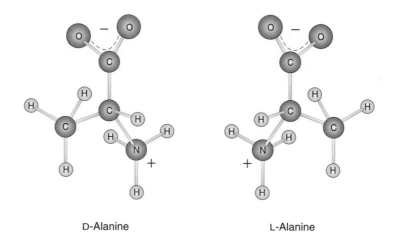

D-Alanine   L-Alanine

FIGURE 6A The covalent structure of alanine, showing the three-dimensional structure of the L- and D-stereoisomeric forms.

Two structures that are related by their mirror images are referred to as *enantiomers*. The two enantiomers for alanine are called L-alanine and D-alanine. The naming by D and L prefixes refers to a convention established a long time ago by Emil Fischer. All commonly occurring amino acids belong to the D series of enantiomorphs.

Not all biomonomers are chiral. Amino acids and sugars are mostly chiral but the purines and pyrimidines that go into nucleic acids are not. The chirality of macromolecules originates in the chirality of the biomonomers from which they are composed. All proteins, polysaccharides, and nucleic acids are chiral. In the case of nucleic acids the purines and pyrimidines are not chiral but the ribose sugars that they contain are.

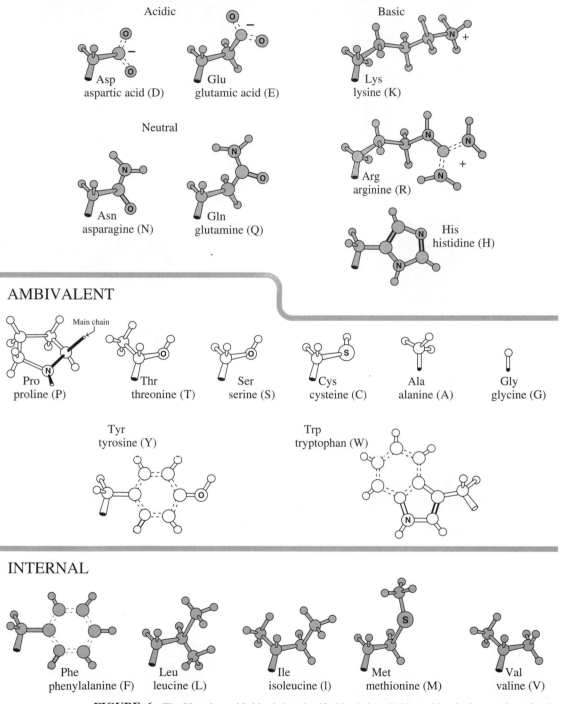

EXTERNAL

Acidic

Asp
aspartic acid (D)

Glu
glutamic acid (E)

Basic

Lys
lysine (K)

Neutral

Asn
asparagine (N)

Gln
glutamine (Q)

Arg
arginine (R)

His
histidine (H)

AMBIVALENT

Main chain

Pro
proline (P)

Thr
threonine (T)

Ser
serine (S)

Cys
cysteine (C)

Ala
alanine (A)

Gly
glycine (G)

Tyr
tyrosine (Y)

Trp
tryptophan (W)

INTERNAL

Phe
phenylalanine (F)

Leu
leucine (L)

Ile
isoleucine (l)

Met
methionine (M)

Val
valine (V)

FIGURE 6 The 20 amino acid side chains classified by their probable position in the protein molecule. Three-letter and one-letter abbreviations are given for each. The forms shown here are the most prevalent at pH 7. Note that histidine can play a dual role: neutral or positively charged.

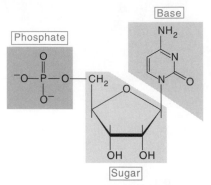

(a) Generalized structure of a nucleotide

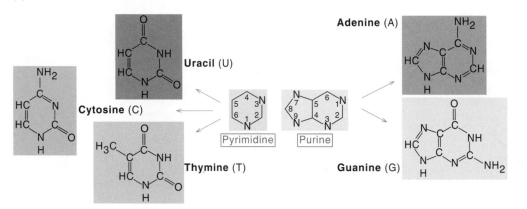

(b) Different bases found in nucleotides

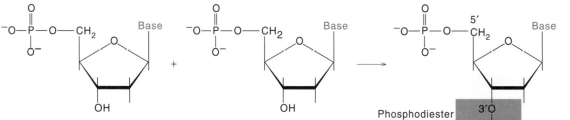

(c) Two nucleotides reacting to form a dinucleotide

Phosphodiester
linkage

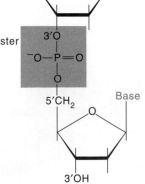

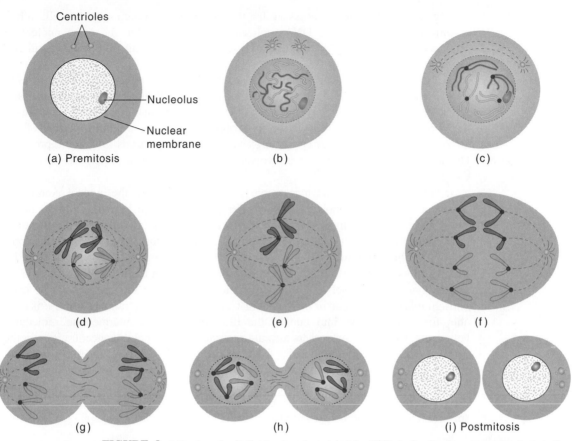

Centrioles

Nucleolus

Nuclear membrane

(a) Premitosis

(b)

(c)

(d)

(e)

(f)

(g)

(h)

(i) Postmitosis

FIGURE 8   Mitosis and cell division in eukaryotes. After DNA duplication has occurred, mitosis is the process by which quantitatively and qualitatively identical DNA is delivered to daughter cells formed by cell division. Mitosis is traditionally divided into a series of stages characterized by the appearance and movement of the DNA-bearing structures, the chromosomes.

and the chloroplasts), also contain DNA, which is usually in the form of a single circular chromosome.

Eukaryotic chromosomes are detectable by microscopic techniques, but they can be seen only in certain cell types, and only at the stage just prior to cell duplication. At this stage, called mitosis, chromosomes appear as elongated refractile structures that can be seen to segregate in equal numbers and types to each of the daughter cells formed as a result of cell division (Fig. 8). Each chromosome carries specific heredi-

FIGURE 7   The structural components of nucleic acids. Nucleic acids are long linear polymers of nucleotides, called polynucleotides. (a) The nucleotide consists of a five-carbon sugar (ribose in RNA or deoxyribose in DNA) covalently linked at the 5′ carbon to a phosphate, and at the 1′ carbon to a nitrogenous base. (b) Nucleotides are distinguished by the types of bases they contain. These are either of the two-ring purine type or of the one-ring pyrimidine type. (c) When two nucleotides become linked, they form a dinucleotide, which contains one phosphodiester bond. Repetition of phosphodiester bond formation leads to a polynucleotide.

tary (genetic) information necessary for the synthesis of proteins of the cell. When chromosomes replicate, the DNA replicates precisely, so that the same nucleotide sequence is passed along to each of the daughter cells resulting from mitosis and cell division.

In the cytoplasm we find another type of nucleic acid known as *ribonucleic acid* (*RNA*). RNAs are smaller than DNAs (usually 10^2 to 10^4 nucleotides in length), and differ in minor chemical respects from their DNA cousins. Each RNA contains a nucleotide sequence that reflects the nucleotide sequence in a specific region of the DNA. Most of the RNAs carry the information for the synthesis of specific proteins. Of the remaining RNAs, most are involved in the mechanics of protein synthesis (see Chapter 9) .

In addition to water and macromolecules and organelles, the cytosol contains a large variety of small molecules that differ greatly in both structure and function. These never make up more than a small fraction of the total cell mass despite their great variety (see Table 4). One class of small molecules consists of the monomer precursors of the different types of macromolecules. These monomers are derived by chemical modification from the nutrients absorbed through the cell membrane. Rarely are the nutrients themselves the actual monomers used by the cell. As a rule each nutrient must undergo one or more enzymatically catalyzed alterations before it is suitable for incorporation into one of the biopolymers. The intermediate molecules between nutrients and monomers also are present in even smaller concentrations in the cytosol. Additional small molecules found in the cytosol include molecules formed as side products in important synthetic reactions and as breakdown products of the macromolecules. Finally, small bioorganic molecules known as *coenzymes* also are present in the cytoplasm. These compounds act in union with the enzymes in a highly specific manner to catalyze a wide variety of reactions.

MACROMOLECULES FORM COMPLEX FOLDED STRUCTURES

In a number of important biochemical reactions that occur within or between biomolecules, weak forces are involved. All these interactions are tempered by the highly polar aqueous environment in which they occur. Thus polar groups in biomolecules prefer associations with other polar groups, especially water. On the other hand, apolar groups in biomolecules prefer contacts with other apolar groups and avoid contacts with water. The result is that a biopolymer adopts a three-dimensional structure in which the polar groups are oriented on the surface, where they can associate with water; on the contrary, the apolar groups are usually buried in the interior of the structure, where they can associate with other apolar groups.

An exception to this rule is observed with polar groups that form *hydrogen bonds*. Hydrogen bonds form between the hydrogen atoms of amino or hydroxyl groups and other nitrogen or oxygen groups that lack covalently bonded hydrogen atoms.

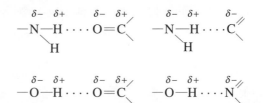

Amino and hydroxyl groups are highly polar, with a partial negative charge (δ^-) on the nitrogen and the oxygen atoms in N—H or O—H functional groups. When the bonding electrons in these functional groups are attracted away from the covalently linked hydrogen, there are no other electrons to shield the positive charge of the bare hydrogen nucleus, a proton. Consequently, there is a very small, highly localized positive charge (δ^+) on the hydrogen atom that seeks to form electrostatic bonds with other nitrogen and oxygen atoms carrying a partial negative charge. Intramolecular and intermolecular hydrogen bonds are frequently formed by complementary polar groups in biomolecules, and often these bonds play a decisive role in determining the three-dimensional structure.

Some structures illustrating these principles of macromolecular interaction are shown in Figs. 9 to 11. Phospholipids (Fig. 9), with a hydrophilic polar group on one end and long hydrophobic side chains attached to it, form multimolecular structures in

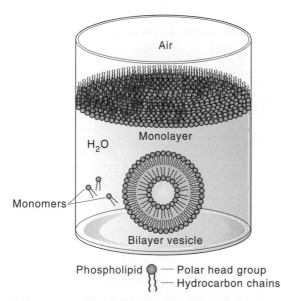

FIGURE 9   Structures formed by phospholipids in aqueous solution. Phospholipids may form a monomolecular layer at the air–water interface, or they may form spherical aggregations surrounded by water. A vesicle consists of a double molecular layer of phospholipids surrounding an internal compartment of water.

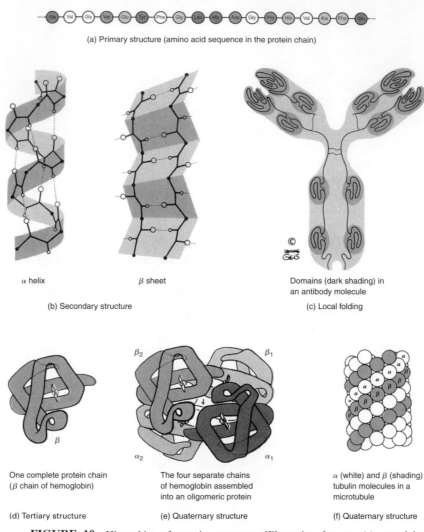

(a) Primary structure (amino acid sequence in the protein chain)

α helix   β sheet   Domains (dark shading) in
                                                                    an antibody molecule

(b) Secondary structure   (c) Local folding

One complete protein chain   The four separate chains   α (white) and β (shading)
(β chain of hemoglobin)   of hemoglobin assembled   tubulin molecules in a
                                  into an oligomeric protein   microtubule

(d) Tertiary structure   (e) Quaternary structure   (f) Quaternary structure

FIGURE 10   Hierarchies of protein structures. [Illustration for part (c) copyright by Irving Geis. Reprinted by permission.]

an aqueous environment (see Fig. 9). These phospholipid aggregates take the form of monomolecular layers at the air–water interface, or of bilayer vesicles within the water. In these structures, the polar head groups of the lipid are in contact with water, whereas the apolar side chains are excluded from the solvent structure.

Polypeptide chains show the greatest versatility in folding. The two most common folds are the α helix in which intramolecular hydrogen bonds give rise to an elongated helix structure, and the β sheet in which interchain hydrogen bonds hold segments of polypeptide chain in the extended configuration (Fig. 10 b). Most globular molecules (see Fig. 10 d to f) contain elements of α-helix and β-sheet structures in a complex folded structure. Overall globular proteins fold in such a way that their hydrophobic amino acid side chains are located on the inside of the structure leaving their hydrophilic amino acid side chains on the surface where they can interact with the water.

DNA forms a complementary structure involving two helically oriented poly-nucleotide chains (Fig. 11). In this structure the polar sugar and phosphate groups are located on the surface, where they can interact with water; the nitrogenous bases from the two chains form intermolecular hydrogen bonds in the core of the structure.

SUMMARY

In this chapter we have discussed the structures and functions of chemicals found in living cells.

1. All living systems, no matter how big or how small, are composed of mem-brane-enclosed objects known are cells. All cells contain chromosomes. In prokaryotes there is only one chromosome. In eukaryotes there are two or more chromosomes and the chromosomes are separated from the remainder of the cell's contents by a nuclear membrane. The eukaryotic cell also usually contains other membrane-bounded organelles.

2. Biochemical reactions are governed by the same principles as ordinary chemical reactions but are more limited in number. This is because biochemical substances are composed, for the most part, of only six elements and because most of the covalent linkages found in biomolecules are of only 16 different types.

3. Cells are their organelles are composed of small molecules and macromolecules.

4. Water is the major component of all cells. Other molecules found in the cell may be classified as hydrophilic if they favor interactions with water or hydrophobic if they avoid such interactions.

5. Aside from water, the mass of cells is made up of four types of macromolecules: lipids, carbohydrates, proteins, and nucleic acids. Each of these molecules has a unique chemistry. Lipids tend to be hydrophobic, while the other three types of macro-molecules tend to be hydrophilic.

6. The three-dimensional structures formed by the four major classes of macromol-ecules are heavily influenced by the way in which the macromolecules interact with water and with themselves.

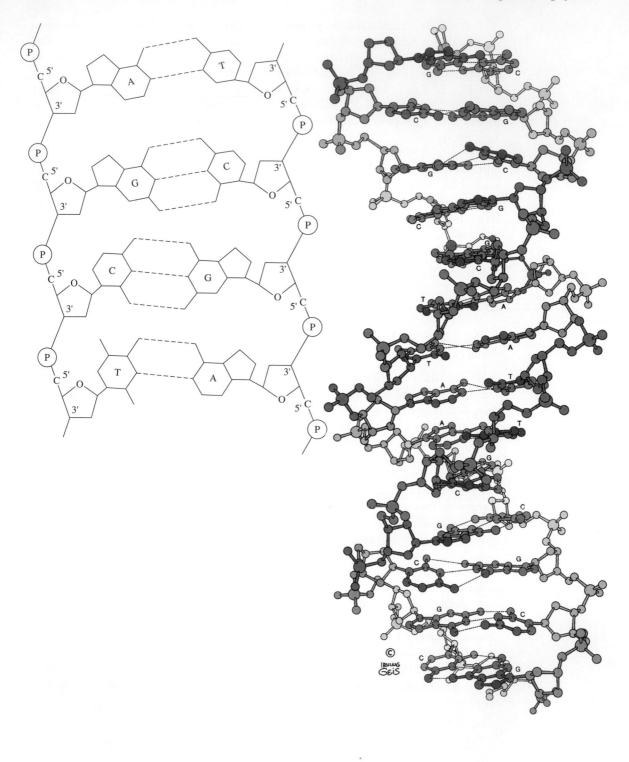

Problems

1. When geologists examine fossil remains embedded in ancient rocks, they are inclined to classify the cells they find as prokaryotic and eukaryotic on the basis of size. Do you see any problems with this criterion?
2. Some meteorites believed to have originated on Mars were carefully analyzed and found to contain fossil-like structures that strongly resembled cell shapes. The only problem was that they are extremely small, too small in fact to contain the structures necessary for protein synthesis. Is this sufficient grounds for rejecting them as fossils of cells?
3. Imagine that there were two types of organisms on Earth, one based on D-sugars and D-amino acids and one based on L-sugars and L-amino acids. If these two types of organisms had been allowed to mingle, does it seem likely that they would have coexisted?
4. If someone suggested to you that the first living things were composed exclusively of carbohydrates, what would you say? Consider the same for RNA.
5. If a protein had a hydrophobic surface, where would you expect to find it in the cell?
6. In addition to water, the living cell contains four types of large molecules: nucleic acids, proteins, carbohydrates, and lipids. Can you imagine life in any form without one of these? Which type of molecule could you eliminate and what type of accommodations would you have to make?
7. Many scientists object to my classifying lipids as macromolecules. They claim that only nucleic acids, proteins, and carbohydrates should be classified as macromolecules. Why do you think they believe this?

References

I give mostly general references for Part II because the material that is dealt with in this section of the text is all well established.

Shultz, H. N., Brinkhoff, T., Ferdelman, T. G., Marine, M. H., Teske, A. and Jorgensen, B. B. Dense Populations of a Giant Sulfur Bacterium in Namibian Shelf Sediments, Science 284:493, 1999. (There are very big prokaryotes after all.)

FIGURE 11   Two representations of the complementary two-chain DNA structure. (a) Segment of DNA, drawn to emphasize the hydrogen bonds (dashed lines) formed between opposing chains. The two chains are oriented antiparallel. One strand (left side) runs $5'$ to $3'$ from top to bottom, and the other strand (right side) runs $5'$ to $3'$ from bottom to top. The planes of the base pairs are turned $90°$ to show the hydrogen bonds between the base pairs. (b) A most realistic model of the DNA structure in its right-handed helical conformation with the planes of the base pairs oriented perpendicular to the helix axis. The hydrogen bonds are represented by dotted lines. (Illustration copyright by Irving Geis. Reprinted by permission.)

For Matters Concerning Structure and Function (Two Outstanding Texts)

Alberts, B., Bray, D., Lewis, J., Raff, M., Roberts, K., and Watson, J. D. *Molecular Biology of the Cell*, 3rd ed. New York: Garland, 1994.

Lodish, H., Baltimore, D., Berk, A., Zipursky, S. L., Matsudaira, P., and Darnell, J. *Molecular Cell Biology*, 3rd ed. New York: Freeman 1995.

For Matters Concerning the Biochemistry (My Own Text)

Zubay, G. *Biochemistry*, 4th ed. New York: McGraw-Hill, 1998.

CHAPTER 7

Metabolic Strategies and Pathway Design

Living systems are designed to thrive and replicate in the surrounding environment. Typically several hundred to several thousand reactions proceed simultaneously in the confines of a living cell. The most distinctive feature of living systems is that all the reactions serve a purpose, the maintenance and propagation of the system. Collectively, the processes involved are referred to as the system's *metabolism*. When new biochemical reactions are discovered, we try to see what role they play in metabolic functioning. Just as frequently, we start with the hypothesis that in light of a known function, a certain reaction must exist, and this conviction gives the impetus that leads to discovery of that reaction.

THERMODYNAMICS GIVES US THE CRITERION TO DETERMINE THE ENERGY STATUS FOR A BIOCHEMICAL REACTION

Photons radiated by the Sun are the ultimate source of energy for driving energy-dependent reactions in living organisms. In photosynthetic plants and microorganisms, sunlight is converted into chemical compounds. Most other organisms obtain their chemical energy by degrading chemical compounds. Thermodynamics provides us with a means for evaluating the potential of an energy-releasing reaction to drive an energy-requiring reaction. The most important quantity for this purpose is the *free energy* designated by the symbol G. In practice we are more concerned with the free energy change, ΔG, for a chemical transformation than with the absolute free energy for any given compound. The free energy change tells us the extent to which a reaction goes; it also tells us how much free energy is available from the reaction to perform useful work. A large negative value for ΔG indicates a spontaneous reaction that releases considerable free energy. A positive ΔG indicates a reaction that does not proceed unless it is coupled to an energy-yielding reaction. When all the reactants and products for a particular chemical transformation are present in their standard state, the free energy difference for a transformation is referred to as the standard free energy, $\Delta G°$, for the reaction. The concentrations of most compounds in their standard state is $1\ M$; for H^+ it is often taken as $10^{-7}\ M$ instead of $1\ M$ because this is close to the H^+ concentration in biological systems. The actual free energy for a chemical reaction, ΔG, differs from the standard free energy $\Delta G°$ by a term that includes the actual concentrations of reactants and products. For the reaction

$$aA + bB \rightleftharpoons cC + dD, \tag{1}$$

the actual free energy change is a composite of the standard free energy change and a term that compensates for the actual concentrations under which a reaction is carried out.

$$\Delta G = \Delta G° + RT \ln(C)^c(D)^d/(A)^a(B)^b. \tag{2}$$

If a reaction is allowed to proceed for a long time in a closed system, it should reach *equilibrium* (Box 7A gives a detailed explanation of the meaning of equilibrium and the equilibrium constant). At this point the rate of conversion of reactants into products is equal to the rate of conversion of products into reactants. As a result there is no net free energy change in going from one state to the other. At equilibrium

$$\frac{(C)^c(D)^d}{(A)^a(B)^b} = K_{eq}, \tag{3}$$

where K_{eq} is the equilibrium constant for the reaction and $\Delta G = 0$. Under these conditions we can see from Eq. 2 that

$$\Delta G° = -RT \ln K_{eq}. \tag{4}$$

This means that if we know K_{eq}, we may calculate $\Delta G°$ or vice versa.

BOX 7A   The Meaning of the Equilibrium Constant for a Reaction

Consider four different molecules A, B, C, and D in which A and B react with each other and give rise to C and D. Let us assume that the reaction is reversible, as most reactions are. In that case C and D can react with each other to produce A and B. In chemistry this reversible situation is written

$$A + B \rightleftharpoons C + D,$$

with conversion arrows pointing in either direction. If we start out with a mixture of A and B, then initially we see only conversions in the rightward direction. Conversely, if we start out with a mixture of C and D, initially we will see only conversions in the leftward direction. If we permit such a reaction to proceed for a long time, it will reach equilibrium. At equilibrium the rates of conversions in the two directions exactly balance. The absolute amounts of the different compounds may vary over a considerable range at equilibrium. It the number of molecules of A and B is equal to the number of molecules of C and D at equilibrium, the equilibrium constant, K_{eq}, is equal to 1

$$K_{eq} = (C + D)/(A + B).$$

If the ratio $(C + D)/(A + B)$ is 10, the $K_{eq} = 10$; or if the ratio is 10^{-1}, the $K_{eq} = 10^{-1}$ and so on.

   It should be noted that we use the number of molecules per unit volume as our standard unit for measuring concentration when computing the equilibrium constant. As a rule this is expressed in units of molarity (M). A 1 M solution of A contains exactly 6.02×10^{23} molecules of A in 1 liter of solution. To make a 1 M solution of A we simply weigh out in grams the equivalent of the molecular weight of A and dissolve it in 1 liter of water. For example, NaCl has a molecular of $23 + 35 = 58$. A solution containing 58 grams of NaCl in 1 litter is a 1 M solution. A solution containing 5.8 grams of NaCl in 1 liter is a 0.1 M solution and so on.

LIVING CELLS REQUIRE A STEADY SUPPLY OF STARTING MATERIALS AND ENERGY

Metabolism depends on factors that are external to the organism. The living system must extract nutrients from the environment and convert then to a biochemically useful form. In the next phase of the metabolism, which is internal, small molecules are synthesized and degraded; this process serves two functions: (1) it supplies the energy needed for the synthesis of macromolecules and other energy-requiring processes; and (2) it furnishes these processes with the necessary starting materials—fatty acids for

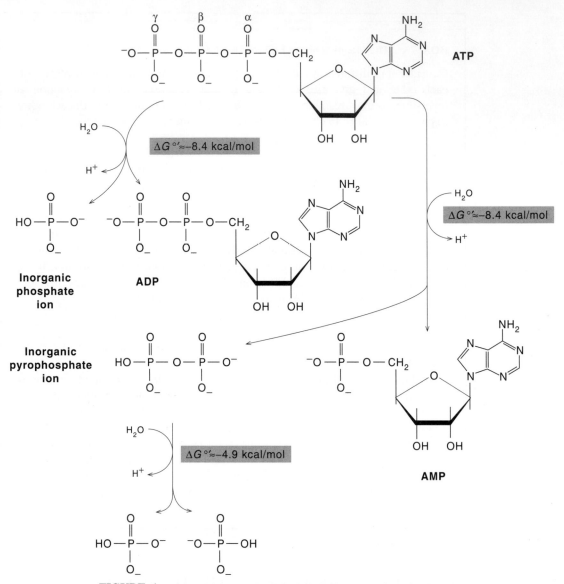

FIGURE 1 Alternative routes of ATP hydrolysis. The charged species shown are the main ones present at physiological pH and ionic strength. The phosphate groups of ATP are referred to as α, β, and γ as indicated. Under physiological conditions, ATP and ADP also bind Mg^{2+} (not shown).

lipid synthesis, sugars for polysaccharide synthesis, nucleoside triphosphates for nucleic acid synthesis, and amino acids for protein synthesis.

The demand for energy and starting materials varies widely in different biological processes. To meet these fluctuating needs, the rates of the reaction sequences must be adjustable over broad ranges.

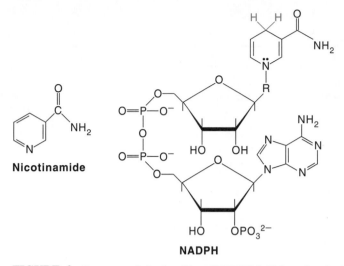

FIGURE 2   Structures of nicotinamide and NADPH (which carries nicotinamide in reduced form).

The compound adenosine triphosphate (ATP) is the main source of chemical energy used by living systems (Fig. 1). Through hydrolysis, ATP is converted into adenosine diphosphate (ADP) and inorganic phosphate ion (P_i), and in the process a great deal of free energy is made available to drive other reactions. We see examples of this later. One of the main compounds that serves as a source of reducing power is reduced nicotinamide adenine dinucleotide phosphate (NADPH), a coenzyme containing nicotinamide (Fig. 2). Reducing power—the capacity to donate electrons—is required because most biosynthesis involves converting compounds in which carbon is in a reduced state.

ORGANISMS DIFFER IN SOURCES OF STARTING MATERIALS, ENERGY, AND REDUCING POWER

Whereas the needs of different organisms are similar, the ways in which organisms satisfy their needs can be quite different. Indeed, the most fundamental metabolic distinction between organisms relates to the ways in which they satisfy their basic metabolic needs (Table 1).

Autotrophs are organisms that exploit the inorganic environment without recourse to compounds produced by other organisms. Thus all their carbon compounds must be synthesized from inorganic compounds of carbon, such as carbonates or carbon dioxide. *Chemoautotrophs* obtain reducing power by oxidation of inorganic materials such as hydrogen or reduced compounds of sulfur or nitrogen. Each species is usually specific as to the electron donors that it can utilize. The same reduced compounds supply electrons for regeneration of ATP by the process

TABLE 1
Metabolic Classification of Organisms

Type of organism	Source of ATP	Source of NADPH[a]	Source of carbon	Examples
Chemoautotroph	Oxidation of inorganic compounds	Oxidation of inorganic compounds	CO_2	Hydrogen, sulfur, iron, and denitrifying bacteria
Photoautotroph	Sunlight	H_2O	CO_2	Higher plants, blue-green algae, photosynthetic bacteria
Photoheterotroph	Sunlight	Oxidation of organic compounds	Organic compounds	Nonsulfur purple bacteria
Heterotroph	Oxidation of organic compounds	Oxidation of organic compounds	Organic compounds	All higher animals, most microorganisms, non-photosynthetic plant cells

NADPH is the main source of reducing power.

known as electron-transfer phosphorylation. *Phototrophs* obtain ATP by electron-transfer phosphorylation during the cycling of photochemically excited electrons. Some *photoautotrophs*, including plants, also are able to use the energy of sunlight to extract electrons from water. Thereby photoautotrophs have achieved independence from all sources of energy except for the Sun and from all sources of electrons and carbon except for water and carbon dioxide. For this reason, photoautotrophic fixation of carbon dioxide is the predominant base of the food chain in the biosphere.

In contrast to autotrophs, the common *heterotrophs* depend on preformed organic compounds for all three primary needs. Although some carbon dioxide is fixed in heterotrophic metabolism, the heterotrophic cell thrives at the expense of compounds formed by other cells, and is not capable of the net conversion (fixation) of carbon dioxide into organic compounds. Some bacteria, referred to as *photoheterotrophs*, are able to regenerate ATP photochemically, but cannot use photochemical reactions to supply electrons to NADP+. Such organisms are like other heterotrophs in their dependence on preformed organic compounds. However, because of their photochemical apparatus, the photoheterotrophs are able to use available organic food more efficiently than common heterotrophs can.

Although all heterotrophs depend on preformed organic compounds, they differ markedly in the numbers and types of compounds they require. Some species can make all required compounds when supplied with a single carbon source; others have lost some or many biosynthetic capabilities. For example, mammals must obtain about half of their amino acids from external sources. They also are unable to make several coenzymes, as evidence by their well-known nutritional requirements for vitamins. Some bacteria and some parasites have even more extensive nutritional requirements than this.

REACTIONS SHOW FUNCTIONAL COUPLING

Metabolism embraces both *catabolism*—the breakdown of complex substances into simpler ones, with a release of energy—and the reverse process of *anabolism*, or *biosynthesis*, which require energy for their completion. The overall operational aspects of metabolism may be clarified by means of block diagrams that omit details and focus on functional relationships. Such a diagram is shown in Fig. 3 for a typical heterotrophic aerobic cell. The metabolism of the cell is symbolized by two functional blocks:

1. A catabolic block occurs when foods are oxidized to carbon dioxide. Most of the electrons liberated in this oxidation are transferred to oxygen, with concomitant production of ATP (electron-transfer phosphorylation). Other electrons are used in the regeneration of NADPH, the most frequently used reducing agent for biosynthesis. The major pathways in this block are the glycolytic sequence and the tricarboxylic (TCA) cycle (see Chapter 11). The same sequences supply carbon skeletons for many of the cell's biosynthetic processes.

2. A biosynthetic block occurs when the starting materials produced by catabolism are converted into hundreds of cell components. In this process NADPH serves as a reducing agent when necessary and ATP serves as the universal energy-transducing compound. Synthesis from starting materials leads to macromolecules and growth.

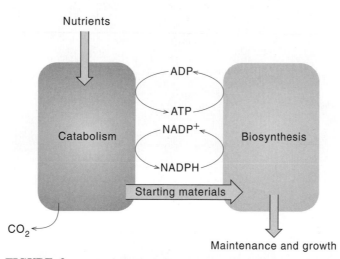

FIGURE 3   A schematic block diagram of the metabolism of a typical aerobic heterotroph. The block labeled catabolism represents pathways nutrients are converted to small-molecule starting materials. Catabolism also supplies the energy (ATP) and reducing power (NADPH) needed for activities that occur in the second block; these compounds shuttle between the two boxes. The block labeled "Biosynthesis" represents the synthesis of low- to medium- molecular-weight components of the cell as well as the synthesis of proteins, nucleic acids, lipids, and carbohydrates; and the assembly of membranes, organelles, and other structures of the cell.

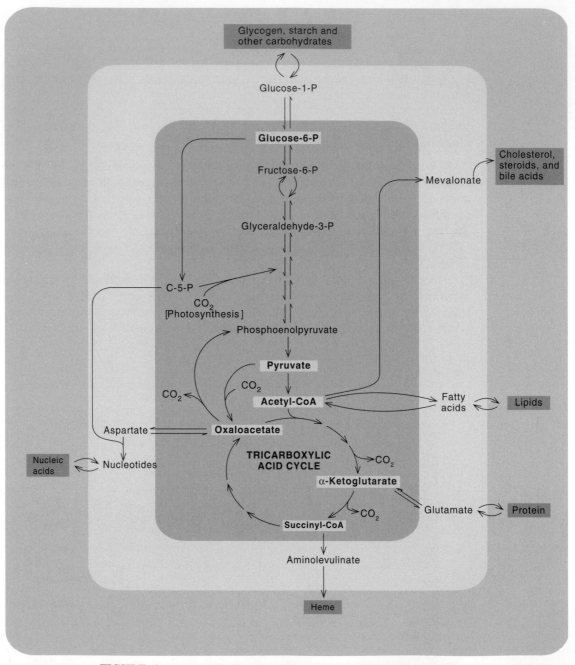

FIGURE 4 The overall metabolism of an aerobic heterotroph represented as three concentric boxes. The central metabolic pathways are represented in the innermost box, within which carbohydrate compounds are degraded to produce energy, reducing power, and starting materials for biosynthesis. Only some key intermediates are shown. In general in moving out from the central box to the outer two boxes, reactions

(Continued)

The starting materials consumed in biosynthetic sequences must be continuously replaced by catabolic processes. The major energy source and reducing source, ATP and NADPH, respectively, are used in very different ways. When they contribute to biosynthesis, ATP is converted to ADP or adenosine 5′-monophosphate (AMP), and NADPH is converted to NADP$^+$. The ATP and NADPH must be regenerated, rephosphorylated, or reduced, at the expense of substrate oxidation, in the catabolic block of reactions.

The block diagram of Fig. 3 represents metabolism in aerobic heterotrophs. The photochemical production of ATP in photoheterotrophs would require the addition of a photochemical block, with ATP production in the catabolic block being deleted. For a photoautotroph two blocks would be needed in addition to those shown in Fig. 3: (1) a photochemical block that regenerates ATP from ADP and reduces NADP$^+$ to NADPH, while consuming water and producing oxygen; and (2) a block that uses the ATP and NADPH for the reduction of carbon dioxide to various products. These photosynthetic products provide the input to the catabolic block, which in this case serves only one function—the production of the starting materials for synthesis.

A more detailed view of metabolism is depicted in Fig. 4. Here metabolism is represented by three concentric boxes. The innermost box shows the central metabolic pathways and the interconversions of various small molecules. These substances serve as the starting materials for all other metabolism. Some of the compounds formed from these intermediates are shown in the middle concentric box. These compounds serve as building blocks for the final products that are indicated in the outermost box.

THERMODYNAMICALLY UNFAVORABLE REACTION CAN BE MADE FAVORABLE BY COUPLING TO THE ADENOSINE TRIPHOSPHATE–ADENOSINE DIPHOSPHATE SYSTEM

The metabolic block diagram of Fig. 3 emphasized the coupling of catabolic sequences and biosynthetic (anabolic) sequences to the ATP–ADP system. In fact, the involvement of the ATP–ADP system is much greater than indicated. Whenever a reaction needs more energy input to make it thermodynamically feasible, it is coupled to the ATP–ADP system.

To appreciate this point we must examine individual sequences. One of the most important interconversions in metabolism is that which involves fructose-1, 6-bisphosphate and fructose-6-phosphate. Sometimes one product needs to be made available and sometimes the other. How is this accomplished? If the same reactants and products

FIGURE 4 *(Continued)*   require energy and reducing power, while moving in the opposite direction leads to the production of energy and reducing power. The middle box shows some of the building blocks for the larger molecules found in the outer box. Metabolites that are maintained at approximately fixed levels are screened in white.

were involved in both reactions, only one of the reactions should be thermodynamically favorable. As the old saying goes, "You can't have it both ways." In fact, the pathways are designed so that the reactions are favorable in either direction. Therefore, you *can* have it both ways. The conversion of fructose-1, 6-bisphosphate to fructose-6-phosphate is an energetically favorable reaction as a simple hydrolysis, so it does not require any ATP-to-ADP conversion to make it go. As for the oppositely directed reaction, conversion of fructose-6-phosphate to fructose-1, 6-bisphosphate, it is coupled to a single ATP-to-ADP conversion to make it thermodynamically favorable.

$$\text{Fructose-6-phosphate} + \text{ATP} \rightarrow \text{fructose-1,6-bisphosphate} + \text{ADP}, K_{eq} \cong 10^4. \quad (5)$$

$$\text{Fructose-1,6-bisphosphate} + \text{H}_2\text{O} \rightarrow \text{fructose-6-phosphate} + \text{P}_i, K_{eq} \cong 10^4. \quad (6)$$

From this example we can see how a reaction that is not thermodynamically favorable can be made favorable by coupling to an appropriate number of ATP-to-ADP conversions. In this case a single ATP-to-ADP conversion is all that is necessary. In other more demanding situations, two or more such conversions can be required.

Because the interconversions of fructose-6-phosphate and fructose-1,6-bisphosphate have dissimilar starting materials and products, it is not surprising to find that the two reactions use different enzymes. In fact, both of the enzymes involved are special because it would serve no useful purpose to have both reactions running simultaneously. From summing Eqs. (5) and (6) it can be seen that this would only result in the net breakdown of ATP as follows:

$$\text{ATP} + \text{H}_2\text{O} \rightarrow \text{ADP} + \text{P}_i. \quad (7)$$

The system is regulated in such a way that when one of the enzymes is active, the other is inactive.

REACTIONS ARE ORGANIZED INTO SEQUENCES OR PATHWAYS

To appreciate how biochemical reactions serve common functions, we must examine the organization of biochemical reactions in some detail. Most of the enzyme-catalyzed reactions in living cells are organized into sequences or *pathways*. In a pathway the overall sequence serves a function, not the individual reactions. For example, the conversion of the organic compound chorismate to tryptophan occurs in five steps, each requiring a specific enzyme catalyst (Fig. 5). The intermediates between chorismate and tryptophan serve no function except as precursors of tryptophan. The end product, tryptophan, has many uses, including its role as a building block in protein synthesis.

A central role of breakdown of the six-carbon sugar, glucose, to the three-carbon compound, pyruvate, is to supply ATP to the cell. However, as in the case of the tryptophan pathway, the function of each reaction in glucose catabolism can be appreciated only by considering the overall sequence. This pathway involves the enzymatically

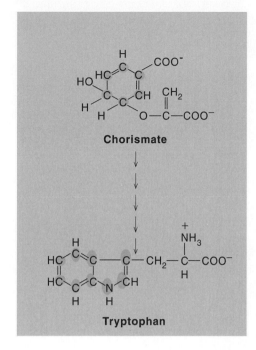

FIGURE 5 The biosynthesis of tryptophan. Tryptophan is synthesized in five steps from chorismate. Each step requires a specific enzyme activity. The four intermediates between chorismate and tryptophan serve no function other than as precursors of tryptophan. The detailed steps of each reaction in this sequence are described in Chapter 24.

catalyzed steps in going from glucose to pyruvate (Fig. 6). In the first and third steps, ATP is actually consumed. Only in the seventh and tenth steps is the starting amount of ATP recovered and is further ATP synthesized. Hence the function of this sequence is not served until each substrate has passed through the entire sequence.

Glucose breakdown and tryptophan biosynthesis illustrate pathways consisting of a linear series of reactions. In Chapter 11 we see that the glycolytic pathway also contains branches so that common intermediates can proceed along diverging routes after the branchpoint. We also see that the glycolytic pathway is organized so that some of the reactions within the pathway can go either in the forward or in the reverse direction.

SEQUENTIALLY RELATED ENZYMES ARE FREQUENTLY CLUSTERED

A fundamental aspect of biochemical organizations is that the enzymes that catalyze a sequence are often clustered in the cell. As a result, an intermediate produced in the first reaction in a pathway is passed directly to the second enzyme in the pathway, and so on. Such an arrangement might be expected to accelerate synthesis of the end product, to minimize the loss of intermediates, and to facilitate regulation of the pathway.

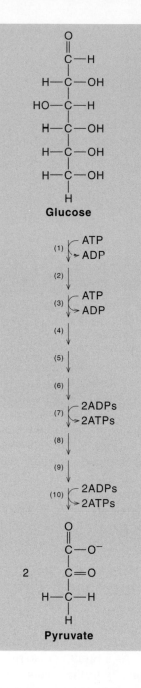

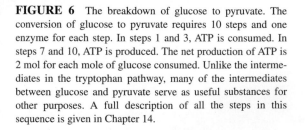

FIGURE 6   The breakdown of glucose to pyruvate. The conversion of glucose to pyruvate requires 10 steps and one enzyme for each step. In steps 1 and 3, ATP is consumed. In steps 7 and 10, ATP is produced. The net production of ATP is 2 mol for each mole of glucose consumed. Unlike the intermediates in the tryptophan pathway, many of the intermediates between glucose and pyruvate serve as useful substances for other purposes. A full description of all the steps in this sequence is given in Chapter 14.

Three types of enzyme clustering are found (Fig. 7). In the simplest situation, all the catalytic activities for a particular pathway are found in proteins that exist as independent soluble proteins in the same cellular compartment. In such cases the intermediates must get from one enzyme to the next in the sequence by free diffusion or by

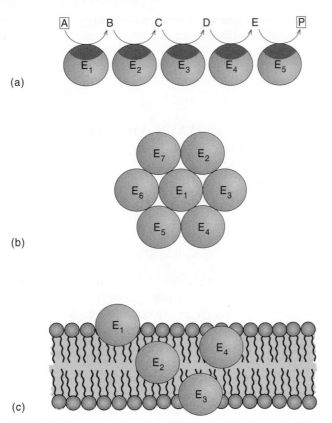

FIGURE 7   The organization of functionally related enzymes. Functionally related enzymes are organized in three ways: (a) as unlinked proteins soluble in the aqueous milieu of the same cellular compartment: (b) as components in a multiprotein complex; or (c) as components on a membrane.

transfer after contact between two sequentially related enzymes, one carrying the reactive intermediate and the other being ready to receive it. Such is the situation for the enzymes involved in the breakdown of glucose to pyruvate.

Independent enzymes with related functions often are localized in specific organelles. This arrangement enhances metabolic efficiency in various ways. First, it lends to higher concentrations of enzymes and intermediates and thus to faster rates of reaction. Second, it facilitates regulation of a pathway by limiting the starting materials or key intermediates to an organelle and by allowing direct interaction of one enzyme with another. Fatty acid metabolism in eukaryotes superbly illustrates this type of arrangement. The enzymes in fatty acid catabolism are all located inside the mitochondrion, whereas all enzymes involved in fatty acid synthesis are located outside the mitochondrion, in the cytosol.

The second type of arrangement observed for sequentially related enzymes is exemplified by the multiprotein complex involved in fatty acid synthesis. This complex contains most of the enzymes involved in fatty acid synthesis. The intermediates in this case are bound to the complex until synthesis is complete.

In the third type of arrangement functionally related enzymes are membrane bound. This is the case for the enzymes that are involved in manipulating the electrons and protons that are extracted when sugars become oxidized so that they can be utilized in the synthesis of ATP (see Chapter 11); these functionally related enzymes are bound to the inner mitochondrial membrane.

ACTIVITIES OF PATHWAYS ARE REGULATED BY CONTROLLING THE AMOUNTS AND ACTIVITIES OF THE ENZYMES

The two most important principles governing the regulation of biochemical pathways are economy and flexibility. Organisms regulate their metabolic activities to avoid deficiencies or excesses of metabolic products. A sudden shift in the concentrations or kinds of nutrients must be met by an equally rapid shift in the cellular enzymes to adjust the metabolism.

Pathways are regulated at the levels of synthesizing the enzyme and of regulating the activity of existing enzymes. The first type of regulation can occur at the levels of RNA or protein synthesis, which are discussed in Chapter 9. The second type of control is exerted by small-molecule factors that interact with the enzyme and thereby alter its activity in an upward or downward direction. The small-molecule factors are metabolic signals that reflect the state of the metabolism. For example, tryptophan is a small-molecule factor that signals the enzymes to stop making tryptophan when there is excess tryptophan already present. ATP is a small-molecule factor that signals the enzymes involved in glucose breakdown to slow down. This is appropriate because glucose breakdown is coupled to ATP synthesis.

An enzyme that responds to regulatory small molecules is called a regulatory enzyme to distinguish it from most enzymes in a particular pathway. As a rule, only the first enzyme in a multienzyme pathway is regulated. As long as the first enzyme in a pathway is inhibited, there is no need to regulate subsequent enzymes because there are no intermediates for the remaining enzymes to act on.

BOTH ANABOLIC AND CATABOLIC PATHWAYS ARE REGULATED BY THE ENERGY STATUS OF THE CELL

It is not always an advantage for the concentration of an amino acid or other end product to be the only factor that determines the rate at which that end product is synthesized. If a cell is low on energy or catabolic intermediates, it may not be able to afford to synthesize amino acids, even if their concentrations are quite low. Continued biosynthesis under such circumstances might deplete the already scarce supply of energy to the point where essential functions, such as maintenance of concentration

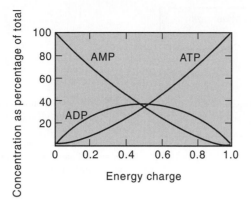

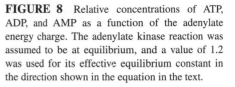

FIGURE 8   Relative concentrations of ATP, ADP, and AMP as a function of the adenylate energy charge. The adenylate kinase reaction was assumed to be at equilibrium, and a value of 1.2 was used for its effective equilibrium constant in the direction shown in the equation in the text.

gradients across membranes, are impaired. It is clearly advantageous for the rate of biosynthesis to be regulated by the general *energy status* of the cell as well as by the need for a specific end product.

Because the adenine nucleotide system ATP–ADP (and occasionally ATP–AMP) couples energy in biosynthetic sequences, we might expect that this system also supplies the necessary signal indicating the energy status of the cell, to be sensed by the kinetic control mechanisms of biosynthesis. It is helpful to have a term that quantitatively expresses the energy status of the cell. The term most often used is the *energy charge*, which is defined as the effective mole fraction of ATP in the ATP–ADP–AMP pool:

$$\text{Energy charge} = \frac{(\text{ATP} + 0.5\,\text{ADP})}{(\text{ATP} + \text{ADP} + \text{AMP})}. \tag{8}$$

In this equation the 0.5 in the numerator takes into account the fact that ADP is about half as effective as ATP at carrying chemical energy. Thus two ADPs can be converted into one ATP and one AMP by a reaction with an equilibrium constant of about 1. The highly active enzyme that catalyzes this reaction is universally distributed as follows:

$$\text{ATP} + \text{AMP} \rightleftharpoons 2\,\text{ADP}, K_{eq} \approx 1. \tag{9}$$

The values for the energy charge could conceivably vary from 0 to 1, as illustrated in Fig. 8. In fact, however, the values for real cells are usually held within very narrow limits. The result is a general stabilizing effect on the cellular metabolism. In much the same way, voltage regulation is necessary for the effective operation of many electronic devices. Now let us see how the limits are maintained.

Several reactions in anabolic and catabolic pathways have been found to respond to variation in the value of the energy charge. As might be expected, the enzymes in catabolic pathways respond in a direction opposite to that of enzymes in anabolic

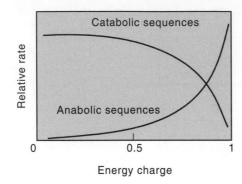

FIGURE 9   Variation in reaction rates as a function of the energy charge. As the energy charge increases, the rate of catabolic reactions decreases. Meanwhile, the rate of anabolic reactions increases. The combined effect is to stabilize the energy charge at a value around 0.9.

pathways. The two responses are compared in Fig. 9. It is clear that if catabolic sequences (which lead to the generation of ATP) respond as shown by the upper curve, and biosynthetic sequences (which use ATP) respond as shown by the lower curve, then the charge is strongly stabilized at a value at which the two curves intersect. A tendency for the charge to fall is resisted by the resulting increases in the rates of catabolic sequences and the decreases in the rates of biosynthetic sequences. A tendency for the charge to rise is resisted by opposing effects.

REGULATION OF PATHWAYS INVOLVES THE INTERPLAY OF KINETIC AND THERMODYNAMIC FACTORS

The general properties of paired unidirectional sequences and their advantages to the organism are in one sense very simple. However, on further consideration we see that they depend on a complex interplay between kinetic and thermodynamic factors and effects.

In any kind of negative feedback system, cause-and-effect relationships are circular. We see the circularity of negative feedback at a simple level in the response of a biosynthetic sequence to the concentration of its end product. The concentration of the end product is a major factor controlling the rate of synthesis, and the rate of synthesis is a major factor controlling the concentration of the end product.

For pairs of oppositely directed metabolic sequences, greater sophistication is needed than for simple feedback control of a synthetic pathway. In our discussion of oppositely directed sequences thus far, we have assumed that the ATP to ADP ratio is held at a value very far from equilibrium. Of course, in the absence of highly effective controls, the ATP to ADP ratio would not remain far from equilibrium; it would rapidly approach its equilibrium value. What are the controls that keep this from happening? The answer is the enzymes that catalyze first steps in oppositely directed sequences—the same controls that we already have discussed.

The rates of reactions in which there is net conversion of ATP to ADP are high when the energy charge for the ATP to ADP ratio is high, and they decrease sharply with a decrease in those parameters. Rates of regulatory enzymes in sequences in which ATP is regenerated are high when the energy charge or ATP to ADP ratio is low, and decrease sharply as the energy charge increases. The curves intersect at an energy charge value of 0.9 (see Fig. 9). The kinetic effects play a major role in stabilizing the energy charge and the ATP–ADP system *in vivo*. Therefore, kinetic control of rates of conversion depends on the value of the ATP to ADP ratio being far from equilibrium (i.e., on thermodynamic factors): however, at the same time the ATP to ADP ratio itself depends on reciprocal regulation of oppositely directed sequences (i.e., on kinetic factors).

As we have seen, the characteristic and essential feature of paired, oppositely directed sequences is that they differ in their ATP stoichiometries. However, that difference is meaningful only because the ATP to ADP ratio is far from equilibrium in the cell. If ATP, ADP, and P_i were at equilibrium, it would not matter how many ATP-to-ADP conversions were coupled to any metabolic sequence. For any system at equilibrium, the free energy change is zero, and coupling to another reaction or sequence would have no effect on the free energy change or the position of equilibrium of the reaction or sequence. Thus the differential ATP stoichiometry would mean nothing if it were not for the kinetic controls that hold the ATP to ADP ratio at a steady-state value far from equilibrium. However, kinetic regulation could not control directions of conversions and thus regulate the balance between ATP utilization and regeneration if it were not for the thermodynamic difference between the sequences of each pair, which depends on the ATP to ADP ratio. Neither the thermodynamic nor the kinetic features can be said to be more fundamental. Metabolic correlation and control, and hence life, depend on an intricate interplay between thermodynamic and kinetic factors.

SUMMARY

In this chapter we have described the major principles that govern the course of chemical reactions in living cells.

1. All cells need energy, reducing power, and starting materials for synthesis. Ultimately, these are supplied by autotrophic organisms, especially green plants. In plants, the starting materials are made from carbon dioxide, and the supply of chemical energy and reducing power is dependent on the absorption of light energy. In a heterotrophic organism, the cells depend on preformed organic compounds to meet their basic needs.

2. Metabolic chemistry is characterized by functionality. Each reaction is important because of its participation in a sequence of reactions and each sequence interacts functionally with other sequences.

3. Sequences may be classified broadly into two main types: biosynthetic or anabolic, and degradative or catabolic. Anabolic sequences are usually energy-requiring and catabolic sequences are usually energy-producing.

4. Metabolic regulatory mechanisms have evolved so as to stabilize concentrations of key metabolites under a broad range of conditions.

5. Energy for the activities of a cell is made available from catabolic sequences by the ATP–ADP system. In a similar manner, reducing power is made available by the NADPH–NADP$^+$ system.

6. ATP-to-ADP conversions contribute to the overall equilibrium constant of a sequence and therefore can determine which reactions are thermodynamically favorable. Any reaction can be made favorable by coupling to an appropriate number of ATP-to-ADP conversions.

7. Metabolic sequences are controlled by regulatory enzymes, which respond to signals in such a way that rates of energy production and biosynthesis are controlled to keep the system operating efficiently and effectively.

Problems

1. What is the primary advantage of subcellular compartments to intermediary metabolism?
2. What is the metabolic advantage of having the "committed step" of a pathway under strict regulation?
3. The conversion of chorismate to tryptophan takes place in five steps (Fig. 5). If the third step instead of the first step were the one to be regulated, how would this affect the distribution of reactants, products, and intermediates. Consider both positive regulation and negative regulation.
4. What is the relationship between anabolism and catabolism?
5. If in a certain anabolic pathway an enzymatic step involves the phosphorylation of the substrate, the process is typically coupled with the conversion of an ATP to an ADP. However, the catabolic version of the pathway often contains a hydrolysis reaction at this point. Why does the organism just not reverse the pattern involved in the anabolic pathway and form an ATP?
6. Living cells must be able to synthesize ATP and to maintain a favorable ratio of ATP to ADP. Explain what is meant by favorable and why this is important to maintain.
7. Explain why a reaction at equilibrium cannot be used to supply energy.

Reference

See references at the end of Chapter 6.

CHAPTER 8

Biochemical Catalysis

In Chapter 6 we noted that biochemical reactions are a subset of ordinary chemical reactions. Whether in a laboratory experiment (*in vitro*) or in a living system (*in vivo*), chemical reactions involve the making and breaking of chemical bonds, which in turn involve the absorption or release of energy.

In theory, all chemical reactions are reversible. By this we mean that the reaction is constantly proceeding both forward, from reactants to products, and backward, from products to reactants. However, the initial rates at which the reaction takes place in the two directions usually are not equal. Sometimes the forward rate is so large in relation to the backward rate that essentially all the reactants are quickly converted to products, and the reaction can be considered to be (under the prevailing conditions) irreversible. Whether or not a chemical reaction goes to completion in this way, the reaction is said to reach *equilibrium* when the rate of forward reaction is equal to the rate of reverse reaction.

Many factors influence the rate at which a reaction proceeds in either direction. Among them are temperature, initial concentrations of reactants, pH level (Box 8A), and the presence or absence of *catalysts* (substances that accelerate a chemical change without being permanently changed themselves).

Biological systems possess structurally complex enzyme catalysts that make them very effective: (1) they make a reaction go faster; (2) by making a particular reaction go faster, they guide the reactant down a specific pathway; and (3) they function under mild conditions compatible with the fragile structure of the living cell.

BOX 8A   Acids, Bases, and pH

An *acid* is any substance that can donate a proton to a reaction, and a *base* is any substance that accepts protons. In this connection, recall that the hydrogen ion, H^+, is in effect a bare proton.

The term pH is a convenient way to express the concentration of hydrogen ions in an aqueous solution. It is defined by the equation

$$pH = -\log[H^+].$$

Water can be considered a weak acid (or a weak base) because it dissociates into hydrogen ions (H^+) and hydroxide ions (OH^-) according to the equilibrium

$$H_2O \rightleftharpoons H^+ + OH^-.$$

The double arrows here indicate an ongoing, reversible reaction in which the product (water) is decomposing into reactants at the same rate as the reactants (hydrogen and hydroxide ions) are combining to form the product.

Pure water is neutral because it contains equal amounts of hydrogen and hydroxide ions. The pH and the pOH are both 7. In an acidic solution the pH is less than 7 and the pOH is greater than 7; in a basic, or *alkaline,* solution the reverse is true. In either case the pH and the pOH sum to 14, just as in a neutral solution.

The pH of an aqueous solution is an extremely important variable for most biological reactions. Some compounds degrade much more rapidly if the pH is not right. Others react much more rapidly at slightly acid or alkaline pH levels. With few exceptions, pH values within living organisms lie in the range between 6 and 8. In the context of possible prebiotic reactions, such as those to be described in Part III, the pH also can have a major effect on the affinity of a small molecule for a clay or mineral support.

In this chapter we consider some features that are common to most catalyzed reactions, and then focus on the distinguishing features of enzyme-catalyzed reactions.

GIVEN FAVORABLE THERMODYNAMICS, KINETIC FACTORS DETERMINE WHICH REACTIONS CAN OCCUR

Reactions can occur only if there is an overall decrease in free energy (ΔG) (see Box 7A). However, more than favorable thermodynamics is needed to ensure the progress of a particular reaction. Some reactions that are highly energetically favorable occur very rapidly; others do not. The reaction of hydrogen ions with hydroxide ions to make water is an example of a highly favorable reaction that occurs rapidly. Thus when we mix equimolar amounts of a strong acid and a strong base in aqueous solution, neutralization occurs as fast as the mixing process and we are left with very small amounts of free hydrogen or hydroxide ions. By contrast, when we mix molecular hydrogen (H_2) and molecular oxygen (O_2) at room temperature, very little if any reaction occurs even though the equilibrium constant is very much in favor of the formation of water. However, if we raise the temperature, the formation of water occurs with explosive violence. This example shows that thermodynamic factors alone do not guarantee reaction. However, if thermodynamic calculations indicate the negative free energy for a reaction, we are likely to be encouraged to find conditions under which the reaction can take place.

Kinetic factors determine the rate at which a reaction can occur. The equilibrium constant is equal to the ratio of the rate constants in both directions. Nevertheless either or both of these may be so small that equilibrium may take a very long time to attain even if the reaction is quite favorable in one direction. When this is the case, we commonly say that there is a kinetic barrier to the reaction.

The gaseous reaction between NO and H_2 to form N_2 and H_2O is an example of a slow reaction. A bimolecular reaction requires that a molecule of NO collides with a molecule of H_2. If all these collisions were effective in forming the product, then N_2 and H_2O would be formed very rapidly. Because the reaction is in fact rather slow, it must be that very few of the collisions between the reacting species are effective. Two factors must be scrutinized to determine why the reaction rate is so limited. First, how many collisions are these two molecules likely to make in a given time? Second, how many of these collisions are effective in leading to products?

The frequency of collisions is a function of the concentration of the two reactants and the speed of the molecules. The probability that a given collision becomes effective is a function of the *free energy of activation* ΔG^* (Box 8B). This quantity is equal to the difference between the free energy of the reactants and the highest free energy state that the reactants must reach before being converted to products. This highest free energy state is referred to as the *transition state*, or the *activated complex* (Fig. 1). The higher the free energy of activation, the lower is the probability that a given collision between reactants can be effective.

For a given concentration of reactants, there are two ways in which a slow reaction can be accelerated. One is by raising the temperature. This causes the molecules to move faster, which increases both the number of molecular collisions and the fraction of effective collisions. The other is by introducing a factor that lowers the activation free energy. In ordinary chemical reactions both methods are used in the laboratory and in industry. In relatively delicate biological systems it is not practical to raise the temperature very much, so the usual means of accelerating a reaction is by introducing a catalyst. The catalyst is a "third party" that lowers the activation energy without itself being consumed. The catalyst does not change the free energy of the reactants or the products, only the activation free energy to achieve the state of the activated complex (see Fig. 1).

BOX 8B   Dependence of Reaction Rate on Activation Energy

The highest free energy state of a reacting species is referred to as the transition state, or the activated complex. Kinetic theory states that the fraction of molecules f that have enough kinetic energy to attain this state is a function of the temperature and the activation energy according to the following equation:

$$f = e^{-\Delta G^*/RT}$$

where e is the base of the natural logarithm, R is the molar gas constant, and T is the absolute temperature. The rate constant for the reaction is related to this quantity by the following equation:

$$k = Ae^{-\Delta G^*/RT} \quad \text{or} \quad \ln k = \ln A - \frac{\Delta G^*}{RT},$$

where A is a constant related to the collision frequency. A also depends on the temperature, because the frequency of collisions increases with temperature. However, this dependence is minimal and can be ignored in most treatments of interest to biochemists, because they work over a narrow range of temperatures.

From the first equation, a rate constant k is 10 times faster at 25°C if the activation energy $\Delta G°$ is lowered by 1.38 kcal/mol. Thus for an enzyme to speed up a reaction by a factor of 10^{10} (ten billion), not an unusual acceleration, the enzyme must lower the activation energy by 13.8 kcal/mol.

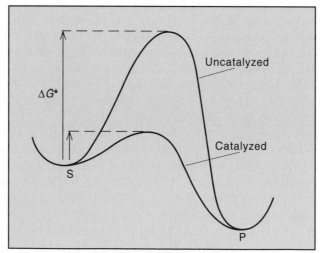

Extent of reaction

FIGURE 1   An enzyme speeds up a reaction by decreasing ΔG^*. The enzyme does not change the free energy of the substrate (S) or product (P); it lowers the free energy of the transition state. The two vertical arrows indicate the activation free energies (ΔG^*) of the catalyzed and uncatalyzed reactions. This figure neglects the enzyme–substrate and enzyme–product complexes that are intermediates in the reaction.

ALL CATALYSTS OBEY THE SAME BASIC SET OF RULES

All catalysts, whether they be small-molecule solution catalysts, clays, or enzymes, obey many of the same basic rules. We now consider some mechanistic features that most catalysts have in common.

All chemical bonds are formed by electrons, and the rearrangement or breakage of these bonds starts with the migration of electrons. In the most general terms, reactive groups can be said to function as either *electrophiles* or *nucleophiles*. The former are electron-deficient substances that are attacked by electron-rich substances. The latter are electron-rich substances that attack electron-deficient substances. Frequently, the main job of the catalyst is to make a potentially reactive center more attractive for reaction (i.e., to potentiate the electrophilic or nucleophilic character of the reactive center of the reacting species). The range of functional groups serving as catalysts for ordinary chemical reactions is much broader than that found in enzymes, but this range serves a similar purpose.

In aqueous solution, protons (actually present as hydronium ions, H_3O^+) or hydroxide ions are the most common catalysts for nonenzymatic reactions. *Generalized acids* or *bases*, defined as compounds capable of yielding protons or hydroxide ions themselves or through interaction with water, also are common forms of solution catalysts. In enzymes, generalized acids and bases always are used in preference to free acids and bases because of the delicacy of biological systems.

The way in which catalysts work is illustrated in Fig. 2 for the hydrolysis of an ester linkage. As a result of the electronegativity of the oxygen atom in the ester $C{=}O$ group, the oxygen has a fractional negative charge δ^- and the carbon has a fractional positive charge δ^+. Hydrolysis of the ester can be accelerated by either acid or base catalysis. In acid catalysis, a proton or a generalized acid (HA) acting as an electrophile is attracted to the oxygen. This leads to an intermediate that accentuates the positive charge on the carbon atom, making it more attractive to a nucleophile, in this case water. Water is a poor nucleophile and would attack the carbon very slowly without such an inducement. This is the key step in the catalysis. The remaining reactions leading to ester hydrolysis and regeneration of the catalyst occur rapidly and spontaneously. In hydroxide ion or general base catalysis of the ester, the carbon atom is attacked more directly by a stronger nucleophile, either OH^- itself, or a water molecule converted into an attacking hydroxide ion by the presence of the generalized base (B:). Again, after hydrolysis, the catalyst is regenerated.

An interesting feature of the general acid and general base mechanisms shown in Fig. 2 is that there is catalysis of both steps. For example, in the first step of general acid catalysis, HA donates a proton, and thus is acting as an acid; however, in the second step, A^- removes a proton and acts as a base. In the general base catalysis mechanism, the reverse sequence is followed. Such sequential catalysis of steps by a general acid or a general base group is much more common in enzymatic reactions than in ordinary chemical reactions.

INITIAL STEP	INTERMEDIATE STEP	PRODUCTS

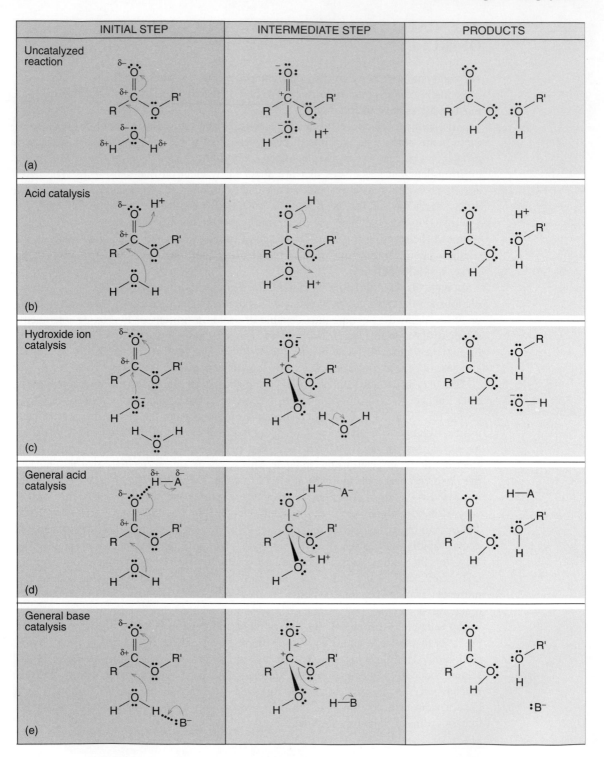

Uncatalyzed reaction (a)

Acid catalysis (b)

Hydroxide ion catalysis (c)

General acid catalysis (d)

General base catalysis (e)

ENZYME CATALYSTS ARE HIGHLY SELECTIVE AND FUNCTION UNDER VERY MILD CONDITIONS

Thus far we have emphasized those aspects of catalysis that are common to ordinary chemical reactions and biochemical reactions. The unique aspects of enzyme catalysis can be summarized as follows:

1. Enzyme catalysis is mediated by a limited number of functional groups found in amino acid side chains, coenzymes, and metal cations. Notably, the amino acid side chains of histidine, serine, tyrosine, cysteine, lysine, glutamate, and aspartate are frequently directly involved in the catalytic process. Coenzymes are organic molecules that work in conjunction with enzymes to enrich the variety of functional groups. Often metals can function in a similar capacity.

2. Enzyme-catalyzed reactions usually have narrow optimum pH ranges for the reactions they catalyze. Extremes of pH cause disruption of the folded three-dimensional structures (denaturation) and consequent loss of enzyme activity, often in an irreversible manner. The narrow optimum pH range results from the involvement of generalized acids or bases that usually function as catalysts in only one ionization state.

3. The catalytic site — the site on the enzyme where the reaction takes place — and the *substrate* binding site on the enzyme are localized and highly specific, both for the type of reaction being catalyzed and for the type of molecule on which the enzyme acts (i.e., the substrate). In some cases the sites are rigid in structure; in other cases they are flexible and change their structure on binding to substrates or during the course of the reaction.

4. Frequently, the catalytic site contains more than one functional group, so that two reactions can occur simultaneously or in very rapid sequence. This phenomenon, known as concerted catalysis, has two advantages. First, it eliminates the need for high-energy intermediates that might slow a reaction. Second, it prevents the loss of reactive intermediates in nonproductive side reactions.

5. The substrate of an enzyme is invariably bound near the catalytic site and is usually oriented in a manner that is highly favorable for reaction with the catalytic site on the enzyme or with the reactive groups of other substrates in a multisubstrate enzyme.

6. Just as the enzyme may sometimes change its conformation on binding, so may the substrate change its structure on binding to the enzyme. Substrates are bound so that a bond strain is created in the substrate that favors the formation of the transition-state complex.

FIGURE 2   Several ways in which the hydrolysis of an ester can occur. The formation of a chemical bond is described by a formal "flow" of an electron pair from an electron donor to an electron acceptor. This electron flow is indicated by a curved arrow drawn from the electron source to the position of the newly formed bond at the electron acceptor.

Oxidoreductases: $A^- + B \rightleftharpoons A + B^-$

Transferases: $A - B + C \rightleftharpoons A + B - C$

Hydrolases: $A - B + H_2O \rightleftharpoons A - H + B - OH$

Lyases:
$$
\begin{array}{cc}
X & Y \\
| & | \\
A & - & B
\end{array}
\rightleftharpoons A = B + X - Y
$$

Isomerases:
$$
\begin{array}{cc}
X & Y \\
| & | \\
A & - & B
\end{array}
\rightleftharpoons
\begin{array}{cc}
Y & X \\
| & | \\
A & - & B
\end{array}
$$

Ligases (synthases): $A + B \rightleftharpoons A - B$

FIGURE 3   The six main classes of enzymes and the reactions they catalyze.

There are thousands of known enzymes and many more catalyzed reactions for which enzymes are still to be discovered. Enzymologists have grouped all known enzymes into six main classes according to the types of reactions they catalyze: *oxidoreductases, transferases, hydrolases, lyases, isomerases,* and *ligases* (Fig. 3). Oxidoreductases catalyze the loss of electrons (oxidation reactions) or the gain of electrons (reduction reactions). Transferases catalyze the transfer of a molecular group from one molecule to another. Hydrolases catalyze bond cleavage by the introduction of water. Lyases catalyze reactions involving removal of a group to form a double bond or addition of a group to a double bond. Isomerases catalyze reactions involving intramolecular rearrangements. Finally, ligases catalyze reactions joining together two molecules.

EACH MEMBER OF THE TRYPSIN FAMILY OF ENZYMES IS SPECIFIC FOR HYDROLYSIS OF A PARTICLE TYPE OF PEPTIDE LINKAGE

The properties of enzyme-catalyzed reactions are best appreciated by considering specific examples. Trypsin, chymotrypsin, and elastase represent a group of closely related digestive enzymes whose role is to hydrolyze polypeptide chains. They are synthesized in the pancreas as inactive proenzymes, and then are secreted into the digestive tract and activated. The three enzymes work as a team. Each cleaves a protein chain at an internal peptide linkage next to a different type of amino acid side group (Fig. 4). Trypsin cuts a chain just past the carbonyl group of a basic amino acid,

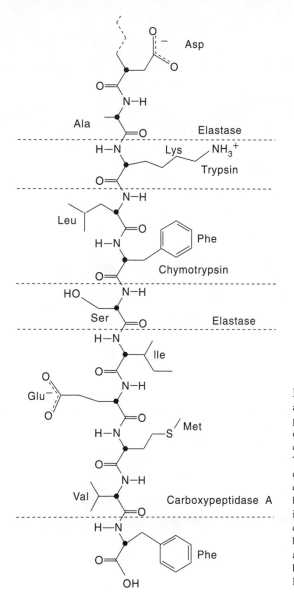

FIGURE 4 Trypsin, chymotrypsin, and elastase—three members of the serine protease family—catalyze the hydrolysis of proteins at internal peptide bonds adjacent to different types of amino acids. Trypsin prefers lysine or arginine residues; chymotrypsin, aromatic side chains; and elastase, small, nonpolar residues. Carboxypeptidases A and B, which are not serine proteases, cut the peptide bond at the carboxyl-terminal end of the chain. Carboxypeptidase A preferentially removes aromatic residues; carboxypeptidase B, basic residues. (Illustration copyright by Irving Geis. Reprinted by permission.)

either lysine or arginine. Chymotrypsin preferentially cuts a polypeptide chain next to an aromatic amino acid. Elastase is less discriminating in its choice of cleavage point, but it tends to make cuts adjacent to small, uncharged side chains.

The fact that the trypsin family of enzymes works together during digestion suggests that these enzymes might be structurally related. Indeed, their amino acid sequences show striking similarities, as illustrated in Fig. 5. They have identical amino

Chymotrypsin

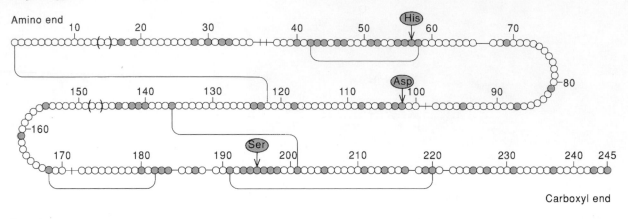

Trypsin

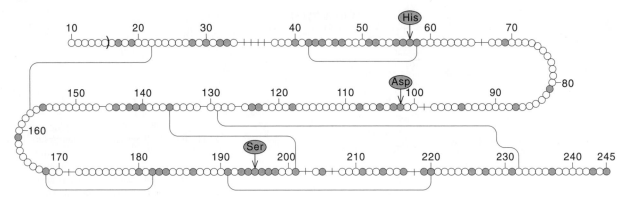

Elastase

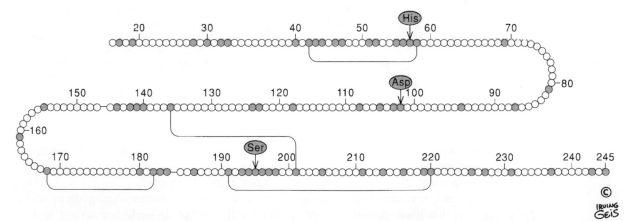

acids at 62 of the 257 positions. Although the proenzymes are different, the native active enzymes in all three cases begin a polypeptide chain at the same place (residue 16). The four disulfide bridges that connect distant parts of the chain in elastase also are present in the other two enzymes, with chymotrypsin having one more disulfide bridge and trypsin having two. All three enzymes have histidine at position 57, aspartic acid at 102, and serine at 195; these three groups constitute the active site of the enzymes.

X-ray crystal structure analyses have revealed that these three proteolytic enzymes are folded the same way. The backbone skeleton for trypsin is shown in Fig. 6. The only amino acid side chains illustrated are His57, Asp102, and Ser195 at the active site and the various disulfide bridges. Because the chains are folded in the same way, positions corresponding to disulfide bridges in one enzyme also are close together even in an enzyme that does not have that particular bridge. The three-dimensional structure reveals another important feature that is not apparent from the amino acid sequences alone. The chain is folded back on itself in such a way that the three catalytic side chains (57, 102, and 195) are brought close together at a depression on the surface of the molecule. This is the active site of the enzyme, where the substrate to be cut during digestion is bound.

Trypsin, chymotrypsin, and elastase share not only a common structure but also a common catalytic mechanism. The mechanism has been determined from chemical and X-ray diffraction studies. A polypeptide substrate binds to the molecule with one portion hydrogen-bonded to residues 215 to 219 in an antiparallel manner. This bonding helps to hold the substrate in place. At the bend in the substrate chain, the NH—CO bond that is to be cut is brought close to His57 and Ser195. There is considerable distortion in this bond in the enzyme–substrate complex, favoring formation of the activated complex. The side group of the substrate just prior to this bond is inserted into a pocket in the surface of the enzyme molecule. The pocket, bordered by residues 214 to 230, a disulfide bridge, and residues 191 to 195, can be seen in Fig. 7. The rim of the pocket from residues 215 to 219 is the binding site for the substrate. The pocket provides an explanation for the different specificities of the three enzymes for the bonds that they cut. In trypsin, the specificity pocket is deep and has a negative charge from aspartic acid 189 at the

---

FIGURE 5 Schematic diagrams of the amino acid sequences of chymotrypsin, trypsin, and elastase. Each circle represents one amino acid. Amino acid residues that are identical in all three proteins are in gray. The three proteins are of different lengths, but have been aligned to maximize the correspondence of the amino acid sequences. All the sequences are numbered according to the sequence in chymotrypsin. Long connections between nonadjacent residues represent disulfide bonds. Locations of the catalytically important histidine, aspartate, and serine residues are marked. The links that are cleaved to transform the inactive zymogens to the active enzymes are indicated by parenthesis marks. After chymotrypsinogen is cut between residues 15 and 16 by trypsin, and is thus transformed into an active protease, it proceeds to digest itself at the additional sites that are indicated; these secondary cuts have only minor effects on the enzyme's catalytic activity. (Illustration copyright by Irving Geis. Reprinted by permission.)

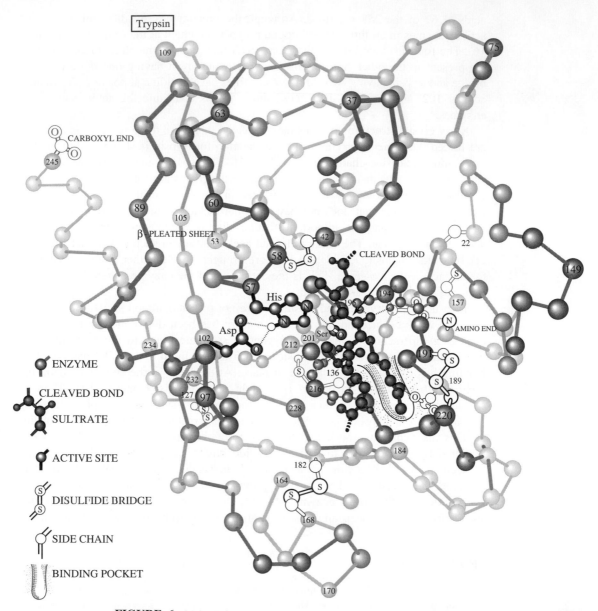

FIGURE 6   Main-chain skeleton of the trypsin molecule. The α-carbon atoms are shown by shaded spheres, with certain ones given residue numbers for identification. The —CO—NH— amide groups connecting the α-carbons are represented by straight lines. Disulfide bridges are shown in outline. A portion of the polypeptide chain substrate is shown darkly shaded near the center. The specificity pocket is sketched in shading, with a lysine side chain from the substrate molecule inserted. The catalytically important aspartate, histidine, and serine are poised for cleavage of the peptide bond marked by an arrow. Activation of trypsinogen involves cleaving six residues from the amino terminal end of the chain. (Illustration copyright by Irving Geis. Reprinted by permission.)

Trypsin

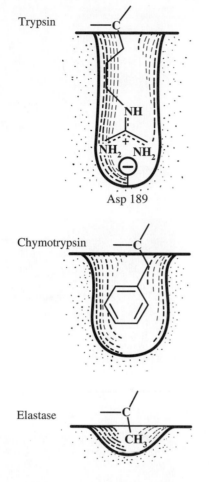

Asp 189

Chymotrypsin

Elastase

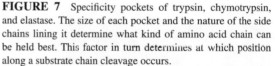

FIGURE 7   Specificity pockets of trypsin, chymotrypsin, and elastase. The size of each pocket and the nature of the side chains lining it determine what kind of amino acid chain can be held best. This factor in turn determines at which position along a substrate chain cleavage occurs.

bottom (see Fig 7). The pocket is designed to accept a long, positively charged basic side chain: lysine or arginine. In chymotrypsin, the corresponding pocket is wider and completely lined with hydrophobic side chains, thereby providing an efficient receptacle for a large, bulky aromatic group. In elastase, the pocket is blocked by valine and threonine at the positions where the other two enzymes have only glycine, which has no side chain. As a result, in elastase no side chain of appreciable size can bind to the enzyme surface.

The proposed mechanism of catalysis is shown in Fig. 8. In steps 1 and 2, a polypeptide chain approaches and binds to the active site of the enzyme, with the proper type of side chain inserted into the specificity pocket. After substrate binding, the three catalytically important groups on the enzyme—Asp102 His57, and Ser195— are connected by hydrogen bonds in what David Blow calls a "charge–relay system." In steps 3 and 4 of the mechanism, the histidine nitrogen acts first as a general base,

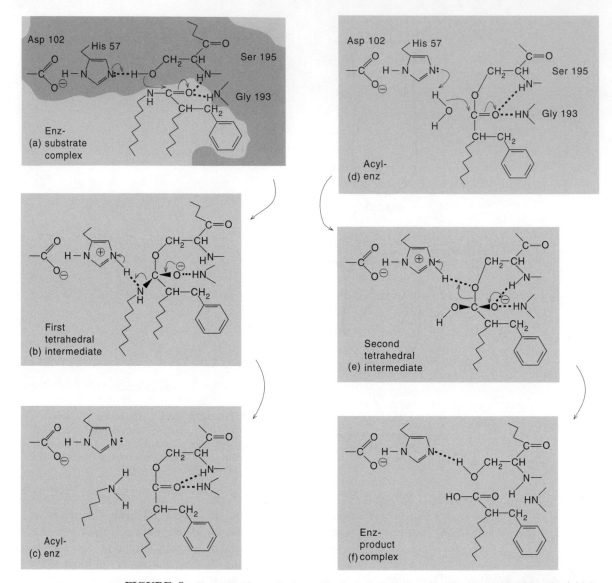

FIGURE 8   The probable mechanism of action of chymotrypsin. The six panels show the initial enzyme–substrate complex (a), the first tetrahedral (oxyanion) intermediate (b), the acyl–enzyme (ester) intermediate with the amine product departing (c), the same acyl–enzyme intermediate with water entering (d), the second tetrahedral (oxyanion) intermediate (e), and the final enzyme–product complex (f). In the transition states between these intermediates, there probably is a more even distribution of negative charge between the different oxygen atoms attached to the substrate's central carbon atom.

pulling the serine proton toward itself, and then as a general acid, donating the proton to the lone electron pair on the nitrogen atom of the polypeptide bond to be cleaved. The aspartic acid group is pictured as helping the histidine to attract the proton by making an electrostatic linkage or hydrogen bond with the other proton in the imidazole ring. The importance of aspartate is shown by the fact that almost all serine proteases have an aspartate in this position. The hydrogen bond between an aspartate carboxylate group and the imidazole also helps aim the imidazole at the serine OH group. With only flexible single bonds holding the imidazole ring to the polypeptide backbone, a histidine needs such an extra hydrogen bond to fix it in place.

As the serine H—O bond is broken in step 3, a bond is formed between the serine oxygen and the carbonyl carbon on the polypeptide chain. The carbon becomes tetrahedrally bonded, and the effect of the negative charge on Asp102 has in a sense been relayed to the carbonyl oxygen atom of the substrate. This negative oxygen in the tetrahedral intermediate is stabilized by hydrogen bonds to N—H groups on the enzyme backbone. The transition state of step 3 is short-lived and cannot be isolated. The enzyme passes quickly through the tetrahedral intermediate to step 4. As the polypeptide N accepts the proton from histidine, the N—C bond is broken. One half of the polypeptide chain falls away as a free amine, RNH_2. The other half remains bound covalently to the enzyme as an acylated intermediate. This acyl–enzyme complex is stable enough to be isolated and studied in special cases where further reaction is blocked .

The steps to deacylate the enzyme and restore it to its original state (steps 5 to 8) are like the first four steps run in reverse, with H_2O playing the role of the missing half-chain. A water molecule attacks the carbonyl carbon of the acyl group in step 5 and donates one proton to His57 to form another tetrahedral intermediate (step 6). This intermediate breaks down when the proton is passed on from histidine to serine (step 7). The second half of the polypeptide chain falls away (step 8), aided by charge repulsion between the newly generated carboxylate anion and the negative charge on aspartic acid 102. The enzyme is restored to its original state.

SUMMARY

In this chapter we have considered the basic properties of catalysis in general and then have described the function of one group of enzyme catalysts.

1. Enzymes do not affect the equilibrium of a reaction but only the rate at which the reaction takes place. They do this by lowering the activation energy.
2. The functional groups on enzymes are limited to a small group of amino acid side chains, coenzymes, and metal cations.
3. Enzyme-catalyzed reactions frequently have narrow optimum pH ranges for the reactions they catalyze.
4. The catalytic site and substrate binding site on the enzyme are localized and highly specific.

5. Enzymes in the trypsin family hydrolyze polypeptide chains. The mechanism of action of chymotrypsin, a member of this family of enzymes, contains eight steps.

Problems

1. Can a catalyst shift the equilibrium for a reaction?
2. What geometric constraints are the amide N—H groups of Ser195 and Gly193 subject to in trypsin, chymotrypsin, and elastase?
3. In a first-order reaction a substrate is converted to the product so that 87% of the substrate is converted in 7 min. Calculate the first-order rate constant. In what time would 50% of the substrate be converted to the product?
4. What is the difference between the enzyme–substrate complex and the transition-state intermediate in an enzymatic reaction?
5. (a) In what ways are the mechanistic features of chymotrypsin, trypsin, and elastase similar?
   (b) If the mechanisms of these enzymes are similar, what features of the enzyme active site dictate substrate specificity?

Reference

See Chapter 6 for references.

CHAPTER 9

Storage, Replication, and Utilization of Biochemical Information

Living systems still defy detailed descriptions due to their complexity. If we were to imagine that these systems arose from chemicals on prebiotic Earth, it seems obvious that they must have been the descendants of far simpler systems that grew, replicated, and evolved. Present-day living systems continue to grow and undergo change via a chain of events that may be thought of as starting from the chromosomal DNA. The chromosomal DNA contains all the information essential for cell maintenance, growth, and replication. This information is transferred from one cell generation to the next, following DNA replication and cell division. It is recorded in the sequence of nucleotides in the DNA. The DNA nucleotide sequence dictates the amino acid sequence in the polypeptide chains of the cellular proteins. The DNA does not transfer sequence information directly. Instead, it passes through RNA intermediaries known as messenger RNAs (mRNAs). Each mRNA is synthesized on a specific region of the DNA template and in turn it serves as a template for the synthesis of a specific

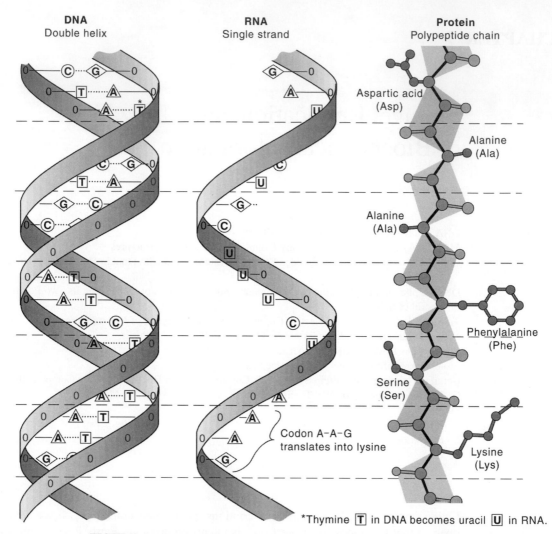

FIGURE 1 Transfer of information from DNA to protein. The nucleotide sequence in DNA specifies the sequence of amino acids in a polypeptide. DNA usually exists as a two-chain helical structure. The information contained in the nucleotide sequence of only one of the DNA chains is used to specify the nucleotide sequence of the messenger RNA molecule (mRNA). This sequence information is used in polypeptide synthesis. A three-nucleotide sequence in the mRNA molecule codes for a specific amino acid in the polypeptide chain. (Illustration copyright by Irving Geis. Reprinted by permission.)

polypeptide chain. Each grouping of three bases in the mRNA encodes a specific amino acid in the mRNA-related polypeptide chain. The overall process of information transfer from DNA to mRNA (*transcription*) and from mRNA to protein (*translation*) is depicted in Fig. 1.

In this chapter we describe the processes of DNA replication transcription and translation. We see that evolutionary change is a natural outgrowth of occasional errors made in DNA replication.

DNA AND RNA HAVE SIMILAR PRIMARY STRUCTURES

Recall that the building blocks of nucleic acids are nucleotides containing a five-carbon sugar with a phosphate group attached on one side and a purine or pyrimidine base attached on the other side (see Chapter 6, Fig. 7). In polynucleotides individual nucleotides are connected by phosphodiester linkages between the 3′ hydroxyl of one sugar and the 5′ hydroxyl of the adjacent sugar. As far as the primary structures of DNA and RNA are concerned, there are two major differences. In RNA the sugar is ribose, while in DNA the sugar is deoxyribose, which has one less hydroxyl group (see Chapter 6, Fig. 7). The second difference between DNA and RNA is in one of the four bases. In DNA the unique base is thymine, which is replaced by uracil in RNA (see Chapter 6, Fig. 7).

Another difference between the two nucleic acids concerns length. The polynucleotides in DNA are extremely long, extending over the full length of the chromosome. In RNA the lengths are quite variable but usually fall in the range of 10^2 to 10^4 monomers.

It should be noted that in both DNA and RNA the polynucleotide chains have a directional sense, with a 5′ end and a 3′ end (Fig. 2). The numbers refer to the 5′ carbon and the 3′ carbon on the sugar (note the use of primes to designate carbons in the sugar). Either of these ends may contain a free hydroxyl group or a hydroxyl group with a phosphate attached. The structure shown in Fig. 2 contains a phosphate group on the 5′ end but none on the 3′ end.

MOST DNAs EXIST AS COMPLEMENTARY DOUBLE-HELIX (DUPLEX) STRUCTURES

Nucleic acids adopt highly organized three-dimensional structures in which complementary forces hold two chains together in a helical complex.

A large body of chemical information that proved vital to understanding DNA structure came from Erwin Chargaff's analyses of the nucleotide composition of DNAs from various sources (Table 1). Although the base compositions for different DNAs were found to vary over a wide range, Chargaff noted that within all the DNAs that he examined, the amount of adenine (A) was very nearly equal to the amount of thymine (T), and the amount of guanine (G) was very nearly equal to the amount of cytosine (C). The two equalities were the first indication that regular complexes exist between A and T and between G and C in DNA.

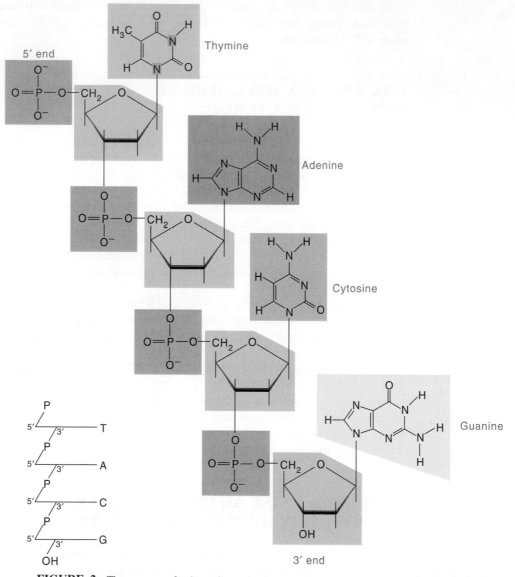

FIGURE 2   The structure of a deoxyribonucleotide. Drawn in abbreviated form at lower left. The illustrated structure is written pTpApCpG.

While searching for the meaning of these equalities, James Watson noted that hydrogen-bonded pairs between A and T and between G and C with the same overall dimensions could be formed (Fig. 3). These hydrogen-bonded pairs could be formed between bases of opposing polynucleotide strands (Fig. 4). With such a structure in

TABLE 1

Base Composition of the DNAs from Different Sources

	(A) Adenine	(G) Guanine	(C) Cytosine	(5-MC) 5-Methyl-cytosine	(T) Thymine	$\dfrac{A + T}{G + C + 5\text{-}MC}$
Human	30.4	19.6	19.9	0.7	30.1	1.53
Sheep	29.3	21.1	20.9	1.0	28.7	1.38
Ox	29.0	21.2	21.2	1.3	28.7	1.36
Rat	28.6	21.4	20.4	1.1	28.4	1.33
Hen	28.0	22.0	21.6		28.4	1.29
Turtle	28.7	22.0	21.3		27.9	1.31
Trout	29.7	22.2	20.5		27.5	1.34
Salmon	28.9	22.4	21.6		27.1	1.27
Locust	29.3	20.5	20.7	0.2	29.3	1.41
Sea urchin	28.4	19.5	19.3		32.8	1.58
Carrot	26.7	23.1	17.3	5.9	26.9	1.16
Clover	29.9	21.0	15.6	4.8	28.6	1.41
Neurospora crassa	23.0	27.1	26.6		23.3	0.86
Escherichia coli	24.7	26.0	25.7		23.6	0.93
T4 Bacteriophage	32.3	17.6		16.7[a]	33.4	1.91

[a] In T bacteriophage all the cytosine exists in the 5-hydroxymethyl form 5-HMC.

mind, Francis Crick pondered the X-ray diffraction pattern produced by DNA. Immediately, it struck him that the diffraction pattern could be interpreted as that produced by a double-helix structure (Fig. 5). The helical aspect of the structure repeats itself every 10 base pairs along the helix axis.

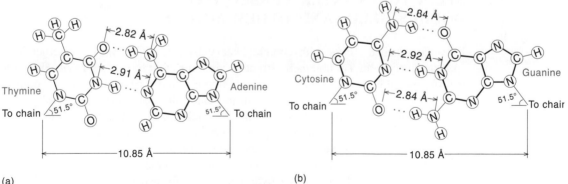

(a)   (b)

FIGURE 3   Dimensions and hydrogen bonding of (a) thymine to adenine and (b) cytosine to guanine. Note that there are two hydrogen bonds formed in the A-T base pair and three in the G-C base pair. The overall dimensions of the base pairs and the same. Consequently, they fit at any position in an otherwise regular polymeric structure. (Academic Press LTD., London, England. Adapted from Arnott, S., Wilkins, M. H. F., Hamilton, L. D., and Langridge, R. Fourier Synthesis Studies of Lithium DNA, Part III: Hoogsteen Models, *J. Mol. Biol.* 11:391, 1965.)

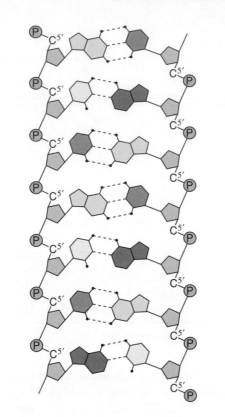

FIGURE 4 Segment of DNA, drawn to emphasize the hydrogen bonds formed between opposing chains. Three hydrogen bonds are present in the G-C pairs and two in the A-T pairs. The two strands are antiparallel: one strand (left side) runs 5′ to 3′ from top to bottom and the other strand (right side) runs 5′ to 3′ from bottom to top. The planes of the base pairs are turned 90° to show the hydrogen bonds between the base pairs. (Adapted from Kornberg, A. The synthesis of DNA, *Sci. Am.* October, 1968.)

CELLULAR RNAs FORM INTRICATE FOLDED STRUCTURES INTERSPERSED WITH DOUBLE-HELIX AND OTHER MOTIFS

If single-stranded RNAs with complementary base pairs are mixed together, they spontaneously form a double-helix structure similar to that formed by DNA. However, RNAs isolated form cells exist as unpaired polynucleotide chains. Instead of a double-helix structure, most cellular RNAs form intricate folded structures that rival protein structures in their complexity. In fact Francis Crick once said that transfer RNAs—a class of RNAs that are involved in protein synthesis—are RNAs trying to look like proteins. Figure 6 illustrates the complex folded structure formed by a typical *transfer RNA (tRNA)*. The structure contains short regions of double helix held together by less regular interactions. Many of the hydrogen bonds formed in the three-dimensional structure are not of the type found in a regular double-helix structure. Larger RNA molecules form even more complex structures.

FIGURE 5 The most common form of the double-helix DNA. The base-paired structure shown in Fig. 4 forms the helix structure shown in (a) and (b) by a right-handed twist. The two strands are antiparallel as indicated by the curved arrows in (a). In (b) a space-filling model depicts the sugar–phosphate backbones as

(Continued)

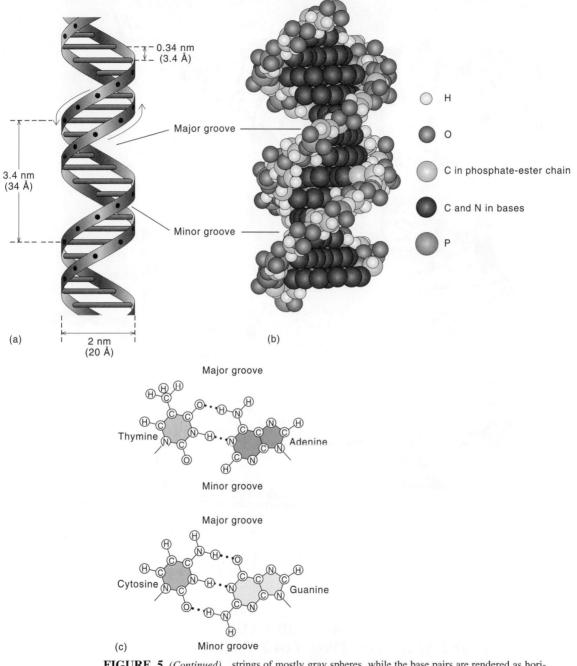

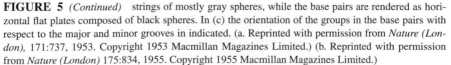

FIGURE 5 *(Continued)*   strings of mostly gray spheres, while the base pairs are rendered as horizontal flat plates composed of black spheres. In (c) the orientation of the groups in the base pairs with respect to the major and minor grooves in indicated. (a. Reprinted with permission from *Nature (London),* 171:737, 1953. Copyright 1953 Macmillan Magazines Limited.) (b. Reprinted with permission from *Nature (London)* 175:834, 1955. Copyright 1955 Macmillan Magazines Limited.)

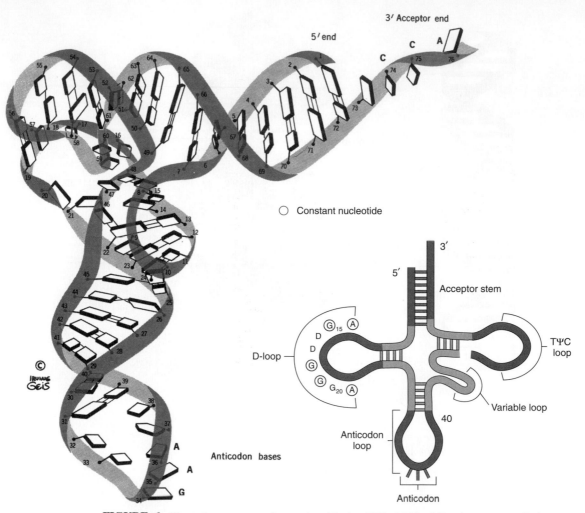

FIGURE 6   The tertiary structure of yeast phenylalanine tRNA. (a) The full tertiary structure. Purines are shown as rectangular slabs; pyrimidines, as square slabs; and hydrogen bonds, as lines between slabs. (b) Nucleotide sequence. Residues that appear in most of the yeast tRNAs and residues that appear to be constantly a purine or a pyrimidine are indicated. Residues involved in tertiary base pairing are shown connected by solid lines. (a. Reprinted from Quigley, G. J. and Rich, A. Structural Domains of Transfer RNA Molecules, *Science* 194:796, 1976 with permission from *Science* © 1976, American Association for the Advancement of Science.)

DNA REPLICATION EXPLOITS THE COMPLEMENTARY STRUCTURE THAT FORMS BETWEEN ITS TWO POLYNUCLEOTIDE CHAINS

Chromosomal DNA replicates once in every cell division. In that way the amount of DNA per cell remains constant. From the complementary duplex structure of DNA it

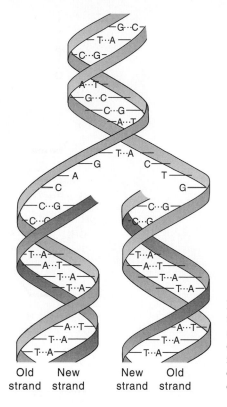

FIGURE 7   Watson–Crick model for DNA replication. The double helix unwinds at one end. New strand synthesis begins by absorption of mononucleotides to complementary bases on the old strands. These ordered nucleotides are then covalently linked into a polynucleotide chain, a process resulting ultimately in two daughter DNA duplexes.

is a short intuitive hop to a model for replication that satisfies the requirement for one round of DNA duplication accompanying every cell division. Such a model was suggested by Watson and Crick when they proposed the double-helix structure for DNA (Fig. 7). First, the two chains of the double helix unwind; next, mononucleotides absorb to complementary sites on each polynucleotide strand; and finally, these mononucleotides become linked to yield two identical daughter DNA double helices. What could be simpler? Subsequent biochemical investigations showed that in many respects this model for DNA replication was correct. Enzymes known as DNA polymerases add nucleoside triphosphates to the growing chain in the 5′ to 3′ direction (Fig. 8). Triphosphates instead of nucleotide monophosphates are used in the polymerization process because energy is needed to make the phosphodiester bonds in the polymer. The splitting of the two terminal phosphates from the triphosphate provides the energy necessary to make a phosphodiester linkage.

Many other enzymes and helper proteins are required to support the DNA synthesis process. There is a large family of enzymes for the synthesis of the nucleotide building blocks. Finally, there are a number of regulatory proteins to ensure that DNA synthesis starts at a well-defined location on the chromosome and that each segment of the DNA is duplicated once and only once for each cell division.

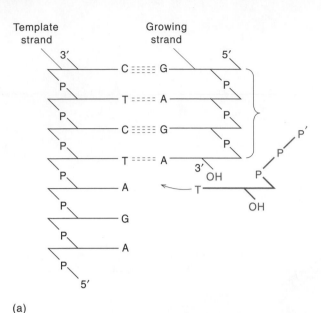

(a)

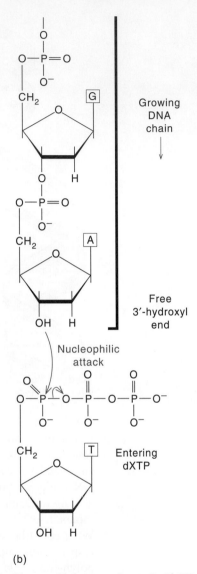

(b)

FIGURE 8   Template and growing strands of DNA. (a) Nucleotides are added one at a time to the 3′-OH end of the growing chain. Only residues that form Watson–Crick H-bonded base pairs with the template strand are added. (b) Covalent bond formation between the 3′-OH end of the growing chain and the 5′ phosphate of the mononucleotide is accompanied by pyrophosphate removal from the substrate nucleoside triphosphate.

Another aspect of DNA replication that is very important deals with the question of faithful replication and repair of damaged DNA. If the wrong base is somehow added during synthesis or if the correct base is somehow damaged after incorporation, there are enzymes that recognize this fault and most of the time remove the aberration and replace it with the correct base, that is, the base that makes a Watson–Crick base pair with the base on the template chain. If the maintenance and repair enzymes cannot rectify an abnormality, the organism may die or alternatively the organism may display a modified function that on rare occasions may actually increase the Darwinian fitness of the organism. An alteration in the sequence of bases in the DNA that leads to a beneficial change to the organism is the stuff that evolution feeds on.

TRANSCRIPTION INVOLVES THE SELECTIVE COPYING OF SPECIFIC REGIONS OF MUCH LONGER DNA CHAINS

RNA polymerases (called transcriptases if they use a DNA template) like DNA polymerases utilize nucleoside triphosphates as substrates in the polymerization process except that in this case ribonucleoside triphosphates are used. In the transcription process one of the DNA strands serves as the template for the growing RNA molecule. For example, a cytosine in the template strand of DNA means that a complementary guanine is to be incorporated at the corresponding location of the RNA. Synthesis proceeds in the 5′ to 3′ direction with each new nucleotide being added onto the 3′—OH end of the growing RNA chain. This is similar to the manner of growth of DNA.

The overall process for RNA synthesis on a duplex DNA template can be conceptually divided into three phases: *initiation*, *elongation*, and *termination* (Fig. 9). In the initiation phase of the reaction, the RNA polymerase binds to a specific site on the DNA called the *promoter*. Here it unwinds and unpairs a small region of the DNA. RNA synthesis begins by the base pairing of ribonucleoside triphosphates to one strand of the DNA, followed by the stepwise formation of covalent bonds from one nucleotide to the next. As elongation proceeds, the DNA unwinds progressively in the direction of synthesis. The short stretch of DNA–RNA hybrid duplex formed during synthesis is prevented from becoming longer than 10 to 20 base pairs by a rewinding of the DNA and the simultaneous displacement of the newly formed RNA. Termination occurs at a sequence recognized by the RNA polymerase as a stop signal. At this point the ternary complex of DNA, RNA, and RNA polymerase breaks up.

The RNA synthesis that takes place from each gene is independently regulated so that the amount of mRNA and mRNA-related protein suits the needs of the cell under a particular set of conditions. For most genes the rate of initiation is proportional to the amount of a particular mRNA that is made and this rate is proportional to the rate of formation of a DNA–RNA polymerase initiation complex. The affinity of the RNA polymerase for a particular promoter is regulated by the sequence of bases in the DNA that bind to the RNA polymerase. In addition to this, there is a broad array of regulatory proteins acting simply or in combination to enhance or to inhibit the formation of the initiation complex. Frequently, the action of these regulatory proteins is influenced by small molecules that reflect the metabolic state of the cell.

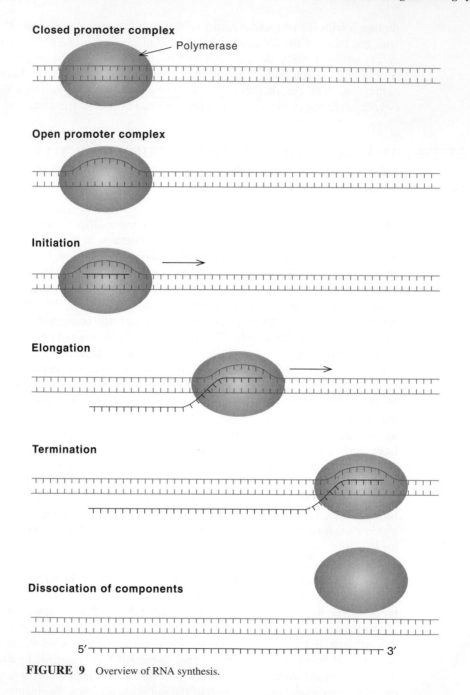

FIGURE 9 Overview of RNA synthesis.

NASCENT TRANSCRIPTS UNDERGO EXTENSIVE CHANGES FOLLOWING SYNTHESIS

The tRNAs in both prokaryotes and eukaryotes undergo extensive changes after transcription. These changes include the alteration of individual bases (e.g., see Fig. 10) and also entails selective cleavages of some of the phosphodiester linkages that were formed in the polymerization reaction.

The mRNAs in eukaryotes undergo extensive changes in the nucleus before they are transported to the cytoplasm. These changes occur both within the chains and at the ends of the chains. They also entail removal of internal segments of the transcripts by a process called *splicing*, because of the resemblance to splicing in film editing. Splicing is particularly interesting for more than one reason: first, because it results in a mature mRNA in which the information from different regions of the nascent RNA molecule have been united; second, because some molecules of RNA catalyze their own splicing.

Protein enzyme-catalyzed splicing is a two-step process. In the first step there is a cleavage of the pre-mRNA at the 5′ splice site that generated a 5′ proximal segment and an RNA species containing a lariat configuration connected to the 3′ proximal segment (Fig. 11). The lariat is formed via a 2′ to 5′ phosphodiester bond, which joins the 5′ terminal guanosine of the internal segment to an adenosine residue that is located 18 to 40 nucleotides upstream of the 3′ splice site. These two RNA species are probably held together in a noncovalent complex until the next step (see Fig. 11, step 2) in a reaction that entails cleavage at the 3′ splice site to generate the free excised segment and ligation of the two preserved segments by a 3′ to 5′ phosphodiester bond. A ribonucleoprotein complex called a *spliceosome* catalyzes the splicing.

SOME RNAs HAVE CATALYTIC PROPERTIES

For more than 50 years since the discovery of the first enzyme urease, growing evidence seemed to confirm that all enzymes were proteins. It came as a giant surprise in the early 1980s when it was found that a number or biochemical activities normally associated with protein enzymes were shared by certain RNAs.

In 1982 Tom Cech discovered that when the pre-rRNA of the protozoan *Tetrahymena pyroformis* was incubated with a magnesium salt and guanosine monophosphate, splicing of the RNA occurred without the involvement of any protein. This ability of RNA to carry out its own splicing gave rise to the term *ribozyme*.

The mechanism of self-splicing in this case is somewhat different from that observed in the spliceosome reaction (Fig. 12). First, the 3′ hydroxyl group of the guanosine attacks the phosphodiester bond at the 5′ splice site. This is followed by another transesterification reaction in which the 3′ hydroxyl group of the upstream RNA attacks the phosphodiester bond at the 3′ splice site, thereby completing the splicing reaction. The final reaction products include the spliced RNA and the excised fragment.

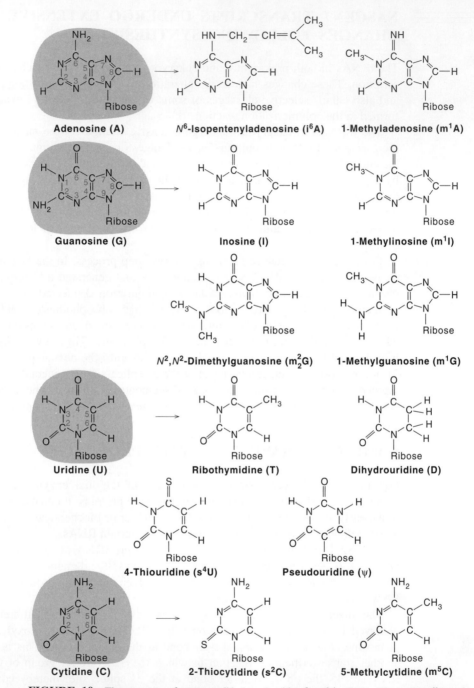

FIGURE 10 The structure of some modified nucleosides found in tRNA. The parent ribonucleosides are shown on the left. The other bases found in RNA result from posttranscriptional modification.

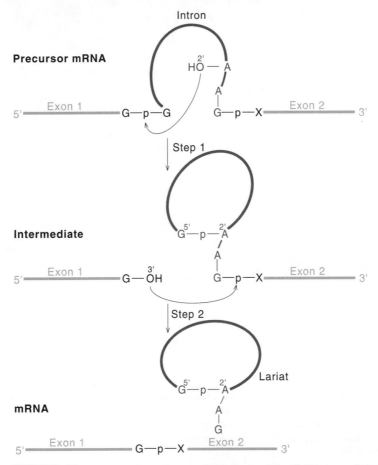

FIGURE 11 Splicing scheme for pre-mRNA. In step 1 the 2′-OH on an adenosine attacks a phosphate that is 5′-linked to a guanine residue. This leads to a lariat configuration connected to the distal exon. The lariat is formed via a 2′–5′ phosphodiester bond, which joins the 5′ terminal guanosine of the intron to an adenosine residue within the intron, 18 to 40 nucleotides upstream of the 3′ splice site. In the next step there is a cleavage at the 3′ splice site to generate the free intron RNA and a ligation of the two exons via a 3′–5′ phosphodiester bond. Bases usually found in the region of the splice sites are indicated. The spliceosome is presumed to hold the reacting components in the proper juxtaposition to facilitate the two-step reaction.

Since its initial discovery, self-splicing has been found to occur for RNAs from a wide variety of organisms. Certain RNAs that exhibit self-splicing produce lariats, just like those seen in the commonly observed splicing reactions that are catalyzed by protein enzymes (see Fig. 11). These findings suggest there was a time when all splicing reactions were RNA-catalyzed.

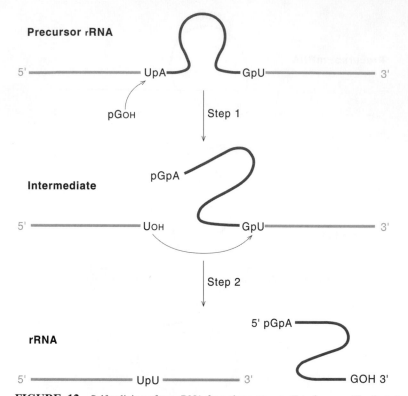

FIGURE 12   Self-splicing of pre-rRNA from the protozoan *Tetrahymena*. The first step is a transesterification reaction in which the 3′ hydroxyl group of guanosine attacks the phosphodiester bond at the 5′ splice site. The second step involves another transesterification reaction in which the 3′ hydroxyl group of the upstream exon attacks the phosphodiester bond at the 3′ splice site and displaces the 3′ hydroxyl group of the intron.

PROTEINS ARE INFORMATIONAL MACROMOLECULES

Proteins are the ultimate heirs of the information encoded in the sequence of nucleotide bases within the chromosomes. Each protein is composed of one or more polypeptide chains and each polypeptide chain is a linear polymer of amino acids. There are 20 amino acids commonly found in polypeptide chains. The order of these amino acids in the polypeptide polymer is determined by the order of nucleotides in the corresponding messenger RNA (mRNA). The process whereby these amino acids are ordered and polymerized into polypeptide chains is called translation because it involves the relationship between the 4-letter code of nucleic acids and the 20-letter code found in proteins.

CELLULAR MACHINERY OF PROTEIN SYNTHESIS IS CONSTRUCTED FROM RNA AND PROTEIN COMPONENTS

Most of the steps in protein synthesis were discovered in the laboratory of Paul Zamecnik in the 1950s. Zamecnik and coworkers injected radioactive amino acids into whole animals and tracked the radioactivity to see what happened to the amino acid on its way to becoming part of a polypeptide chain. It was shown that the amino acid first becomes

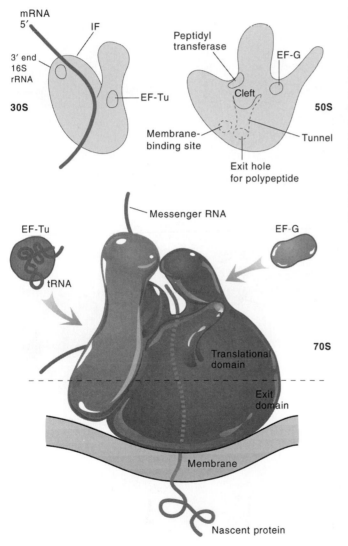

FIGURE 13 Functional sites on the prokaryotic ribosome. This figure shows a ribosome in the process of elongating a polypeptide chain. The mRNA and the EF–Tu–aminoacyl–tRNA complex are more closely associated with the 30S subunit. The peptidyl transferase and the EF-G elongation factor are associated with the 50S subunit, as in the nascent polypeptide chain. IF and EF are standard abbreviations for proteins that function as initiation factors and elongation factors, respectively. (From Bernabeu, C. and Lake, J. A. Nascent Polypeptide Chains Emerge from the Exit Domain of the Large Ribosomal Subunit: Immune Mapping of the Nascent Chain, *Proc. Natl. Acad. Sci. USA* 79:3111–3115, May 1982. Reprinted by permission.)

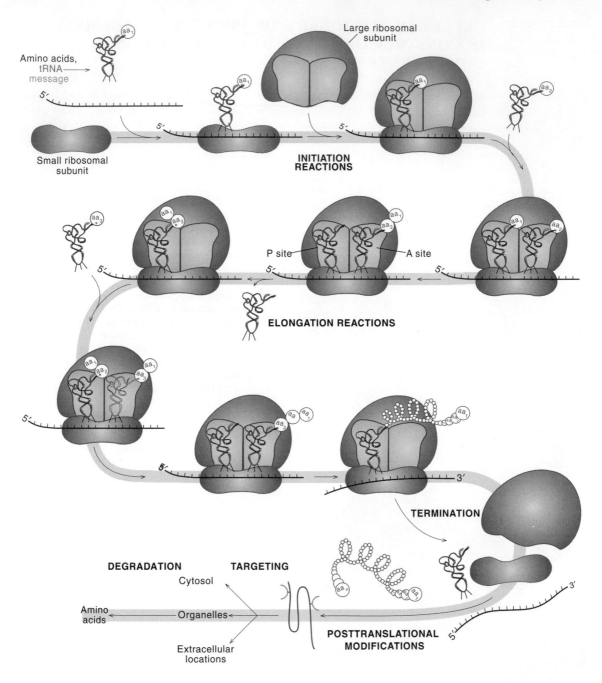

Amino acids,
tRNA
message

Small ribosomal
subunit

Large ribosomal
subunit

INITIATION
REACTIONS

P site   A site

ELONGATION REACTIONS

TERMINATION

DEGRADATION   TARGETING

Cytosol

Amino
acids   Organelles

Extracellular
locations

POSTTRANSLATIONAL
MODIFICATIONS

linked to a low-molecular-weight RNA molecule, the tRNA, which then migrates to a large ribonucleoprotein particle, the *ribosome*, where the amino acid becomes peptide-linked. During translation the mRNA associates with the ribosome. These key components of the translation system are illustrated in Fig. 13.

In addition to tRNAs for each amino acid, ribosomes, and mRNAs, protein synthesis requires a number of other protein factors, adenosine triphosphate (ATP), and guanosine 5'-triphosphate (GTP). All the steps in protein synthesis are illustrated in Fig. 14. First, the tRNAs become linked to amino acids: then, the initiation complex on the ribosome is formed. Peptide synthesis does not start until a second amino acid linked tRNA becomes bound to the ribosome. The tRNAs bind to sites on the mRNA template by complementary base pairing. Elongation reactions involve peptide bond formation, dissociation of the discharged tRNA, and translocation. The elongation process is repeated many times until the termination point on the message is reached. Termination is marked by dissociation of the two ribosomal subunits and release of the newly formed polypeptide chain. The polypeptide chain sometimes folds into its final form without further modifications. Frequently, the folded polypeptide chain is modified by removal of part of the polypeptide chain or by addition of various groups to specific amino acid side chains. The completed polypeptide chain migrates to different locations in the cell according to its structure and function.

CODE USED IN TRANSLATION WAS DECIPHERED WITH THE HELP OF SYNTHETIC MESSENGERS

A solution to the coding problem resulted from experiments initiated by Marshall Nirenberg and Heinrich Matthaei in 1961 while they were attempting to synthesize proteins in cell-free extracts prepared from bacteria. These extracts, containing the essential cellular components for protein synthesis, were supplemented with nucleotides, salts, and amino acids. For mRNA Nirenberg and Matthaei chose to use viral RNA from the tobacco mosaic virus (TMV) because their goal was to make viral protein. As a control for these first incorporation experiments, these biochemists

---

FIGURE 14   Overview of reactions in protein synthesis (aa$_1$, aa$_2$, aa$_3$ = amino acids 1, 2, 3). Protein synthesis requires transfer RNAs (tRNAs) for each amino acid, ribosomes, messenger RNA (mRNA), and a number of dissociable protein factors in addition to ATP, GTP, and divalent cations. First, the tRNAs become charged with amino acids; then, the initiation complex is formed. Peptide synthesis does not start until the second aminoacyl-tRNA becomes bound to the ribosome. Elongation reactions involve peptide bond formation, dissociation of the discharged tRNA, and translocation. The elongation process is repeated many times until the termination codon is reached. Termination is marked by the dissociation of the mRNA from the ribosome and the dissociation of the two ribosomal subunits. The polypeptide chain sometimes folds into its final form without further modifications. Frequently, the folded polypeptide chain is modified by removal of part of the polypeptide chain or by addition of various groups to specific amino acid side chains. The completed polypeptide chain migrates to different locations according to its structure. It may remain in the cytosol or it may be transported into one of the cellular organelles or across the plasma membrane into extracellular space. Proteins have different lifetimes. Eventually a protein is degraded usually down to its component amino acids.

needed an mRNA that would not be expected to code for a protein. For this purpose their choice was an artificial polynucleotide, polyuridylic acid [poly (U)], which contained only uridine residues. When Nirenberg and Matthaei added poly (U) to their cell-free system, substantial incorporation was evident but only the amino acid phenylalanine was incorporated.

This result sent a wave of excitement through the biochemical community. As so often happens in research, a major discovery had resulted from serendipity. A control experiment that was supposed to yield a null result had in fact led to the spectacular incorporation of large quantities of a single amino acid. It appears that poly(U) had stimulated the synthesis of polyphenylalanine. Soon thereafter it was demonstrated

TABLE 2
The Genetic Code

First position (5′ end)	U		C		A		G		Third position (3′ end)
U	UUU	} Phe	UCU	}	UAU	} Tyr	UGU	} Cys	U
	UUC		UCC		UAC		UGC		C
	UUA	} Leu	UCA	} Ser	UAA	} STOP	UGA	STOP	A
	UUG		UCG		UAG		UGG	Trp	G
C	CUU	} Leu	CCU	}	CAU	} His	CGU	}	U
	CUC		CCC		CAC		CGC		C
	CUA		CCA	} Pro	CAA	} Gln	CGA	} Arg	A
	CUG		CCG		CAG		CGG		G
A	AUU	} Ile	ACU	}	AAU	} Asn	AGU	} Ser	U
	AUC		ACC		AAC		AGC		C
	AUA		ACA	} Thr	AAA	} Lys	AGA	} Arg	A
	AUG	Met	ACG		AAG		AGG		G
G	GUU	} Val	GCU	}	GAU	} Asp	GGU	}	U
	GCU		GCC		GAC		GGC		C
	GUA		GCA	} Ala	GAA	} Glu	GGA	} Gly	A
	GUG		GCG		GAG		GGG		G

that polyadenylic acid [poly (A)] promotes polylysine synthesis and [poly (C)] promotes polyproline synthesis. From these observations it seemed clear that the code words UUU, AAA, and CCC correspond to the amino acids phenylalanine, lysine, and proline, respectively. A solution to the coding problem was well underway. The remaining code words were determined in the following 3 years by a painstaking analysis of the amino acids incorporated with other synthetic and natural mRNAs with a known nucleotide sequence. The results are presented in Table 2.

SUMMARY

In this chapter we discuss the structures, functions, and biochemical reactions of the two major nucleic acids found in biochemical systems, DNA and RNA.

1. All nucleic acids consist of covalently linked nucleotides. Each nucleotide contains a phosphate group linked to the sugar on one side and a purine or pyrimidine base on the other side. The purine bases in both DNA and RNA are always adenine (A) and guanine (G). The pyrimidine bases in DNA are thymidine (T) and cytosine (C); in RNA they are uracil (U) and cytosine.

2. DNA usually exists as a double-stranded molecule. The continuity of each strand is maintained by repeating 3′ to 5′ phosphodiester linkages formed between the sugar and the phosphate groups; they constitute the covalent backbone of the macromolecule. The side chains of the covalent backbone consist of the purine or pyrimidine bases. In double-stranded DNA the two chains are oriented antiparallel to one another.

3. Specific pairing occurs between bases on one strand and bases on the other strand in such a way that A always pairs with T and G always pairs with C. The pairs are held together by hydrogen bonds. The duplex is stabilized by these hydrogen bonds and also by stacking interactions between adjacent base pairs.

4. Most cellular RNAs exist as complex folded single-stranded polynucleotides. The conformation of these folded structures is sequence dependent. Hydrogen bonds and stacking forces are important in stabilizing the folded structures of RNA. In addition to the Watson–Crick hydrogen bonds found in DNA, a wide variety of additional hydrogen bonds are found in folded RNA structures.

5. DNA replication proceeds by the synthesis of one new strand on each of the parental strands.

6. Many proteins are required for DNA synthesis. Some enzymes that act on DNA are involved in processes other than DNA synthesis, such as DNA repair.

7. RNA is synthesized in the 5′ direction by the formation of 3′ to 5′ phosphodiester linkages between the nucleotide substrates as in DNA. The sequence of bases in RNA transcripts catalyzed by DNA-dependent tRNA polymerases is specified by the complementary DNA template strand.

8. Some newly synthesized transcripts are immediately functional; others must be processed or modified before they can function. One of the more interesting

processing reactions of nascent transcripts involves the removal of intervening sequences, or introns. This type of processing is referred to as splicing. Most splicing reactions appear to require host proteins. However, it is clear that certain splicing reactions require only RNA, a fact demonstrating that some RNAs are capable of functioning like protein enzymes in catalyzing the formation and breakage of phosphodiester linkages in polyribonucleotides.

9. There are three types of RNA required for protein synthesis: messenger RNA (mRNA), ribosomal RNA (rRNA), and transfer RNA (tRNA). The tRNA transports amino acids in an activated form to the template for protein synthesis.

10. Ribosomes, which are composed of ribosomal RNA and ribosomal proteins, are the site of protein synthesis.

11. The mRNA is the template for protein synthesis.

12. During translation of the mRNA, aminoacyl–mRNAs complex with specific trinucleotide sites on the mRNA template. The order of these trinucleotide sites, or codons, determines the order of amino acids in the resulting polypeptide chain.

13. The genetic code is the relationship between three-base sequences and the amino acids they represent.

Problems

1. Why do we refer to proteins as informational macromolecules?
2. If you had a cell-free system for polypeptide synthesis that simulated the *in vivo* system for protein synthesis, what polypeptides would you expect to see synthesized if you supplied the system with the following synthetic mRNAs:
   (a) A repeating sequence . . . CUCUCU . . . ?
   (b) A repeating sequence . . . GAUGAU . . . ?
3. Why do proteins make superior catalysts to nucleic acids? See Problem 2 in Chapter 17.
4. In what sense is every nucleic acid a potential catalyst?
5. If life were to originate in another biosphere and follow the same general lines of nucleic acid and protein development with the same four bases in DNA and RNA, do you think the same codon–amino acid relationship would have developed? If you think not, in what ways might the relationship be similar and in what ways might it have been different?

Reference

See references at the end of Chapter 6.

PART III

Biochemical and Prebiotic Pathways

A Comparison

In Part III we explore the chemistry that could have given rise to living systems on primitive Earth. This section of the text deals directly with the question of the origin of life. I take the view that biochemical pathways give us strong clues as to how life originated. Accordingly, this section of the text seesaws between considerations of how the chemistry for a particular pathway works in biochemical systems now and how they evolved in the prebiotic world and in very primitive forms of life.

In Chapter 10 we consider the environment around the time of the origin of life. This discussion includes a description of the substances that were present and the types of energy sources that were available to drive reactions of potential prebiotic significance. The subsequent chapters present detailed biochemical pathways and their prebiotic precursors.

Chapter 11 deals with some of the detailed biochemistry of the most primitive biochemical pathways: the glycolytic pathway, the pentose phosphate pathway, and the tricarboxylic acid (TCA) cycle. In Chapter 12 we attack the question of how sugars were made in the prebiotic world. The focus is on ribose because ribose is an essential component of nucleic acids and for that reason was probably the most important sugar around the time of the origin of life.

In Chapter 13 we turn our attention to the synthesis of nucleotides. Both biochemical and possible prebiotic pathways are considered. Strong parallels are seen between the biochemical pathways and the prebiotic pathways for purine synthesis in the later steps of both pathways.

In Chapter 14 RNA metabolism and prebiotic synthesis of RNA are discussed. The major classes of RNA found in biochemical systems are described and the basic aspects of transcription are considered. It is pointed out that there are RNA viruses in which RNA serves as a template for RNA synthesis in much the same way as DNA serves as a template for most cellular RNAs. Next experiments are discussed that deal with attempts to synthesize RNA in a prebiotic world. The extent to which these experiments have attained success is emphasized, as are those aspects of the process still to be reckoned with. In the final section of Chapter 14 the evidence for ribozymes in the current biochemical world is reviewed and current efforts to make additional ribozymes by *in vitro* methods are described.

Chapter 15 sketches the main mechanisms for amino acid biosynthesis and goes on to suggest that amino acids in the prebiotic world also were synthesized from carbohydrate precursors. It seems obvious that only a few amino acids were needed to get life underway. It is not so easy to make choices (of which amino acids were needed) but it seems highly likely that the amino acids that were utilized by the first living organisms must have been relatively easy to synthesize.

Chapter 16 discusses protein biosynthesis and Chapter 17 suggests how the current system for protein synthesis evolved. In this text we take the view that the first living organisms did not contain proteins. In Chapter 17 a realistic scenario for the evolution of the protein-synthesizing machinery is presented. In particular, it is suggested that the first transfer ribonucleic acids (tRNAs) were used for other types of synthesis in cells dominated by the directive forces of RNA and the ribozymes.

Chapter 18 describes fatty acid degradation and synthesis, and lipid synthesis before considering how lipids might have been synthesized in the prebiotic world. In Chapter 19 we go on to explain how lipids were assembled to make membranes, how proteins integrate into the membranes, and the functions that they serve. This is followed by a consideration as to how membranes were involved in the first living cells. Strategically it seems likely that membranes encompassed living systems at some point prior to the evolution of the protein-synthesizing machinery.

Chapter 20 describes the basic structures and properties of clays. Although the notion of living clays seems somewhat fanciful, there is little doubt that clays played crucial roles as surfaces where the first RNA life could assemble and flourish.

CHAPTER 10

General Considerations Concerning the Origin of Life on Earth

It seems highly likely that a very special environment was required for the origin and survival of living things. To verify this, all we have to do is to look our own Solar System. There is no indication that life exists or ever existed on any of the planets or planetesimals in our Solar System except for Earth. Although the possibility that living organisms were brought to Earth from another source cannot be ignored, the possibility that life originated on Earth is the only one we can reasonably test.

Even so, we face a difficult task, because living organisms with which we are familiar are immensely complex; it is hard to conceive of how even the most primitive living cells imaginable could have arisen from nonliving organic substances. Any scheme that is proposed involves a great deal of speculation, because the records for the origin of life have been erased during the course of time. Many biologists would even question that this is a fruitful activity, one amenable to scientific investigation, because biologists usually have more tangible matters to investigate.

However, even if the precise pathway that was followed cannot be ascertained, feasible pathways can be proposed and tested experimentally. In the process it seems likely that new principles governing chemical evolution are to be discovered. I believe that the origin of life on Earth was a likely outcome of an extended period of chemical evolution. This belief has become more and more firm as I pursue my scientific interests in this subject. In this chapter I address the problems that must be resolved and the chemical principles that must be considered.

EARTH AND THE ORIGIN OF LIFE

What Is Life?

If we are going to examine reactions that led to the origin of life, we must first decide when that ill-defined boundary between the nonliving and the living was crossed. If you assemble a group of biologists in an attempt to arrive at a definition of life, you immediately spark controversy. Most insist that a cellular structure is essential to the description of all living systems or that nucleic acids and proteins are essential attributes of all living systems. That is because they are locked into the characteristics of known living systems. However, we must be more broad-minded. If we focus on functions instead of specific chemical characteristics, we are more likely to arrive at a broader definition for a living system. There are two functions that are invariably associated with living organisms: (1) the ability to replicate and (2) the capacity to undergo gradual change. The National Aeronautics and Space Administration (NASA) is a little more specific than this and defines life as a self-sustained chemical system capable of undergoing Darwinian evolution. It is implicit in the NASA definition that variants (mutants) that arise after the replication process provide the raw material for Darwinian selection. The most important aspect to note about this definition is that it frees us from being committed to specific structures or molecules. Having said that I now would like to impose a limitation by suggesting that we restrict our inquiry to systems that are very likely to have preceded the living systems currently on Earth. This greatly simplifies the problem and we are more likely to recognize the solution if and when we find it.

Did Life Originate on Earth or Was It Transported Here from Elsewhere?

The second question that we must ask at the start of our inquiry is did life originate on this planet or was it transported here from elsewhere? Considerations about the complexity of living systems have led some evolutionists to reject the idea that life on Earth could have originated here. The hypothesis that life on Earth began as a result of being transported here from elsewhere is known by the term *panspermia*. This hypothesis was first proposed by Svante Arrhenius (1908), who suggested that living microorganisms in a dehydrated state (spores) could have been driven here by the pressure of light from the star of another planetary system. Theories like this that are based on accidental transport have been criticized because the long journey of a living organism through space would have resulted in massive doses of radiation that might preclude survival.

Francis Crick and Leslie Orgel have aired the possibility that panspermia was directed by an intelligent force transporting life to Earth from another star system. The late astrophysicist Carl Sagan proposed that life has almost certainly arisen *de novo* many times throughout the Universe; he speculated that our Solar System has been visited many times by various galactic civilizations. In fact, there are several thousand stars within a hundred light years of Earth that could be reached within as few as a million years by a spaceship traveling at 60,000 miles per hour, or within as few as 10,000 years at a speed one-hundreth that of light. Although the technology required to do this is not currently available, it seems very likely that within the forseeable future it will be. Consequently, it is easy to imagine that the necessary technology could have been developed by an intelligent force located elsewhere within the Universe. A properly designed spaceship could provide the necessary shielding for organisms traveling through space so that they would be able to grow and even replicate. The panspermia hypothesis cannot be totally dismissed. Even though all known organisms appear to be descended from a common source, this source could have been an organism that originated in another world, instead of on Earth itself.

The fossil record provides only a sketchy chronicle of evolutionary events. It does, however, indicate that the earliest recognizable forms of life on Earth were microorganisms. Records for the existence of microorganisms go back about 3.5 billion years. The environment at that time was such that it would have favored the existence of resilient anaerobic microorganisms over more complex forms of eukaryotic life.

If life was transported here from elsewhere, perhaps it was not from another galaxy, but simply from another planet within our Solar System. The most likely planets would be either Mars or Venus. Both of these planets currently have surface environments hostile to life, but present conditions could be due to changes that occurred after organisms were transplanted to Earth. Of the two planets, Mars would appear to have a friendlier environment, and the possibility that living forms currently inhabit the interior or Mars cannot be excluded. Indeed, a meteorite of Martian origin that was scooped recently from the Antarctic ice cap was thought to contain fossils of bacterial life. However, the evidence for life has been seriously questioned by most experts.

From the foregoing discussion, it should be clear that we are not rejecting the hypothesis of panspermia. However, exploring it further would be most difficult. By contrast, the alternative notion—that life arose on Earth independently of extraterrestrial beginnings—immediately opens fertile ground for scientific inquiry with present-day technology. In the remainder of this chapter we explore this possibility in general terms. In the subsequent chapters of Part III, we attack specific problems related to this scientific inquiry.

Can We Extrapolate from Familiar Living Systems to the First Living Systems?

Our next task is to see if somehow we can propose a plausible description of the first living systems. Obviously, this would be conjectural because there are no records. However, our vast understanding of the structure and components found in living cells should help us to make an intelligent guess. A description of the first living systems would be immensely valuable because it would focus our inquiries. On the one hand, prebiotic studies could be pursued to see if a feasible pathway to the first living systems could be found. On the other hand, additional studies could be pursued to see if feasible pathways from the first living systems to more familiar living cells could be found. Whatever we suggest is highly speculative and we cannot take any stock in it until the feasible pathways referred to have been found.

How are we going to do this? The best way I can think of is by extrapolation. It is like budget cutting. If you find yourself in a financial bind, you ask yourself what you can do without. Current living systems are the very elaborate products of billions of years of evolution. It is very clear that they could be greatly simplified without losing the fundamental attributes of a living system. Nevertheless, how far can we go? We do this step by step until we can do no more without losing the essence of a living system.

To begin our extrapolation we first note that in terms of both size and number of components bacteria are the simplest organisms that can be found on today's Earth. The simplest known bacteria is *Mycoplasma genatilium*, which contains 580,070 bp of DNA with 470 translation reading frames for encoding polypeptides. It is highly likely that under more primitive conditions bacteria with far fewer genes were viable. Organisms with as few as 100 to 200 carefully chosen genes probably could conduct all the essential biosynthetic processes in a suitable environment.

If we retrace the steps of evolution even farther, we can imagine a time when chromosomes were represented by RNAs instead of DNA. DNA and RNA are very similar chemically, but the deoxyribose sugar in DNA is more difficult to synthesize than the ribose found in RNA. In fact the reduction of ribose to deoxyribose does not occur unless ribose is part of a nucleotide. This biochemical observation is consistent with the notion that RNA preceded DNA as the cellular genome in living systems. The main disadvantage in using RNA as an information-bearing chromosome would appear to be that RNA is less stable than DNA. However, in a living system with far fewer genes, stability would not have been as important as it is in more complex genetic systems.

Now we must consider a much more profound evolutionary leap. Nucleic acids are key macromolecules that bear biochemical information in the form of genes while proteins are the ultimate heirs of this genetic information. Proteins are responsible for catalyzing most of the reactions in living cells. Life without either of these macromolecules at first thought seems inconceivable. Yet it is even more difficult to see how these two systems could have coevolved. If one of these systems evolved before the other, then it must have been capable of carrying out both types of functions: that is, the information storage and replication functions normally possessed by nucleic acids and the catalytic functions normally possessed by proteins. The findings of Tom Cech and Sid Altman at the beginning of the 1980s that certain RNAs possessed enzymelike activities raised the possibility that there was a time when most enzymes were parts of RNA molecules. This line of thinking has led to the currently popular proposal of a preexisting RNA-only world.

Living systems without proteins could have been much simpler. Far fewer enzymes (or ribozymes) would have been essential and most if not all of these could have been contained within the RNA molecules that also served as the genome.

Can we extrapolate back any farther? In particular, could the requirement for a cellular membrane be lifted? A membrane is a useful way of enclosing the components needed for cell growth and function, but in the most primitive living systems most of the components needed for growth and function might have been produced in the surrounding environment. Under such circumstances a membrane might have been more of an impediment than an asset. For these reasons it seems possible that membranes would not have been needed by the first "living" RNA molecules, although they would have become a major asset in time.

At this point our hypothetical living system consists of little more than an RNA molecule. Under favorable circumstances an RNA molecule should be capable of satisfying the NASA definition for a living system. Thus RNA with the help of an appropriate ribozyme should be able to replicate by a template mechanism that involves complementary base pairing between template strand and growing strand. It also has the capacity to change by occasional mispairings during the replication process. This clearly does not solve all problems with respect to the origin of the first living systems. Above all we must wonder how it got nucleotides for its own replication. We also might ask if RNA itself could be simplified any further. This possibility is raised in Chapters 13 and 14.

Conditions on Prebiotic Earth Favored the Origin of Life

Earth was formed by a process of accretion that resulted in a molten mass lacking the gravitational force to retain its gases at elevated temperatures. Within 1.1 billion years of its formation the isotopic and fossil record of sediments indicates that life existed. If we are ever to appreciate the chemical reactions leading to the origin of life, we must reproduce the conditions that existed around that time, at least on a miniscale. We must ask many questions. What raw materials were available? What sources of energy were available to drive reactions? Where did the reactions take place? Was the medium for reaction in the gas, liquid, or solid phases (or in more than one phase)? Within each of these phases, was a particular microenvironment favored? Whatever the specific answers are to these questions, it seems clear that sufficiently favorable conditions did prevail, because all indications are that life has existed on this planet for over 3 billion years.

First, let us focus on a description of the solid portions of the planet in prebiotic times. In Chapters 2, 3, and 4 we have seen that a combination of nuclear physics (which sets the relative abundances of the elements) and inorganic chemistry (which sets the chemical form of the elements in the planetary nebula) dictated that rocky planets like Earth should consist primarily of the four elements—oxygen, magnesium, silicon, and iron. In Chapter 6, Table 1, we enumerate the major elements found in the Universe and in various regions of Earth. From a comparison of the composition of the same elements found in Earth's crust, ocean waters, and the human body, two facts emerge. First, the composition of Earth's crust differs most significantly from the composition of the Universe as a whole. Second, the composition of the human body, which is reasonably representative of a large segment of living organisms, is closest to that of Earth's crust; yet there are significant differences in relative amounts of the different elements. The four most abundant elements in the human body are hydrogen, carbon, nitrogen, and oxygen. Of these, only oxygen belongs to the highest abundance class of elements making up Earth's crust. Despite this quantitative discrepancy, most elements needed to make living things were present in significant quantities in Earth's crust.

The composition of the atmosphere and the oceans also has a direct bearing on the origin of life, because many of the early reactions in the prebiotic pathways probably took place in a gaseous or liquid medium. Even though the total mass of the atmosphere is only about one-millionth the mass of Earth itself, it is highly likely that this is where most of the precursor compounds containing nitrogen, carbon, and oxygen were formed. These compounds must have been chemically altered and subsequently incorporated into larger and larger organic molecules that served as the building blocks of living organisms.

Recall from Chapter 5 that the composition of the atmosphere today is quite different from that around the time of the origin of life. The current atmosphere primarily is composed of molecular nitrogen (N_2) and molecular oxygen (O_2), with smaller amounts of carbon, primarily being in the form of carbon dioxide (CO_2). In prebiotic times there was virtually no gaseous oxygen and there was probably far less nitrogen; there may have been some carbon in the form of carbon dioxide, but much of the carbon probably existed in more reduced states such as carbon monoxide (CO) and methane (CH_4). Thus primitive Earth probably had a reducing atmosphere, as contrasted with today's highly oxidizing atmosphere (see Chapter 5). This difference was extremely important, because organic compounds that are found in living organisms are mostly in a reduced state and they would be much less stable in an oxidizing atmosphere—if, indeed, they could even form under such circumstances.

The composition of seawater (see Chapter 6, Table 1) also has undergone great changes in the course of geologic time. For example, the current content of iron in seawater is very low. Before molecular oxygen gas abundant in the atmosphere, the ferrous ion content of seawater is believed to have been much higher. Supporting this belief are the giant banded iron formations that were deposited in sedimentary layers during the period when the oxygen content of the atmosphere was believed to be on the rise (see Chapter 22, Fig. 3). Ferrous salts are highly soluble in water, while ferric salts are highly insoluble. Contact with molecular oxygen would result in the rapid conversion of ferrous ions to ferric ion, which should precipitate soon after formation. Hence there is a direct correlation between the rise of atmospheric oxygen and the formation of iron-rich sedimentary layers.

In contrast to its mineral composition, the average pH of the oceans probably has not changed much with time. The current pH of the oceans is between 8.0 and 8.5. It is believed that this pH has been maintained over the course of geologic time, because the pH is determined by the combined buffering action of vast deposits of clays and minerals that are in contact with the oceans. Because the pH inside cells is about 8.1, it appears that the ocean pH is in the ideal range for biochemical reactions.

GETTING THE CHEMISTRY OF LIFE UNDERWAY

Before we focus on specific reactions there are basic considerations that need airing: the relevance of thermodynamics and kinetics, the available sources of energy and starting materials, and the importance of location.

Both Thermodynamic and Kinetic Factors Were Important in Determining Which Reactions Would Be Favored

As we have seen in Chapter 7, a favorable free energy is a necessary but not a sufficient condition for a reaction to take place. This point can be illustrated by considering four possible routes to the synthesis of the amino acid glycine (Table 1). The free energy of formation of glycine from methane and ammonia [reaction (1)] is highly unfavorable. Synthesis of glycine from carbon monoxide, hydrogen, and ammonia [reaction (2)] is possible but requires heating. The synthesis of glycine from formaldehyde, ammonia, and hydrogen cyanide [reaction (3)] is also possible and occurs readily at 25°C with a greater release of free energy. The hypothetical synthesis of glycine from methane, ammonia, and oxygen [reaction (4)] also would give rise to a large release of free energy, but this reaction does not, in fact, take place. This simple comparison shows us that of four conceivable routes for making glycine, reaction (1) is excluded on thermodynamic grounds, and reaction (4) is excluded for kinetic reasons. This leaves reactions (2) and (3) as possible routes for glycine synthesis.

TABLE 1

Free Energies Involved in Different Routes to Glycine

	Starting materials	Reaction	Free energy of formation, $\Delta G°$
(1)	Methane and ammonia	$2CH_4(g) + NH_3(g) + 2H_2O(1) \rightleftarrows H_2N\text{—}CH_2\text{—}COOH(aq) + 5H_2(g)$	+53 kcal/mol
(2)	Carbon monoxide, hydrogen, and ammonia	$2CO(g) + NH_3(g) + H_2(g) \rightleftarrows H_2N\text{—}CH_2\text{—}COOH(aq)$	−21 kcal/mol
(3)	Formaldehyde, ammonia, and hydrogen cyanide	$H_2CO(g) + HCN(g) + NH_3(g) + H_2O(1) \rightleftarrows H_2N\text{—}CH_2\text{—}COOH(aq) + NH_3(g)$	−35 kcal/mol
(4)	Methane, ammonia, and oxygen	$2CH_4(g) + NH_3(g) + \frac{5}{2} O_2(g) \rightleftarrows H_2N\text{—}CH_2\text{—}COOH(aq) + 3H_2O(1)$	−232 kcal/mol

Left out of these considerations is the possibility that a third-party catalyst, which does not influence the thermodynamics, could have provided a kinetically feasible pathway where one did not exist before. The possibilities for catalysis are so numerous that it is virtually impossible to rule out a reaction that is thermodynamically favorable.

If the thermodynamics for a reaction is unfavorable, an alternative pathway for producing the desired reaction product or products must be sought. Sometimes only minor alterations in the pathway or reaction conditions are all that is needed. For example, in some reactions merely changing the concentrations of one or more of the reactants or products can create a thermodynamically favorable situation. In many important prebiotic reactions, water is one of the reaction products, and in such cases a reaction that at first sight seems thermodynamically unfavorable can be made favorable simply by removing the water from the system by gradual evaporation. For example, peptide bond formation between two amino acids involves the removal of a water molecule (see Chapter 6, Fig. 5). In aqueous solution, peptide bond formation is thermodynamically unfavorable, but if a solution containing amino acids is evaporated, peptide bond formation becomes thermodynamically favorable. The thermodynamics of a reaction also can be appreciably altered if one of the reacting species is susceptible to absorption of light. The resulting high-energy intermediate might make an otherwise unfavorable reaction highly favorable. Indeed, this is exactly what happens in photosynthesis.

Most Energy for Driving Chemical Reactions Originates from the Sun

The efficiency of most energy sources for living organisms can be measured in terms of their capacity to produce adenosine triphosphate (ATP), the main source of chemical energy directly utilizable to drive biochemical reactions. In prebiotic systems it is unlikely that ATP was available, so we must think in somewhat different terms. In spite

TABLE 2

Present Sources of Energy Averaged Over Earth[a]

Source	Energy (cal/cm^{-2}/year)
Total radiation from the Sun	260,000
Ultraviolet (UV) light	
< 3000 Å	3,400
< 2500 Å	563
< 2000 Å	41
< 1500 Å	1.7
Electric discharges	4
Cosmic rays	0.0015
Radioactivity (to 1.0-km depth)	0.8
Volcanoes	0.13
Shock waves	1.1

[a]The figure for electric discharges includes 0.9 cal/cm^{-2}/year from lightning and about 3 cal/cm^{-2}/year from corona discharges from pointed objects (from Schonland, B. *Atmospheric Electricity*, London: Methuen, 1953, pp. 42, 63). The value for radioactivity 4×10^9 years ago was 28 cal/cm^{-2}/year (Reprinted with permission from *The Earth as a Planet*, E. Bullard, G. P. Kuiper (ed.). Chicago: University of Chicago Press, © 1954, p. 110). The volcanic energy is based on the emission of 1 km^3 of lava per year at 1000°C (Cp $= 0.25$ cal/gm, $\rho = 3.0$ cal/gm). The energy from shock waves includes 0.1 cal/cm^{-2}/year from meteorites and meteors and 1 cal/cm^{-2}/year from the shock or pressure wave of a lightning bolt (Reprinted from Bar-Nun, A. *et al.*, *Science* 168:470, 1970 with permission from *Science* © 1976 American Association for the Advancement of Science).

of this the same sources of energy that are used to drive the synthesis of ATP must be considered as potential sources of energy for prebiotic reactions leading to the origin of life. These sources are sunlight and chemical compounds. In addition, we must not neglect other, more exotic sources of energy; these include electric discharges, cosmic rays, radioactivity, volcanoes, steam vents, and hydrothermal vents.

The largest source of energy available for organic synthesis on primitive Earth was direct sunlight. This energy comprises a mixture of wavelengths that have different potentials. Ultraviolet (UV) light comprises wavelengths that can be absorbed by most organic compounds. When this light is absorbed, electrons within the organic molecules are catapulted to higher energy levels so that the molecules become more reactive chemically. Some of the reactions resulting from UV light exposure may be useful, while others may be destructive.

As one goes to longer wavelengths, the list of molecules (inorganic and organic) that can absorb light becomes shorter but the total amount of energy available increases, because the Sun emits more radiation at these higher wavelengths (Table 2).

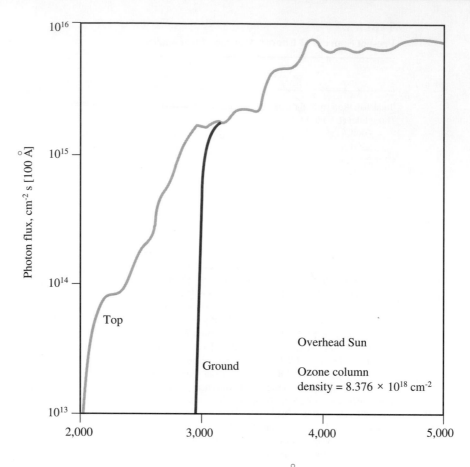

FIGURE 1   Flux of solar radiation at the top of the atmosphere and at the ground. The loss of radiation at the ground level is primarily due to the ozone layer. (From Walker, J. C. G., *Evolution of the Atmosphere*, New York: Macmillan, 1977, p. 57. Reprinted by permission of Prentice-Hall, Inc., Upper Saddle River, NJ.)

Most of the low-wavelength UV (<3000 Å) does not impinge on Earth's surface at the present time because of the UV-absorbing ozone shield in the stratosphere (Fig. 1). This shield was not present on primitive Earth, when there was little atmospheric oxygen; it was formed later by irradiation of the much more abundant oxygen in the atmosphere.

Much of the energy of the sunlight supplied by longer wavelengths could have been useful in supplying heat that would have accelerated chemical reactions, because most reactions go faster at higher temperatures. In addition, many important bioorganic reactions that involve dehydration could have been driven by the simple removal of water resulting from a heating process. Finally, sunlight has been critical to maintaining a favorable temperature on the planet, for it is sunlight that has kept the average temperature above the water-freezing point over most of the course of geologic time.

The Sun's rays have provided Earth with a tremendous surplus of energy. Given this abundance of energy, it is reasonable to suppose that chemical systems that could take advantage of solar energy were likely to be favored. Indeed, one of the major reasons why biological systems have flourished is that they have developed the means to convert the Sun's rays into useful chemical energy by the process of photosynthesis (see Chapter 23). There is little doubt that prebiotic systems also took advantage of this abundant source of energy.

Among the remaining nonchemical sources of energy we find electric discharges that arise from lightning and corona discharges. Historically, electric discharges have been used in the laboratory to produce a number of simple organic compounds, including amino acids. We discuss some of these experiments in Chapter 15. Given the amounts of energy available from electric discharges, however, it seems unlikely that they were as important as solar sources for supplying the energy for important prebiotic reactions.

Energy Was Also Obtainable from Chemical Sources

In Earth's early history the oceans and the crust were richly endowed with reduced ferrous iron, Fe^{2+}, which could have supplied the chemical energy for the synthesis of a wide range of organic compounds. Ferrous ion and sulfide, in the form of iron sulfide (FeS) and hydrogen sulfide (H_2S), have been shown to provide an efficient means of producing reduced organic compounds under mild conditions (see Chapter 18). One advantage of such inorganic sources of energy is that they could be useful in sheltered subterranean environments where direct exposure to the Sun or electric discharges is not possible. In cases where exposure to the Sun occurs, the potential of ferrous ion could have been augmented by the absorption of UV, which would have increased its reducing power. We may think of such systems as a primitive form of photosynthesis.

High-energy phosphate compounds are omnipresent in biological systems. Simple polyphosphates are used as a form of energy storage in many microorganisms. Such compounds could have been produced in the prebiotic world by mere heating of inorganic orthophosphates to dryness.

Organic heterotrophic metabolism may have had its counterpart in the prebiotic world. A broad range of organic compounds could have been produced in the prebiotic world as a result of reactions driven by sunlight, iron sulfides, or other earthly sources of energy. Some of these organic compounds could have served as direct precursors, for the first living systems. Others could have supplied an organochemical source of energy for driving important prebiotic reactions, in much the same way as heterotrophs now use the catabolism of organic compounds to fulfill their energy needs.

Observations of Extraterrestrial Chemistry Provide Clues about the Chemistry of Prebiotic Earth

If our goal is to understand what Earth was like over 4 billion years ago, then why be concerned about the chemistry of the cosmos? The reason is that living organisms have had such a profound effect on Earth's environment that there is no hope of finding areas on Earth's surface that have not been altered by them. On the other hand, if we explore the cosmos, we may find that the types of molecules that occurred on Earth around the time of its formation are still to be found in outer space or in other celestial bodies. Spectroscopic analyses of the chemistry of the molecular species present in interstellar gases in our galaxy indicate an abundance of carbon compounds, such as CO, CH_4, CH_3OH, HCN, and $HC{\equiv}CCN$, together with water and ammonia (Table 3). Many of these compounds have good potential uses in chemical processes leading to the origin of life. Although most of these carbon and nitrogen compounds were probably not transferred intact to the atmosphere of primitive Earth, their presence in the solar nebula increases the likelihood that they were also available on primitive Earth.

Meteorites also provide clues concerning the chemistry of primitive Earth. The largest influx of meteorites came during the first 0.5 billion years after the accretion process. Meteorites still are striking Earth today, and those containing carbon possess organic molecules like those present in biological molecules. Carbon compounds present in carbonaceous meteorites include alcohols, aldehydes, amines, amino acids, hydrocarbons, ketones, purines, pyrimidines, and other heterocyclic compounds. These compounds probably resulted from photochemical reactions or other reactions of simple precursors on the surfaces of interstellar dust particles and from the subsequent thermal reaction of these organics at the high temperatures and pressures of the contracting solar nebula.

TABLE 3
Some Interstellar Molecules

HCN	CH_3OH
H_2O	$HCONH_2$
H_2S	CH_3CN
SO_2	CH_3CHO
$H_2C{=}O$	$CH_2{=}CHCN$
$HN{=}C{=}O$	CH_3NH_2
$H_2C{=}S$	$CH_3C{\equiv}CH$
NH_3	$HC{\equiv}CC{\equiv}CCN$
$CH_2{=}NH$	HCO_2CH_3
$HC{\equiv}CCN$	CH_3CH_2OH
HCO_2H	CH_3OCH_3
NH_2CN	CH_3CH_2CN
CH_4	$HC{\equiv}CC{\equiv}CC{\equiv}CCN$
	$HC{\equiv}CC{\equiv}CC{\equiv}CC{\equiv}CCN$

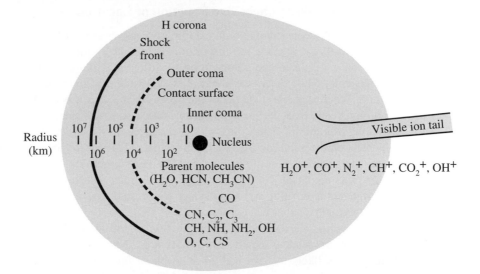

FIGURE 2 A model for the chemical processes that occur when a comet approaches the Sun.

Some meteorites also contain clays; because clays are believed to require water for their formation (see Chapter 20), these meteorites may be chips from much larger bodies such as asteroids or even other planets.

Comets probably were formed in the colder regions of the solar nebula at the time of the birth of the Solar System. They have highly eccentric orbits that often extend beyond the outer planets. In the past many comets must have collided with Earth, thereby depositing their contents on the planet. The volatile molecular species that are contained in comets were probably picked up in the colder regions of the nebula at points far more distant from the Sun than our Earth (Fig. 2).

Starting Materials for Synthesis Were Available in the Atmosphere and the Lithosphere

Four of the six most common elements found in biological systems, carbon, hydrogen, oxygen, and nitrogen, make up the bulk of Earth's atmosphere. This fact, together with the obvious accessibility of the atmosphere to the Sun's rays, has led to the suggestion that the major starting materials for prebiotic synthesis came from the atmosphere. Sulfur also may have been present to some extent in the atmosphere (e.g., as hydrogen sulfide), but most sulfur probably existed in the form of metal sulfides present in Earth's crust. The phosphorus needed for prebiotic synthesis also was present in Earth's crust probably mostly in the form of mineral phosphates that have slowly but surely released their phosphates to the oceans over eons of time.

Most Favorable Conditions Were a Sensitive Function of Location

We have spent considerable time extolling the advantages of Earth compared with the other planets, as a place for the origin of life. On Earth itself there are enormous variations in climate and composition in different regions. Some parts of Earth are always cold, while others are always hot. Most regions go through seasonal hot–cold cycles. Whereas the average pH of the oceans probably has been close to 8 for the past 4 billion years, there are local variations that could result in considerably more acid or alkaline conditions. There are wide variations in the types of clays and mineral deposits on the solid portions of Earth's crust. There are regions that are always wet, and others that are almost always dry. Most reactions would not take place in an area where it is perpetually dry. At the same time, a region such as a tidal flat, where there are frequent alternating wet and dry periods, would favor one class of reactions when it was wet and another when it was dry.

Even when we consider the composition of the atmosphere, it seems likely that there were tremendous local variations. An area densely populated by volcanoes belching gases is likely to have had an atmosphere far richer in the carbon and nitrogen compounds that serve as starting materials for key prebiotic compounds.

Only the atmosphere and the crust are exposed to direct sunlight. Thus for reactions where direct sunlight is required, we are limited to the atmosphere, the surface of the land masses, and the topmost layers of the oceans. It seems likely that the formation of very simple organic compounds took place in the atmosphere. These compounds would eventually have settled in freshwaters, in oceans, or on land masses. There they would have undergone further reactions and some of these compounds would have become bound to the clays or other minerals that abound on Earth's surfaces and in its oceans.

It is tempting to try to think of one location where all the reactions essential to the origin of life occurred. However, there are virtues to having different locations for the synthesis of different types of compounds. For example, iron sulfide deposits are likely to supply the right type of chemistry for the synthesis of hydrocarbons and other highly reduced organic compounds, but clays in tidal regions might provide a superior location for the synthesis of nucleotides or polynucleotides. As we approach the question of the chemical reactions leading to the origin of life, we must be flexible in considering the location or locations where different reactions might have flourished.

Concentration Mechanisms Were Essential for Key Reactants

It is not enough that a particular process exists for the synthesis of a compound crucial to the overall prebiotic pathway. There must be enough of the compound produced for it to be useful. Because synthetic mechanisms are limited in how much material of any sort can be produced in a given time, mechanisms for concentrating reactants and products would be very useful and probably essential in some cases. This is particularly true for bimolecular or multimolecular reactions. We encounter two such cases in the following chapters: (1) the purine pathway in its early steps requires successive condensations of several molecules of hydrogen cyanide to produce key intermediates (see Chapter 13); and (2) the formation of ribose from formaldehyde requires successive condensations of formaldehyde (see Chapter 12).

There are several possible mechanisms for concentrating key intermediates of these types. Selective binding to a clay or nonclay mineral is one mechanisms. Concentration by evaporation or by freezing from aqueous solution provides another possible mechanism. In the case of hydrogen cyanide, for example, freezing to $-21°C$ permits the formation of very concentrated hydrogen cyanide solutions (up to 75%) from dilute solutions because the water freezes first. As the water freezes, the solution becomes more and more concentrated in hydrogen cyanide.

Chemical Stability Is Most Important for Compounds That Are Not Rapidly Utilized

In biochemical systems, pathways often are extended multistep processes and many of the intermediates often are quite unstable. The problem of stability is minimized by the rapid transformation of one intermediate into the next before it has time to decompose or react in a nonproductive way. In prebiotic pathways such a tight coupling of reactions seems unlikely in most cases. Indeed, in many cases it is easy to imagine long pauses (hours, days, or even years) before intermediates are converted into final products. For these reasons stability of compounds takes on added importance.

Let us consider the stability of some well-known biochemical intermediates. The most stable amino acids are the aliphatic amino acids—glycine, alanine, isoleucine, leucine, and valine. They decompose by decarboxylation as shown:

$$R-CH-COO^- \rightarrow R-CH_2-NH_2 + CO_2.$$
$$\underset{NH_3{}^+}{|}$$

The rate of decarboxylation has been studied in detail for alanine. This amino acid has a half-life of 3×10^9 years at $25°C$ and 10 years at $150°C$. From this information we can say that alanine is stable on a prebiotic timescale as long as the temperature is not too high. Of course, these numbers do not take into account the problem of stability in the presence of other compounds that might have a destabilizing influence.

Most other amino acids commonly found in proteins have alternative mechanisms of decomposition in addition to decarboxylation. These generally lead to more rapid decomposition. For example, glutamine decomposes rapidly to give pyroglutamic acid; it is therefore doubtful that significant concentrations of glutamine could have accumulated in primitive oceans. On the other hand, in peptide linkages glutamine is considerably more stable. Thus if glutamine occurred in prebiotic polypeptides, it must have been incorporated fairly readily into peptide linkages. Other amino acids that are particularly unstable include tryptophan, asparagine, and arginine.

Fatty acids are among the most stable of organic compounds. This could be very important, because the slow accumulation of considerable amounts of fatty acids may have preceded the formation of membranes in the prebiotic milieu. These compounds could have accumulated for tens of millions of years before they were utilized in prebiotic systems.

Sugars are very unstable. No doubt they must have been utilized shortly after synthesis, and correspondingly there must have been synthetic pathways so that they could be readily supplied when needed.

The half-lives of the purines and pyrimidines have not been resolved but it is clear that these bases are considerably more stable than neutral ribose is.

As we have noted, it is fortunate that there was very little gaseous oxygen in prebiotic times, because almost all organic compounds are unstable in the presence of molecular oxygen. In fact, it seems unlikely that life could have originated on any planet that started with an oxygen-rich atmosphere.

Compounds that exist in two or more *enantiomeric* forms (see Box 6A) pose an additional problem as far as stability is concerned—the problem of utilizing the compound before it changes from one enantiomer to the other (*racemizes*). This problem could be serious for amino acids and sugars, because they racemize with half-lives in the range of hundreds to thousands of years. The problem does not exist for purines and pyrimdines, because they do not give rise to enantiomers.

EVOLUTIONARY ASPECTS OF THE ORIGIN OF LIFE

Most biologists tend to equate evolution with Darwinian evolution, that is, the changes that have taken place in various life forms over time through natural selection. However, evolution can be defined more broadly as any change in a certain direction. According to this definition the Universe has been evolving at least since the Big Bang. In Part I we have seen that after the Big Bang the first elements to appear were hydrogen and helium. Vast clouds of these two elements condensed to form stars in which other elements formed. Planets were formed in our Solar System, one of which was ideally suited for the origin of life. In this part of our text we are concerned with the chemical changes that took place between the formation of Earth and the origin of life. In this pursuit we search for descriptions of how prebiotic systems might have evolved and by what mechanisms.

Kinetic Factors Were Probably Decisive in the Evolution of Chemical Reactions Leading to the Living State

At first thought it might appear that the evolution of chemical reactions to the highly organized state found in living cells could never have occurred because it would have been a violation of the second law of thermodynamics, which states that a closed system spontaneously goes in the direction of increasing entropy (i.e., to a more disordered state). However, Earth is not a closed system; it interacts strongly with other parts of the Universe, particularly the Sun, which constantly showers it with energy-filled light rays. The Sun guarantees an excess of energy on Earth's surface, a supply that is more than adequate to overcome the negative entropy problem.

If the chemical reactions on Earth were not limited by the available energy, what was decisive in the evolution of chemical reactions to living systems? There is only one possible answer. Kinetic factors must have been decisive in this evolution.

The kinetic efficiency of a highly organized biological system for processing matter far exceeds that of almost any other chemical system. Just consider the chemical efficiency of a photosynthetic autotroph, the cyanobacterium, which converts the Sun's rays into chemical energy and requires only carbon dioxide, water, nitrogen, and trace elements for growth. Contrast the chemical efficiency of this microorganism with that of a ground homogenate containing precisely the same chemicals. It is a foregone conclusion that the organized system is much more likely to flourish. To explain the evolution of living from nonliving systems, it is merely necessary to assume that at each step in the evolution of chemical systems to living systems, the same principle applies. Selection pressure favors any chemical system that can process matter more rapidly and make more of its kind.

First Living Systems Must Have Been the Product of a Multistep Process

From the comparison just given, it should be obvious that a highly organized cyanobacterium would have tremendous advantages over a disorganized chemical system. It is far more difficult to see how such an intricate system could have arisen from nonliving substances. A great many seemingly improbable events would have had to take place to create the living system. Each step in the direction of making a living cell seems improbable, because there are so many possible reactions that a disorganized chemical system could undergo. Could all these steps have taken place simultaneously? This is most unlikely, because the probability of the occurrence of many reactions—each with a low probability of occurring—would be the product of all the individual probabilities.

We are driven to the conclusion that the origin of life must have been a multistep process in which each step must have resulted in a chemical system with some advantage over the one that preceded it. In gross terms we may divide this complex evolutionary process into three overlapping phases (Fig. 3):

1. Initially, there was a phase during which the chemicals needed for the synthesis of instructional polymers were assembled.

2. Following the synthesis of the first instructional polymers, possibly resembling nucleic acids, modifications began to take place. Changes would have been favored that led to more efficient polymers for replication and the catalytic activities required to make the precursors for polymer synthesis. At an early stage in their evolution such polymers should display the two key properties of living systems: the ability to

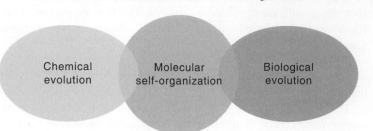

FIGURE 3   The origin of life probably occurred in three overlapping phases. In phase I, chemical evolution, involved the noninstructed synthesis of biological macromolecules. In phase II, biological macromolecules self-organized into systems that could reproduce. In phase III, organisms evolved from simple genetic systems to complex multicellular organisms. The arrow pointing from left to right emphasizes the unidirectional nature of the overall process.

replicate and the ability to evolve into more efficient polymers. At an advanced stage in their evolution, the polymers with a future would develop the capacity to direct the synthesis and organization of other types of polymers, and of membrane phospholipids: only then would the first living cells emerge.

3. In the third phase, the divergent processes of biological evolution would commence, leading to further developments of single-celled organisms, including the differentiation of some of them into complex multicellular organisms.

Overlaps between these preceding three phases of evolution reflect the difficulty in determining exactly when the transition from one form of evolution to another was initiated. The arrow in Fig. 3 is intended to indicate that there is a unidirectional character to this evolutionary process.

Instructional Polymers Carry the Information for Their Own Synthesis and the Means to Undergo Further Evolution

It is much easier to visualize how evolutionary steps proceeded once there were instructional polymers of nucleic acids. A nucleic acid polymer carries information in the form of a linear sequence of bases. Infrequent changes in sequence would be expected to occur during replication. Such changes would have a high probability of propagation by the template mode of replication. If a change resulted in a more efficient system, then the altered nucleic acid would be expected to replicate more effectively than its unaltered predecessor would. It might even become the dominant nucleic acid molecule in the population. Although we are accustomed to thinking about nucleic acids as the information-bearing component of chromosomes in whole cells, the arguments about selection and survival also should apply to nucleic acids in a favorable prebiotic situation.

We Are Making Headway in Understanding the Chemical Evolution That Preceded the Origin of Nucleic Acids

The main difficulty with the notion that nucleic acids were the first information-bearing polymers is that nucleic acid molecules are delicate, and the building blocks that are required for nucleic acid synthesis are quite complex. The primitive prebiotic environment contained a broad array of organic compounds, only a few of which would have been useful to the origin of nucleic acids. What sort of evolutionary processes leading to nucleic acids can be imagined in such an environment? If a useful step in chemical evolution were achieved, how would it become stabilized so that it would be an integral part of the developing system? There are no sure answers to such questions.

The outline of a possible route to the first nucleic acids is sketched in Fig. 4. In this diagram the initial reactions leading to RNA-like molecules are pictured as taking place in the atmosphere. This stage is followed by the formation of purines and sugars on Earth's surface and finally by the formation of oligonucleotides and self-replicating polymers. We elaborate on the steps suggested by this figure in Chapters 13 and 14.

The complexity of nucleic acids has led many evolutionists to doubt that nucleic acids could have been the first informational polymers to be synthesized. However, except for a proposal that clays may have been the first informational polymers (discussed in Chapter 20) no reasonable alternative proposals have been made.

Most of the doubts about whether nucleic acids could have evolved directly from simpler chemical systems have arisen because we know so little about the evolution of simpler chemical systems. If we attempt to build model systems for studying chemical evolution prior to the time when information-bearing templates were part of the developing prebiotic system, we immediately encounter three problems: (1) very few experimental systems have been developed for such studies; (2) a complex chemical system has no obvious repository for the storage of useful chemical information; (3) the rules governing the evolution of purely chemical systems are not known.

A model for testing that I find attractive is one in which clay surfaces provide a location for chemical evolution. For example, imagine an environment that has an abundance of the precursors necessary for the synthesis of two compounds, A and B (Fig. 5). Now consider three examples of the relationship between A and B and their affinity for the clay. First, assume that A and B both bind strongly to the clay, and that A catalyzes the synthesis of B, while B catalyzes the synthesis of A. In this situation it seems likely that there will be a rapid buildup of both A and B on the clay surface as long as there are ample precursors for their synthesis. Alternatively, assume that A catalyzes B synthesis as before, but that B does not catalyze A synthesis. In this event there will be a slower buildup of A and B on the clay and the concentration of B will greatly exceed that of A. Finally, assume that B neither catalyzes A synthesis nor does it bind strongly to the clay surface. In this event there will be no buildup of B on the clay surface and only a relatively slow buildup of A. This simple set of hypothetical examples depicts how chemical evolution on

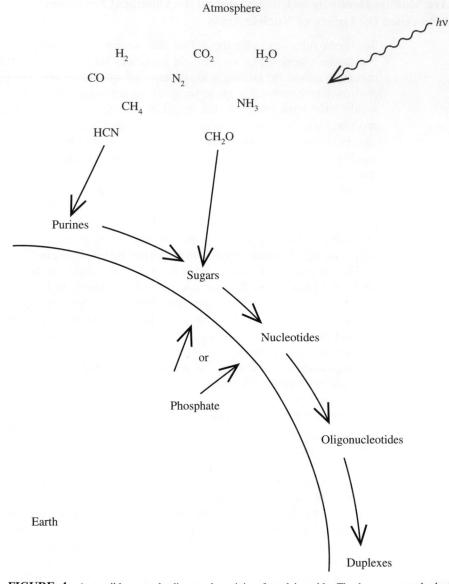

FIGURE 4   A possible route leading to the origin of nucleic acids. The key compounds, hydrogen cyanide (HCN) and formaldehyde (CH_2O), were made in the atmosphere in reactions involving simpler gaseous compounds and sunlight. The HCN and CH_2O rained down to Earth's surface where they served as precursors of purines and ribose, respectively. Inorganic phosphate became linked to the ribose either before or after nucleoside formation. The mononucleotides became activated prior to oligonucleotide formation. The first oligonucleotides served as templates for the formation of complementary nucleotides.

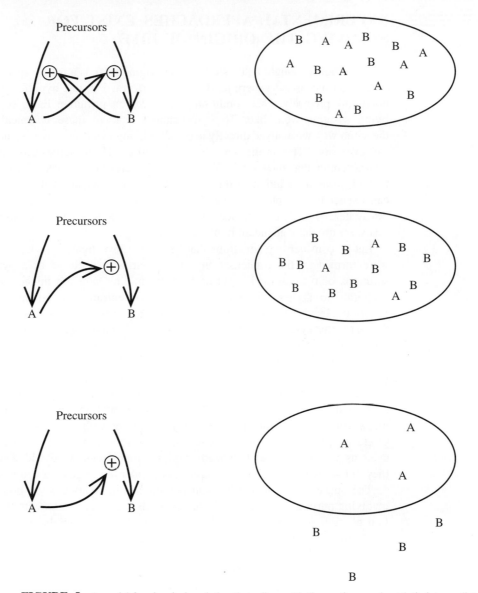

FIGURE 5   A model for chemical evolution that relies on binding surfaces and catalytic intermediates.

surfaces might depend on the affinity of molecules for the surface as well as their catalytic properties. It would not be difficult to extend this type of hypothetical example to more complex situations. However, real progress in this direction depends on direct experimental demonstrations that such systems can be constructed and shown to be effective.

EXPERIMENTAL APPROACHES EXIST FOR STUDYING THE ORIGIN OF LIFE

It is commonly thought that a study of the origin of life is a purely speculative endeavour, where no useful experiments can be performed to test hypotheses. True, we do not know precisely what conditions were like on primitive Earth or what the first living systems were like. Thus we cannot analyze these chemical processes in the same way we analyze directly accessible living systems. It is very unlikely that we can ever say, "This is the way that life originated." To some, this realization is so disheartening that they say, "Why bother?" Nevertheless even if we cannot expect to find a unique solution to the puzzle of how life began, in the next few years we can expect to find plausible answers. A plausible beginning is one that starts from raw materials likely to have been present in prebiotic times and uses sources of energy that were probably abundant in prebiotic times.

Let us consider two situations that have been investigated. To test that nucleosides were formed by the condensation of a sugar with a purine or a pyrimidine base, solutions of D-ribose and a purine or a pyrimidine base are mixed and allowed to evaporate to dryness. Simple evaporation concentrates the potential reactants and makes nucleoside formation thermodynamically favorable by removal of water. The finding is that certain purine nucleosides can be formed in this way but that pyrimidine nucleosides cannot.

A more complex experimental system has been used by Orgel to investigate prebiotic systems for polyribonucleotide synthesis. Starting from polycytidylic acid and activated monomers of guanine. Orgel has been able to synthesize polyguanylic acids with nothing more than a divalent cation catalyst. This result is considered a major triumph even though it is not yet clear how the reactants could have been produced prebiotically. It is interesting to note that when polyguanylic acid was used as a template and activated cytidylic monomers were added to the system, they did not react to form polycytidylic acid under the same incubation conditions.

The successes in nucleoside and polynucleotide synthesis are both for reactions involving purines. What are we to say about the negative results with the pyrimidines? Is it possible that by modifying conditions or starting materials, we would be able to synthesize pyrimidine nucleosides and polynucleotides containing pyrimidines? Alternatively, is it possible that the first prebiotic systems did not utilize pyrimidines, but only purines? What criteria do we apply to decide between these alternatives? Questions of this sort are be explored in succeeding chapters.

SUMMARY

In this chapter we have dealt with some preliminary issues that pertain to research into the origin of life, we have discussed basic constraints on the reactions that might have been involved, and we have identified possible stages in the evolution of living systems from nonliving ones.

1. Living systems as we know them could have originated elsewhere and been transported to Earth from another planet or another Solar System. However, we note two facts in favor of the idea that they originated here on Earth. First, all records indicate that there has been a natural progression from the simplest microorganisms to the most complex multicellular organisms, which would be consistent with life having originated on Earth. Second, the conditions on Earth around the time of the appearance of living forms seem to be quite compatible with the notion that life originated on Earth.

2. The types of organic compounds that would have been required for the origin of life on Earth are found elsewhere in the cosmos. Therefore, it is likely that the formation of such compounds also could have taken place on Earth in its early stages of development.

3. The chemical composition of primitive Earth and the physical conditions on Earth around the time of the origin of life are those that would have been most satisfactory for the origin of life.

4. The necessary energy for driving chemical reactions would have resulted directly or indirectly from sunlight. This puts a premium on chemical reactions that can be driven by or harness sunlight and convert it into chemical energy.

5. Given this abundant source of energy, the chemical systems that would prosper the most would be those with favorable kinetics. Thus both thermodynamic and kinetic factors must have been important in determining which reactions would be favored.

6. Many of the simplest organic compounds required for the origin of life could have been formed in the atmosphere, where simple gaseous compounds to serve as precursors were available and where there was ample sunlight and electric discharge to provide the energy necessary to drive the chemical reactions.

7. Compounds made in the atmosphere would have been expected to rain down into the oceans. The concentrations of the important chemical compounds in the oceans would have been too low to drive further reactions. This means that there must have been a means of concentrating these organic compounds. Several mechanisms exist for concentrating organic compounds in the oceans (e.g., eutectic freezing, evaporation, or absorption at a solid–liquid interface such as a clay).

8. The first living organisms must have been the product of an extensive multistep process. At each stage in this multistep process, developments must have proceeded in a direction that presented a kinetically favorable situation for synthesis of a mixture of products that somehow was sufficiently stable to presist until the next step was taken.

9. Once instructional polymers had been made, it is relatively easy to see how further evolution could have taken place, because the instructional polymer has a means of self-reproduction and a means for recording evolutionarily significant information.

10. The biggest problem in the origin of life is determining the evolutionary events that led up to the first instructional replicating polymers. Because the only polymers of this type that we know about are nucleic acids, the first question to ask is were there simpler instructional polymers that preceded nucleic acids? Clays have been considered a candidate for such polymers.

11. Chemical stability is an important consideration in the origin of life, because many of the compounds required for this process might have had to wait until the time was appropriate for their being needed.

12. Although there is a lot of speculation involved in the science of the origin of life, there are important experimental approaches to evaluating various possibilities. Once an idea has been proposed as to how a particular bioorganic molecule was first made, experiments can be set up in the laboratory to see if a feasible route can be demonstrated for the synthesis of the compound in question.

Problems

1. Most "lifers" think that nucleic acids came before proteins. Is it possible that proteins came first, or is it possible that nucleic acids and proteins coevolved? Explain why or why not.

2. If Earth had an atmosphere rich in oxygen 4 billion years ago, would this have interfered with the origin of life? See Problem 5 in Chapter 5.

3. Carbon and silicon are in the same column in the Periodic table and they both can adopt the tetravalent state as is the case with dioxides. Because silicon is so much more abundant than carbon, why do you suppose that more use was not made of silicon in biosystems?

4. In addition to water, the living cell contains four types of large molecules: nucleic acids, proteins, carbohydrates, and lipids. Can you imagine life in any form without one of these? Which type of molecule could you eliminate and what type of accommodations would have to be made?

5. The three polymers, nucleic acids, proteins, and carbohydrates, all are composed of monomers that in each case link to one another to form identical linkages. Inspection of the structures of the monomers and the polymers indicates that they differ by the loss of a water molecule for each polymeric linkage that is made. What significance could be attached to this observation as it concerns the origin of life?

6. Imagine that there were two types of organisms on Earth, those based on D-sugars and D-amino acids and those based on their enantiomorphs L-sugars and L-amino acids. If these two types of organisms were allowed to mingle, does it seem likely that they would coexist for long? See Problem 3 in Chapter 6.

7. Which do you think came first, anabolism or catabolism? Do you think the living world could have existed without catabolism?

8. Because all living systems are composed of cells encompassed by membranes, does life without membranes seem possible?

9. What roles did iron play in influencing the composition of the atmosphere? See Chapter 5.

10. If a particular chemical reaction has unfavorable thermodynamics, is there any point in searching for a good catalyst to drive the reactions?

11. Why is it dogma that nucleic acids came before proteins?

12. Why did DNA take over the job of keeper of the genetic information from RNA? At what stage in the course of evolution is it likely that this event took place?

13. If someone suggested to you that the first living organisms were composed exclusively of carbohydrates, what would you say? See Problem 4 in Chapter 6.

14. If life had not developed on Earth, in what ways would Earth be different now? See Problem 4 in Chapter 5.

15. If Earth had an atmosphere rich in oxygen 3.8 billion years ago, would this have interfered with the origin of life? See Problem 5 in Chapter 5.

16. Why are volcanoes beneficial?

References

General Books

Bernal, J. D. *The Origins of Life*. London: Weidenfeld & Nicholson, 1967. (This is mainly of historical interest.)

Calvin, M. *Chemical Evolution*. New York: Oxford University Press, 1969. (This also is mainly of historical interest.)

Cairns-Smith, A. G. *Genetic Takeover and the Mineral Origins of Life*. Cambridge: Cambridge University Press, 1982. (This text is written by the champion of the clay theory for the origin of life.)

Deamer, D. W. and Fleischaker, G. R. *Origins of Life*. Boston: Jones & Bartlett, 1994. (This is mostly a collection of papers. Pay special attention to the material on lipids and membranes.)

DeDuve, C. Blueprint for a Cell. In *The Nature and Origins of Life*. Burlington, NC: Neil Patterson Publishers/Carolina Biological Supply, 1991. (Focus on Chapter 7 where the author presents his theories on harnessing energy.)

Gesteland, R. F. and Atkins, J. F. (Eds.). *The RNA World*. Cold Spring Harbor, NY: Cold Spring Harbor Laboratory Press, 1993. (This gives a broad perspective on the potential of RNA but is somewhat dated.)

Miller, S. L. and Orgel, L. E. *The Origins of Life*. Englewood Cliffs, NJ: Prentice-Hall, 1973. (This is the classic predecessor of this text.)

Schopf, J. W. (Ed.). *Earth's Earliest Biosphere: Its Origin and Evolution*. Princeton, NJ: Princeton University Press, 1983. (Information from many disciplines are brought together in this multiauthor book.)

Theng, B. K. G. *The Chemistry of Clay-Organic Reactions*. New York: John Wiley & Sons, 1974. (This is dated but still the best general text on this subject.)

Journals

Origins Life Evol. Biosphere and *J. Mol. Evol.* (for keeping up with original research); *Science* and *Nature (London)* (for occasional articles on topics of current interest).

Specific

Eigen, M., Gardiner, W., Schuster, P., and Winkler-Oswatitsch, R. The Origin of Genetic Information, *Sci. Am.* 244(4):188–118, 1981.

Eigen, M. and Schuster, P. The Hypercycle: A Principle of Self-Organization, Part C, *Naturwissenschaften* 65:341–369, 1978.

Ferris, J. P. Chemical Markers of Prebiotic Chemistry in Hydrothermal Systems, *Origins Life Evol. Biosphere* 22:109–134, 1992.

Henning, Th. and Salama, F. Carbon in the Universe, *Science* 282:2204, 1998.

Horowitz, N. H. On the evolution of Biochemical Synthesis, *Proc. Natl. Acad. Sci. USA* 31:153–158, 1945. (This is a must reading.)

Joyce, G. F. RNA evolution and the origin of life, *Nature (London)* 338:217–224, 1989. (This review gives an excellent overview of the subject up to 1989.)

Kerr, R. A. Requiem for Life on Mars? Support for Microbes Fades, *Science* 282:1398, 1998.

Orgel, L. E. Molecular Replication, *Nature (London)* 358:203–209, 1992. (This presents the view that simple replicating molecules are likely to have played a critical role in the origins of life.)

Orgel, L. E. *The Origin of Life—How Long Did It Take? Origins Life Evol. Biosphere* 28:91–96, 1998.

Oro, J., Mills, T., and Lazcano, A. Comets and the Formation of Biochemical Compounds on the Primitive Earth—a Review. *Origins Life Evol. Biosphere* 21:267–277, 1992.

Siskin, M. and Kartritzky, A. R. Reactivity of Organic Compounds in Hot water: Geochemical and Technological Implications, *Science* 254:231–238, 1991.

CHAPTER 11

Biochemical Pathways Involving Carbohydrates

One of our main goals in Part III of this text is to emphasize the similarities between reactions in the contemporary biochemical world and related reactions in the prebiotic world. This is particularly hard to do with carbohydrates because they have assumed so many roles in addition to those few that were crucial in prebiotic times. In spite of this problem, we think it is worthwhile to briefly survey the role of carbohydrates in the biochemical world and to point out the similarities and differences with the prebiotic world as we move along.

In this chapter we discuss how hexoses are broken down into smaller molecules. The reverse process of how hexoses are synthesized from two or three carbon compounds also is considered. Next we consider how pentoses are derived from hexoses. Finally, we consider the process whereby the three carbon units are broken down to carbon dioxide and water. This last process is considered in greater detail in Chapter 22.

BREAKDOWN OF SUGARS (GLYCOLYSIS) FOLLOWS A LINEAR PATHWAY WITH MANY BRANCHPOINTS

Glycolysis is the process by which glucose or other hexoses are converted into the three-carbon compound pyruvate. Glycolysis is the most ubiquitous pathway in all energy metabolism, occurring in almost every living cell. It is regarded as a primitive process because it probably arose early in biological history long before there was a significant amount of molecular oxygen in the atmosphere.

The glycolytic pathway is multifunctional. Thus it provides the cell with energy [adenosine triphosphate (ATP)] from glucose catabolism and also can serve an anabolic function by yielding C-3 precursors for the synthesis of amino acids, fatty acids, and cholesterol. We focus on the catabolic energy-producing function.

The 10-step glycolytic pathway is illustrated in Fig. 1. Glycolysis involves the splitting of the C-6 hexose into two molecules. The splitting occurs in step 4 (starting from glucose). At this point, a six-carbon sugar is cleaved to yield 2 three-carbon compounds, one of which, glyceraldehyde-3-phosphate, is the only oxidizable molecule in the entire pathway. After the cleavage in step 4, two successive ATP-generating steps occur: one at step 7 and the other at step 10.

One oxidation step in the pathway, step 5, results in the conversion of nicotinamide adenine dinucleotide (NAD^+) to reduced NAD (NADH). This NAD^+ must be regenerated if glycolysis is to continue. Under anaerobic conditions some reaction must occur that permits the regeneration of NAD^+ if the pathway is to remain operative.

Except for glycerate-1,3-bisphosphate all the intermediates in the pathway are pictured as belonging to one of three *metabolic pools*. Within each metabolic pool the intermediates are readily interconvertible and usually present in relative concentrations close to their equilibrium values. Between the pools, the concentrations of the intermediates can be very different because of the lack of rapid interconversion and also because of equilibrium values often being very large or very small.

The 10 steps from glucose to pyruvate are reviewed in Table 1. It can be seen that overall, glycolysis involves a significant negative free energy, about -25 kcal/mol (104 kJ) of glucose consumed. In calculating the free energy drop at each step, we do not get a realistic picture from the standard free energies. This is because the standard free energies are all based on concentrations of 1 *m*. If we use more realistic concentrations, we can see that a sizable drop in free energy occurs at only three steps. These steps are all outside the metabolic pools.

The glycolytic pathway resembles a series of "lakes" (metabolic pools) connected by short "rivers" (the reactions between the pools). This pattern is reflected in the ways that functional metabolic relationships have evolved. Reactions involving ATP and adenosine diphosphate (ADP) occur in the interconnecting reactions, or rivers. Clearly, this is where they are expected, because an ATP-linked reaction within a metabolic pool makes no more sense than a hydroelectric power plant in the middle of a lake.

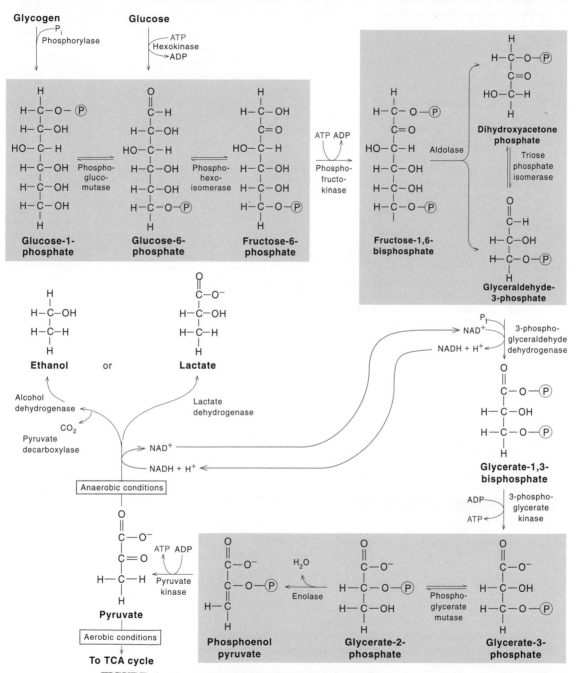

FIGURE 1 The glycolytic pathway from glucose to pyruvate, indicating two anaerobic options (ethanol or lactate) and one aerobic option (TCA cycle). The long curved arrow indicates how NADH formed in glycolysis could be reoxidized to NAD^+ when pyruvate is converted to an end product of ethanol or lactate under anaerobic conditions.

TABLE 1

Reactions, Enzymes, and Standard Free Energies for Steps in the Glycolytic Pathway

Step	Reaction	Enzyme	$\Delta G^{\circ\prime a,b}$	$\Delta G'^{a,b}$
1	Glucose + ATP $\longrightarrow$ glucose-6-phosphate + ADP + H$^+$	Hexokinase	−4.0	−8.0
2	Glucose-6-phosphate $\rightleftharpoons$ fructose-6-phosphate	Phosphohexose isomerase	+4.0	−0.60
3	Fructose-6-phosphate + ATP $\longrightarrow$ fructose-1,6-bisphosphate + ADP + H$^+$	Phosphofructokinase	−3.4	−5.3
4	Fructose-1,6-bisphosphate $\rightleftharpoons$ dihydroxyacetone phosphate + glyceraldehyde-3-phosphate	Aldolase	+5.7	−0.31
5	Dihydroxyacetone phosphate $\rightleftharpoons$ glyceraldehyde-3-phosphate	Triose phosphate isomerase	+1.8	+0.60
6	Glyceraldehyde-3-phosphate + P$_i$ + NAD$^+$ $\rightleftharpoons$ glycerate-1,3-bisphosphate + NADH + H$^+$	Phosphoglyceraldehyde dehydrogenase	+1.5	−0.41
7	Glycerate-1,3-bisphosphate + ADP $\rightleftharpoons$ glycerate-3-phosphate + ATP	3-Phosphoglycerate kinase	−4.5	+0.31
8	Glycerate-3-phosphate $\rightleftharpoons$ glycerate-2-phosphate	Phosphoglyceromutase	+1.1	+0.19
9	Glycerate-2-phosphate $\rightleftharpoons$ phosphoenolpyruvate + H$_2$O	Enolase	+0.4	−0.79
10	Phosphoenolpyruvate + ADP + H$^+$ $\longrightarrow$ pyruvate + ATP	Pyruvate kinase	−7.5	−4.0

Net reaction:

$C_6H_{12}O_6$ + 2NAD$^+$ + 2ADP + 2P$_i$ $\longrightarrow$ 2C$_3$H$_4$O$_3$ + 2NADH + 2H$^+$ + 2ATP + 2H$_2$O $\qquad$ −27.8[c] −23.0

[a]ΔG values are for the human erythrocyte using approximate values for concentrations.

[b]Values are given in units of kcal/mol; to convert this to kJ, merely multiply the given values by 4.18.

[c]To obtain this sum for the standard free energy of the overall reaction per mole of glucose, the indicated free energies for reactions 6 to 10 must all be doubled. This is because each of these reactions occurs twice for every glucose consumed.

TABLE 2

Some Properties of Reactions Associated with Glycolysis

Reaction	Enzyme	Precursor	Product	Mechanism
1	Phosphorylase	P$_i$	Glucose-1-P	?
2	Phosphoglucomutase	Glucose-1-P	Glucose-6-P	*bis*-Phosphorylated intermediate
3	Hexokinase	ATP	Glucose-6-P	Nucleophilic attack on γ-phosphate of ATP
4	Phosphofructokinase	ATP	Fructose-1,6-*bis* P	Nucleophilic attack on γ-phosphate of ATP
5	3-Phosphoglyceraldehyde dehydrogenase	P$_i$	Glycerate-1,3-*bis* P	Thiol ester intermediate
6	3-Phosphoglycerate kinase	Glycerate-1,3 *bis* P	ATP	Substrate level phosphorylation of ADP
7	Phosphoglyceratemutase	Glycerate-3-P	Glycerate-2-P	*bis*-Phosphorylated intermediate
8	Pyruvate kinase	Phosphoenol-pyruvate	ATP	Substrate level phosphorylation of ADP
9	Phosphohexoisomerase	Glucose-6-P	Fructose-6-P	Tautomerization
10	Triose phosphate isomerase	Dihydroxyacetone P	Glyceraldehyde-3-P	Tautomerization
11	Enolase	Glycerate-2-P	Phosphoenolpyruvate	Dehydration
12	Aldolase	Fructose-1,6-*bis* P	Dihydroxyacetone P, glyceraldehyde-3-P	Reverse aldol condensation

Except for the starting material (glucose) and the end product (pyruvate) all the inter-mediates in the glycolytic pathway are phosphorylated. Because negatively charged phosphorylated compounds are resistant to passage through cell membranes, this leads to the suggestion that one of the functions of phosphorylation is to prevent intermediates in the glycolytic pathway from getting lost by passage across the cell membrane. There are others. In fact most of the reactions in the glycolytic pathway involve phosphorylations (see reactions 1 to 8 in Table 2). This leaves just four reactions that do not directly involve phosphate (see reactions 9 to 12 in Table 2). Two of these reactions, phosphohex-oisomerase and triose phosphate isomerase are tautomerizations that involve enediol intermediates. One, the enolase reaction, involves a dehydration; and the other, the aldolase reaction, involves a reverse aldol condensation. In the following chapter we see that tautomerizations and aldol condensations are believed to have played a major role in sugar synthesis in the prebiotic world. The mechanisms for these reactions are discussed in Chapter 12.

MOST OF THE ENZYMES USED IN GLYCOLYSIS ARE USED IN THE REVERSE PROCESS OF SUGAR SYNTHESIS

Glycolysis and gluconeogenesis constitute a set of oppositely directed conversions. The organization of glycolysis as a series of connected metabolic pools makes it possi-ble for most of the same enzymes to function in both directions (Fig. 2).

Only at three points, all outside the metabolic pools, do we find reactions in gluconeogenesis that use different enzymes: (1) the conversion of pyruvate to phosphoenolpyruvate (PEP), (2) the conversion of fructose-1,6-bisphosphate to fructose-6-phosphate, and (3) the conversion of hexose phosphate to storage polysac-charide or hexose phosphate to glucose. At these three points we find sizable energy drops in the glycolytic direction (see Table 1). Clearly, if cells are to conduct these reactions in the reverse direction, the three reactions must have a different ATP-to-ADP stoichiometry and accordingly different enzymes are required (see Fig. 2).

PENTOSE PHOSPHATE PATHWAY SUPPLIES RIBOSE AND REDUCING POWER

Many kinds of organisms and some mammalian organs, notably the liver, possess an alternative pathway for the oxidation of hexoses that results in a pentose phosphate and carbon dioxide (Fig. 3). This pentose can be used as a precursor for the ribose found in nucleic acids or other sugars containing from three to seven carbon atoms, which are needed in smaller amounts. The first and third reactions in the pentose phosphate path-way generate NADPH, which is a major source of reducing power in many cells.

Cells differ considerably in their use of the pentose phosphate pathway. In muscle, a tissue in which carbohydrates are utilized almost exclusively for generation of

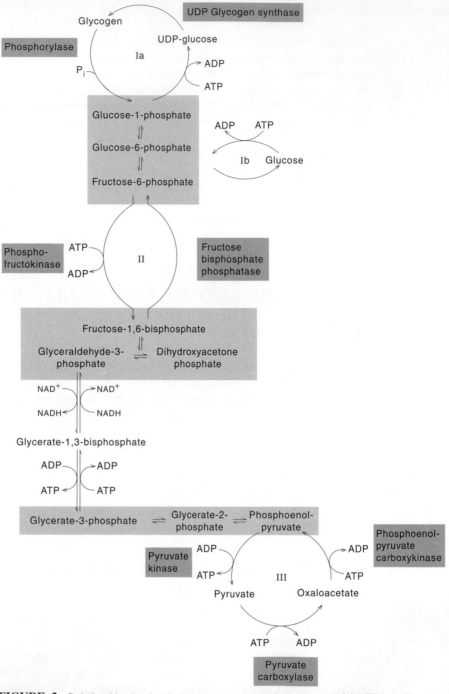

FIGURE 2 Relationships in glycolysis and gluconeogenesis. Points at which ATP is produced or consumed are indicated. Compounds in the same metabolic pools are indicated by light shading. Three small pseudocycles (Ia, II, III) in the paired sequences occur between glycogen and pyruvate, or between glycogen and glucose (Ib, II, III). Only enzymes that are unique to either glycolysis or gluconeogenesis are indicated (dark shading).

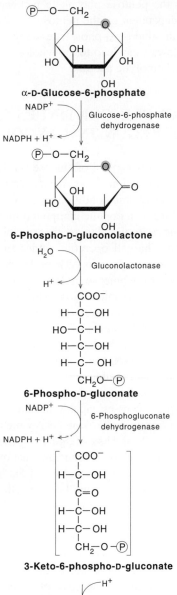

α-D-**Glucose-6-phosphate**

NADP$^+$

Glucose-6-phosphate
dehydrogenase

NADPH + H$^+$

6-Phospho-D-gluconolactone

H$_2$O

Gluconolactonase

H$^+$

6-Phospho-D-gluconate

NADP$^+$

6-Phosphogluconate
dehydrogenase

NADPH + H$^+$

3-Keto-6-phospho-D-gluconate

H$^+$

CO$_2$

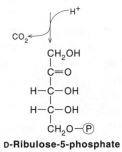

D-**Ribulose-5-phosphate**

FIGURE 3 Stage 1 of the pentose phosphate pathway: the oxidation of glucose-6-phosphate ribulose-5-phosphate and CO$_2$ and the reduction of NADP$^+$. Net Reaction: Glucose-6-phosphate + 2 NADP$^+$ ⟶ ribulose-5-phosphate + CO$_2$ + 2NADPH.

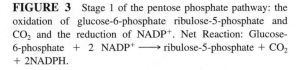

mechanical energy, the enzymes of the pentose phosphate pathway are lacking. By contrast, red blood cells are totally dependent on the pentose phosphate pathway as a source of NADPH. A deficiency in glucose-6-phosphate dehydrogenase, the first enzyme in the pentose phosphate pathway, can lead to the wholesale destruction of red blood cells and a condition known as hemolytic anemia.

TRICARBOXYLIC ACID CYCLE CONTINUES THE DEGRADATION PROCESS BEGUN IN GLYCOLYSIS

Only a small fraction of the total free energy content of glucose is released under anaerobic conditions. This is because no net oxidation of organic substrates can occur in the absence of oxygen. Catabolism under anaerobic conditions means that every oxidative event in which electrons are removed from an organic compound must be accompanied by a reductive event in which electrons are returned to another organic compound, often closely related to the first compound. The cell operating under anaerobic conditions must content itself with the generation of only two ATP molecules per molecule of glucose fermented. Most of the energy of the glucose molecule remains untapped. Given access to oxygen, however, the cell can do much more with the oxidizable organic molecules available to it, and the energy yield increases dramatically. With oxygen available as the electron acceptor, the carbon atoms of glucose (or another substrate) can be oxidized fully to CO_2, and all the electrons that are removed during the multiple oxidation events are transferred ultimately to oxygen. In the process, the ATP yield per glucose is close to 15 times greater than that possible under anaerobic conditions. There lies the advantage of the aerobic way of life.

It is not surprising then, that aerobic processes capable of extracting further energy from pyruvate have come to play such a prominent role in energy metabolism. The overall process of aerobic respiratory metabolism and its distinguishing characteristics are (1) the use of oxygen as the ultimate electron acceptor, (2) the complete oxidation of organic substrates to CO_2 and water, and (3) the conservation of much of the free energy as ATP.

Under aerobic conditions the glycolytic pathway becomes the initial phase of glucose (carbohydrate) catabolism. The other three components of respiratory metabolism are the tricarboxylic acid (TCA) cycle (also known as the Krebs cycle). This cycle is responsible for further oxidation of pyruvate, the electron-transport chain, which is required for the reoxidation of coenzyme molecules at the expense of molecular oxygen, and the oxidative phosphorylation of ADP to ATP, which is driven by a proton gradient generated in the process of electron transport (see Chapter 22, Fig. 2). Overall, this leads to the potential formation of approximately 30 molecules of ATP per molecule of glucose in the typical eukaryotic cell.

In this chapter we discuss the reactions of the TCA cycle, leaving for later the discussion of how the reduced coenzymes (NADH and $FADH_2$) produced by the cycle stimulate ATP synthesis (see Chapter 22).

The TCA cycle begins with acetyl-coenzyme A (acetyl-CoA) that results from the oxidative decarboxylation of pyruvate or the oxidative cleavage of fatty acids (Fig. 4).

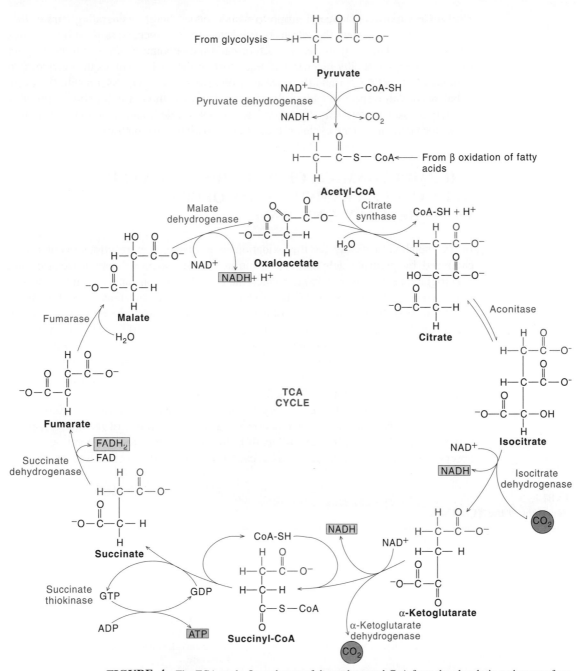

FIGURE 4   The TCA cycle. In each turn of the cycle, acetyl-CoA from the glycolytic pathway or from β oxidation of fatty acids enters and two fully oxidized carbon atoms leave (as CO_2). ATP is generated at one point in the cycle, and coenzyme molecules are reduced. The two CO_2 molecules lost in each cycle originate from the oxaloacetate of the previous cycle instead of incoming acetyl from acetyl-CoA.

Acetyl-CoA transfers its acetyl group to oxaloacetate, thereby generating citrate. In a cyclic series of reactions the citrate is subjected to two successive decarboxylations and several oxidative events, leaving a four-carbon compound from which the starting oxaloacetate is eventually regenerated. Each turn of the cycle involves the entry of two carbons from acetyl-CoA and the release of two carbons as CO_2. As a result, the cycle is balanced with respect to carbon flow and functions without the net consumption or buildup of oxaloacetate (or any other intermediate) unless side reactions occur that either feed carbon into the cycle or drain it off into alternative pathways.

THERMODYNAMICS OF THE TRICARBOXYLIC ACID CYCLE PERMITS IT TO OPERATE IN MORE THAN ONE WAY

In the metabolic pathway for the oxidation of pyruvate, four reactions occur (those catalyzed by pyruvate dehydrogenase, citrate ligase, isocitrate dehydrogenase, and α-ketoglutarate dehydrogenase) for which the equilibrium constants are large. For four others (those catalyzed by aconitase, succinate thiokinase, succinate dehydrogenase, and fumarase) the equilibrium constants are fairly close to 1; and for one reaction (catalyzed by malate dehydrogenase) it is much less—about 10^{-5} (Table 3). At first it seems strange that such an unfavorable reaction should occur in a sequence (the TCA cycle) that has one of the largest fluxes in metabolism. This is possible only because the following reaction, catalyzed by citrate ligase, is so thermodynamically favorable. The equilibrium constant for this reaction is about 5×10^5.

The highly unfavorable free energy for the conversion of malate to oxaloacetate may actually serve an important function in facultative aerobes (a facultative aerobe can operate aerobically or anaerobically). In facultative aerobes operating in the absence of O_2, the TCA or Krebs cycle cannot operate in the normal way. In spite of

TABLE 3
Reactions of the TCA Cycle

Reaction	Enzyme	$\Delta G^{\circ\prime}$ (kcal/mol)
1. Acetyl-CoA + oxaloacetate + H_2O $\longrightarrow$ citrate + CoA	Citrate synthase (or ligase)	-7.7 (-32 kJ)
2. Citrate $\rightleftharpoons$ cis-aconitase $\rightleftharpoons$ isocitrate	Aconitase	$+1.5$ (6.3 kJ)
3. Isocitrate + NAD^+ $\longrightarrow$ α-ketoglutarate + NADH + CO_2	Isocitrate dehydrogenase	-5.0 (-20.9 kJ)
4. α-Ketoglutarate + NAD^+ + CoA $\longrightarrow$ succinyl-CoA + NADH + CO_2	α-Ketoglutarate dehydrogenase	-8.0 (-33 kJ)
5. Succinyl-CoA + GDP + P_i $\longrightarrow$ succinate + GTP + CoA	Succinate thiokinase	-0.7 (-2.9 kJ)
6. Succinate + FAD $\longrightarrow$ fumarate + $FADH_2$	Succinate dehydrogenase	0.0 (0.0 kJ)
7. Fumarate + H_2O $\longrightarrow$ L-malate	Fumarase	-0.9 (-3.7 kJ)
8. L-Malate + NAD^+ $\longrightarrow$ oxaloacetate + NADH + H^+	Malate dehydrogenase	$+7.1$ (29.7 kJ)
Acetyl-CoA + $2H_2O$ + $3NAD^+$ + FAD + ADP + P_i $\longrightarrow$ $2CO_2$ + 3NADH + $2H^+$ + $FADH_2$ + CoA + ATP		-13.7 (-57.3 kJ)

this the organism still has a need for the intermediates for biosynthesis normally supplied by the Krebs cycle. This situation is remedied by operating the first part of the cycle to α-ketoglutarate or succinyl-CoA in the forward direction while operating the last part of the cycle from oxaloacetate to succinate or succinyl-CoA in the reverse direction. In this way the organism is able to synthesize the intermediates for biosynthesis while balancing its needs for oxidation and reduction. The highly unfavorable free energy for the conversion of malate to oxaloacetate becomes a highly favorable free energy for the conversion of oxaloacetate to malate so that the reactions between oxaloacetate and succinyl-CoA have a favorable overall free energy for operating in the reverse directions (see Table 3).

GLYOXYLATE CYCLE PERMITS GROWTH ON A TWO-CARBON SOURCE

Usually, condensation of acetyl-CoA with oxaloacetate to form citrate is a signal that the metabolic fate of the acetyl carbons is sealed; the inevitable result, by means of the TCA cycle, is their oxidation and eventual release as CO_2. However, the glyoxylate cycle, shown in Fig. 5 represents an alternative pathway that also begins with citrate formation but results in anabolism to the four-carbon level instead of catabolism to the one-carbon level. Comparison of the glyoxylate cycle with the TCA cycle (see Fig. 5) reveals that two of the five reactions of the glyoxylate cycle are unique to this pathway, while the other three also are part of the TCA cycle. Specifically, the glyoxylate cycle effectively bypasses the two steps of the TCA cycle in which CO_2 is released. Furthermore, two molecules of acetyl-CoA are taken in per turn of the cycle instead of just one, as in the TCA cycle. The net result is the conversion of two molecules of two carbons each (i.e., the acetate of acetyl-CoA) into succinate, a four-carbon compound.

The glyoxylate cycle is an indispensable metabolic capability for those species of bacteria, protozoans, fungi, and algae that grow on a two-carbon substrate such as acetate or ethanol. It is also an essential reaction sequence for seedlings of fat-storing plant species that must effect net synthesis of sugars and other cellular components from the acetyl-CoA produced by oxidation of storage triglycerides.

SUMMARY

In this chapter we have considered some aspects of carbohydrate metabolism.

1. Glycolysis involves the breakdown of energy-storage polysaccharides or glucose to the three-carbon acid, pyruvate. Its function is to produce energy in the form of ATP and three-carbon intermediates for further metabolism. Glycolysis occurs under both aerobic and anaerobic conditions.

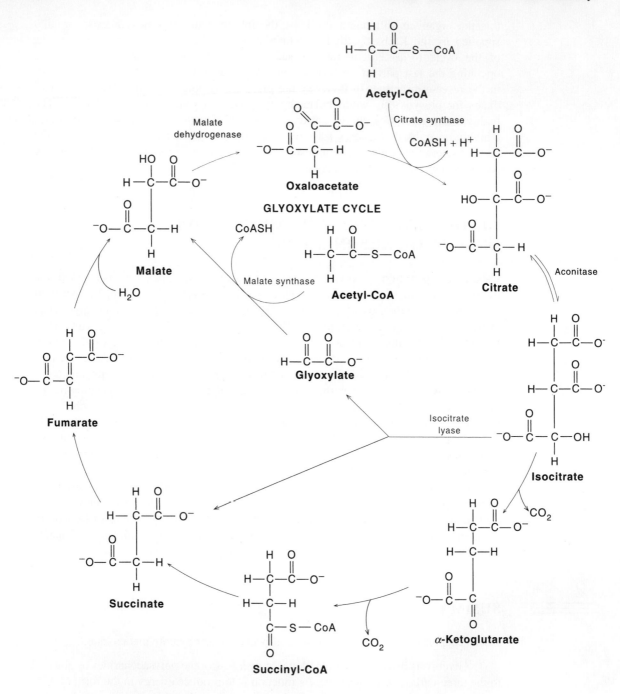

2. Glycolysis and the oppositely directed process gluconeogenesis share common intermediates and many enzymes. Only at three steps are unique enzymes used in different directions.

3. An alternative pathway for glycolysis is provided by the pentose phosphate pathway. In this pathway glucose is degraded to a pentose phosphate and carbon dioxide, and in the process $NADP^+$ is reduced to NADPH. This pathway serves a variety of functions in different cell types: (1) the production of NADPH for biosynthesis, (2) the production of ribose required mainly for nucleic acid synthesis, and (3) the interconversion of a variety of phosphorylated sugars.

4. The TCA (Krebs) cycle begins with acetyl-CoA, which is obtained either by oxidative decarboxylation of pyruvate available from glycolysis or by oxidative cleavage of fatty acids.

The acetyl-CoA transfers its acetyl group to oxaloacetate, thereby generating citrate. In a cyclic series of reactions, the citrate is subjected to two successive decarboxylations and four oxidative events, leaving a four-carbon compound malate from which the starting oxaloacetate is regenerated.

Only a single ATP is directly generated by a turn of the TCA cycle. Most of the energy produced by the cycle is stored in the form of reduced coenzyme molecules, NADH and FADH. Reoxidation of these compounds liberates a large amount of free energy, which is captured in the form of ATP.

5. The glyoxylate cycle permits growth on a two-carbon source. The glyoxylate cycle bypasses the two steps per turn of the cycle of the TCA cycle in which CO_2 is taken in instead of just one, as in the TCA cycle. The net result is the conversion of two molecules of two carbons each into a four-carbon compound, succinate. Two additional enzymes are needed for operation of the glyoxylate cycle: isocitrate lyase and malate synthase.

Problems

1. Indicate three major roles carbohydrates play in living cells. Which of these roles is likely to have been the most important in the first living organisms?

2. 2-Phosphoglycerate and phosphoenolpyruvate (PEP) differ only by dehydration between C-2 and C-3, yet the difference in $\Delta G^{o\prime}$ of hydrolysis is about -12 kcal/mol. How does dehydration "trap" so much chemical energy?

3. Summarize in the simplest words the portion of the tricarboxylic acid (TCA) cycle that is bypassed by the glyoxylate cycle.

FIGURE 5   Comparison of the TCA cycle and the glyoxylate cycle. In the TCA cycle, one molecule of acetyl-CoA is oxidized to two molecules of CO_2. In the glyoxylate cycle (gray), two molecules of acetyl-CoA are converted to one molecule of oxaloacetate. As indicated, the glyoxylate cycle uses some of the enzymes of the TCA cycle. Only enzymes operative in the glyoxylate cycle are shown. In plant cells, enzymes of the glyoxylate cycle are located in specialized organelles called glyoxysomes. In yeasts and other eukaryotic organisms these enzymes are located in the cytosol.

4. What would you expect to be the metabolic consequences of the following mutations in yeast:
   (a) Inability to synthesize malate synthase?
   (b) Pyruvate carboxylase that is not activated by acetyl-CoA?
   (c) Pyruvate dehydrogenase that is inhibited by acetyl-CoA more strongly than is the wild-type enzyme?

Reference

See references at the end of Chapter 6.

CHAPTER 12

Prebiotic Pathways
Involving Carbohydrates

Most organisms are heterotrophs that live by degrading carbon compounds produced by other organisms. They also have the capacity to synthesize pentoses and hexoses from three-carbon compounds and in some cases also from two-carbon compounds (see Chapter 11). Heterotrophs could not survive unless there were some organisms that fix carbon from one-carbon compounds. Hence the food chain starts with the photoautotrophs that are able to harness the energy of the Sun to synthesize sugars from CO_2. Plants and other photoautotrophs convert about 10^{11} tons of carbon from CO_2 into organic compounds annually.

The overall equation for CO_2 fixation as it occurs in plants is

$$6CO_2 + 6H_2O + light \rightarrow C_6H_{12}O_6 + 6O_2, \qquad (1)$$

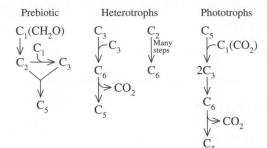

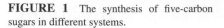

FIGURE 1 The synthesis of five-carbon sugars in different systems.

or, more generally,

$$CO_2 + H_2O + light \rightarrow (CH_2O) + O_2, \tag{2}$$

where (CH_2O) represents the basic unit that participates in a carbohydrate molecule. Electrons are removed from H_2O, O_2 is evolved, and carbon from CO_2 is incorporated into carbohydrate. In the process the carbon is reduced from the $+4$ valence state to the 0 valence state.

In the prebiotic world it seems likely that one-carbon chemistry also dominated the synthesis of carbon compounds that were crucial to the origin of life. However, in this case the key precursor carbon was probably formaldehyde (CH_2O) formed in the atmosphere or in the lithosphere.

An overall equation for the fixation of CH_2O into carbohydrates is

$$nCH_2O \rightarrow (CH_2O)_n. \tag{3}$$

In contrast to the fixation of carbon dioxide by photoautotrophs, the incorporation of formaldehyde into organic compounds does not involve any change in the valence state of the carbon and does not require any other molecules for incorporation except as catalysts.

It seems likely that formaldehyde gave rise to a wide array of carbohydrates and other organic molecules. However, among these the most important sugar that was required to get the nucleic acid world underway was probably ribose (Fig. 1). Although simpler sugars than ribose have been proposed as preceding ribose as the sugar component in nucleic acids, the arguments in support of such proposals are far from convincing. A major stumbling block for a long time was that no feasible routes from formaldehyde to ribose had been found. Yields always were very small and ribose always constituted a very minor component of the mixture of sugars that was usually formed. In this chapter we present new information and new possibilities suggesting that we may be closer to finding a feasible prebiotic pathway to ribose.

SYNTHESIS OF SUGARS IN THE PREBIOTIC WORLD IS LIKELY TO HAVE STARTED WITH FORMALDEHYDE

In the 1860s Butlerov discovered that when an aqueous solution of formaldehyde is warmed in the presence of a calcium hydroxide suspension, a mixture of sugars is produced (Fig. 2). The process is referred to as the *formose reaction*. Since Butlerov's discovery, a wide variety of conditions have been used, but calcium hydroxide remained the favorite catalyst for sugar synthesis for more than a century. The process even has been proposed as a way of producing food but some toxic compounds that are formed make this impossible. The formose reaction starts from a concentrated solution of formaldehyde (usually 1 to 2%). Within a matter of minutes at temperatures around 55°C a broad array of sugars and other products is formed. Could

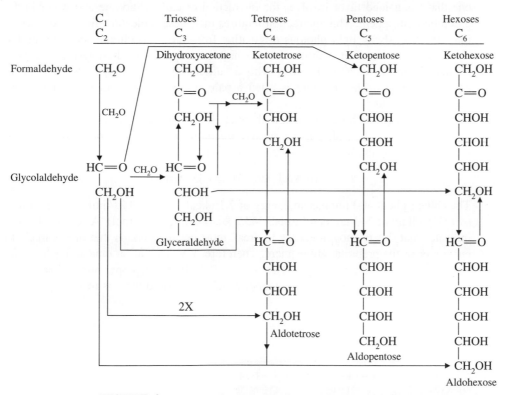

FIGURE 2   Some of the main routes followed by formaldehyde in the formose reaction according to Melvin Calvin. Note that formaldehyde only is directly involved in the synthesis of glycolaldehyde or trioses or tetroses. Pentoses are made from a combination of glycolaldehyde and a triose. Hexoses are made from a combination of two trioses.

this procedure be used without modification for the prebiotic synthesis of ribose? This is very unlikely because ribose usually constitutes 1% or less of the reaction products. Furthermore, ribose is very unstable at the high pHs that prevail in a calcium hydroxide suspension.

Considerable effort has been exerted to find milder conditions under which the formaldehyde molecule could be channeled down a pathway to ribose.

FORMALDEHYDE WAS PROBABLY SYNTHESIZED IN THE PREBIOTIC ATMOSPHERE

The results from the formose reaction suggested that formaldehyde might be a most feasible precursor for ribose under some circumstances. Because of its potential importance in carbohydrate synthesis, considerable effort has been exerted on the study of possible routes for the synthesis of formaldehyde in the prebiotic world. The one that has gained favor involves the interaction of carbon dioxide and water in the prebiotic atmosphere. This interaction requires that these molecules first be converted into free radicals by light photons originating from the Sun. Photons can elevate the electrons in an atom or a molecule to a higher energy level, thereby making them chemically more reactive. Photons with a sufficiently high energy also can break covalent linkages. The resultant is often a pair of free radicals, that is, nuclei with unpaired electrons.

For a quantum of light to break a chemical bond, its energy must equal or exceed that of the chemical bond. The energy of a quantum of light is reciprocally related to its wavelength by the formula:

$$E \text{ (in kcal/mol)} = 28{,}635/\lambda(\text{nm}).$$

Thus blue light at 400 nm has an energy of 7.16 kcal/mol (299 kJ/mol), while ultraviolet (UV) light at 200 nm can furnish 143.2 kcal/mol (599 kJ/mol). A scan of Table 1 indicates that this is enough energy to break most covalent bonds that one is likely to encounter in the prebiotic atmosphere. Therefore, UV irradiation emanating from the Sun is very likely to produce a shower of free radicals in the upper atmosphere. Such considerations led Pinto to experiments in which he exposed various mixtures of gases

TABLE 1
Bond Energies[a]

C—C 82	C—H 99	S—H 81
O—O 34	N—H 94	C—N 70
S—S 51	O—H 110	C—O 84
C≡C 147	C≡N 147	C≡S 108
O=O 96	C=O 164	N≡N 226

[a]Expressed in kcal/mol.

including water and carbon dioxide to the kind of UV radiation they should be exposed to in unfiltered sunlight. Low-wavelength UV dissociates both of these molecules into more reactive species as shown:

$$H_2O + h\nu \rightarrow OH\cdot + H\cdot \quad \lambda \leq 240 \text{ nm} \tag{4}$$

$$CO_2 + h\nu \rightarrow CO + O, \quad \lambda \geq 230 \text{ nm}. \tag{5}$$

The H and CO so produced can react further to make the formyl radical CHO·

$$H\cdot + CO \rightarrow HCO\cdot, \tag{6}$$

and two formyl radicals can combine to produce formaldehyde and CO

$$CHO\cdot + CHO\cdot \rightarrow CH_2O + CO. \tag{7}$$

Gaseous formaldehyde is very water soluble and would be expected to rain out of the atmosphere. It has been estimated that approximately 10 million years would have been required for this scheme to produce concentrations of formaldehyde in the oceans sufficient for synthesis of more complex condensation products. The unduly long time and the high reactivity of formaldehyde with other substances makes it unlikely that such a slowly developing situation could ever have reached the necessary formaldehyde concentrations. Thus more efficient pathways must have been available for synthesis; alternatively, some means of concentrating the formaldehyde must have existed. In fact dilute solutions of formaldehyde may be concentrated by evaporation of aqueous solutions containing formaldehyde.

GLYCOLALDEHYDE CATALYZES THE INCORPORATION OF FORMALDEHYDE INTO SUGARS

At high levels of calcium hydroxide (20 to 50 mg/0.5 ml) the formose reaction proceeds with a short delay and reaches completion in 10 to 20 min at temperatures between 45 and 75°C. When sparing amounts of calcium hydroxide are used (5 mg/0.5 ml), the reaction requires small amounts of specific organic catalysts to proceed. Glycolaldehyde makes an excellent catalyst.

Because glycolaldehyde is alleged to be the first intermediate formed when two formaldehydes condense (see Fig. 2) as well as a catalyst, there must be a mechanism whereby each glycolaldehyde incorporated results in the generation of one or more glycolaldehydes. To explain this Ron Breslow proposed a series of reactions in which every glycolaldehyde that is incorporated into the reaction results in two glycolaldehydes being produced.

$$CH_2O + HOCH_2-CHO \rightarrow HOCH_2-CHOH-CHO \tag{8}$$

$$HOCH_2O-CHOH-CHO \rightleftharpoons HOCH_2-CO-CH_2OH \tag{9}$$

$$CH_2O + HOCH_2-CO-CH_2OH \rightarrow HOCH_2-CHOH-CO-CH_2OH \qquad (10)$$

$$HOCH_2-CHOH-CO-CH_2OH \leftrightharpoons HOCH_2-CHOH-CHOH-CHO \qquad (11)$$

$$HOCH_2-CHOH-CHOH-CHO \rightarrow 2HOCH_2-CHO\cdot \qquad (12)$$

This series of reactions is intended to explain only the first phase of the reaction because no pentoses or hexoses are produced. These five reactions can be arranged in a circular pathway with five members: glycolaldehyde, glyceraldehyde, dihydroxyace-tone, a ketotetrose, and an aldotetrose (Fig. 3). Step 1 [Eq. (8)] involves an aldol condensation between formaldehyde and glycolaldehyde to form glyceraldehyde.

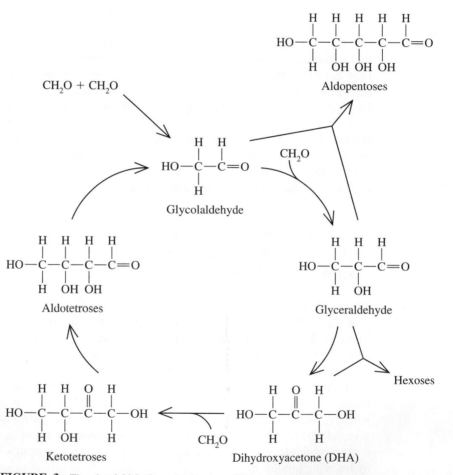

FIGURE 3 The glycolaldehyde cycle showing additional reactions that feed into and tap the cycle. It is presumed that once the first glycolaldehyde is formed the formation of glycolaldehyde from two formaldehyde molecules is no longer rate significant. As can be seen the fomation of aldopentoses and hexoses removes intermediates from the cycle.

Step 2 [Eq. (9)] involves a tautomerization in which glyceraldehyde is reversibly converted into dihydroxy acetone (DHA). Step 3 [Eq. (10)] involves another aldol condensation, this time between DHA and formaldehyde to form a ketotetrose. Step 4 [Eq. (11)] involves another tautomerization resulting in the conversion of the ketotetrose to an aldotetrose. Finally, step 5 [Eq. (12)] involves a reverse aldol condensation in which an aldotetrose splits into two glcolaldehydes. Support for this cycle, *the glycolaldehyde cycle,* comes from the observation that any member of the cycle makes an equally fit catalyst to glycolaldehyde. Other reactions shown as branching from the cycle include the reaction of two formaldehyde molecules to form glycolaldehyde. This is a very slow reaction and explains why a catalyst is needed under conditions of low calcium hydroxide. Also shown are likely routes to the aldopentoses and hexoses. The route to hexoses is very similar to the one used in biosystems.

In reactions leading to living systems, chemical catalysts must have played a major role in determining which out of the seemingly endless cluster of chemical reactions that could occur were the reactions that actually did occur. We suspect these reactions to be unusual with many of the pathway intermediates or end products doubling as catalysts. Nowhere is this possibility more in evidence that it is in the glycolaldehyde cycle.

The chemistry of the glycolaldehyde cycle is simple and repetitious involving reactions that are very familiar from organic chemistry. The *aldol condensation* and *tautomerization* reactions also are found in the glycolytic–gluconeogenetic pathways (see Chapter 11 Fig. 1). The mechanisms for a typical aldol condensation and a typical tautomerization are described in Box 12A.

Reactions that involve tapping the intermediates in the glycolaldehyde cycle such as those consumed in aldopentose synthesis are considerably slower than the reactions within the glycolaldehyde cycle. Only as the formaldehyde supply approaches exhaustion and as the intermediate concentrations build do the rates of these reactions become competitive. If it were not for this, the cycle might shut down prematurely before the formaldehyde is used up. In fact our experience is that no pentoses or hexoses appear until the formaldehyde concentration level is down to miniscule levels.

STRONGLY BASIC CONDITIONS USED IN THE FORMOSE REACTION ARE NOT CONDUCIVE TO HIGH YIELDS OF THE ALDOPENTOSES

Even at low calcium hydroxide concentrations (5 mg/0.5 ml) the pH is very high (above 11.0). These strongly basic conditions result in a broad array of products and unstable conditions for survival of the aldopentoses. Of the four common aldopentoses (Fig. 4), ribose is usually the least abundant in the reaction products. This is due at least partly to the fact that ribose is the most unstable aldopentose to alkali.

Finding low yields, 5% or less of the aldopentoses based on the formaldehyde input, led us to search for factors that might adversely affect the outcome of the

BOX 12A   Key Mechanisms That Are Used in the Proposed Prebiotic Pathways to Ribose

(a) A tautomerization is shown using as an example the reversible conversion of glyceraldehyde to dihydroxyacetone.

(b) An aldol condensation is shown using as an example the reversible conversion of a glycolaldehyde to an aldotetrose. Note that in both of these reactions the conversions do not involve any gain or loss of atoms. Both reactions are base catalyzed.

(a) Tautomerization

(b) Aldol Condensation

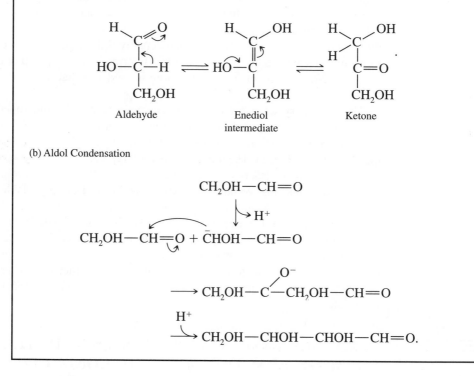

reaction. Our investigations showed that all four aldopentoses are unstable in calcium hydroxide. After 6 hr at 67°C, over 90% of the aldopentoses are lost. This may be due to the Cannizzaro reaction that involves the formation of an acid and an alcohol from two aldehydes. High pHs are known to favor the Cannizzaro reaction. In addition to losses of ribose that may result from the Cannizzaro reaction, ribose tautomerizes to arabinose. Although the reverse reaction also occurs, the equilibrium according to our measurements is about 3:1 in favor of arabinose.

```
    CHO            CHO            CHO            CHO
     |              |              |              |
   HCOH           HOCH           HCOH           HOCH
     |              |              |              |
   HCOH           HCOH           HOCH           HOCH
     |              |              |              |
   HCOH           HCOH           HCOH           HCOH
     |              |              |              |
   CH₂OH          CH₂OH          CH₂OH          CH₂OH
  D-Ribose*     D-Arabinose*    D-Xylose*      D-Lyxose*
   (Rib)          (Ara)          (Xyl)          (Lyx)
```

FIGURE 4   The four common straight-chain aldopentoses.

UNDER MILDLY BASIC CONDITIONS FORMALDEHYDE INCORPORATION INTO PENTOSES AND HEXOSES IS GREATLY REDUCED

The low yields of aldopentoses at the high pHs generated by calcium hydroxide led us to seek milder reaction conditions. Lowering the temperature did not improve the situation in the presence of calcium hydroxide; it just increased the reaction time. In general, conditions that favored aldopentose synthesis also favored their degradation, as long as we were working in the high pH milieu of calcium hydroxide suspensions. Several alternative catalysts were examined. The formose reactions were studied in some detail using magnesium hydroxide suspensions. The pH of a saturated solution of magnesium hydroxide at room temperature is about 9.4. Although magnesium hydroxide is a strong base like calcium hydroxide, it is much less soluble. This accounts for the lower pH of magnesium hydroxide suspensions. In magnesium hydroxide the aldopentoses were much more stable, with a less than 10% loss in 12 hr at 67°C. Over the same time period only about 10% of the ribose isomerizes to arabinose. Unfortunately, no aldopentoses were formed from formaldehyde alone and only matching amounts were formed in the presence of glycolaldehyde. Clearly, the latter compound was not operating catalytically as it had done at the higher pHs.

Despite the limitations of the magnesium hydroxide system, we used this system to scrutinize some of the formose-related reactions. We found that the yields of aldopentoses formed from glycolaldehyde and glyceraldehyde were up severalfold from what they were in parallel calcium hydroxide experiments. Moreover, ribose was no longer underrepresented but usually constituted somewhat more than one-quarter of the aldopentose fraction. Once formed, the aldopentoses were stable for much longer times than they were at the higher pHs that exist in calcium hydroxide suspensions.

We also used the magnesium hydroxide system to explore the possibility that already formed sugars might serve as an alternative starting point for ribose. Just as heterotrophic organisms feed off of the carbon compounds generated by autotrophic organisms, a means might have existed so that many of the carbohydrates present in the prebiotic world could be utilized as an alternative source to formaldehyde for ribose synthesis.

How could these carbohydrates be "recycled" for ribose synthesis? We need to look no farther than the reactions of the glycolaldehyde cycle. Just as erythrose generates glycolaldehyde by a reverse aldol condensation, so might we expect other carbohydrates to generate compounds that could be used for ribose synthesis. For example, two successive reverse aldol condensations should produce three glycolaldehydes from one glucose. We have tested glucose and found that in the presence of formaldehyde and magnesium hydroxide at 67°C, glucose breaks down; in this process, sizable amounts of the four aldopentoses are formed. We have not studied these reactions in any depth and have not found optimum conditions for them. Nevertheless, the use of other carbohydrates like glucose for ribose synthesis presents exciting possibilities for how sugars that may not have any other function in the prebiotic world might be remodeled to make ribose.

LEAD (PLUMBOUS) SALTS CATALYZE ALDOPENTOSE SYNTHESIS UNDER MILDLY BASIC CONDITIONS

The frustrations of working at the lower pHs generated by magnesium hydroxide suspensions led to a search for additional catalytic agents. First, reports were pursued that α-hydroxyacetophenone could catalyze sugar formation from formaldehyde and that it might be even superior to most other carbohydrate catalysts. When α-hydroxyacetophenone was added to a magnesium hydroxide suspension containing formaldehyde, it stimulated sugar synthesis but the yields were small and positive results required prolonged incubations of 10 hr or more at elevated temperatures. Despite the limitations experienced in working with α-hydroxyacetophenone, these were the first positive results we had witnessed for making sugars from formaldehyde at lower pHs so we sought a means for potentiating the catalytic action of α-hydroxy-acetophenone. An encouraging observation had been made by Langenbeck in 1954 that plumbous hydroxide in conjunction with benzoin catalyzes the formation of gly-colaldehyde from formaldehyde. When plumbous hydroxide was added together with α-hydroxyacetophenone, there was a pronounced stimulation effect. In short order conditions were found under which this new concoction increased the yields of aldopentoses from formaldehyde by 20-fold over those found in the absence of lead. This made it possible to convert formaldehyde into the four aldopentoses with an efficiency of 20 to 30%. Hexoses were not detectable under these conditions. Actually, we used lead nitrate in which lead is present as the doubly-charged plumbous ion; this converts to the soluble plumbite ion ($HOPbO^-$) at the pH that prevails in a magnesium hydroxide suspension because the pK_a for lead hydroxide is 7.7.

The stimulatory action of lead has not been duplicated with any other metallic cations despite an extensive search. So far negative results have been obtained with

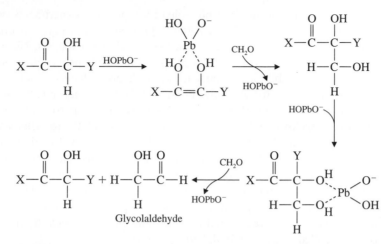

FIGURE 5 Proposed mechanism whereby lead in concert with an organic molecule catalyzes the synthesis of glycolaldehyde from formaldehyde. The organic catalyst is drawn in a general form. It could be a sugar, benzoin or α-hydroxyacetophenone. In benzoin, X and Y would be phenyl groups. In α-hydroxyacetophenone, X would be a phenyl group and Y would be a hydrogen.

Mn^{2+}, Fe^{2+}, Zn^{2+}, Co^{2+}, Sn^{2+}, Tl^+, Mg^{2+}, and Ca^{2+}. We speculate that this unique property of plumbous ion is due to a high affinity for *cis*-hydroxyls and an extraordinarily low pK_a, which makes the lead soluble at pH 9.4. A mechanism for its action based on Langenbeck's proposed mechanism for the catalysis of glycolaldehyde synthesis by benzoin and lead is shown in Fig. 5. According to this mechanism lead hydroxide in the form of plumbite ion stabilizes the enediol form of α-hydroxyacetophenone. In this form one of the carbons of the enediol–lead complex attacks a formaldehyde carbon. Following the formation of a formaldehyde adduct the lead shifts to another location resulting in reaction with a second formaldehyde molecule. Overall this leads to the formation of one glycolaldehyde from two formaldehyde molecules and the return of α-hydroxacetophenone to its original form. Even though this mechanism was designed to explain how α-hydroxyacetophenone catalyzes the formose reaction, it should be apparent that the constellation of key atoms in the organic catalyst is found in most sugars. In fact we have observed that α-hydroxyacetophenone can be replaced by any of the glycolaldehyde cycle intermediates or for that matter almost any sugar as long as lead is present. There are two notable exceptions to this. First, ribose at high lead concentrations (0.08 M) does not function as a catalyst. At this concentration ribose is almost quantitatively precipitated by the lead. The other three aldopentoses only partially are precipitated at this lead concentration. Normally we use lead at a concentration of 0.04 M. Negative results also were obtained with the sugar sedoheptulose for unknown reasons. In the process of obtaining the negative results with sedoheptulose, we discovered that when incubation times are extended to 10 hr or more, lead catalyzes the conversion of formaldehyde into sugars in the absence of any organic catalyst.

In view of the seemingly unique role of lead in these experiments, we pondered the legitimacy of lead as a prebiotic catalyst. Obviously, lead is less abundant than many other metals that are considered to be likely prebiotic catalysts. Nevertheless, lead is the most abundant element in the solar spectrum for elements above atomic number 56 and it even might be more abundant in Earth's crust. Indeed, lead is a component of numerous minerals, for example, galena (PbS), boulangerite ($Pb_5Sb_4S_{11}$), bournonite ($PbCuSbS_3$), cerussite ($PbCO_3$), anglesite ($PbSO_4$), pyromorphite ($Pb_5(PO_4)_3Cl$), mimetite ($Pb_5(AsO_4)$ Cl), and vanadinite ($Pb_5(VO_4)_3Cl$). So whereas lead is not as abundant as metals like Mg^{2+}, Ca^{2+}, and Fe^{2+}, it should be found in numerous locales. The fact that lead has been shunned by most biosystems is a weak argument against its consideration as a prebiotic compound. If we were to use this as a criterion for excluding lead, we also should have to exclude toxic compounds like formaldehyde and cyanide, which are likely to have been very important prebiotic compounds. All three of these compounds are avoided in biosystems probably because they are too reactive to be controlled by protein modulators.

Lead also has been implicated in reactions involving polynucleotides and RNA. For example, Sawai has described lead-catalyzed, non-template-assisted oligomerization of adenosine-5'-phosphorimidazolide. Following this Orgel and colleagues observed that plumbous ion catalyzes polyadenylic and polyguanylic acid synthesis on complementary polynucleotide templates. There also is extensive literature on the Pb^{2+}-catalyzed hydrolysis of RNA.

In the new incubation system containing both lead and catalytic amounts of glycolaldehyde or other cycle intermediates, the optimum temperature for aldopentose synthesis is between 55 and 75°C. Kinetic studies indicate that riobose is the first aldopentose to appear. This is followed by arabinose and finally by the other two aldopentoses. Lyxose is usually present in the smallest amounts. In prolonged incubations the amount of arabinose exceeds the amount of ribose. The significance of these kinetic observations is discussed in the next section.

LEAD SALTS ALSO CATALYZE THE INTERCONVERSION OF ALDOPENTOSES AND THE SYNTHESIS OF ALDOPENTOSES FROM TETROSES AND HEXOSES

In magnesium hydroxide suspensions without lead, arabinose and ribose slowly interconvert so that 4 days of incubation at 67°C results in a ratio of arabinose to ribose of 3:1. The same ratio is obtained whether one starts from arabinose or from ribose, indicating that this is an equilibrium value.

In the lead-containing system any one of the four aldopentoses converts readily into a mixture of all four aldopentoses. A kinetic study shows that arabinose is the first interconversion product observed when starting from ribose; xylose and lyxose follow. This result is remarkably similar to what is observed when starting from formaldehyde and incubating for several hours. Reconsideration of the results on *de novo* synthesis in the light of these interconversion results is consistent with the notion that ribose is the main if not the only sugar synthesized directly, the other sugars arising as interconversion products. Further work is necessary to reach a firm conclusion on this issue and to determine the mechanisms for some of these interconversions.

In addition to the interconversions of aldopentoses that are catalyzed by lead, we were quite surprised to see that small amounts of the aldopentoses were produced when the tetrose erythrose or the hexoses (mannose, glucose, or fructose) were incubated under the same conditions. We do not understand the mechanisms for these remarkable reactions. However, these observations could be of considerable importance because they reveal another possible route whereby existing sugars in the prebiotic world might be converted into ribose.

HIGH YIELDS OF RIBOSE 2,4-BISPHOSPHATE CAN BE SYNTHESIZED UNDER CONTROLLED CONDITIONS FROM GLYCOLALDEHYDE AND FORMALDEHYDE

Before the plumbite system for making ribose from formaldehyde was discovered, there was a wave of pessimism about the prebiotic synthesis of ribose because of the poor yields. This wave of pessimism gave way to limited optimism when Alfred Eschenmoser and coworkers found an organochemical approach to making a derivative of ribose in high yields. In their procedure glycolaldehyde was phosphorylated and then incubated with 0.5 M equivalents of formaldehyde in strongly basic solution. Under optimum conditions this resulted in ribose 2,4-bisphosphate as the major product. Unfortunately, the 2 and 4 positions are just where the phosphates do not normally occur in RNA. Although this procedure gives high yields of a ribose derivative, the conditions used are contrived and it is not possible at this point to see how they could be translated into prebiotic conditions.

$$
\begin{array}{c}
\text{H}\!\!\diagdown \\
\text{C}\!\!=\!\!\text{O} + 2\text{HC}\!\!-\!\!\text{C}\!\!=\!\!\text{O} \longrightarrow \text{HC}_5\!\!-\!\!\text{C}_4\!\!-\!\!\text{C}_3\!\!-\!\!\text{C}_2\!\!-\!\!\text{C}_1\!\!=\!\!\text{O}. \\
\text{H}\!\!\diagup
\end{array}
\qquad (13)
$$

WE STILL FACE PROBLEMS WITH THE SYNTHESIS OF RIBOSE

You might say that things had improved since the time-worn $Ca(OH)_2$ procedure was replaced for ribose synthesis. The lead-catalyzed system developed in our laboratory certainly has led to a large increase in ribose yields. However, even if the total yield of aldopentoses is 30%, ribose only accounts for about a quarter of this and we have the problem of eliminating compounds from the mixed product that also might react in the next step of nucleoside or nucleotide synthesis. There is also the problem of the instability of neutral ribose. It must be modified quickly; otherwise it converts to other aldopentoses or other carbohydrate derivatives.

The Eschenmoser procedure results in ribose 2,4-bisphosphate as the major product. This is certainly quite interesting but it does not provide a solution to our problem of producing ribose in a form suitable for nucleoside or nucleotide synthesis either. Not only the conditions of synthesis are highly contrived and therefore not imaginable in a prebiotic world but also the product contains phosphates in abnormal locations.

Thus despite the fact that some interesting results have been obtained from two laboratories, we do not regard either of them as providing a final answer to the ribose synthesis problem.

SUMMARY

In this chapter we have considered potential prebiotic pathways to carbohydrates with a particular focus on ribose. Because much of the material presented represents work done in my laboratory to find a better pathway to ribose, the material is presented in chronicle form.

1. Formaldehyde always has been considered to be the most likely precursor of carbohydrates in the prebiotic world. It has the advantage that it is a high-energy compound for which routes of synthesis in the atmosphere have been found.

2. The favoured route for formaldehyde synthesis in the prebiotic atmosphere involves water and carbon dioxide and exposure of these compounds to low-wavelength UV light. The formaldehyde formed in a series of free-radical reactions would have rained down because formaldehyde is very soluble in water. Formaldehyde in small bodies of water could have been concentrated by simple evaporation.

3. In a calcium hydroxide suspension, formaldehyde rapidly undergoes a series of condensation reactions that lead to a braod array of carbohydrates of which ribose constitutes only a small fraction. At low concentrations of calcium hydroxide, formaldehyde does not react even in heated solution unless an organic catalyst is added. Glycolaldehyde makes an excellent catalyst. Because glycolaldehyde is alleged to be the first intermediate formed when two formaldehyde molecules condense as well as a catalyst, there must be a mechanism whereby each glycolaldehyde molecule incorporated into carbohydrate results in the generation of one or more glycolaldehyde molecules. To explain this a cycle of reactions was proposed in which the net reaction entails the formation of one glycolaldehyde from two molecules of formaldehyde. The chemistry of the glycolaldehyde cycle is simple and repetitious, involving reactions that are very familiar from organic chemistry: aldol condensations and tautomerizations. Strong support for the glycolaldehyde cycle comes from the observation that any member of the cycle makes an equally fit catalyst to glycolaldehyde. The cycle keeps

going as long as there is formaldehyde. When the supply of formaldehyde is exhausted, condensation reactions between members of the cycle take place to give larger
products. For example, the aldopentoses including ribose can be formed by an aldol
condensation reaction between glycolaldehyde and glyceraldehyde.

4. Only small amounts of ribose are formed from formaldehyde in calcium hydroxide suspensions. This is because the high pH conditions in calcium hydroxide
suspensions encourage many competing reactions. In search of milder conditions that
would favor ribose synthesis, it was found that lower pHs generated by a magnesium
hydroxide syspension combined with a lead catalyst led to greatly improved yields of
ribose. A mechanism for the lead-catalyzed reaction is proposed.

5. Ribose 2,4-bisphosphate can be made in high yields by reacting phosphorylated
glycolaldehyde with half molar equivalents of formaldehyde in strongly basic
conditions. It is not clear if this reaction has the same prebiotic potential as the lead-
catalyzed reaction because the phosphorylated derivative is not the normal one found
in RNA and the conditions for its synthesis are highly contrived.

Problems

1. Is it likely that the prebiotic synthesis of sugars occurred in the oceans? If it did
   not, why not?
2. In many prebiotic reactions it is advantageous to proceed from the liquid to the dry
   state with heating. In the synthesis of ribose from formaldehyde this strategy does
   not seem to be useful or necessary. Explain.
3. If you have a concentrated solution of formaldehyde, the addition of a very small
   amount of glycolaldehyde triggers the formose reaction. Can you suggest why glycolaldehyde appears to function as a catalyst?
4. Which of the following reactions is likely to proceed more readily:

$$CH_2O + CH_2O \rightarrow CH_2OHCHO?$$

   or

$$CH_2O + CH_2OHCHO \rightarrow CH_2OHCHOHCHO?$$

   By using reaction arrows, draw likely mechanisms for these two reactions.
5. Calculate the $\Delta G°$ for the following reaction (one way of estimating this $\Delta G°$
   would be to find a biochemical reaction that involves similar known chemistry):

$$CH_2OHCHOHCHOHCHO \rightarrow 2CH_2OHCHO.$$

6. Donors and acceptors occur in the aldol condensation reaction.
   (a) If glycolaldehyde is the donor, indicate the reaction products when
       glycolaldehyde is the acceptor in an aldol condensation reaction. Do the same
       when formaldehyde or glyceraldehyde is the acceptor. Can you draw the mech-
       anism with reaction arrows?

(b) If dihydroxyacetone reacts with glycolaldehyde, which is the donor and which
is the acceptor? Can you draw the mechanism and indicate the products?

References

Bolli, M., Micura, R. and Eschenmoser, A. Pyranosyl-RNA: Chiroselective Self-Assembly of Bases
Sequences by Ligative Oligomerization of Tetranucleotide-2′,3′-Cyclophosphates, *Chem. Biol.* 4:
309–320, 1997. (This has a commentary concerning the origin of biomolecular homochirality.)

Muller, D., Pitsch, S., Kittaka, A., Wagner, E., Wintaer, C. E. and Eschenmoser, A. Chem von a-Aminoni-
trilen Adlomerisierung von Glycolaldehyd-phosphat zu racemischen Hexose-2,4,6-triphosphaten and (in
Genenward von Formaldehyd) racemischen Pentose-2,4-diphosphaten: rac, Allose, 2,4,6-triphosphat und
rac-Ribose, 2,4-diphosphat sind die Reaktions hauptprodukte, *Helv. Chim. Acta* 73:1410, 1990.

Pinto, J.P. and Gladstone, G.R. Photochemical Production of Form aldehyde in Earth's Primitive Atmosphere, *Science* 210:183–185, 1980.

Weber, A.L. Nonezymatic Formation of "Energy-Rich" Lactoyl and Glyceroyl Thioesters from Glyceraldehyde and a Thiol, *J. Mol. Evol.* 20:157–166, 1984.

Zubay, G. Studies on the Lead-Catalyzed Synthesis of Aldopentoses, *Origins Life Evol. Biosphere*
28:12–26, 1998.

CHAPTER 13

Similarities and Differences between the Biosynthesis of Nucleotides and the Prebiotic Synthesis of Nucleotides

In Chapter 10 we describe a possible scenario for the origin of life in which RNA is the key component of the first living organisms. Once the means for synthesizing and propagating RNA has evolved, further evolution would be expected to follow a standard pattern in which nucleic acid replication results in the occasional synthesis of variants that are the raw material for Darwinian selection. It is unclear how evolutionary events were directed prior to the first RNAs. In my view this is the most challenging problem in all of evolution. The first nucleotides were probably synthesized in simple steps by a process that was ideally suited to the raw materials that were present in Earth's atmosphere and on its crust.

Our major goal in this chapter is to describe likely routes to the first nucleotides. How does one commence the search for such a system? First, the existing biochemical pathways are scrutinized for clues. Then experiments are attempted using materials and conditions that seem appropriate to the Archean world. We begin with a description of the biochemical pathways of nucleotides. Following this we give a 50-year progress report on what has been learned about the origins of the first nucleotides through laboratory studies.

OVERVIEW OF NUCLEOTIDE METABOLISM

Nucleotides are composed of three components: ribose, phosphate, and a purine or pyrimidine base. We focus on ribonucleotides instead of deoxyribonucleotides because RNA is believed to have preceded DNA. The most common bases in RNA are adenine, guanine, uracil, and cytosine. The nucleotides with uracil, guanine, and hypoxanthine are illustrated in Fig. 1, along with inosine, the nucleotide for hypoxanthine, which we suspect played a more significant role than guanine in the first RNAs (see Chapter 14).

From *in vivo* isotope labeling experiments on animals we know that the precursor atoms that go into the purine and pyrimidine rings come from a few simple molecules (Fig. 2). A Single amino acid, glycine, is required for purine synthesis, and a single amino acid, aspartic acid, is required for pyrimidine synthesis. The remaining nitrogen atoms in both of these heterocyclic compounds originate as single atoms from the α-amino group of aspartate or the amide nitrogen of glutamine. The remaining carbon atoms originate from formate or carbon dioxide. The overall pathways to the five most common ribonucleotides [adenosine triphosphate (ATP), guanosine 5'-triphosphate (GTP), uridine 5'-triphosphate (UTP), cytidine 5'-triphosphate (CTP), and inosine 5'-monophosphate (IMP)] are illustrated in Fig. 3.

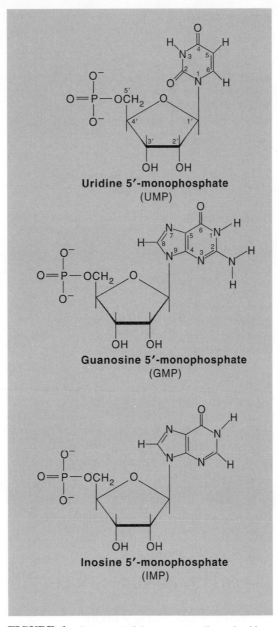

Uridine 5'-monophosphate
(UMP)

Guanosine 5'-monophosphate
(GMP)

Inosine 5'-monophosphate
(IMP)

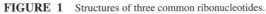

FIGURE 1   Structures of three common ribonucleotides.

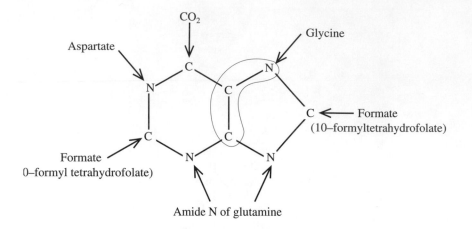

Precursors of the atoms in the purine ring

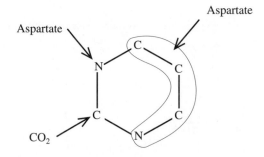

Precursor of the atoms in the pyrimidine ring

FIGURE 2   Sources of atoms found in the purines and pyrimidines in living cells.

BIOSYNTHESIS OF PURINE RIBONUCLEOTIDES

De novo Biosynthesis of Purine Ribonucleotides Starts with an Activated Ribose

The most striking difference in the pathways to the purines and pyrimidines is the timing of ribose involvement. In *de novo* purine synthesis the purine ring is built on the ribose in a stepwise fashion. In pyrimidine synthesis the nitrogen base is synthesized prior to attachment of the ribose. In both instances the ribose-5-phosphate is first

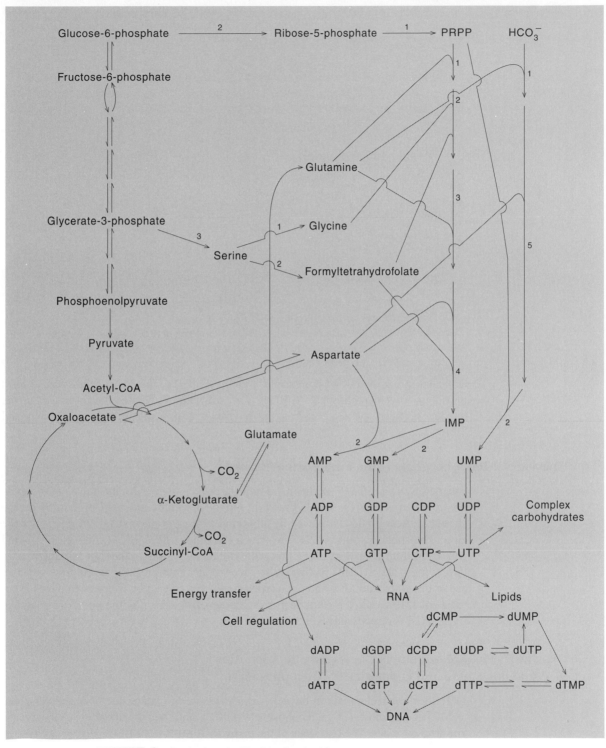

FIGURE 3 Synthesis and utilization of nucleotides.

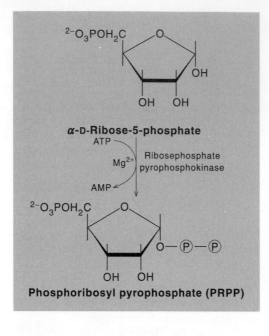

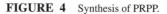

FIGURE 4   Synthesis of PRPP.

activated by addition of a pyrophosphate group (Fig. 4) to the C'-1 of the sugar to form phosphoribosyl pyrophosphate (PRPP). This activation facilitates the formation of the linkage between the C'-1 carbon of the ribose and a nitrogen of the purine and pyrimidine bases.

Inosine 5′-Monophosphate Is the First Purine Formed

The pathway from PRPP to the first complete purine nucleotide, inosine 5′-monophosphate, involves 10 steps and is shown in Fig. 5. It would seem logical that the purine should be built up first, followed by addition of ribose-5-phosphate, but this is not the case. The starting point is PRPP, to which the imidazole is added; the six-membered ring is built up afterwards. Reaction with glutamine results in the first intermediate in the purine pathway. This pathway also incorporates into purines, atoms from glycine, aspartate, glutamate, CO_2 and one-carbon fragments carried by folate coenzymes. Synthesis of IMP requires a great deal of energy as can be seen by the number of steps in which ATP is directly involved.

Inosine 5′-Monophosphate Is Converted into Adenosine 5′-Monophosphate and Guanosine 5′-Monophosphate

IMP does not accumulate in the cell but is converted to adenosine 5′-monophosphate (AMP) and guanosine 5′-monophosphate (GMP), and the corresponding diphosphates and triphosphates. The two steps of the pathway from IMP to AMP (Fig. 6) are typical

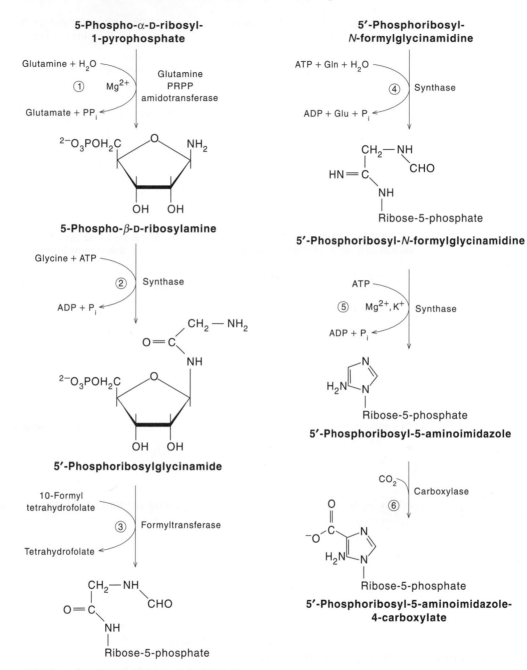

FIGURE 5 Biosynthetic pathway to inosine 5′-monophosphate. The names on the right of the reaction arrows indicate the enzymes that catalyze the individual steps in this 10-step pathway. The curved arrows on the left of the vertical reaction arrows indicate coreactants and coproducts for each step.

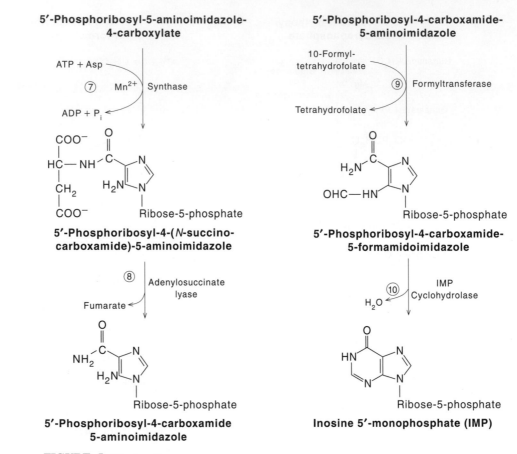

FIGURE 5 (*Continued*)

reactions by which the amino group from aspartate is introduced into a product. The 6-hydroxyl group of IMP (tautomeric with the 6-keto group) is first displaced by the amino of aspartate to give adenylosuccinate, and the latter is then cleaved nonhydrolytically by adenylosuccinate lyase to yield fumarate and AMP. In the condensation of aspartate with IMP, cleavage of GTP to guanosine diphosphate (GDP) and phosphate provides energy to drive the reaction.

Conversion of IMP to GMP also proceeds by a two-step pathway as shown in Fig. 6.

Conversion of Nucleoside Monophosphates to Nucleoside Triphosphates Goes through the Diphosphates

The products of the biosynthetic pathways discussed in the preceding section are mononucleotides. In cells a series of kinases (phosphotransferases) converts these mononucleotides to their metabolically active diphosphate and triphosphate forms.

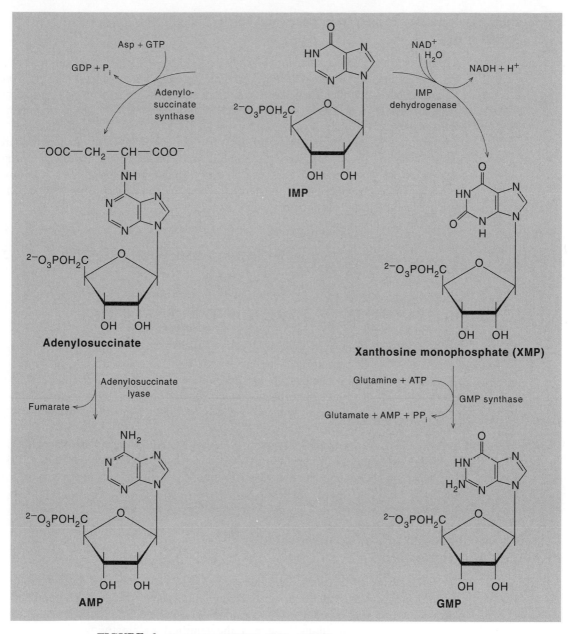

FIGURE 6   Conversion of IMP to AMP and GMP.

Nucleotides Also Can Be Formed from Bases and Nucleosides (Salvage Pathways)

In addition to the pathways for *de novo* synthesis, mammalian cells and microorganisms readily can form mononucleotides from purine bases and, to a lesser extent, from pyrimidine bases and their nucleosides. In this way, bases and nucleosides formed by constant breakdown of messenger RNA (mRNA) and other nucleic acids can be reconverted to useful nucleotides.

In mammals, specific enzymes for converting purine bases to nucleotides are present in many organs. In some tissues this may be the main source of purine nucleotides. The most important enzyme for carrying out these conversions in hypoxanthine-guanine phosphoribosyl transferase, which catalyzes the formation of IMP from hypoxanthine and GMP from guanine as follows:

$$\text{Hypoxanthine} + \text{PRPP} \rightleftharpoons \text{IMP} + \text{PP}_i, \tag{1}$$

$$\text{Guanine} + \text{PRPP} \rightleftharpoons \text{GMP} + \text{PP}_i. \tag{2}$$

BIOSYNTHESIS OF PYRIMIDINE NUCLEOTIDES

The biosynthetic pathway to pyrimidine nucleotides is simpler than that for purine nucleotides, reflecting the simpler structure of the base. In contrast to the biosynthetic pathway for purine nucleotides, in the pyrimidine pathway the pyrimidine ring is constructed before ribose-5-phosphate is incorporated into the nucleotide. The first pyrimidine mononucleotide to be synthesized is orotidine-5′-monophosphate (OMP); and from this compound, pathways lead to nucleotides of uracil, cytosine, and thymine. OMP thus occupies a central role in pyrimidine nucleotide biosynthesis, somewhat analogous to the position of IMP in purine nucleotide biosynthesis. Like IMP, OMP is found only in low concentrations in cells and is not a constituent of RNA.

Uridine 5′-Monophosphate Is a Precursor of Other Pyrimidine Mononucleotides

The pathway for UMP synthesis is shown in Fig. 7. It starts with the synthesis of carbamoyl phosphate, catalyzed by carbamoyl phosphate synthase. The carbamoyl phosphate reacts with aspartate to form carbamoyl aspartate. The reaction is catalyzed by aspartate carbamoyl transferase and the equilibrium greatly favors carbamoyl aspartate synthesis. In the third step the pyrimidine ring is closed by dihydroorotase to form L-dihydroorotate. Dihydroorotate then is oxidized to orotate by dihydroorotate dehydrogenase. In the final two steps of the pathway, orotate phosphoribosyl transferase yields orotidine-5′-phosphate, and a specific decarboxylase then produces UMP.

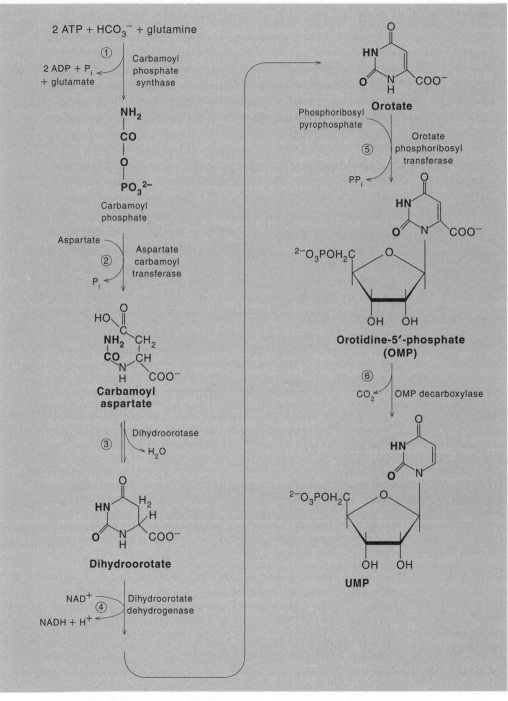

FIGURE 7   Biosynthesis of UMP.

Deoxyribonucleotides Are Formed by the Reduction of Ribonucleotides

Tracer studies with isotopically labeled precursors have shown that both in mammalian tissues and in microorganisms, deoxyribonucleotides are formed from the corresponding ribonucleotides by replacement of the 2′-OH group with hydrogen.

It is interesting to see that there is no independent way in which deoxyribose is produced in living cells. This finding is consistent with the notion that ribonucleotides came before deoxyribonucleotides.

PREBIOTIC SYNTHESIS OF NUCLEOTIDES

In the prebiotic world there are a variety of ways in which the three components of the nucleotides could have been synthesized and connected to one another. For example, the sugar and the base could have been linked first or the sugar or sugar phosphate could have served as a foundation on which the nitrogenous base was synthesized. Because the nitrogenous base is much more stable than the ribose, it seems highly likely that the nitrogenous bases were synthesized first so that when the ribose was synthesized the linkage of the two entities could be made before the sugars decomposed. It also seems likely that the phosphate was added after nucleoside formation. This is because conditions for phosphate addition are rather harsh, but once the C-1 of the ribose has been linked to the base, the sugar is stabilized so that it can tolerate the rigorous conditions necessary for phosphate addition.

In Chapter 12 we discuss possible routes for the prebiotic synthesis of ribose. In this section we start with a discussion of the synthesis of the base, then we consider the attachment of the preformed base to ribose, and finally we consider the conversion of the nucleoside to a nucleotide and its necessary activation prior to polynucleotide synthesis.

Hydrogen Cyanide Was Probably the Starting Point for the Prebiotic Synthesis of Purines

In 1961 John Oro discovered that small amounts of adenine could be synthesized after prolonged incubations of hydrogen cyanide (HCN). From that time on, HCN has been considered to be the key precursor for purine ring synthesis in the prebiotic world. HCN has a high free energy, making it an ideal starting compound for a multistep reaction. The situation here is parallel to that for formaldehyde in the prebiotic pathway.

Many claims have been made that HCN can be synthesized from mixtures of small molecules in the atmosphere. A spark discharge produces some HCN from a mixture of CO, N_2, and H_2. A spark discharge also triggers some HCN synthesis in a gaseous mixture of CO_2, N_2, and H_2. However, there are reports that the presence of CO_2 encourages the formation of NO. It also should be noted that producing nitrogen atoms needed for the formation of HCN is not easy because the N molecular bond is very stable (see Chapter 11, Table 1). Solar radiation less than 100 nm, which is only available in the very upper regions of the atmosphere, is needed to supply the dissociation energy, K. Zahnle has proposed a mechanism that starts with the breakage of N_2 in the ionosphere by ionization followed by dissociative recombination. This reaction is carried out in a weakly reducing atmosphere containing small amounts of CH_4.

$$N_2 + h\nu \rightarrow N_2^+ + e \tag{3}$$

$$N_2^+ + e \rightarrow N + N \tag{4}$$

Most of the N atoms should return to the ground state and flow down into the stratosphere where they react with the by-products of methane photolysis as follows:

$$CH_4 + h\nu \rightarrow CH_3 + H \tag{5}$$

$$CH_3 + N \rightarrow HCN + H_2 \tag{6}$$

CH_2, which also is produced by photolysis, can react with N to form additional HCN. The HCN made by photolysis or other high-energy sources in the atmosphere would be expected to rain down because of its high solubility in water.

In dilute solutions hydrogen cyanide slowly hydrolyzes to formamide and then to formic acid and ammonia, which readily convert into the very soluble salt ammonium formate

$$HCN \xrightarrow{H_2O} \underset{\text{Formamide}}{HCNH_2} \xrightarrow{H_2O} \underset{\text{Formic acid}}{HCOH} + \underset{\text{Ammonia}}{NH_3} \longrightarrow \underset{\text{Ammonium formate}}{HCO^- + NH_4^+} \tag{7}$$

It is hard to see how hydrogen cyanide concentrations sufficient to result in productive polymerization reactions could have been achieved in the ocean at large. The relatively rapid hydrolysis of cyanide to ammonium formate would make it difficult to slowly accumulate the hydrogen cyanide from atmospheric precipitation. Some means of concentrating the HCN would have been necessary to facilitate the process. Concentration of hydrogen cyanide by evaporation is not possible because it is more volatile than water. On the other hand, partial freezing provides an efficient means for concentrating hydrogen cyanide. At $-12°C$ a dilute solution of hydrogen cyanide is converted into a very concentrated solution because most of the water freezes out, leaving the hydrogen cyanide to become increasingly concentrated in the remaining aqueous solution.

Four Hydrogen Cyanide Molecules Condense to Form Diaminomaleonitrile

In concentrated solutions of hydrogen cyanide, a series of interactions takes place over a period of several months that leads to the formation of *diaminomaleonitrile* (*DAMN*) (Fig. 8). The reaction proceeds fastest at a pH of 9.2, which is the pK_a for HCN, suggesting that in the slow first step involving a dimerization a neutral HCN and an anionic CN^- are the reacting species as shown in Fig. 8.

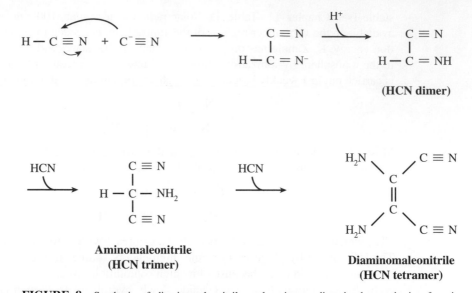

FIGURE 8   Synthesis of diaminomaleonitrile, a key intermediate in the synthesis of purines and pyrimidines from hydrogen cyanide.

Diaminomaleonitrile Is Readily Converted into Aminoimidazolecarbonitrile by Ultraviolet Light

Brief exposure (90 min) to ultraviolet (UV) light results in the conversion of DAMN of *aminoimidazolecarbonitrile* (*AICN*). The reaction is monomolecular and probably involves a four-membered ring intermediate (Fig. 9). Ordinary sunlight or a UV lamp rich in 310-nm radiation make ideal radiation sources. The reaction may be carried out in a dilute aqueous solution of DAMN or on a dried film of DAMN. If done properly, yields of AICN of 90% may be obtained.

Aminoimidazolecarbonitrile Is the Gateway to the Purines

The view that most purines can be formed directly or indirectly from AICN arose from observations made in the 1960s (Fig. 10). It was established that AICN can be convert-

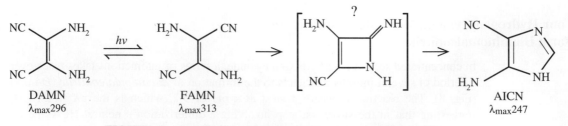

FIGURE 9   Established mechanism for the photochemically driven synthesis of aminimidazolecarbonitrile (AICN) from diaminomaleonitrile (DAMN).

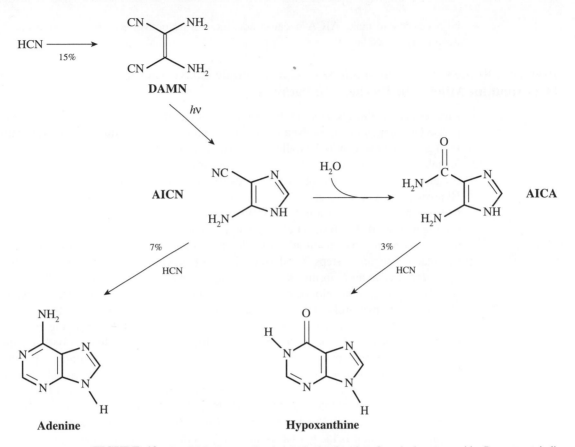

FIGURE 10   Established routes for the synthesis of purines from hydrogen cyanide. Percentages indicate the approximate maximum yields that have been observed for the different reaction steps.

ed directly to adenine by exposure to hydrogen cyanide. Alternatively, AICN can be hydrolyzed to *aminoimidazolecarboxamide* (*AICA*). Subsequently, AICA can be converted to hypoxanthine by treatment with hydrogen cyanide. These reactions are not particularly efficient. Thus the yields are low and the reaction conditions involve high temperatures (100°C) and long times (months). All this suggested that HCN was not the correct reagent at this stage but that it was being converted into something else before reaction with AICA. Further investigations in our laboratory led to superior conversions for this reaction. These are discussed later.

Aminoimidazolecarbonitrile Can Be Converted into Aminoimidazole Carboxamide by Limited Hydrolysis

AICN can be hydrolyzed to AICA by limited exposure to base. Prolonged exposure to base must be avoided or the amide converts to a carboxylic acid and further decomposition results. If the goal is to make the purines adenine and hypoxanthine, it may not be necessary to make AICA because adenine can be converted into hypoxanthine by deamination (see later).

Preferred Route for the Aminoimidazole Carboxamide Conversion to Hypoxanthine Mimics the Biochemical Pathway

Biochemical pathways tend to be conserved so that once an effective route to a particular compound has evolved it tends to be preserved. Sometimes this conservation is so great that one can find parallels between the biochemical pathway and the prebiotic pathway. The parallels are more likely to occur in the later steps of a pathway. This is because the early steps often of necessity start from quite different materials. Hypoxanthine synthesis is a case in point. The *de novo* biochemical pathway for hypoxanthine synthesis starts from PRPP; the purine ring of hypoxanthine is built on the sugar (Fig. 5). The product of step 8 in the biochemical pathway is 5-phosphoribosyl-4-carboxamide-5-aminoimidazole. This compound is identical to AICA except for the attached sugar. In steps 9 and 10 of the biochemical pathway, this intermediate is converted first to an *N*-formyl derivative and then by cyclization to the final product, IMP (see Fig. 5). In the biochemical pathway the formyl donor is a complex coenzyme (10-formyltetrahydrofolate). Let us assume that hypoxanthine synthesis in the prebiotic world occurred in a similar manner except for the absence of the ribose, the enzymes, and the complex coenzyme. In the prebiotic pathway it seems highly likely that a simpler molecule would have served as the formyl group donor. The simplest possible formyl donor is formate itself. Formate is highly likely to have been abundant in the prebiotic world because it is readily formed from the complete hydrolysis of HCN.

When an aqueous solution of AICA and ammonium formate are heated at 75°C and then allowed to evaporate to dryness, most of the AICA is converted to hypoxanthine. In this experiment it is likely that formate adds as in step 9 and that cyclization occurs as in step 10 of the biochemical pathways (Fig. 11). These combined reactions are much more efficient than those used in the previously described conversion using hydrogen cyanide (see Fig. 10), which resulted in only 3% conversion of AICA to hypoxanthine after many months. This finding underscores the value of scrutinizing the biochemical pathway for leads in exploring the corresponding prebiotic pathway. It also illustrates the potential importance of evaporation as a driving force for prebiotic reactions that entail the elimination of water. Both the formylation and cyclization reactions are driven by the removal of water because it is one of the reaction products when ammonium formate is used as one of the reactants.

Conversion of Aminoimidazolecarbonitrile to Adenine Parallels the Aminoimidazole Carboxamide to Hypoxanthine Conversion

Quite remarkably, when AICN is treated with ammonium formate under the same conditions that were used in the AICA–hypoxanthine conversion, adenine is produced in 90% yield. This conversion is more complex in that both the formate carbon and the ammonium nitrogen contribute to the six-membered ring of the purine. The likely steps in this reaction are depicted in Fig. 12.

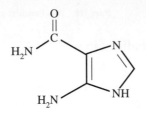

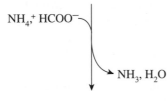

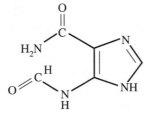

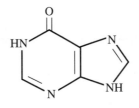

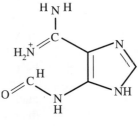

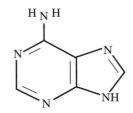

▲

FIGURE 11 New pathway from AICA to hypoxanthine. Reaction is carried out in an ammonium formate solution that is evaporated to dryness at 75°C. The chemical changes occurring in this transformation are identical to those that occur in steps 9 and 10 of the biochemical pathway. See Fig. 5.

FIGURE 12 New pathway from AICN to adenine. Reaction is carried out in an ammonium formate solution that is evaporated to dryness at 75°C.

Reactions Leading to Purine Synthesis Can Be Visualized in a Prebiotic Setting

The favored reactions from hydrogen cyanide to the two purines adenine and hypoxanthine are illustrated in Fig. 13. Although it is not possible to see how all these reactions could take place rapidly within the confines of a very small space, it is possible to imagine the reactions being localized in one geographic area that included, for example, a mountain and the atmosphere surrounding it. Formation of hydrogen cyanide in the atmosphere would probably have proceeded best in a region of intense volcanic activity, because this is where the atmospheric concentrations of the precursor molecules — carbon monoxide, hydrogen, nitrogen, and ammonia — would have been the greatest. The hydrogen cyanide formed in the atmosphere would be expected to rain down because of its high solubility in water. A freshwater pond on a mountainside might have served as a convenient catch basin for the hydrogen cyanide; here the compound could have become concentrated in the winter months by partial freezing. In a cold concentrated solution, DAMN would be expected to form slowly over a period of many months. The conversion of DAMN to AICN might occur in the spring after a thaw as a stream containing DAMN irradiated by the sunlight flows down the mountainside to a second location, where a fraction of the AICN might be converted to AICA by contact with a clay at somewhat elevated temperatures (e.g., 75°C). Following this, the stream now containing both AICN and AICA could continue to flow until it reached warmer waters containing ammonium formate. The slow evaporation of this pool to dryness over the summer months should result in the efficient conversion of the two imidazoles into the corresponding purines.

While AICN may be hydrolyzed to AICA to create the necessary precursor for hypoxanthine, hypoxanthine also could be formed by deamination of adenine by nitrite. This reaction is very efficient and selective, and might be the preferred route from adenine to hypoxanthine provided that nitrite was available. Nitrite often has been considered as a likely prebiotic form of nitrogen.

Once synthesized, the two purines could react with ribose to form nucleosides or nucleotides. However, because the purines are quite stable, they need not react immediately. Thus they could have accumulated for a long time until the conditions were right for going on to the next step in nucleotide synthesis.

In the conversions just described, ammonium formate plays a crucial role. As already indicated, the complete hydrolysis of hydrogen cyanide results in ammonium formate. Prior to the studies just presented, this hydrolysis might have been considered a wasteful reaction; however, in the light of these findings, ammonium formate can be seen to serve an important function.

FIGURE 13   Proposed prebiotic pathway for the synthesis of the purines adenine and hypoxanthine ▶ from hydrogen cyanide. By one route adenine is formed first from AICN and hypoxanthine is subsequently formed by partial deamination of the adenine. By an alternative route AICN is a branchpoint in which adenine is formed directly while hypoxanthine is formed via AICA.

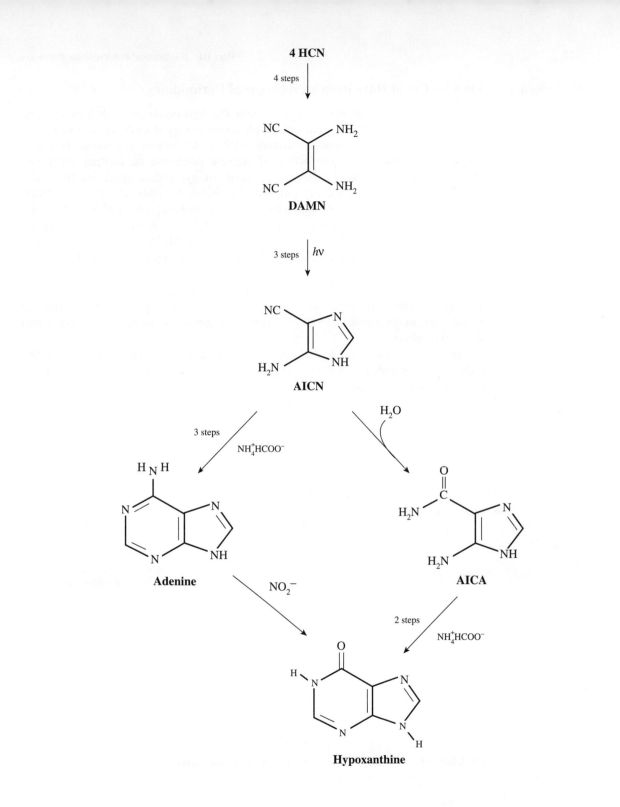

Hydrogen Cyanide Also Could Have Been a Precursor of Pyrimidines

Pyrimidines as well as purines result from the hydrolysis of hydrogen cyanide oligomers (Fig. 14). Orotic acid can be photochemically decarboxylated to uracil, one of the two major pyrimidines found in RNA. Alternatively, cyanoacetylene, a major product formed by the action of electric discharge on methane–nitrogen mixtures, could serve as the starting material for pyrimidine synthesis (Fig. 15). This route, which was originally proposed by Ferris, Sanchez, and Orgel in 1968, entails the hydrolysis of cyanoacetylene to cyanoacetaldehyde, followed by reaction with guanidine in aqueous solution to give 2,4-diaminopyrimidine. Cyanoacetaldehyde also can react with urea to give cytosine. M. Robertson and S. Miller have shown that high concentrations of urea result in greatly expanded yields of cytosine.

The efficiency of pyrimidine synthesis from cyanoacetylene far exceeds that from hydrogen cyanide. However, because hydrogen cyanide was probably formed in greater yield under primitive Earth conditions, it is difficult to assess the relative merits of the two pathways.

Uracil results from the hydrolysis of cytosine (Fig. 16). The half-time for this hydrolysis is 19 days at 100°C or an estimated 100 years at 30°C. Thus, provided cytosine can be made, it is possible to obtain uracil in one additional step.

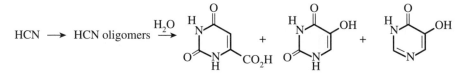

Orotic acid   5-Hydroxyuracil   4, 5-Dihydroxy-
                                          pyrimidine

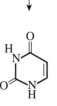

$h\nu$

Uracil

FIGURE 14   Synthesis of various pyrimidines from hydrogen cyanide oligomers.

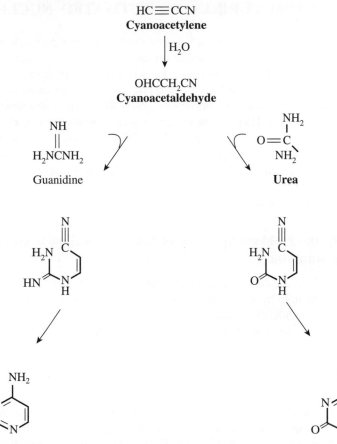

2,4 – Diaminopurine

Cytosine

FIGURE 15   Sythesis of cytosine from cyanoacetylene.

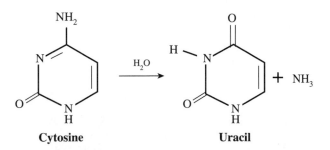

Cytosine   **Uracil**

FIGURE 16   Formation of uracil by hydrolysis of cytosine. As in the case of the deamination of adenine to hypoxanthine, nitrite is the reagent of choice for this hydrolysis.

FROM PURINES TO ACTIVATED NUCLEOTIDES

The biggest difference between the biochemical pathway and the prebiotic pathway to hypoxanthine just described relates to the time of attachment of ribose. The purine is built on the sugar in the *de novo* biochemical pathway, whereas in the prebiotic pathway the purine is fully formed prior to sugar involvement.

Why should the purines be synthesized before attachment to ribose in the prebiotic pathway? There are at least two good reasons for this. First, the difference in starting materials makes it impossible to follow the same route as in the biochemical pathway Second, because the purines are much more stable than ribose is, it is a better strategy to have the purines available first so that the sugars can be attached as soon as they are synthesized. Once the sugars are attached to form nucleosides the sugars are much more stable.

Conditions for the Prebiotic Synthesis of Nucleosides Have Been Found Only for Hypoxanthine

Fuller, Sanchez, and Orgel found that small amounts of inosine could be formed by heating hypoxanthine and ribose to dryness in the presence of Mg^{2+}. Divalent cation is absolutely essential and Mg^{2+} is the best one found so far. It should be noted that there are two possible isomers that could be formed when a purine reacts with ribose: one in which the purine and the C'-5-OH are on the same side of the ribose (β configuration) and one where the purine and the C'-5-OH are on opposite sides of the ribose (α configuration). The β configuration is the one found in natural products and also predominates in the prebiotic simulation experiment. A possible mechanism for formation of the β-isomer involving nucleophilic attack by the N9 of the purine on the C'-1 of the ribose is shown in Fig. 17. In this mechanism the divalent cation forms a transient complex with the C'-1 and C'-2 hydroxyl groups, which increases the nucleophilicity of the C'-1 carbon. The formation of the nucleoside mimics the biochemical salvage

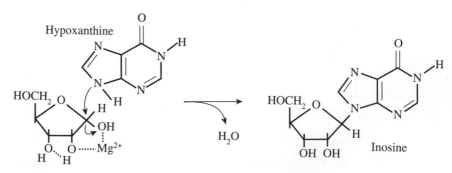

FIGURE 17   Formation of inosine nucleoside from hypoxanthine purine and D-ribose sugar by dehydration.

pathway except that neutral ribose is used in place of PRPP. The prebiotic reaction with neutral ribose would not be expected to occur in solution for energetic reasons. However, removal of water by dehydration makes the reaction energetically favorable by shifting the equilibrium in favor of complex formation.

The procedure used for inosine does not work for adenosine synthesis or for pyrimidine nucleoside synthesis.

The current procedure for making nucleosides has several shortcomings. First, the yields of inosine are low. Second, the procedure only works with hypoxanthine. Finally, there is no selectivity in the reaction; thus all four aldopentoses work equally well in the reaction with hypoxanthine. Perhaps the prebiotic route to linking ribose and base would be more effective if an activated ribose were used as in the biochemical pathway where PRPP is used. As in the case of ribose synthesis, more experiments are clearly necessary to resolve this important issue.

Nucleosides Can Be Phosphorylated by Orthophosphate or Trimetaphosphate

If ribose-5′-phosphate is used in the purine condensation experiment, the product is a mononucleotide. If ribose is used instead, the resulting nucleoside must be phosphorylated to convert it to a nucleotide. In biochemical reactions a high energy phosphate compound is used in such a reaction. However, in a prebiotic situation inorganic orthophosphate can be used. This was shown first by Lohrmann and Orgel who discovered that nucleosides can be phosphorylated in a mixture containing urea, ammonium chloride, ammonium bicarbonate, and sodium phosphate at temperatures between 65 and 100°C. A solution containing these ingredients is evaporated to the dry state. The removal of water provides the thermodynamic pull for the reaction because water is one of the reaction products. Unfortunately, the main product in this system is the nucleoside 2′,3′-cyclic phosphate. In our laboratory we have found that if the experiments with orthophosphate are carried out in the presence of urea and sodium phosphate, the main derivative is the 5′-phosphomononucleoside in yields of up to 60%. Under our conditions, the 5′-phosphomononucleoside is the main product with traces of the 2′,3′-cyclic derivative also being formed. Derivatives of 2′- or 3′-phosphomononucleoside may account for as much as 30% of the mononucleotide formed under our conditions.

Nucleoside-5′-polyphosphates are formed by the dry-phase heating of nucleoside-5′-phosphates with trimetaphosphates in the presence of Mg^{2+} (Fig. 18). On hydration these polyphosphates degrade rapidly to the corresponding triphosphates, which in turn decompose much more slowly to diphosphates and monophosphates.

As a component in a prebiotic pathway for making phosphorylated nucleotides, meta- or polyphosphates raise some concern because of the scarcity of accessible phosphate in Earth's crust. Trimetaphosphates are readily produced by heating calcium dihydrogen phosphates to about 300°C but neither trimetaphosphates nor dihydrogen phosphate are currently found in any abundance in mineral deposits. Most of the phosphates not found in living organisms or their recent decomposition products

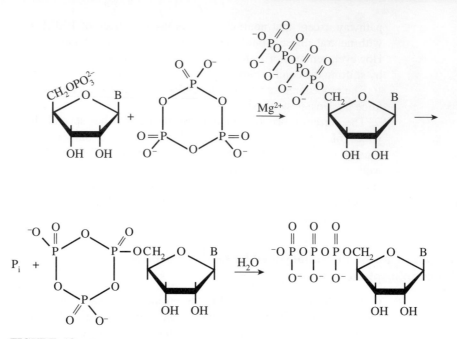

FIGURE 18   Formation of nucleoside-5'-triphosphates by dry-phase heating with trimetaphosphate.

exist in the form of the calcium phosphate mineral hydroxylapatite. Phosphate in this form is considered difficult to access. These concerns may not be all that serious if the distribution of phosphates has changed substantially in the last 3.5 billion years. It is entirely possible that in prebiotic times there existed much more accessible phosphates in the mineral reservoirs, but that these have been depleted as a result of transfer to living organisms. Most of what is found in living cells has a rapid turnover time and appears to be transferred from one living form to another as birth follows death.

Prebiotic Synthesis of Activated Nucleotides Is Poorly Understood

Once the mononucleotide has been made, it must be converted to an activated derivative suitable for incorporation into a polynucleotide chain. In biochemical pathways the nucleoside triphosphate derivative is usually used. The triphosphate derivative has more than enough chemical energy to power the formation of the phosphodiester linkages found in polynucleotides so there is no thermodynamic problem here. However, these compounds are not very reactive. In biosystems sophisticated polymerases are essential to catalyze the polymerization of nucleoside triphosphates.

Orgel and others have searched for other forms of activated nucleotides that would be reactive under mild conditions and would not require any more than a divalent cation catalyst. Their extensive search has led them to the use of imidazole-activated mononucleotides (Fig. 19, and see Chapter 14, Fig. 6). Compounds of this type can be synthesized very efficiently by organochemical methods but a satisfactory prebiotic route for their synthesis has not been discovered. Because the reaction of imidazole and a mononucleotide involves the loss of a water molecule, a remote possibility is that the phosphorimidazolide is formed under dehydrating conditions (see Fig. 19).

The formation of activated nucleotides by a prebiotically plausible route remains a most challenging problem. The first problem to be resolved is the form of activation. For 20 years imidazole activation of nucleotide monophosphates has been a favorite. We wonder if more conventional triphosphates or tetraphosphates, which are easier to make, might be more appropriate. Their use has not gained favor because triphosphates do not appear to be reactive under mild prebiotic conditions that would be necessary for template reactions. In this regard very little actually has been done. It has been pointed out by Tom Steitz that all polymerases contain two divalent cations strategically spaced so that they function in the polymerization growth step (Fig. 20). To my knowledge no such model system ever has been tested.

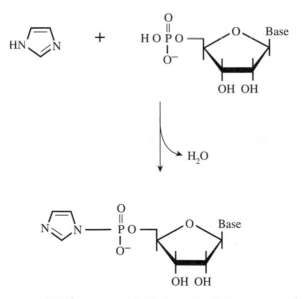

FIGURE 19   A possible route for nucleoside phosphorimidazolide synthesis from imidazole and a 5'-mononucleotide. The secondary pK for the phosphate group on a 5'-mononucleotide is about 6.2, meaning that the nucleotide would be approximately 50% protonated at this pH.

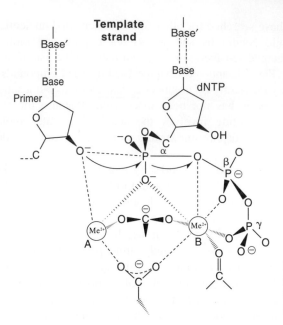

FIGURE 20 All known nucleic acid polymerases employ two strategically positioned divalent cations in the polymerization step.

SUMMARY

In this chapter we have reviewed the major pathways for nucleotide biosynthesis and have considered how these nucleotides might have been synthesized in the prebiotic world.

1. The ribose for nucleotide synthesis is made from glucose by the pentose phosphate pathway as ribose-5-phosphate. This is converted to PRPP, the starting point for purine synthesis. This pathway also incorporates into purines atoms from glycine, aspartate, glutamate, CO_2, and one-carbon fragments carried by folates. IMP synthesized by this route is converted by two-step pathways to AMP and GMP, respectively.

2. The biosynthetic pathway to UMP starts from carbamoyl phosphate and results in the synthesis of the pyrimidine orotate, to which ribose phosphate is subsequently attached.

3. Conversion of nucleoside monophosphates to nucleoside triphosphates goes through the diphosphates.

4. Nucleotides also can be formed from bases and nucleosides by the so-called salvage pathways.

5. Adenine can be synthesized in low yields from HCN. This result plus the fact that there are a number of ways in which HCN could have been synthesized in the prebiotic atmosphere with the help of UV radiation suggests that HCN was the major precursor of purines and possibly also of pyrimidines.

6.   HCN made in the atmosphere would be expected to rain down because of its high solubility in water. At subzero temperatures a dilute solution of HCN becomes more concentrated as the water freezes out of it. In the cold, HCN interacts to form diaminomaleonitrile, a tetramer of HCN. Diaminomaleonitrile is efficiently converted into AICN by exposure to UV light. AICN is most efficiently converted to adenine by evaporating dilute solutions of AICN in the presence of ammonium formate. Partial hydrolyisis of AICN to AICA and evaporation of dilute solutions of AICA in the presence of ammonium formate lead to the purine hypoxanthine. This pathway mimics the biochemical pathway very closely.

7.   It is possible that HCN was also an important precursor to pyrimidines.

8.   Effective prebiotic conditions for nucleoside synthesis only have been found for hypoxanthine. A solution of hypoxanthine and ribose in the presence of a magnesium salt is heated to dryness and then heated as a dry film for many hours. This results in inosine nucleoside in yields of up to 50%.

9.   Nucleosides can be converted into nucleotides by heating to dryness in a solution containing inorganic orthophosphate and urea.

10.   No satisfactory pathways exist for the conversion of nucleoside monophosphates into phosphorimidazolides, which make excellent precursors for oligonucleotide synthesis in nonenzymic systems.

Problems

1.   When PRPP is treated with a base, 5-phosphoribose-1,2-cyclic phosphate is formed. Produce a mechanism for this reaction. What is the second product in this reaction?

2.   Compare and contrast the pathways for the biosynthesis of IMP with that of UMP. What are the two most striking features of this comparison?

3.   Contrast the synthesis of purines by the biochemical route with the most likely prebiotic route.

4.   Adenine can be synthesized most efficiently from an aminoimidazole carbonitrile (AICN) and ammonium formate in an evaporite. What was the most likely source of ammonium formate in the prebiotic world?

5.   Describe three reactions in the prebiotic pathway to activated nucleotides in which heating to dryness is absolutely essential for reaction. Explain.

6.   Ribose is very sensitive to heating. This sensitivity is relieved when it becomes linked in a nucleoside. Can you suggest why this makes a big difference?

7.   In the prebotic environment at the time when nucleoside synthesis is underway, which compound is more likely to be present in excess, adenine or ribose? Explain. Is this likely to create any special problems?

8.   Starting with HCN and CH_2O, diagram the reactions that are essential for 5'-AMP synthesis and indicate where divalent cation catalysis is likely to be involved. Indicate when heating and dehydration are required. Also indicate the

reactions that you think need further work before it can be said that we have an effective prebiotic pathway.

References

Concerning the Biochemistry of Nucleotides

Zubay, G. *Biochemistry*. 4th ed. New York: McGraw-Hill, 1998.

Concerning Possible Prebiotic Pathways to the Nucleotides

Ferris, J. P., Sanchez, R. A., and Orgel, L. E. Studies in Prebiotic Synthesis. III. Synthesis of Pyrimidines from Cyanoacetylene and Cyanate, *J. Mol. Biol.* 29:693–704, 1968.

Fuller, W. D., Sanchez, R. A., and Orgel, L. E. Studies in Prebiotic Synthesis. VI. Synthesis of Purine Nucleosides, *J. Mol. Biol.* 67:25–33, 1972.

Lohrman, R. and Orgel L.E. Urea-Inorganic Phosphate Mixtures as Prebiotic Phosphorylating Agents, *Science* 171:490–494, 1971.

Oro, J. Mechanism of Synthesis of Adenine from Hydrogen Cyanide under Possible Primitive Earth Conditions, *Nature (London)* 191:1193–1194, 1961. (This marks a turning point in our understanding of the origin of purines.)

Oro, J. and Kimball, A. P. Synthesis of Purines under Possible Primitive Earth Conditions. I. Adenine from Hydrogen Cyanide, *Arch. Biochem. Biophys.* 94:217–227, 1961. (This paper focuses attention on the potential of HCN as a precursor of purines.)

Robertson, M. P. and Miller, S. L. An Efficient Prebiotic Synthesis of Cytosine and Uracil, *Nature (London)* 375:772–774, 1995.

Sanchez, R. A., Ferris, J. P,. and Orgel, L. E. Studies in Prebiotic Synthesis. II. Synthesis of Purine Precursors and Amino Acids from Aqueous Hydrogen Cyanide, *J. Mol. Biol.* 30:223–253, 1967

Sanchez, R. A., Ferris, J. P., and Orgel, L. E. Studies in Prebiotic Synthesis. IV. Conversion of 4-Aminoimidazole-5-carbonitrile Derivatives to Purines, *J. Mol. Biol.* 38:121–128, 1968.

Zubay, G. To What Extent Do Biochemical Pathways Mimic Prebiotic Pathways? *Chemtracts-Biochem. Mol. Biol.* 4:317–323, 1993.

Zubay, G. A feasible prebiotic pathway to the purines, *Chemtracts-Biochem. Mol. Biol.* 5:179–198, 1994.

CHAPTER 14

RNA Metabolism and Prebiotic Synthesis of RNA

In Chapter 9 some aspects of RNA structure and function are briefly considered. In this chapter we elaborate on the chemistry of the basic transcription process and then proceed to the question of how RNAs were synthesized in the early stages of biological evolution.

RNA METABOLISM

In prokaryotes DNA, RNA, and protein synthesis all take place in the same cellular compartment, the cytosol. In eukaryotes the DNA is compartmentalized in the cell nucleus. Experiments with radioactively labeled compounds were used to demonstrate that eukaryotic RNA synthesis takes place in the cell nucleus. Much of the newly synthesized RNA is degraded quickly; most of the portion that survives is transferred to the cytosol where protein synthesis takes place. These observations are consistent with the proposition that RNA is the carrier of genetic information for the synthesis of proteins.

First RNA Polymerase to Be Discovered Did Not Use a DNA Template

The first RNA polymerase to be discovered was a bacterial enzyme called *polynucleotide phosphorylase*. This enzyme catalyzed the formation of 5′, 3′-linked polyribonucleotides from nucleoside diphosphates. It seemed unlikely that this could be the major cellular RNA polymerase, however, because there was no requirement for DNA to direct synthesis. In fact the nucleotide sequence was uncontrollable except in a crude way by adjusting the relative concentrations of different nucleotides in the starting materials.

Stimulated by these studies, Sam Weiss began a search for an RNA polymerase that had a requirement for DNA. He found such an enzyme in liver cell-free extracts. Unlike polynucleotide phosphorylase this polymerase used nucleoside triphosphates for substrates instead of nucleoside diphosphates.

DNA–RNA Hybrid Duplexes Indicate That Most Cellular RNAs Carry DNA Sequences

Sol Spiegelman reasoned that if RNA was made on a DNA template, it should be complementary to one of the DNA chains just as two DNA chains in a standard double helix are complementary to one another. In this event it should be possible to make a specific DNA–RNA duplex complex by mixing the two nucleic acids together under conditions favoring complex formation. By using ^{3}H-labeled T2 bacteriophage DNA and ^{32}P-labeled RNA that was synthesized after T2 infection of *Escherichia coli* cells, he was able to show that the newly synthesized RNA forms a specific complex with the viral DNA. It was presumed that the specific complex must involve Watson–Crick-like base pairing between an RNA and a DNA chain, with uracil playing the role of thymine in the RNA. Such a complementary interaction was a strong indication that the RNA was a product of DNA-directed synthesis. Although most RNAs are synthe-

TABLE 1

Types of RNA in *Escherichia coli*

Type	Function	Number of different kinds	Number of nucleotides	Percentage of synthesis	Percentage of total RNA in cell	Stability
mRNA	Messenger	Thousands	500–6000	40–50	3	Unstable ($t_{1/2} = 1$–3 min)
rRNA	Structure and function of ribosomes	3 { 23S / 16S / 5S	2800 / 1540 / 120	50	90	Stable
tRNA	Adapter	50–60	75–90	3	7	Stable

Biochemistry 2/E by Zubay, © 1988. Reprinted by permission of Prentice-Hall, Inc., Upper Saddle River, NJ.

sized as a result of transcription from a DNA template, different strategies are used in their synthesis and in their posttranscriptional modification.

There Are Three Major Classes of Cellular RNA

There are three major types of RNA that are transcribed from a nuclear DNA template: messenger RNA (mRNA), ribosomal RNA (rRNA), and transfer RNA (tRNA). These three RNAs work together in protein synthesis. Some of the properties of the three RNAs as they are found in *E. coli* bacteria are summarized in Table 1. mRNA carries the message from the DNA for the synthesis of a polypeptide chain with a specific sequence of amino acids. tRNA carries amino acids to the template for protein synthesis. The bulk of the cellular RNA is rRNA. There are three ribosomal RNA molecules (see Table 1), which differ substantially in size. The three rRNAs are always found in a complex with proteins in a functional component known as the ribosome. The ribosome is the site where mRNA and tRNAs meet to engage in protein synthesis.

Transcription Process Is Divided into an Initiation Phase, an Elongation Phase, and a Termination Phase

All DNA-dependent RNA polymerases carry out the following reactions:

$$NTP + (NMP)_n \overset{\underset{\text{DNA}}{Mg^{2+}}}{\longrightarrow} (NMP)_{n+1} + PP_i$$

The subsequent breakdown of PP_i ensures the irreversibility of this reaction, which explains why RNA polymerases utilize nucleoside triphosphates (NTPs) instead of nucleoside diphosphates (NDPs). The DNA template strand determines which base is added to the growing RNA molecule. For example, a cytosine in the template strand of DNA means that a complementary guanine is incorporated at the corresponding location of the RNA. Synthesis proceeds in a $5' \rightarrow 3'$ direction, with each new nucleotide being added onto the 3'-OH end of the growing RNA chain.

The overall process for RNA synthesis on a duplex DNA, called transcription, can be conceptually divided into initiation, elongation, and termination (see Chapter 9, Fig. 9). In the initiation phase of the reaction, RNA polymerase binds at a specific site on the DNA called the promoter. Here it unwinds, exposing the bases in a limited region of the DNA. Once the polymerase binds to the promoter and strand separation occurs, initiation usually proceeds quite rapidly. The first or initiating NTP, which is usually adenosine triphosphate (ATP) or guanosine 5′-triphosphate (GTP), binds to the enzyme. The binding is directed by a complementary base in the DNA template strand at the start site. A second NTP binds, and initation occurs on formation of the first phosphodiester bond by a reaction involving the 3′-hydroxyl group of the initiating NTP with the inner phosphorus atom of the second NTP. Inorganic pyrophosphate derived from the second NTP is a product of the reaction.

After initiation has occurred, chain elongation proceeds by the successive binding of the nucleoside triphosphate complementary to the base at the growth point on the template strand, bond formation with pyrophosphate release, and translocation of the polymerase one base farther along the template strand (Fig. 1). As elongation proceeds, the DNA unwinds progressively in the direction of synthesis. The short stretch of DNA–RNA hybrid formed during synthesis is prevented from becoming longer than 10 to 20 bp by a rewinding of the DNA and the simultaneous displacement of the newly formed RNA. Transcription proceeds in the $5′ \rightarrow 3′$ direction, antiparallel to the $3′ \rightarrow 5′$ direction of the templating DNA strand.

Termination of transcription involves stopping the elongation process at a region on the DNA template that signals termination and release of the RNA product and the RNA polymerase. The termination signal is a sequence of bases on the DNA that causes the RNA polymerase to pause, terminate, and detach.

Transcripts Are Frequently Subject to Posttranscriptional Processing and Modification

Most RNAs are not made in their final functional froms as they peel off the DNA template. They must undergo backbone phosphodiester bond cleavages into smaller molecules (processing) and individual base changes (modification). The types of alterations that pre-tRNA and pre-rRNA transcripts undergo are very similar in prokaryotes and eukaryotes.

In prokaryotes, most mRNAs function in translation with no prior alterations. Indeed, little opportunity arises for any alterations because translation starts on the nascent transcript before transcription has been completed. By contrast, in eukaryotes transcription occurs in the nucleus, and a transcript undergoes extensive changes before being transported to the cytoplasm where translation takes place. One of the most interesting types of processing in eukaryotic pre-mRNAs involves the removal of interinal segments by a process called *splicing* because it mimics film splicing. Splicing was discussed briefly in Chapter 9 and we elaborate on it later in this chapter because of the involvement of RNA enzymes (ribozymes) in the process.

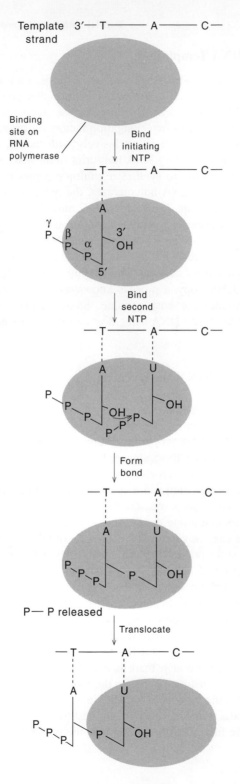

FIGURE 1 Details of phosphodiester bond formation. The α, β, and γ phosphates are indicated on the initiating NTP, which in this case is ATP. The shaded ovals represent NTP binding sites on the RNA polymerase. The biochemistry of bond formation in RNA synthesis is very similar to that in DNA synthesis.

Some Viruses Transcribe RNA from RNA Templates

The RNA genomes of single-stranded RNA bacterial viruses, such as Q_B, are themselves mRNAs. Bacteriophage Q_B codes for a polypeptide that combines with three host proteins to form an RNA-dependent RNA polymerase (*replicase*). The three host proteins are ribosomal protein S1 and two elongation factors for protein synthesis: EF-Tu and EF-Ts. The Q_B replicase functions exclusively with the Q_B RNA plus-strand template. It first makes a complementary RNA transcript (minus-strand) and ultimately uses the minus-strand as a template to synthesize multiple copies of viral RNA plus-strands. Like the DNA-dependent RNA polymerases, the replicase utilizes rNTPs and transcribes in the $5' \rightarrow 3'$ direction. The phage RNA must first act as an mRNA to direct the synthesis of the previously mentioned protein component of the replicase, because uninfected cells do not have an RNA-dependent RNA polymerase or replicase.

RNA tumor viruses (retroviruses) that infect animal cells exhibit a different replication strategy. In their virions they carry an enzyme that uses the viral RNA as a template to synthesize a DNA copy, a process known as *reverse transcription*. This DNA becomes integrated into the host genome. Subsequently, the viral RNA is transcribed from the integrated viral DNA using host cell RNA polymerase.

PREBIOTIC SYNTHESIS OF RNA

The main polymers of biological importance are polysaccharides, polypeptides, and polynucleotides. Although the connectivities between monomeric units with each polymer type are virtually identical, other properties distinguish individual monomeric units. Polypeptides and polynucleotides almost always form strictly linear structures, while polysaccharides also can form branched-chain structures. Polysaccharides that contain more than one type of monomeric unit use a specific enzyme to catalyze each type of intermonomeric linkage. Thus, in large measure the enzymes and the availability of activated monomeric substrates determine the sequence for polysaccharides. By contrast, with polypeptides a single enzyme catalyzes the addition of all monomers to the growing chains; the same is true for polynucleotides as we have just seen. For polypeptides and polynucleotides the order of monomers is dictated not by the enzyme but by the precise positioning of activated monomers on the templates employed in their synthesis.

Nucleic acids in different forms function as the templates for DNA, RNA, and protein synthesis. Despite this fact, until 1982 it was commonly thought that polypeptides and polynucleotides somehow coevolved. This long-held view was all but dismissed when it was discovered that some RNAs possessed catalytic activity for processing their own sequences. Currently, there is a widespread belief that nucleic acid enzymes (ribozymes) were the main enzymes in the most primitive living systems and that their existence predated the process of translation.

In this section we discuss template-independent and template-directed synthesis of polyribonucleotides. In the final section we consider the properties of ribozymes and their isolation.

Could Monomer-to-Polymer Conversion Have Occurred by Simple Dehydration?

Formally, the growth of all three types of biopolymers—polysaccharides, polypeptides, and polynucleotides—could be represented by a simple condensation reaction entailing the loss of a water molecule. For example, when two amino acids form a dipeptide, a water molecule is released (see Chapter 6, Fig. 5). The resulting dipeptide contains a carboxyl group at one end and an amino group at the other end. Further reaction could occur between another amino acid and the dipeptide or between two dipeptides. By repeating this process, polypeptides could eventually be formed.

In polysaccharide formation the situation is similar. Hydroxyl groups from two monomers could form a glycosidic bond by the release of a water molecule (see Chapter 6, Fig. 4).

In the case of nucleic acids, dehydration condensations can occur at three different levels (Fig. 2): first, in the formation of the nucleoside from the heterocyclic base and the ribose sugar; second, in the formation of a phosphate ester between the primary alcohol group of the terminal C'-5 carbon atom of the ribose and one of the hydroxyl groups of inorganic phosphate; and third, in the formation of the internucleotide linkage in the polymerization process. At successive stages in the polymerization process there is a phosphoric acid group at one end of the molecule and a hydroxyl

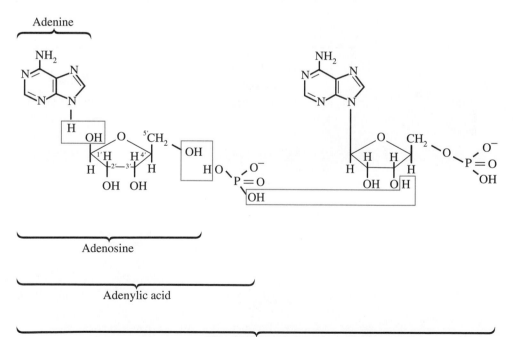

FIGURE 2   Dehydration condensation reactions in the formation of ribonucleic acid.

group at the other end; as a result the molecule can be extended farther at either end by additional dehydrations.

Synthesis of First Oligonucleotides May Have Occurred on Wet Clay Surfaces

The polymerization of monomeric nucleotides can be accomplished by dry heating or by addition of a condensing agent such as a polyposphate or cyanamide.

A more effective method of making oligonucleotides is to first activate the phosphate group on the mononucleotides and then to allow reaction to take place in aqueous solution. Nucleoside phosphorimidazolides have been used for this purpose. When a nucleoside-5'-phosphorimidazolide (5'-ImpN) is used, a dinucleotide is formed by reaction with another nucleotide (Fig. 3a); the reaction can continue if the receptor nucleotide is also 5' activated. If a 3'-phosphorimidazolide is used instead, it undergoes an intramolecular reaction, leading to a nucleoside-2',3'-cyclic phosphate (see Fig. 3b).

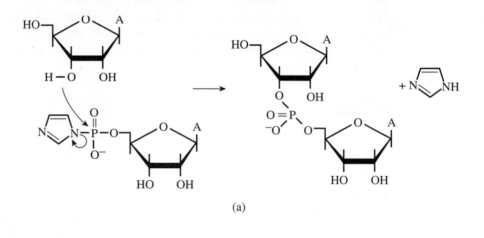

(a)

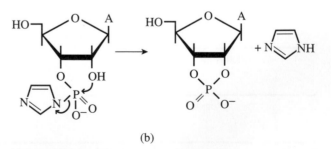

(b)

FIGURE 3 The formation of an internucleotide bond from a nucleoside and a nucleoside-5'-phosphorimidazolide (a). A 3'-phosphorimidazolide would rapidly give a 2',3'-cyclic phosphate (b).

Polymerization of imidazole-activated 5'-nucleotides is not very efficient in aqueous solution because of the competing hydrolysis reaction (Fig. 4). Furthermore, when activated nucleotides react with one another, there is a tendency for them to form pyrophosphate-linked products. For example, 5'-ImpA forms P,P'-diadenosine-5',5'-pyrophosphate (AppA). If the clay montmorillonite is added to the reaction mixture, the polymerization reaction is strongly favored. James Ferris and coworkers have shown that oligonucleotides of up to 55 nucleotides, mostly in 3',5' linkages, can be synthesized in this way. Here, the clay may perform a role similar to its alleged role in the formation of polypeptides from aminoacyl AMPs (see Chapter 17). As we soon see, reaction of activated nucleotides is far more efficient if they are first bound to a polynucleotide template. Nevertheless, observations on oligonucleotide synthesis in the absence of such templates are very important to studies on the origin of life, because the first oligonucleotides to be synthesized clearly did not have the advantage of nucleic acid templates. It seems likely that the first oligonucleotides were rather short and peppered with imperfections, as in the experiments just described.

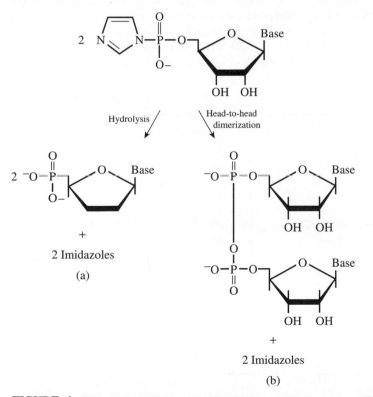

FIGURE 4   Side reactions of phosphorimidazolides in solution that prevent efficient polymerization. (a) Hydrolysis in which the activating group detaches; (b) head-to-head dimerization that leads to a pyrophosphate linkage.

Thereafter, the more regular oligonucleotides so formed were probably preferred as templates for further oligonucleotide synthesis. This form of Darwinian selection led to a larger number of regular oligonucleotides in a template-conducted second round of oligonucleotide synthesis. As the supply of regular templates increased, the rate of synthesis is also likely to have increased, as long as there was an adequate supply of activated monomers to feed the growth of new chains.

Template-Directed Polynucleotide Synthesis Follows Base-Pairing Rules of Watson and Crick

The first successful template-catalyzed reaction for oligonucleotide formation was conducted by Naylor and Gilham in 1966 (Fig. 5). They showed that a poly(dA) template catalyzed the formation of dodecamers of dT from the corresponding hexamers. Subsequently, Leslie Orgel and coworkers embarked on an extensive study of polymerization that could be carried out with select combinations of polymer template and activated monomers.

Template-catalyzed reactions appear to be most effective when a stable complex forms between template and activated monomers. A mixture of polycytidylic acid [poly(C)] and monomers of guanylic acid forms helical complexes that are stable at 0°C. Below pH 7 a triple helix is formed between two poly(C) chains and pG monomers. In this complex the cytosine residues are partially protonated. Above pH 8, a poly(C): pG Watson-Crick-like double helix is obtained. Polyuridylic acid [poly(U)] and pA form a 2-poly(U): pA triple helix that is stable at 0°C over a wide range of pH values. Polyguanylic acid [poly(G)] and pC do not form a helical complex because poly(G) tends to self-associate and the pyrimidine monomers have a low-stacking

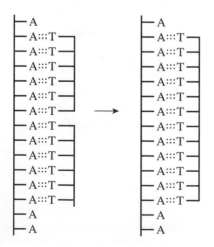

FIGURE 5 Mechanism proposed to explain how poly(dA) could direct the synthesis of dodecamers of dT from hexamers of dT.

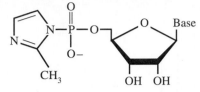

FIGURE 6 Structure of the activated phosphorimidazole derivative used in template-directed condensation. In biochemistry, nucleotide triphosphates are the activated building blocks for template-directed polynucleotide synthesis. These derivatives are not sufficiently reactive to work in prebiotic systems.

energy compared with purine monomers. Poly(A):pU complexes are unstable, most likely because of the small stacking energy of the pU monomers. We see that both the successes and failures in making template–monomer complexes are reflected in polymerization events.

Both the efficiency and the specificity of the oligomerization reaction are sensitive to the stereochemical orientation of template-bound reactants. For example, in the absence of template, adenosine 5′-phosphorimidazolide (5′-ImpA) undergoes an inefficient dimerization reaction that yields 2′,5′- and 3′,5′-linked isomers in a ratio of about 6:1. In the presence of poly(U) the overall yield increases more than fivefold and the 2′5′:3′5′ ratio rises to about 18:1. Comparing the poly(U):ImpA reaction with the poly(C):ImpG reaction, we find that overall yields are roughly the same, but in the latter reaction 3′,5′ linkages are favored by about 2:1.

Addition of divalent metal cations has a big effect on template-directed reactions involving ImpA or guanosine 5′-phosphorimidazolide (ImpG). For example, the addition of 1 to 10 mM Pb^{2+} to a reaction mixture containing 100 mM poly(U) and 50 mM of ImpA monomers results in a fourfold increase in the yield of oligomeric products of pentamer length or longer. When Pb^{2+} is present, 3′,5′-phosphodiester linkages predominate in the oligo(A) product. Pb^{2+} also enhances the formation of oligo(G) on a poly(C) template, but in this case the products are mainly 2′,5′ linked. If 10 to 100 mM Zn^{2+} is used in place of Pb^{2+} in the poly(C):ImpG reaction, the yield of oligo (G) is still enhanced but in this case the formation of 3′,5′ linkages is favored. Thus in both the poly(U): ImpA and poly(C): ImpG reactions, it is possible to obtain regularly 3′,5′-linked oligomers simply by addition of the appropriate divalent metal cation. In this connection it is notable that most nucleic acid polymerases contain bound Zn^{2+} and usually require Mg^{2+} as a cofactor.

A large number of ImpG derivatives have been tested in the poly(C):ImpG reaction. The two-methyl ImpG derivative (2-MeImpG) is the most active compound that has been found (Fig. 6). In 0.2 M MgCl$_2$ and 1M NaCl, 2-MeImpG forms oligomers with four or more residues after 7 days of incubation at 0°C. The product distribution is conveniently characterized by high-pressure liquid chromatography (HPLC) on an RPC-5 column. On such a column oligomers are separated mainly on the basis of chain length and secondarily by phosphodiester-linkage isomerism (Fig. 7). The major peaks in the HPLC profile contain 3′,5′-linked oligo(G)s of successive lengths. The minor peaks preceding each of the major peaks correspond to oligo (G)s that contain one or more 2′,5′ linkages.

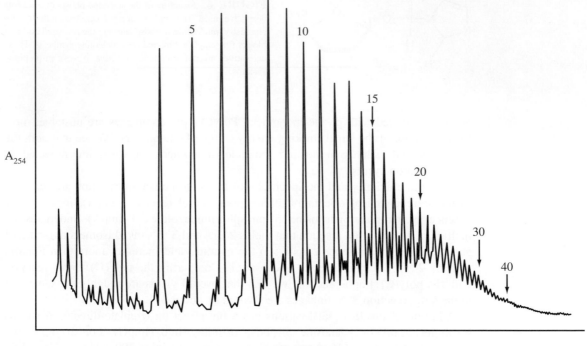

FIGURE 7 HPLC elution profile of 2-MeImpG condensation on a poly(C) template. The numbers above the peaks indicate the lengths of the corresponding 3′-5′-linked oligo(G)s. The materials were eluted from the RPC-5 column using a linear NaClO$_4$ gradient (pH 12, 0 to 60 mM, 90 min), and UV absorption was monitored at 254 nm. (From Joyce, G. F. Nonenzymatic Template-Directed Synthesis of Informational macromolecules, *Cold Spring Harbor Symp. Quant. Biol.* 52:43, 1987. Copyright © 1987 Cold Spring Harbor Laboratory Press, Cold Spring Harbor, NY. Reprinted by permission.)

The poly(C)-directed oligomerization of 2-MeImpG proceeds vigorously in the presence of divalent cation. To ensure the stability of the poly(C)–oligo(G) helical complex during polymerization, the reaction temperature is kept below 37°C; typically a temperature of 0°C is used. The preferred pH is in the range of 7.6 to 9.0. Below pH 7.6 the cytosine residues protonate and above 9.0 the guanine residues lose protons. Optimum polymerization occurs when the concentration of the 2-MeImpG relative to that of poly(C) is high enough to permit monomer–monomer interaction in solution.

The interaction of monomers in free solution results in the formation of dimers. Activated dimers so formed can bind to the template and become covalently linked to one another. Elongation of small template-bound oligo(G) chains proceeds in the 5′ → 3′ direction with the addition of template-bound monomers to the 3′ terminus of the growing oligo(G) chains. The relative rates of initiation and elongation, which are dependent on complex stability and concentration of both monomer and template, determine the overall yield and the distribution of products in the oligomerization reaction. The faster the rate of elongation is relative to initiation, the greater the yield of oligomers (beyond the dimer) and the longer the chain length of the average oligomer.

When the $MgCl_2$–NaCl or $ZnCl_2$ mixtures of cations are used, the 2-methyl imidazole derivative of guanosine 5′-phosphate undergoes efficient condensation in the presence of poly(C) but the corresponding derivative of adenosine 5′-phosphate does not oligomerize in the presence of poly(U). However when a poly(C,U) random copolymer template is used, 2-MeImpG and 2-methyl ImpA (2-MeImpA) are cooligomerized to yield a variety of oligo(G, A)s. In general, any random, mixed polymer template that contains a substantial excess of cytosine residues can be used to direct the synthesis of a complementary polynucleotide chain. Monomers are incorporated into newly synthesized oligomers if and only if their complement is present in the template strand. This result suggests that the specificity of Watson–Crick pairing is sufficient to provide the basis for information transfer in template-directed oligomerization reactions. Confirmation of this notion comes from studies of defined sequences that catalyze the synthesis of products with precisely complementary sequences.

A major limitation of the poly(C,N)-directed reactions is the requirement that the template should contain an excess of cytosine residues. If the ratio of cytosine to noncytosine residues in the template is decreased, the template begins to take on a different character, and the efficiency of the reaction drops sharply. Clearly, this limitation presents a serious obstacle to chemical self-replication, because any cytosine-rich polynucleotide that serves as a good template produces a complementary chain that is cytosine-poor and consequently unable to direct further rounds of synthesis. Thus this system does not provide a solution to the ultimate goal of replicating nucleic acid in an enzyme-free system.

Self-Replication of Nucleic Acids in "Prebiotic Systems" Is Possible with Specially Designed Oligonucleotides

Nucleic acid self-replication in a template-directed reaction system has been demonstrated in cases where chemically modified nucleotide substrates and a self-complementary (palindromic) template are used. The first, developed by G. von Kiedrowski, utilizes two trideoxynucleotide substrates d(5′-MeCCGp-3′) and d(5′-CGGp-O-PhCl-3′), and hexadeoxynucleotide template d(5′-MeCCGCGGp-O-PhCl-3′). The two trinucleotide substrates bind to the template by complementary pairing, and, in the presence of water-soluble carbodiimide, undergo condensation to form a new template molecule (Fig. 8a). The hexamer duplex can dissociate to yield complementary hexamers, each of which is then able to bind additional substrate and begin a new round of synthesis.

Another example of self-replication was developed by Zielinski and Orgel. In this example the 3′-OH of ribose has been replaced by a 3′-amine (see Fig. 8b). This modification enhances the nucleophilicity of the 3′ terminus of the sugar. As a result the

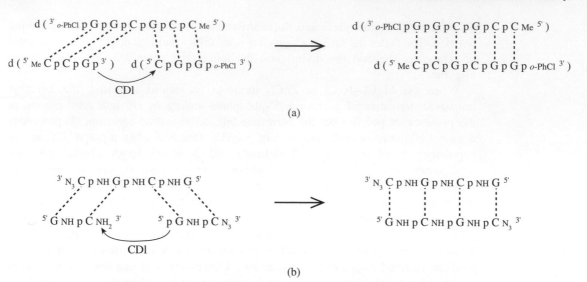

FIGURE 8 Mechanism of two self-replication reactions involving palindromic oligonucleotide–analog templates. (a) The 3′-terminal phosphate of CCG is activated by water-soluble carbodiimide (CDI); and in the presence of a template, it reacts with the 5′-terminal hydroxyl of CGG to produce 5′-CCGCGG-3′. (b) The 5′-terminal phosphate of GC is activated by CDI, and in the presence of a template, it reacts with the 3′-terminal amine of GC to produce 5′-GCGC-3′. (From Joyce, G. F. Nonenzymatic Template-Directed Synthesis of Informational Macromolecules, *Cold Spring Harbor Symp. Quant. Biol.* 52:43, 1987. Copyright © 1987 Cold Spring Harbor Laboratory Press, Cold Spring Harbor, NY. Reprinted by permission.)

3′-deoxy-3′-amino derivative of ImpG condenses with high efficiency. The resulting oligomers contain 3′,5′-phosphoramidate linkages, instead of the conventional 3′, 5′-phosphodiester linkages. This self-replication reaction employs a tetranucleoside triphosphoramidate template, $5'-G_{NHP}C_{NHP}G_{NHP}C-3'$, and a dimeric substrate, $5'-G_{NHP}C_{NH_2}-3'$. This reaction scheme is similar to the one previously described, except that a 5′-phosphate instead of a 3′-phosphate is activated by water-soluble carbodiimide.

In both of these template-catalyzed reactions the kinetic data indicate a square-root dependence of the initial reaction rate on the concentration of the template, consistent with autocatalytic synthesis. In the previously mentioned self-replication reactions, the products have the same sequence as the templates. Kiedrowski has extended these experiments to a system employing two hexameric templates and four trimeric substrates. The product hexamer is complementary to the template but does not appear to be self-complementary.

Could Asymmetrical Synthesis Have Been Initiated at the Level of Nucleic Acid Synthesis?

One of the most intriguing features of biomolecules is the highly specific way in which they deal with asymmetry. Most amino acids and most sugars are asymmetrical

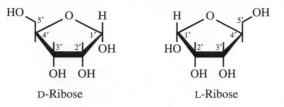

FIGURE 9 Ribose exists in two stereoisomeric forms that are mirror images of each other. Only D-ribose is found in naturally occurring nucleic acids. Locations of carbon atoms are numbered.

or chiral. They come in pairs of enantiomers that are related through their mirror images. An enantiomer cannot be converted into its enantiomeric partner without covalent bond breakage. Biosystems have developed so that only one chiral form is used by its sugars and amino acids. For example, in nucleic acids all the ribose is in the D-ribose configuration. The D-ribose and its mirror image L-ribose are illustrated in Fig. 9. When we discussed sugar synthesis in Chapter 13, we passed over the chirality question because we knew that the synthetic methods we described there must have resulted in equal amounts of both chiral forms. This is because asymmetrical synthesis requires asymmetrical catalysis. This is a fundamental law of nature. In biosystems all enzymes are asymmetrical because of their amino acids and their tertiary structures. Asymmetrical protein catalysts are very discriminating in selecting between chiral pairs of substrates. For evolutionists the paramount question is when and how chiral selection was introduced into prebiological systems. It is particularly appropriate that we discuss this matter here because the helical factor in nucleic acid synthesis hints at a possible answer to this question. Forgetting the chirality of the individual bases for the moment it should be clear that a helix has an intrinsic chirality. A right-handed helix is the mirror image of a left-handed helix. The chiral monomeric nucleotides with their D-ribose components are known to strongly favor folding into two-chain right-handed helices. It is presumed that if all the monomers contained L-ribose instead of D-ribose, thay would form left-handed helices instead.

Orgel and coworkers studied the template-directed incorporation of a pair of enantiomers. In his system for the poly (C)-directed synthesis of poly(G). As we have seen, the poly(D-C)-dependent oligomerization of the D-enantiomer of guanosine 5'-phospho-2-methylimidazole (D-2-MeImpG) yields 3',5'-linked oligo(G)s ranging in length from dimers to 30 mers. Far less efficient is the poly (D-C)-directed oligomerization of a racemic mixture of the two 2-MeImpGs. No products greater than octamers were observed. This is because the L-2-MeImpG was incorporated much less readily than the D-2-MeImpG but once it was incorporated it acts as a chain terminator.

The inhibition of the L-guanosine residues can be understood in terms of the stereochemistry of mononucleotides when bound to a complementary template. Guanosine mononucleotide binds to poly(C) by forming a Watson–Crick base pair. The orientation of the sugar–phosphate component of the monomer relative to the template is determined by the handedness of the ribose sugar (D or L) and the rotational conformation of the purine about the glycosidic bond (*syn* or *anti*; Fig. 10).

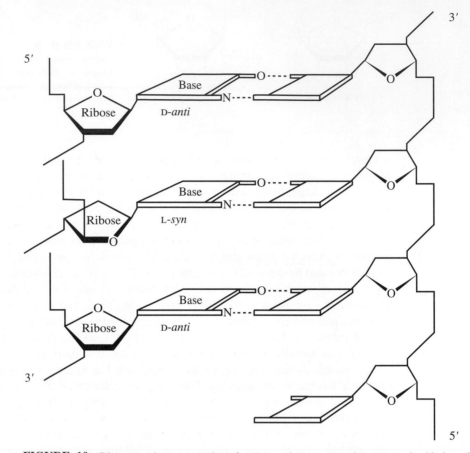

FIGURE 10   Diagrammatic representation of *anti*-D- and *syn*-L-guanosine mononucleotide bound to a poly(D-C) template. The orientation of a template-bound monomer is fixed by Watson–Crick base pairing and by base-stacking interactions. The position of the sugar–phosphate component of the monomer relative to the template is determined by the handedness of the ribose sugar (D versus L) and by the rotational conformation about the glycosidic bond (*syn* versus *anti*). The D-isomer in the *anti* conformation and the L isomer in the *syn* conformation are oriented antiparallel to the poly(D-C) template. The 5′-phosphate of *syn*-L-guanosine mononucleotide lies in close proximity to the 3′-hydroxyl of an adjacent *anti*-D-guanosine residue. (From Joyce, G. F., Schwartz, A. W., Miller, S. L., and Orgel, L. E. The Case for an Ancestral Genetic System Involving Simple Analogues of the Nucleotides, *Proc. Natl. Acad. Sci. USA* 84:4399, 1987. Reprinted by permission of the author.)

Monomer addition to the 3′ terminus of a template-bound oligo(D-C) can involve either the *anti*-(D)-isomer or the *syn*-L-isomer. Thus when bound to a poly (D-C) template, there is occasional incorporation of an L-guanosine residue in a reaction designed to optimize incorporation of D-guanosine (see Fig. 10). However, the addition of a single L-guanosine residue inhibits the addition of further monomers to the growing chain and therefore serves as a chain terminator.

When Orgel and colleagues first reported this *enantiomeric cross-inhibition effect*, they seemed to be discouraged by the outcome. My opinion is just the opposite because I see this, with some modification, as providing a possible answer to how chiral synthesis got its start in biosystems. Imagine a prebiotic situation where there was a racemic mixture of activated mononucleotides. After an initial burst of oligonucleotide synthesis we would expect to see the first oligonucleotides serving as templates for complementary oligonucleotide synthesis. The chirally pure oligonucleotides would be favored in both template and complementary oligonucleotide products. Missing at present is a means for removing the inhibitory terminal residues of the opposite chiral form. Finding such a factor would open the door to the synthesis of substantially longer chirally pure oligonucleotides.

The eventual dominance of D-ribose polymers over L-ribose polymers in such a system could have resulted from a favorable mutation providing the D-polymers with a big selective edge.

An All-Purine Nucleic Acid May Have Preceded a Purine–Pyrimidine Nucleic Acid

Aside from the RNA synthesis problems that relate to ribose, we have seen that there are frustrations concerning the prebiotic replication of RNA. Most of these problems center around the synthesis and polymerization of pyrimidine nucleotides. These problems with pyrimidines would be nonexistent if the first RNAs were made exclusively from purine nucleotides. In fact, two proposals of this sort have been made. In 1968 Crick suggested that the first RNA may have contained only two bases, adenine and hypoxanthine (Fig. 11). Another proposal made by Gunter Wachtershauser in 1982, has the two pyrimidines, cytosine and uracil, replaced by the two purines, isoguanine and xanthine, these purines have the same hydrogen-bonding potential as the two pyrimidines. Crick's proposal is preferable for three reasons. First, it results in a simpler first system. Second, it presents no new problems in the synthesis of mononucleotides. Finally, the transition from the Crick all-purine

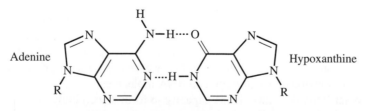

FIGURE 11 The only base pair in an all-purine duplex first proposed by Crick. Hypoxanthine is the name of the purine found in the purine nucleoside inosine (I).

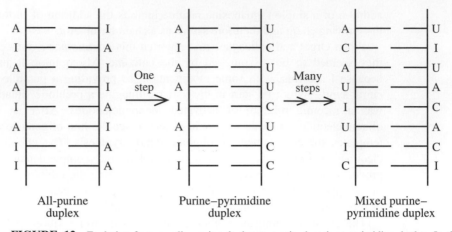

| All-purine | Purine–pyrimidine | Mixed purine– |
| duplex | duplex | pyrimidine duplex |

FIGURE 12 Evolution from an all—purine duplex to a mixed purine–pyrimidine duplex. In the all-purine duplex the enzyme activity is visualized as being confined to a single purine chain, which is probably overproduced so as to give rise to single strands with catalytic activity as well as duplex. After the system has evolved to the stage where it can synthesize and incorporate activated pyrimidine nucleotides, the complementary purine strand could be replaced by a complementary pyrimidine chain. In time this could change so that both chains become a mixture of purines and pyrimidines. The purine–purine duplex has a larger diameter than the purine–pyrimidine duplex, which makes it impossible for a stable duplex to contain both purine–purine and purine–pyrimidine base pairs.

duplex to a Watson–Crick purine–pyrimidine duplex could occur smoothly in two steps (Fig. 12).

EFFECTIVENESS OF RNA TO FUNCTION AS A RIBOZYME IS DEPENDENT ON ITS CAPACITY TO FORM COMPLEX FOLDED STRUCTURES

Beyond the immediate problems of synthesizing and replicating RNA-like polymers in the prebiotic environment, there are concerns that an RNA polymer with four bases or possibly only two bases could have provided the necessary functional groups to satisfy the variety that would be needed to operate a smoothly running living system. In Table 2 we list some of the ribozymes that would have been needed in the RNA-only world and in making the transitition to an RNA–protein world. Even this small list suggests activities and specificities that we would be hard pressed to allocate to any specific functional groups within an RNA molecule.

What RNA does have is the capacity to form a seemingly endless variety of folded structures that rely on the hydrogen-bonding possibilities generated within its limited variety of functional groups. Any doubts about this should be erased by inspection of the complex folded structures found in tRNAs, rRNAs, or viral RNAs.

TABLE 2

Some Ribozymes That Would Have Been Useful in the Early Evolution of Living Systems

Ribozymes for the synthesis of purines, pyrimidines, ribose, nucleosides, nucleotides, and activated nucleotides

Ribozymes for RNA synthesis and processing

Ribozymes for cellular membrane synthesis

Ribozymes for the synthesis of a limited number of coenzymes

Ribozymes for the synthesis of a limited number of amino acids

Ribozymes for the synthesis of aminoacyl tRNAs, linking amino acids to tRNAs, and for the synthesis of mRNAs and primitive ribosomes

In the next section we see that intricate folding patterns also play a crucial role in the structures of known ribozymes. Folding permits a precise orientation of the limited number of functional groups that arise from the nucleotides in the RNA. It also creates potential binding sites for coenzymes, metal cations, and other molecules that may assist in the catalytic process or that may serve as the substrates for catalyzed reactions. Folding may even create situations in which certain bases within the RNA structure are favorably juxtaposed for base modification to give them additional catalytic potential.

PROPERTIES OF KNOWN RIBOZYMES

In 1981 Tom Cech and coworkers reported that a mitochondrial RNA from a protozoa had the capacity for self-cleavage. This self-cleavage was found to be associated with a naturally occurring splicing reaction. Splicing had been discovered 4 years before as a major mechanism for processing eukaryotic mRNAs but the degree of involvement of RNA in the splicing process was not appreciated. Shortly after Cech's discovery, Sid Altman found that RNase P, a tRNA-processing enzyme, also contained a catalytically active RNA. The concept of a ribozyme was born.

Extensive investigations done since these two monumental discoveries have uncovered many additional enzymes in which the catalytic center resides partially or solely in the RNA component of a ribonucleoprotein complex. In this section we address a description of known ribozymes. These ribozymes originate from two sources: biological systems and *in vitro* systems. We consider first the proven ribozymes that have been found in biological systems.

Naturally Occurring Ribozymes May Be Divided into Large Ribozymes and Small Ribozymes

Most naturally occurring ribozymes may be roughly divided into large ribozymes that contain 90 bases or more and small ribozymes that contain substantially fewer bases.

TABLE 3
Some Characteristics of Naturally Occuring Catalytic RNAs

Ribozyme species	Nucleophile	Reaction products
Group I intron	3'-OH of guanosine	5' to 3' Joined exons and intron with 5'-guanosine and 3'-OH
RNase P	H_2O	5'-Phosphate and 3'-OH
Group II intron	2'-OH of adenosine	5'To 3' Joined exons and intron with 2'-3' lariat joined at A and 3'-OH tail: also acts as a DNA endonuclease when bound to a protein
Hammerhead ribozyme	Divalent metal hydroxide, e.g. $[Mg(H_2O)_5(OH)]^+$	5'-OH and 2',3' cyclic phosphatase
Hairpin ribozyme	Divalent metal hydroxide, e.g. $[Mg(H_2O)_5(OH)]^+$	5'-OH and 2',3' cyclic phosphatase
Hepatitis δ virus ribozyme	Divalent metal hydroxide, e.g. $[Mg(H_2O)_5(OH)]^+$	5'-OH and 2',3' cyclic phosphatase
tRNA^{Phe}	Divalent lead hydroxide, e.g. $[Pb(H_2O)-(OH)]^+$	5'-OH and 2',3' cyclic phosphatase

The large ribozymes (Group I splicesosomes, Group II splicesosomes and RNase P) cleave RNA to generate 3'-hydroxyl termini. The small ribozymes cleave RNA to generate 2',3'-cyclic phosphate termini (Table 3).

Group I ribozymes are located in the sequences of group I *introns*. Introns are contiguous sequences that interrupt genes for mRNA, rRNA, and tRNA; introns are removed from preRNA in the splicing process. They are found in mitochondria, chloroplasts, and nuclear genomes of diverse eukaryotes. The sequences within group I introns provide the active site for their own splicing reactions. They fold into a secondary structure with a common "core" region (centered on paired regions P3, P4, P6, and P7 in Fig. 13a). Group I introns have a common splicing pathway that involves a two-step transesterification mechanism that uses an exogenous molecule of guanosine (Fig. 14a). The guanosine serves as an attacking nucleophilic group in the first step. The binding site for guanosine and other elements of the catalytic apparatus are all contained in the intron region of the unspliced RNA. Many group I introns are self-splicing *in virto*. This entails excision of the intron from a precursor RNA followed by the ligation of the flanking exon sequences in the absence of protein. *In vivo*, proteins that facilitate an intrinsically RNA-catalyzed reaction are usually complexed with the splicing RNAs.

Group II introns have distinct structural features (see Fig. 13b). Although they also undergo splicing by a two-step transesterification mechanism (see Fig. 13b), the initial attacking group is an adenosine covalently integrated into the intron and located several bases upstream from the 3'-splice site (see Fig. 14b). All known group II introns are found in eukaryotic organelles.

Several group II introns have been shown to undergo self-splicing *in vitro*. However, self-splicing in this case requires extreme conditions that are clearly

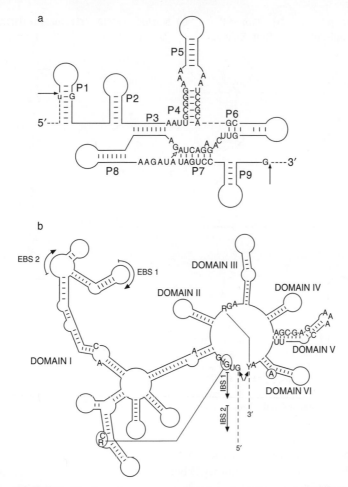

FIGURE 13   Secondary structures of (a) group I and (b) group II introns as proved by comparative sequence analysis. Dashed lines and lowercase *u* indicate exons. Solid lines and uppercase letters indicate introns. Filled arrows indicate 5′ and 3′ splice sites. (a) Group I structure showing most common nucleotide at each position in core region. Dashed line between P4 and P6 was added to make the diagram less crowded. Open arrow indicates site of insertion of extra stem–loops in a subgroup of these introns. Reprinted from *Gene* 73, Thomas R. Cech, Secondary structures of group I introns, 59–271, © 1988, with permission from Elsevier Science. (b) Group II secondary structure for a subgroup that includes two self-splicing introns from yeast mitochondria. Indicated nucleotides are present in all introns with at most one exception; R = purine, Y = pyrimidine. Each exon binding site (EBS) pairs with the corresponding intron binding site (IBS). (Dept. of Chemistry & Biochemistry, Univ. of Colorado, Boulder, Colorado, 80309-0215) Reprinted from *Gene* 82, F. Michael, K. Omesono, and H. Ozeki, Secondary structures of group II introns, 5–30, © 1989, with permission from Elsevier Science.

nonphysiological (e.g., 100 m*M* MgCl$_2$ and 500 m*M* (NH$_4$)$_2$SO$_4$ at 45°C. Even under these conditions splicing proceeds at a rate 10 times slower than group I self-splicing. Genetic evidence has shown that splicing of group II introns *in vivo* requires specific proteins.

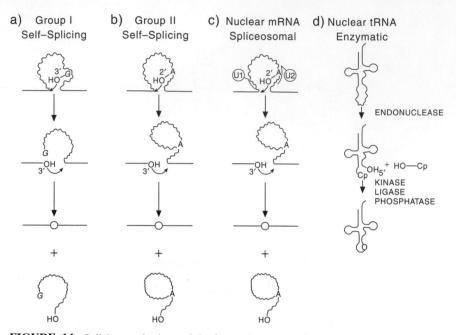

FIGURE 14 Splicing mechanisms of the four major groups of precursor RNAs. Wavy lines indicate introns. Smooth lines indicate flanking exons. For nuclear mRNA splicing, only two of the small nuclear RNPs and other factors that assemble with the pre-mRNA are shown. ("Splicing mechanisms of the four major groups of precursor RNAs" from Thomas R. Cech, "Self-Splicing of Group I Introns" from Annual Review of Biochemistry 59 (1990), Figure 1, page 544. Copyright © 1990 by Annual Reviews. Reprinted with the permission of Annual Rreviews, http://www.AnnualReviews.org.)

RNase P is the only catalytic RNA known to act *in vivo* as a true enzyme, in the sense that it turns over without itself being altered. This ribozyme acts as a ribonucleoprotein in cells.

The reaction catalyzed by RNase P results in the removal of a block of nucleotides from the 5′ end of a tRNA precursor (Fig. 15). RNase P has a catalytic subunit composed of RNA and an accessory polypeptide subunit. *In vitro* the polypeptide allows the reaction to proceed efficiently under physiological conditions. The maximum velocity of the intact ribozyme is about 20 times higher than that for the RNA alone.

The second class of naturally occurring catalytic RNAs, the small ribozymes, are generally found in the genomes of RNA viruses and in virus-related RNAs that replicate by forming tandem repeats of the entire genome. A prominent member of this class of ribozymes, the *hammerhead motif*, bears three base-paired stems flanking a central core of 15 conserved nucleotides (Fig. 16). The conserved central bases are essential for ribozyme activity. Most of these conserved bases cannot form Watson–Crick base pairs, but instead form more complex structures that mediate RNA folding. The relative effectiveness of different divalent cations in influencing cleavage rates is indicated in Table 4. The key role of the divalent cations in influencing the

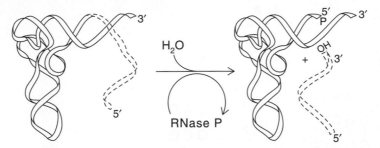

FIGURE 15 Maturation of tRNA catalyzed by RNase P, an enzyme with a catalytic RNA subunit. The 5′-leader sequence removed by processing may have variable structure, so it is shown as a dashed ribbon. The eubacterial consenus structure of RNase P RNA with phylogenetically proven helices designated by filled rectangles, invariant nucleotides in uppercase letters, >90% conserved nucleotides in lowercase letters, and nucleotides less conserved in sequence shown as dots. (Reprinted, with permission, from J. W. Brown and N. R. Pace 1992 by permission of Oxford University Press.)

folded structure of the RNA is underscored by the finding that high levels of monovalent cations are equally catalytically proficient.

It should be mentioned that metal ions often play a crucial role in RNA folding. While most proteins fold without metal ions, RNAs usually require them for folding. As a rule divalent cations are usually effective in catalyzing folding at much lower concentrations than monovalent cations.

The crystal structure of one of the small catalytic RNAs, the hammerhead ribozyme, has been determined. This gives us our first direct look at the tertiary structure of a ribozyme. The conformation of the hammerhead ribozyme is depicted in Fig. 17a as a roughly γ-shaped fold. Stem II and stem III are nearly coaxial, with stem I and the catalytic pocket branching away from this axis. Stem II stacks directly upon stem III, forming a discontinuous helix. The lack of continuity arises from a three-strand junction where the active site cytosine (C-17) is squeezed out of the helix and is forced into a four-nucleotide catalytic pocket, which is formed by a sharp turn in the hammerhead enzyme strand. The phosphate backbone strands, which diverge at the three-strand junction, subsequently reunite to form stem I. These structural features are

TABLE 4

Relative Cleavage Rates for Hammerhead Ribozyme with Various Divalent Metals

Metal	Relative rate
Ca^{2+}	1/16
Mg^{2+}	1.0
Mn^{2+}	~ 10
Co^{2+}	~ 10
Cd^{2+}	~ 6–10
Pb^{2+}	No cleavage

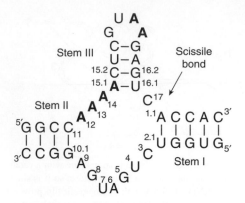

FIGURE 16   Hammerhead RNA secondary structure and cleavage site. The secondary structure of the all-RNA hammerhead ribozyme used for structural determination, consisting of a 16-nucleotide enzyme strand and a 25-nucleotide substrate strand. The conserved bases shown in bold letters are required for catalytic activity. The cleavage site is indicated. (From Scott, W. G. and Klug, A. Ribozymes: Structure and Mechanism in RNA Catalysis, *TIBS* 21: 220–224, 1996.)

illustrated schematically in (Fig. 17b). The uridine turn in the hammerhead-RNA smoothly connects stem I to the stem II helix by bending the enzyme strand of the ribozyme molecule, forming a highly structured pocket into which the cleavage base is positioned.

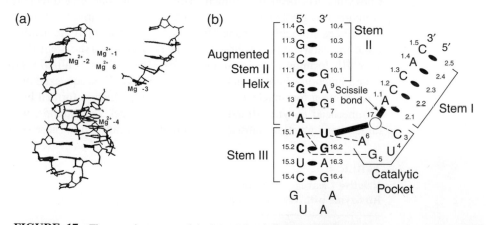

FIGURE 17   The crystal structure of the hammerhead ribozyme. (a) The three-dimensional structure of the all-RNA hammerhead ribozyme, showing the enzyme strand and the substrate strand. The cleavage-site base is (C-17). Difference electron density interpreted as $Mg(H_2O)_6^{2+}$ sites contain spheres corresponding to the complex ion center of mass. (b) A corresponding schematic diagram indicating the location of stems I, II, and III; the catalytic pocket; the augmented stem II helix, and the tetraloop. The essential nucleotides are shown as shadow letters. The universal numbering scheme is indicated. (Scott, W. G. and Klug, A. Ribozymes: Structure and Mechanism in RNA Catalysis, *TIBS* 21: 220–224, 1996.)

Most Ribozymes Found in Biosystems Are Involved in Transesterification or Hydrolytic Reactions

Many ribozymes have been isolated from cells that can catalyze transesterification or hydrolysis of phosphodiester linkages in RNA (see Table 3). As we have seen, the details of the chemistry can vary. Thus while group I and group II ribozymes and RNase P leave $3'$-OH groups at the cleavage site, the small ribozymes form $2'$, $3'$-cyclic phosphates, using the $2'$-hydroxyl group adjacent to the cleavage site as the nucleophile to cleave the phosphodiester linkage. The chemistry of small ribozymes also is employed by common protein ribonucleases as well as random and nonrandom cleavage of RNA by metal ions.

All the reactions just described involve reaction at phosphorus centers (Fig. 18). Other results show that ribozymes also can catalyze reactions at carbon centers. An oligoribonucleotide substrate that can pair at the internal binding (guide) sequence of the *Tetrahymena* ribozyme and that terminates in a $3'(2')$-aminoacyl ester is deacylated at a rate up to 15 times faster than the uncatalyzed rate (see Fig. 18). Like the RNA cleavage reaction, this aminoacyl esterase reaction requires Mg^{2+} and sequence elements of the ribozyme catalytic core. Although the observed catalytic rate enhancement is quite modest, one would not expect an active site that has been selected in nature for phosphoryl transfer to be optimum for catalyzing carboxylate ester hydrolysis. Thus it seems likely that a ribozyme selected to catalyze reactions at carbon centers would be considerably more efficient. In fact, the 23S ribosomal RNA of the thermophile *Thermus acquaticus* may be a case in point. Thus extensively deproteinized *T. aquaticus* ribosomes catalyze peptide bond formation with a high efficiency. This result has led to the exciting possibility that the 23S rRNA may be directly involved in the catalysis of peptide bond formation in protein synthesis.

Many RNAs with Enzymelike Activities Have Been Made *in vitro*

One of the great frustrations of the prebiotic scientist is that there are no sources of evolutionary intermediates available for investigations. Everything is by inference. The conventional evolutionist can search for fossil remains, but there are no fossil remains of the prebiotic molecules that led to the origin of life. This is the greatest challenge faced by the prebiotic investigator.

Consider the potenital of ribozymes. It seems most likely that such enzymes originated before translation systems existed. After translation and protein enzymes became a reality, the majority of ribozymes were presumably replaced by more efficient protein enzymes. For example, the first enzymes for catalyzing the synthesis of activated nucleotides and RNA were probably ribozymes, but we know that most of the enzymes that now play these roles are protein enzymes.

To get a better notion of the full potential of RNA to function as an enzyme, several laboratories have begun developing systems in which random sequences are tested for catalytic activities of various sorts. Once such an activity is detected, a cyclic process

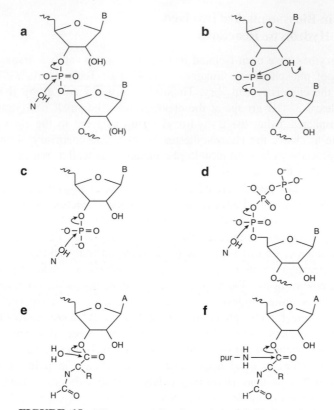

FIGURE 18   Ribozyme-catalyzed reactions described in the text. (a) Cleavage of RNA or DNA generating 3'-OH terminus at cleavage site, reactions catalyzed by group I and group II introns and RNase P. Nucleophiles (N-OH) vary as listed in Table 1. (b) Cleavage of RNA generating 2',3'-cyclic phosphate end, the reaction catalyzed by hammerhead, hairpin, and hepatitis δ-ribozymes as well as many common protein ribonucleases. (c) Phosphotransferase and phosphatase activities on a 3'-phosphomonoester by a group I ribozyme. (d) Chain extension by attack on α-phosphate releasing inorganic pyrophosphate, as catalyzed by a group II intron and RNAs selected by *in vitro* evolution as well as protein polymerases and replicases. (e) Hydrolysis of N-formylmethionine (R = side chain of met) from a hexanucleotide comprising the 3' end of f-met tRNA$^{f\text{-met}}$ by a group I ribozyme. (f) Same substrate as (e); amino acid is transferred to puromycin (pur-NH$_2$) by extensively deproteinized *Tetrahymena aquaticus* ribosomes. (Cech, T. R., 1988.)

that stimulates Darwinian selection is used to enrich for RNA molecules with increased activity (Fig. 19). There are two variations of this process that have been introduced. In the first variation, mutations are introduced at each cyclic step to generate more variety in the polynucleotides that have been enriched in the first round of selection. In the second variation, known ribozymes serve as the starting point. These are reproduced under conditions where mutations are introduced and selection pressures are used that are designed to favor ribozymes with somewhat different properties

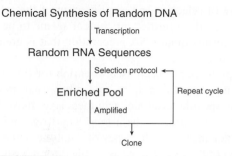

FIGURE 19   Scheme for *in vitro* selection. A chemically synthesized pool of random DNA is transcribed to generate a pool of RNA sequences. The RNA sequences are then subject to a selection protocol to create an enriched pool. This enriched pool is amplified, and the cycle of selection and amplification is repeated. This process is repeated until the remaining sequences all pass the selection procedure. At this point individual sequences can be cloned and characterized.

from the initial ribozyme. For example, a ribozyme that functions with a Mg^{2+} cocatalyst is used as a starting point for isolating a ribozyme that functions with a Mn^{2+} cocatalyst. Are there any indications that these *in vitro* procedures for creating new or altered ribozymes are effective? The answer is emphatically yes. Many more ribozyme activities have been found in this way than have been found in existing biological systems. The range of activities found thus far includes ribozymes with the following functions: cleavage of DNA–RNA hybrid, DNA cleavage, RNA ligation, transesterification, peptidyl transfer, DNA ligation, *N*-alkylation, amide cleavage, aminoacylation, acyl transfer, and porphyrin methalation.

In the simplest procedure for *in vitro* selection, one first prepares a large pool of DNA with random sequences. This pool is transcribed to give rise to a comparable pool of random RNA sequences. The RNA pool is subjected to a predesignated selection procedure. For example, a simple selection procedure might be set up to select for RNA that binds the side chain of arginine. For this selection arginine might be covalently linked to a cellulose column. When the unassorted RNA mixture is passed through the column, the RNA that has an affinity for the amino acid would be expected to bind to the column while the vast majority of the RNA should pass through the column unimpeded. The bound RNA can then be removed from the column by passing a solution containing arginine through it. The release of column-bound RNA is likely because of the competitive binding of the free arginine in the solution. In practice this procedure would have to be repeated many times to enrich for the RNA that binds the amino acid. Also, because only a very small fraction of the RNA in the original pool should have this property, it would be necessary to amplify this RNA after each column pass. For this purpose the RNA would be reverse transcribed into DNA. This DNA would be amplified by conventional DNA recombinant procedures. A second pool of RNA sequences would be produced by transcription from the amplified DNA. This pool is no longer random because it is enriched in sequences that bind the amino acid. This cyclic procedure of selection and amplification of the selected product can be repeated many times or until no further enrichment seems to be occurring. Partial success by this procedure results in an RNA preparation that is greatly enriched in its capacity to bind the amino acid.

The binding selection protocol has been used to enrich for RNAs that bind to various amino acids, coenzymes, specific proteins, and other small molecules. Selection protocols that enrich for specific catalytic activities also have been successfully employed (earlier). There still appears to be a long way to go before this procedure results in purified ribozymes that carry the types of catalytic activities that would be most relevant to replication of the first RNAs (see Table 2 for a wish list).

An All-Purine RNA Should Facilitate the *in vitro* Selection Process

We have presented arguments for an all-purine predecessor of contemporary RNA. If this judgment is correct, it would have a major impact on the *in vitro* selection process and would probably greatly facilitate its use. In this section we consider some of the potential advantages.

By current *in vitro* procedures it is impossible to produce a starting pool of RNA that contains all possible sequences. This can be seen from a simple calculation and the assumption based on current knowledge that to be effective ribozymes require a length of 30 to 60 monomers in a specific sequence. In an all-purine RNA, there would be about 10^9 possibile 30 mers and 10^{18} possible 60 mers. A more conventional four-base RNA has a much greater variety; about 10^{18} possible 30 mers and about 10^{36} possible 60 mers. Prior to selection, a pool of random sequence polymers must be synthesized. Ideally, this pool should be large enough so that there is a high probability that every possible sequence is represented at least once, preferably more than once. A simple calculation using Avogadro's number indicates that a 10-g sample of 60 mers contains about 3×10^{21} molecules; there would be twice this number of molecules in a 10-g sample of 30 mers. This is probably larger than the largest sample size that it would be practical to make *in vitro*. Even if this is so, it is clear that a sample of this size is not nearly large enough to contain all possible sequences of 60 mers for four-base RNA. However, for an all-purine RNA, all possible 60 mers are likely to be represented many times and the initial pool size could be cut back 10- to 30-fold without much loss.

This is one very obvious advantage of the two-base all-purine RNA to the *in vitro* investigator. Another advantage accrues at the level of detection or isolation. The smaller the polydispersity is for a given sample size, the better the detection limit and the fewer the number of cycles that would be required for purification. Incidentally, the same advantage that would accrue to researchers in the laboratory would accrue in a natural selection process taking place in the biosphere.

There is another advantage of the all-purine system. Most known ribozymes have extensive regions of secondary structure (e.g., see Fig. 13). The typical ribozyme has regions of intra- or interstrand base pairing that alternate with looped or single-strand regions. The chances of a region of regular secondary structure forming are far greater for a two-base system than for a four-base system. For example, consider a stem–loop structure with a perfectly paired 10-base stem. The probability of this forming from a random pool in the two-base system is $1/2^{10} \simeq 10^{-3}$, because the probability of a single pair forming is one-half. In calculating this probability, it is assumed that the system contains equal quantities of the two bases. In the case of a four-base system the probability of a perfect stem forming is $1/4^{10} \simeq 10^{-6}$ because the probability of a single pair forming is just one-fourth. Clearly, because regular base pairing is of major importance in most ribozymes, the two-base RNA would have a tremendous advantage over the four-base RNA in the early stages of selection, where positive selection had to rely on these infrequent structures in a pool of nearly random sequences.

The possibility that natural selection of ribozymes from random sequence pools may have begun at a time when RNAs contained only purines raises serious doubts about the relevance of *in vitro* selection procedures based on a four-base RNA. If the first RNAs were all purine, then it is likely that the first pyrimidines did not enter the picture by random selection but instead as complements of established polypurine sequences. In this event, there may never have been a stage in the evolution of RNA when selection started from random sequence pools containing all four bases.

SUMMARY

In this chapter the basic properties of the transcription process have been reviewed and then experiments that deal with prebiotic synthesis have been described. Finally, naturally occurring ribozymes and methods for synthesizing and enriching ribozymes have been considered.

1. Polynucleotide phosphorylase polymerizes ribonucleoside diphosphates into long chains in a nontemplate enzyme-catalyzed reaction.

2. Most cellular RNAs reflect the sequences of the DNA from which they are transcribed.

3. The transcription process is divided into three phases: initiation, elongation, and termination. Synthesis is in the $5'$ to $3'$ direction and the substrates are nucleoside triphosphates.

4. Frequently, transcripts are processed and modified after synthesis.

5. Certain viruses can synthesize RNA on viral RNA templates. Others can synthesize DNA on an RNA template.

6. Formally monomer-to-polymer conversion could be represented by a simple condensation reaction entailing the loss of a water molecule.

7. The noninstructed synthesis of polynucleotides can be accomplished by dry heating or by adding a condensing agent such as a polyphosphate or cyanamide. Imidazole-activated nucleotides can form short oligomers; the reaction is more productive in the presence of clays.

8. Template-directed polynucleotide synthesis also uses imidazole-activated monomers. Reactions usually are conducted under conditions where the activated monomers absorb to polynucleotide templates. This occurs best at low temperatures in the presence of divalent cations. Although activated purine nucleotides polymerize on a polypyrimidine template, the reciprocal is not the case.

9. Another potenitally serious problem associated with the template-directed system is enantiomeric cross-inhibition. Poly(D-C) serves as a template for poly(D-G) synthesis as long as only D-guanine monomers are presented to the template. In the presence of a racemic mixture of activated guanine nucleotides, the occasional incorporation of L-guanine residues leads to chain termination.

10. A possible solution to the enantiomeric cross-inhibition problem would be to have a proofreading meachnism that would eliminate occasionally incorporated residues that block elongation. This is an attractive possibility that awaits being tested experimentally.

11. An all-purine duplex composed of adenine and hypoxanthine residues may have preceded the purine–pyrimidine duplex.

12. The versatility of RNA enzymes (ribozymes) in large measure results from complex folded structures that create a great variety of sites for binding coenzymes and substrates.

13. *In vitro* systems have been developed to select for RNAs with catalytic activities.

Problems

1. Why does mRNA have a much more rapid turnover time than that of rRNA and of tRNA?

2. Why do you suppose that nucleoside triphosphates instead of nucleoside diphosphates are used in RNA synthesis?

3. Human RNA polymerase II generates RNA at a rate of approximately 3000 nucleotides per minute at 37°C. One of the largest mammalian genes known is the 2000 kbp [1 kbp = 1 kilobase pair = 1000 base pairs (bp)] gene encoding the muscle protein dystrophin. How long would it take one RNA polymerase II molecule to completely transcribe this gene?

4. By assuming that RNA preceded DNA in evolution, for what reasons was there a need of DNA? See Problem 12, Chapter 10.

5. In the 1960s Crick proposed that the first nucleic acids exclusively contained purines. We now have reasons to believe that this was a good idea. Give three reasons for considering this to be a good possibility.

6. Indicate three types of ribozymes that probably existed in early evolution that probably do not exist anymore. Explain your choices.

7. What property of the first ribozyme discovered by Tom Cech prevents us from regarding it as a true enzyme?

8. If someone proposed that the first living organisms were composed exclusively of carbohydrates, how would you respond? See Problem 4, Chapter 6.

9. Jim Ferris has found that it is possible to make extended polyribonucleotide chains in the presence of the clay montmorillonite. If Ferris were able to make an effective ribozyme for the polymerase in this way, how would this change our approach to the RNA synthesis problem? Even if Ferris did make some polymerase ribozyme in this way, how would he detect it and how would he be able to amplify it in the complex mixture that it would be contained in?

10. On page 260 of the text it states that below pH 7 cytosine residues in polycytidylic acid [poly(C)] are partially protonated. Because the pK (basic) for cytosine is 4.0, how can you explain the partial protonation at pH 7.0?

11. Why is folding considered to be a crucial aspect to ribozyme formation? In what way would folding be easier if the first RNAs were made from the two purine bases, A and I, than if they were composed of the usual two purines and two pyrimidines?

12. Table 2 lists some ribozymes that would probably have been useful in an RNA-only world. Can you name others?

References

Evolution of Catalytic Function, *Cold Spring Harbor Symp. Quant. Biol.* 52: 1987. (Many articles of interest are on this subject.)

Bartel, D. P. and Szostak, J. W. Isolation of a New Ribozyme from a Large Pool of Random Sequences, *Science* 261:1411–1418, 1993.

Baudry, A. A. and Joyce, G. F. Directed Evolution of an RNA Enzyme, *Science* 257: 635–641, 1992.

Breaker, R. R. and Joyce, G. F. Directed Evolution of an RNA Enzyme. Self-Incorporation of Ribozymes, *J. Mol. Evol.* 40: 551–558, 1995.

Breslow, R., Anslyn, E., and Huang, D.-L. Ribonuclease Mimics. *Tetrahedron* 47: 2365–2376, 1991.

Cech, T. R. Ribozyme Self-Replication? *Nature (London)* 339: 507–508, 1989

Cech, T. R. RNA as an Enzyme, *Sci. Am.* 255 (5): 64–75, 1986.

Cech, T. R. Structure and Mechanism of the Large Catalytic RNAs: Group I and Group II Introns and Ribonuclease P, In *The RNA World*, Cold Spring Harbor, NY: Cold Spring Harbor Laboratory Press, 1993, pp. 239–269.

Cuenoud, B. and Szostak, J. W. A DNA Metalloenzyme with DNA Ligase Activity. *Nature (London)* 375: 611–614, 1995.

Dai, X., DeMesmaeker, A., and Joyce, G. F. Cleavage of an Amide Bond by a Ribozyme, *Science* 267: 237, 1995.

Doudna, J. A., Usman, N., and Szostak, J. W. Ribozyme-Catalyzed Primer Extension by Trionucleotides; a Model for RNA-Catalyzed Replication of RNA, *Biochemistry* 32: 2111–2115, 1993.

Eigen, M., Biebricher, C. K., and Gebinoga, M. The Hypercycle. Coupling of RNA and Protien Biosynthesis in the Infection Cycle of an RNA Bacteriophage, *Biochemistry* 30: 11007–11011, 1991.

Ekland, E. H., Szostak, J. W., and Bartel, D. P. Structurally Complex and Highly Active RNA Ligases Derived from Random RNA Sequences, *Science* 269: 364, 1995.

Ellington, A. D., Robertson, M. P., and Bull, J. Ribozymes in Wonderland, *Science* 276: 546, 1997.

Ferris, J. P. and Ertem, G. Oligomerization of Ribonucleotides on Montmorillonite: Reaction of the 5′-Phosphorimidazolide and Adenosine, *Science* 257: 1387, 1992. (Finally, this is a reasonable approach to making the first oligonucleotides.)

Ferris, J. P., Hill, A. R., Liu, R., and Orgel, L. E. Synthesis of Long Prebiotic Oligomers on Mineral Surfaces. *Nature (London)* 381: 59, 1996. (This extends the use of montmorillonite referred to in the Ferris and Ertem paper to making longer polymers of RNA on montmorillonite and polypeptides on hydroxylapatite. It also refers to minerals as presenting a library of surfaces for different reactions.)

Green, R. and Szostak, J. W. Selection of a Ribozyme That Functions as a Superior Template in a Self-Copying Reaction. *Science* 258: 1910–1915, 1992.

Illangasekare, M., Sanchez, G., Nickles, T., and Yarus, M. Aminoacyl–RNA Synthesis Catalyzed by an RNA, *Science* 267: 643–647, 1995.

Joyce, G. F. Directed Molecular Evolution, *Sci. Am.* December: 90–97, 1992.

Joyce, G. F. Ribozymes: Building the RNA world, *Curr. Biol.* 6: 965, 1996. (The isolation of ribozyme with replicase activity, at present only a hypothetical molecule, is considerably closer following the recent demonstration of RNA-catalyzed polymerization of nucleoside triphosphates.)

Joyce, G. F., Schwartz, A. W., Miller, S. L., and Orgel, L. E. The Case for an Ancestral Genetic System Involving Simple Analogues of the Nucleotides. *Proc. Natl. Acad. Sci. USA* 84: 4398–4402, 1987.

Kazakov, S. and Altman, S. A Trinucleotide Can Promote Metal Ion-Dependent Specific Cleavage of RNA, *Proc. Natl. Acad. Sci. USA* 89: 7919–7943, 1992.

Lohrman, R. Evidence That ImpNs Could Have Been Formed Prebiotically, *J. Mol. Evol.* 10: 137, 1977.

Lorsch, J. R. and Szostak, J. W. *In vitro* Evidence of New Ribozymes with Polynucleotide Kinase Activity, *Nature (London)* 372: 31–36, 1994.

Murphy, F. L., Wang, Y.-H., Griffith, J. D., and Cech, T. R. Coaxially Stacked RNA Helices in the Catalytic Center of the Tetrahymena Ribozyme, *Science* 265: 1709–1712, 1994.

Murray, J. B., Seytian, A. A., Walter, N. G., Burke, J. M., and Scott, W. C. The Hammerhead, Hairpin and V5 Ribozymes Are Catalytically Proficient in Monovalent Cations alone, *Chem. Biol.* 5: 587, 1998.

Noller, H. F., Hoffarth, V., and Zimniak, L. Unusual Resistance of Peptidyl Transferase to Protein Extraction Procedures, *Science* 256: 1416–1419, 1992.

Orgel, L. E. The Origin of Polynucleotide-Directed Protein Synthesis, *J. Mol. Evol.* 29: 465–474, 1989.

Orgel, L. E. RNA Catalysis and the Origins of Life. *J. Theor. Biol.* 123: 127–149, 1986.

Piccirilli, J. A., McConnell, T. S., Zaug, A. J., Noller, H. F., and Cech, T. R. Aminoacylesterase Activity of the *Tetrahymena* Ribozyme, *Science* 256: 1420–1424, 1992.

Scott, W. G. and Klug, A. Ribozymes: Structure and Mechanism in RNA Catalysis, *TIBS* 21: 220–224, 1996.

Sievers, D. and von Kiedrowski, G. Self-Replication of Complementary Nucleotide-Based Oligomers. *Nature (London)* 369: 221–224, 1994.

Stribling, R. and Miller, S. L. Template-Directed Synthesis of Oligo Nucleotides under Eutectic Conditions. *J. Mol. Evol.* 32: 289–295, 1991.

Tohidi, M. and Orgel, L. E. Polymerization of the Cyclic Pyrophosphates of Nucleosides and Their Analogues. *J. Mol. Evol.* 30: 97–103, 1990.

Tsang, J. and Joyce, G. F. Specialization of the DNA-Cleaving Activity of a Group I Ribozyme through *in vitro* Evolution, *J. Mol. Biol.* 262: 31, 1996.

Wilson, C. and Szostak, J. W. *In vitro* Evolution of a Self-Alkylating Ribozyme. *Nature (London)* 374: 777–782, 1995.

Wright, M. C. and Joyce, G. F. Continuous *in Vitro* Evolution of Catalytic Function, *Science* 276: 614, 1997. (A population of RNA molecules that catalyze the template-directed ligation of RNA substrates was made to evolve in a continuous manner in the test tube. This looks like an important advance in technique for evolving ribozymes).

Zubay, G. Arguments in Favor of an All-Purine RNA First, *Chemtracts-Biochem. Mol. Biol.* 6: 251, 1996.

CHAPTER 15

Amino Acid Synthesis Now and Then

Amino acids are the building blocks of proteins. There are 20 L-amino acids directly incorporated into polypeptide chains by an elaborately constructed template mechanism (see Chapter 6, Fig. 6). In addition to their major role in protein synthesis, amino acids serve as precursors for many low-molecular-weight compounds, including peptides, nucleotides, porphyrins, lipid head groups, and coenzymes. Amino acids also provide the vehicles for converting nitrogen and sulfur from inorganic to bioorganic forms. In this chapter we see that the carbon skeletons for most amino acids are derived from the central metabolic pathways of carbohydrate metabolism. After a discussion of the main routes for amino acid biosynthesis, we turn to a consideration of amino acid chemistry in the prebiotic world and in early living systems that predated the advent of translation systems.

AMINO ACID BIOSYNTHESIS

Pathways to Amino Acids Arise as Branchpoints from the Central Metabolic Pathways

Inspection of amino acid biosynthetic pathways shows that most amino acids arise from one of six intermediates in the central metabolic pathways for carbohydrate metabolism (Fig. 1). Amino acids derived from a common carbohydrate are said to be in the same family. For example, the serine family of amino acids, which includes serine, glycine, and cysteine, all arise from glycerate-3-phosphate (see Fig. 1). Serine is made first and serves as a common intermediate for glycine and cysteine synthesis. The carbon flow from the central metabolic pathways to amino acids is a regulated process that provides amino acids in the amounts needed for maintenance and growth. The flow is regulated by end-product inhibition, meaning that the amino acid end product inhibits the first enzyme in the pathway for its synthesis.

Glutamate Family of Amino Acids Includes Glutamate, Glutamine, Proline, Lysine, and Arginine

Glutamate is the cardinal member of the glutamate family of amino acids. It is made in one step from α-ketoglutarate. The most direct route to glutamate is that exhibited by many bacteria when grown in a medium containing an ammonium salt as the sole nitrogen source. The reaction entails a reductive amination catalyzed by glutamate dehydrogenase (Fig. 2, top right). In *Escherichia coli* bacteria this enzyme is specific for the oxidation–reduction coenzyme reduced nicotinamide adenine dinucleotide phosphate (NADPH) as the hydrogen donor, as would be expected for a biosynthetic reaction involving a reductive step.

Studies of the glutamate dehydrogenases from green plants indicate that they require a very high concentration of NH_3 to be effective. In fact an alternative pathway is responsible for glutamate biosynthesis in most plants. Thus, in green plants, as well

---

FIGURE 1 Outline of the biosynthesis of the 20 amino acids found in proteins. The *de novo* biosynthesis of amino acids starts with carbon compounds found in the central metabolic pathways. Some key intermediates are illustrated, and the number of steps in each pathway is indicated alongside the conversion arrow. All common amino acids are emphasized by boxes. Dashed arrows from pyruvate to both diaminopimelate and isoleucine reflect the fact that pyruvate contributes some of the side-chain carbon atoms for each of these amino acids. Note that lysine is unique in that two completely different pathways exist for its biosynthesis. The six amino acid families are screened.

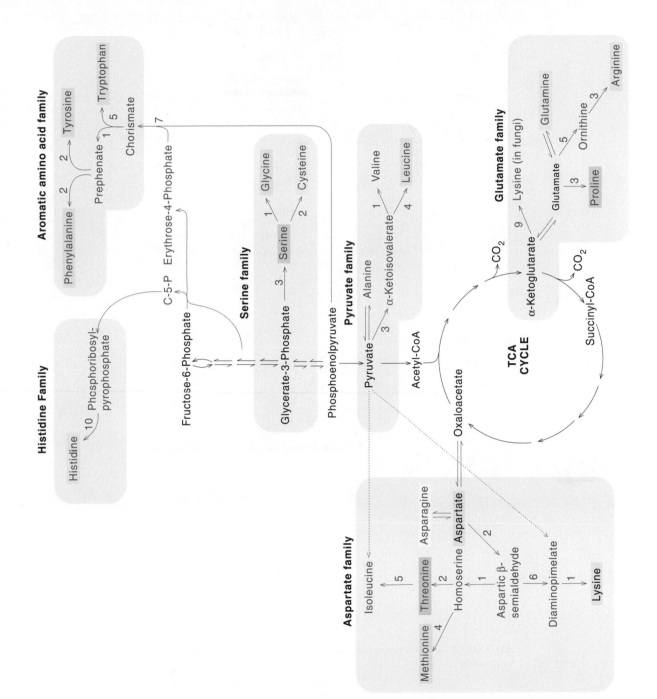

Histidine Family

Histidine ←10— Phosphoribosyl-pyrophosphate

Aromatic amino acid family

Phenylalanine —2→ Prephenate —2→ Tyrosine
Prephenate —1→ 5→ Tryptophan
Chorismate
—7→

C-5-P   Erythrose-4-Phosphate

Fructose-6-Phosphate

Serine family

Glycerate-3-Phosphate —3→ Serine —1→ Glycine
Serine —2→ Cysteine

Phosphoenolpyruvate

Pyruvate family

Pyruvate —1→ Alanine
Pyruvate —3→ α-Ketoisovalerate —1→ Valine
α-Ketoisovalerate —4→ Leucine

Acetyl-CoA

Oxaloacetate

TCA CYCLE

α-Ketoglutarate ←9— CO₂
α-Ketoglutarate → CO₂ Succinyl-CoA

Glutamate family

Glutamate —5→ Glutamine
Glutamate —3→ Proline
Ornithine —3→ Arginine
Lysine (in fungi)

Aspartate family

Isoleucine ←5— Threonine ←2— Homoserine ←1— Aspartate
Methionine ←4— Threonine
Aspartate —2→ Asparagine
Aspartate → Aspartic β-semialdehyde —6→ Diaminopimelate —1→ Lysine

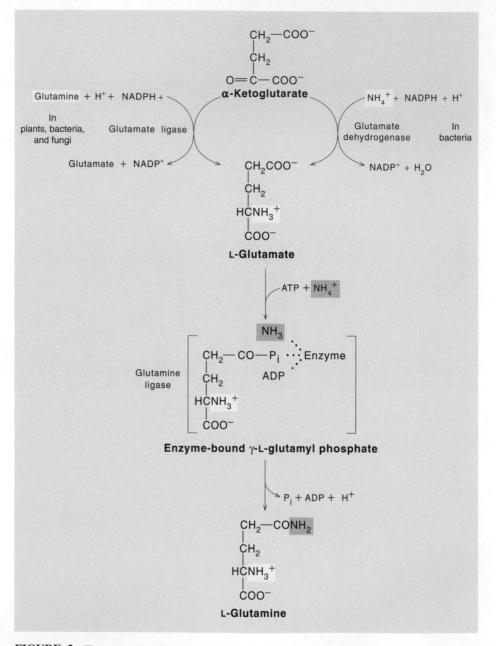

FIGURE 2   The conversion of ammonia into the α-amino group of glutamate and into the amide group of glutamine.

as many bacteria and fungi, the formation of the amino group of glutamate usually originates from the amide group of glutamine. The reaction is catalyzed by glutamate ligase (see Fig. 2, top left).

Glutamine Synthesis Brings Nitrogen into the Biological World

Glutamate is a branchpoint compound from which three of the four remaining amino acids in the glutamate family are derived (see Fig. 1). We discuss only the synthesis of glutamine because of its importance in nitrogen fixation.

A common element in all protein-bound amino acids is the amino group. Directly or indirectly, the α-amino groups of most amino acids are derived from ammonia by way of the amide group of L-glutamine. The uptake of NH_3 by glutamate to make glutamine represents a key step in the process whereby reduced nitrogen becomes incorporated into organic molecules (see Fig. 2, bottom half). The reaction is catalyzed by glutamine ligase.

Glutamine ligase is just one of the links in the nitrogen cycle whereby nitrogen passes through several chemical forms. This passage involves a chain of widely distributed organisms (Fig. 3).

NH_3 is the form in which inorganic nitrogen is incorporated into organic materials, but it is often less available to plants or bacteria for biosynthesis than other forms of nitrogen are, as previously explained. When present for any length of time in the free state, NH_3 is likely to be oxidized by nitrifying bacteria (such as *Nitrosomonas* and *Nitrobacter*) to nitrite (NO_2^-) and nitrate (NO_3^-). Reduction of nitrate by plants and bacteria to NH_3 is vital to maintaining the nitrogen cycle. Nitrogen fixation whereby atmospheric nitrogen is reduced to NH_3 occurs in a limited number of microorganisms. Some of these microorganisms exist in the free state, and some exist as symbionts in the root nodules of certain plants.

Serine Family of Amino Acids Includes Serine, Glycine, and Cysteine

The serine family includes three amino acids: serine, glycine, and cysteine (see Fig. 1). We focus on cysteine synthesis, which funnels sulfur into the biochemical world and supplies the cysteine needed for biosynthesis.

The biosynthesis of L-cysteine entails the sulfhydryl transfer to an activated form of serine. This pathway to L-cysteine has been most thoroughly studied in *E. coli*. In the first step an acetyl group in transferred from acetyl-coenzyme A (acetyl-CoA) to serine to yield *O*-acetylserine (Fig. 4). The reaction is catalyzed by serine transacetylase. The formation of cysteine itself is catalyzed by *O*-acetylserine sulfhydrylase. In the Archean world before the advent of atmospheric oxygen, most sulfur was probably present in the reduced state amenable to direct incorporation into cysteine. Currently, most sulfur is present as sulfate that must undergo an elaborate eight-electron transfer to H_2S. This reduction reaction is found in plants and microorganisms but not in the animal kingdom.

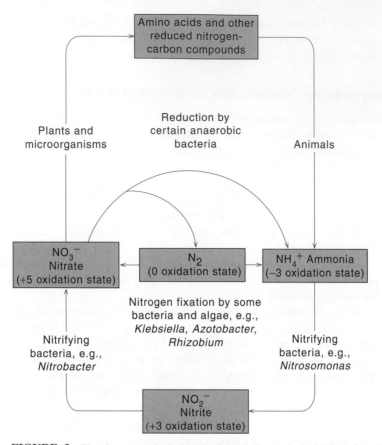

FIGURE 3 The nitrogen cycle depicts the flow of nitrogen in the biological world. The proteins of animals and plants are cleaved by many microorganisms to free amino acids from which ammonia (or ammonium ion) is released by deamination. Urea, the main nitrogen excretion product of animals, is hydrolyzed to NH_3 and CO_2. *Nitrosomonas* soil bacteria obtain their energy by oxidizing NH_3 to nitrite, NO_2^-. *Nitrobacter* bacteria obtain their energy by oxidizing nitrite, NO_2^-, to nitrate, NO_3^-. Plants and many microorganisms reduce nitrate for incorporation into amino acids, completing the cycle. Other microorganisms reduce nitrate partly to NH_3 and partly to N_2, which is lost to the atmosphere. Atmospheric nitrogen, N_2, can be recaptured, reduced, and converted into organic substances by a limited number of nitrogen-fixing bacteria and algae.

Aspartate Family of Amino Acids Includes Aspartate, Asparagine, Methionine, Lysine, Threonine, and Isoleucine

Aspartate is the cardinal member of the aspartate family of amino acids, which has six members. Aspartate itself is derived from oxaloacetate in a transamination reaction (see Fig. 1). All transamination reactions employ the coenzyme pyridoxal-5′-phosphate, and the α-amino group of glutamate is the most common amino group donor. The full series of intermediates in the transamination reaction leading to aspartate is

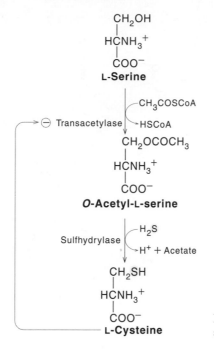

FIGURE 4   The biosynthesis of cysteine by direct sulfhydrylation.

illustrated in Fig. 5. The aldehyde group of the coenzyme first reacts with the α-amino group of an amino acid to produce an aldimine or Schiff's base, which is internally stabilized by H bonding. After protonation at the aldimine carbon of pyridoxal-5′-phosphate (step 3), hydrolysis (step 4) forms an α-keto acid and pyridoxamine-5′-phosphate. The reverse of this sequence with a second α-keto acid (steps 5 through 8) completes the transamination reaction.

The complex chemistry of the remaining amino acids in the aspartate family are not covered because it has little bearing on the primitive pathways for amino acids that are discussed in the latter half of this chapter.

Pyruvate Family Includes Alanine, Valine, Leucine, and to a Limited Extent Lysine and Isoleucine

As can be seen from inspection of Fig. 1, pyruvate itself is the branchpoint compound for the pyruvate family of amino acids. Alanine is produced in one step from pyruvate by a transamination reaction. By another pathway pyruvate is coverted into α-ketoiso-valerate, which is a branchpoint intermediate leading to valine by one branch and to leucine by the other branch. Pyruvate also makes contributions to the side-chain struc-tures of isoleucine and lysine. Lysine is unique because there are two completely dif-ferent pathways for lysine found in different species. We forego a discussion of the complex biochemistry of these pathways.

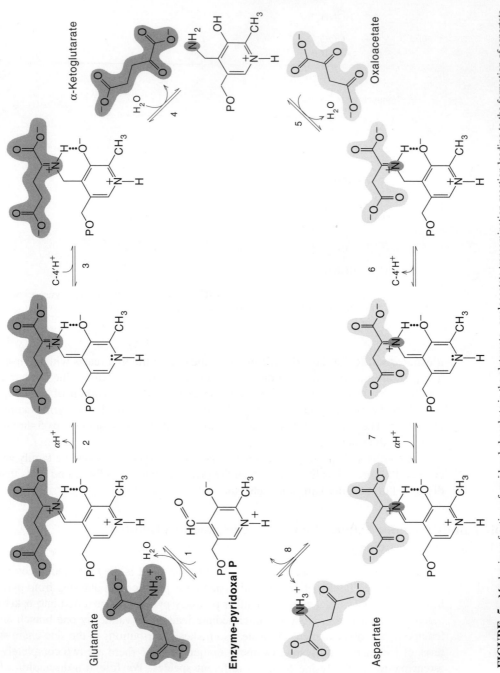

FIGURE 5 Mechanism of action of pyridoxal phosphate in the glutamate–oxaloacetate transamination reaction leading to the formation of aspartate from oxaloacetate.

Aromatic Amino Acid Family Includes Phenylalanine, Tyrosine, and Tryptophan

Aromatic amino acid biosynthesis proceeds via a long series of reactions, most of them concerned with the formation of the aromatic ring before branching into the specific routes to phenylalanine, tyrosine, and tryptophan. Chorismate, the common intermediate of the three aromatic amino acids (Fig. 6), is derived in eight steps from erythrose-4-phosphate and phosphoenolpyruvate. We focus on the biosynthesis of tryptophan, which has been intensively studied by both geneticists and biochemists.

In *E. coli* and *Salmonella* bacteria the details of the enzymatic steps, the correlation between DNA sequence and the protein products, and the factors controlling the transcription of the structural genes exceed those known for most other sets of related genes. Comparative studies of other bacteria and fungi have revealed a variation on the themes found in *E. coli*, particularly with respect to the distribution on one protein or another of the sequence of enzyme activities, which are identical in all forms. In addition, these studies also have revealed differences in the way genes are arranged in the DNA and in the way expression of these genes is controlled.

The first specific step in tryptophan biosynthesis is the glutamine-dependent conversion of chorismate to the simple aromatic compound anthranilate. Like most other glutamine-dependent reactions, this reaction also can occur with ammonia as the source of the amino group. However, high concentrations of ammonia are required. Thus far, almost all the anthranilate ligases examined have the glutamine amidotransferase activity (component II) and the chorismate-to-anthranilate activity (component I) on separate proteins.

Anthranilate is transferred to a ribose phosphate chain in a phosphoribosyl-pyrophosphate-dependent reaction catalyzed by anthranilate phosphoribosyltransferase. Phosphoribosyl anthranilate undergoes a complex reaction known as the Amadori rearrangement, in which the ribosyl moiety becomes a ribulosyl moiety. The product, 1-(*O*-carboxylphenylamino)-1-deoxyribulose-5′-phosphate, is cyclized to indole-glycerol phosphate by removal of water and loss of the ring carboxyl by indole-glycerol phosphate ligase. The final step in tryptophan biosynthesis is a replacement reaction, catalyzed by tryptophan ligase, in which glyceraldehyde-3-phosphate is removed from indole-glycerol phosphate and the enzyme-bound indole undergoes a β-replacement reaction with serine.

Histidine Constitutes a Family of One

Histidine biosynthesis is unusual in two respects: it is in a family by itself, and both its structure and its pathway show a strong interplay with the purine pathway. The starting point for histidine biosynthesis is phosphoribosyl pyrophosphate (PRPP) as in the purine pathway (see Chapter 13). The first specific step in histidine biosynthesis entails a condensation reaction between PRPP and adenosine triphosphate (ATP) leading to phosphoribosyl ATP (Fig. 7). In the fifth step most of the purine nucleotide donated in the first step is returned to the purine pathway while the histidine precursor, imidazole

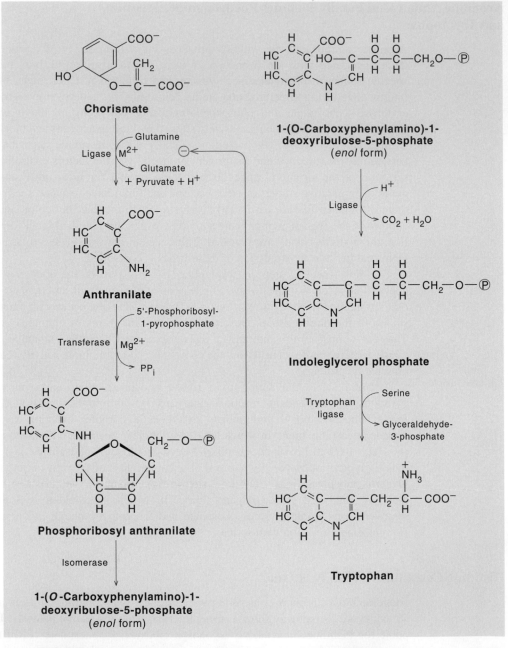

FIGURE 6   The biosynthesis of tryptophan from branchpoint compound chorismate in *E. coli*. The arrow from tryptophan to the first reaction indicates that the first reaction is inhibited by tryptophan.

glycerol phosphate, which now contains a newly synthesized imidazole ring, undergoes additional conversions leading to the formation of histidine. Not only does interplay between the histidine pathway and the purine pathway occur but also the final products of the two pathways, histidine and purine nucleotide, both contain imidazole rings.

Many Amino Acids Found in Proteins Are Formed by Modification after Incorporation into the Polypeptide Chain

The side chains of many amino acids are modified in proteins by specific modifying enzymes. Because not all amino acids of a given type show the same modification, it is clear that the location of the amino acid in the polypeptide chain or folded protein is an important part of the signal for the modification. Phosphorylation of serine—OH groups is an important modification that is observed in regulatory proteins. Similar phosphorylations are observed for threonine and tyrosine. The phosphorylations are observed for threonine and tyrosine. The phosphorylation is usually reversible so that the enzyme is active in one form and inactive in the other form. Hydroxylation of proline and lysine is commonly seen in collagen. Up to one-half of the prolines can be hydroxylated. The most common repeating sequence in collagen is glycine-proline-hydroxyproline. Histones, the positively charged proteins most commonly found complexed with negatively charged chromosomal DNA in eukaryotes, are subject to acetylation of their lysine side chains. This causes them to bind less strongly to DNA, which is consistent with the notion that histone acetylation is a regulatory mechanism that triggers gene expression in that region of the chromosome where the histones become acetylated. Other types of amino acid side-chain midifications include glycosylation and methylation.

Nonprotein Amino Acids Are Usually Derived from Protein Amino Acids

In addition to the amino acids most frequently found in proteins, a large group of amino acids occur in plants, bacteria, and animals that are not found in proteins. Some are found in peptide linkages in compounds that are important as cell wall or capsular structures in bacteria or as antibiotic substances produced by bacteria and fungi. Others are found as free amino acids in seeds and other plant structures. Some amino acids are never found in proteins. These nonprotein amino acids, numbering in the hundreds, include precursors of normal amino acids, and amino acid analogs that might be formed by unique pathways or by modification of normal amino acid biosynthetic pathways.

Certain D-amino acids are commonly found both in microbial cell walls and in many peptide antibiotics. For example, the peptidoglycans of bacteria contain both D-alanine and D-glutamate. The latter is present in a γ-glutamyl linkage. In some forms, the α-carboxyl of the D-glutamyl residue is either amidated or peptide linked with glycine. D-lysine or D-ornithine is found in the glycopeptide of some Gram-positive organisms. The capsule of the dreaded anthrax bacillus is composed of nearly

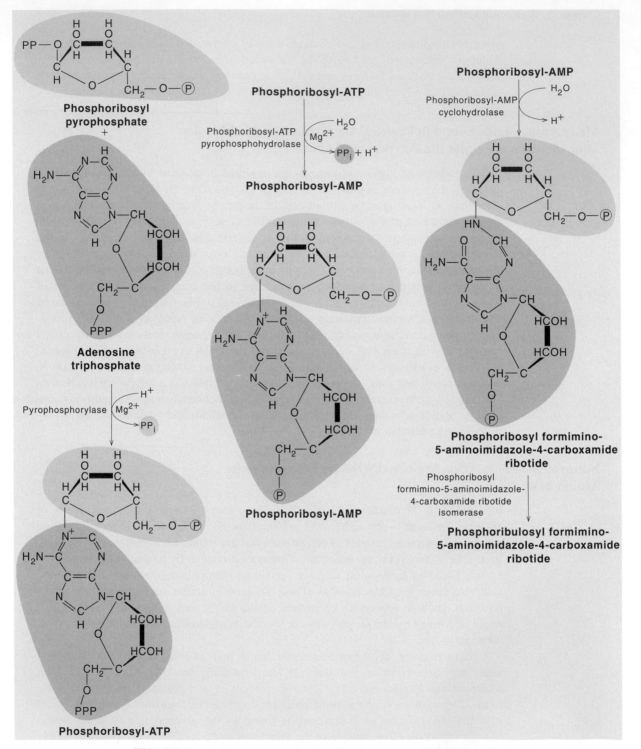

FIGURE 7 The biosynthesis of histidine. The 5-aminoimidazole-4-carboxamide ribotide formed during the course of histidine biosynthesis is also an intermediate in purine nucleotide biosynthesis. Therefore it can be readily regenerated to an ATP, thus replenishing the ATP consumed in the first step in the histidine biosynthetic pathway.

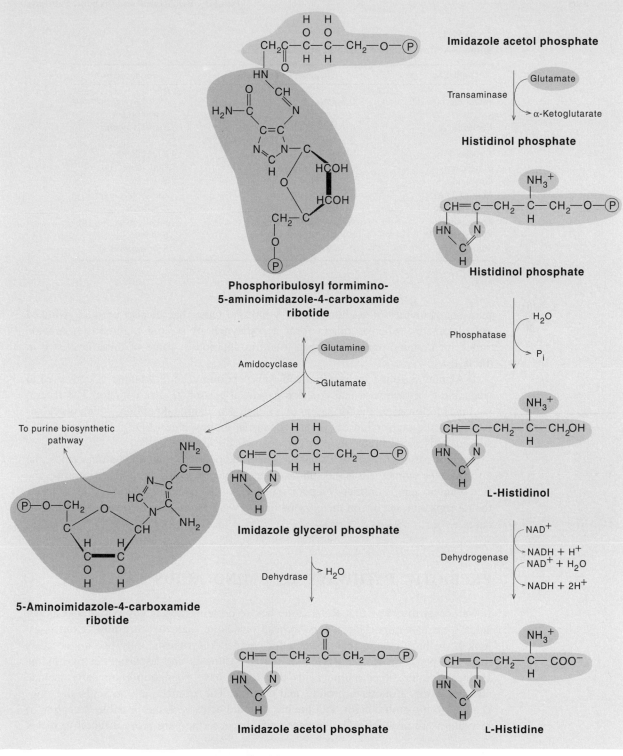

Imidazole acetol phosphate

Phosphoribulosyl formimino-
5-aminoimidazole-4-carboxamide
ribotide

5-Aminoimidazole-4-carboxamide
ribotide

To purine biosynthetic
pathway

Amidocyclase

Glutamine

Glutamate

Imidazole glycerol phosphate

Dehydrase

H₂O

Imidazole acetol phosphate

Transaminase

Glutamate

α-Ketoglutarate

Histidinol phosphate

Histidinol phosphate

Phosphatase

H₂O

Pᵢ

L-Histidinol

Dehydrogenase

NAD⁺

NADH + H⁺

NAD⁺ + H₂O

NADH + 2H⁺

L-Histidine

FIGURE 7 (*Continued*)

295

TABLE 1

Some D-Amino Acids Found in Peptide Antibiotics

Antibiotic	D-Amino acids present	Produced by
Antinomycin C$_1$(D)	D-Valine	*Streptomyces parralus* and others
Bacitracin A	D-Asparagine, D-glutamate, D-ornithine, D-phenylalanine	*Bacillus subtilis*
Circulin A	D-Leucine	*B. circulans*
Fungisporin	D-Phenylalanine, D-valine	*Penicillium* spp.
Gramicidin S	D-Phenylalanine	*B. brevis*
Malformin A$_1$, C	D-Cysteine, D-leucine	*Aspergillus niger*
Mycobacillin	D-Aspartate, D-glutamate	*B. subtilis*
Polymixin B$_1$	D-Phenylalanine	*B. polymyxa*
Tyrocidine A, B	D-Phenylalanine	*B. brevis*
Valinomycin	D-Valine	*S. fulrissimus*

pure homopolymer of D-glutamate in γ linkage. Other bacilli also produce γ-linked polyglutamates, whereas others form a copolymer of D- and L-glutamate. A wide variety of D-amino acids have been found in antibiotics; some of these are listed in Table 1.

D-Alanine is found in bacterial cell wall peptidoglycan. L-Alanine is converted to D-alanine by a racemase that contains pyridoxal phosphate as a cofactor. The racemization is followed by the formation of a D-alanyl-D-alanine dipeptide, which is accompanied by the conversion of ATP to adenosine diphosphate (ADP). The dipeptide is subsequently incorporated into the glycopeptide.

In most cases of peptide formation where the peptides contain D-amino acids, the L form of the amino acid is the substrate for the incorporating enzyme. In contrast, the free D-amino acid is ordinarily a poor substrate for the incorporation reaction. Whether the racemization occurs on the enzyme or afterward remains to be determined in most cases.

PREBIOTIC PATHWAYS TO AMINO ACIDS

If we accept the view that RNA came before protein, then it is obvious that amino acids were less important in the Archean biosphere than they are now. Even if we look at the biosynthetic requirements for nucleotide synthesis, only two amino acids play indispensable roles: glycine in purine synthesis and aspartate in pyrimidine synthesis. Some other amino acids that might have been important early include alanine, serine, glutamate, proline, and cysteine. The aromatic amino acids, the long-chain aliphatic amino acids, and the basic amino acids with the possible exception of histidine were probably relative latecomers because they are more difficult to make.

PIONEERING EXPERIMENTS OF MILLER AND UREY SUGGEST A PREBIOTIC ROUTE TO AMINO ACIDS THAT STARTED IN THE ATMOSPHERE

In 1953 Stanley Miller, working in Harold Urey's laboratory, showed that when an electric discharge was passed through a mixture of methane, ammonia, water, and hydrogen, a variety of amino acids, α-hydroxyacids, and other products resulted. The starting mixture of gaseous compounds was intended to simulate the strongly reducing atmosphere that Urey and Oparin had proposed to exist in prebiotic times; the electric discharge was intended to simulate lightning. An apparatus was designed in which the four compounds and their reaction products could be recycled again and again through a reaction vessel, and from which samples could be periodically removed without disturbing the system (Fig. 8).

The electric discharge was operated continuously for 1 week in Miller's experiments, as the gaseous compounds and some of their reaction products were cycled repeatedly through the electric discharge reaction vessel. Finally, the gases were pumped out and the accumulated products were analyzed in detail by chromoatography. The results of this analysis are given in Table 2. Identified compounds account for only 15% of the carbon that was added to the apparatus; most of the carbon ends up as a poorly defined polymer tar. In addition, some of the methane remained undecomposed in the gas phase, along with carbon monoxide and nitrogen formed in the spark discharge. In spite of the low yields, the results were considered exciting because some amino acids that occur in proteins were found in the reaction mixture. The list of amino acids included small amounts of the acidic amino acids. A number of other organic molecules were found, some of which were closely related to these amino acids and most probably originated from divergent pathways.

In an attempt to study the mechanism of amino acid synthesis in this system, the concentrations of ammonia, hydrogen cyanide (HCN), and aldehydes were monitored periodically during the experiment (Fig. 9). It was found that concentrations of hydrogen cyanide and aldehydes rose for the first 25 hr to levels that were maintained for about 80 hr before starting to fall. The concentration of the reactant ammonia fell steadily from the beginning of the experiment. The concentration of amino acids rose steadily and plateaued after about 100 hr. These kinetic observations were consistent with a relationship in which the hydrogen cyanide and aldehydes were the reaction intermediates and the amino acids were the final products. It appears that Miller had rediscovered the so-called Strecker synthesis for amino acids, which starts from the addition of ammonia to an aldehyde to give an imine [Fig. 10, Eq. (1)]. Cyanide then adds to the imine to give an aminonitrile [Eq. (2)]. Amino acid formation is completed by the hydrolysis of the nitrile [Eq. (3)]. The addition of cyanide to an imine is analogous to the formation of a cyanohydrin from an aldehyde [Eq. (2′)]. The corresponding hydroysis product in this case is a hydroxyacid [Eq. (3′)]. The finding that the hydroxyacids, glycolic acid and lactic acid, were present in amounts

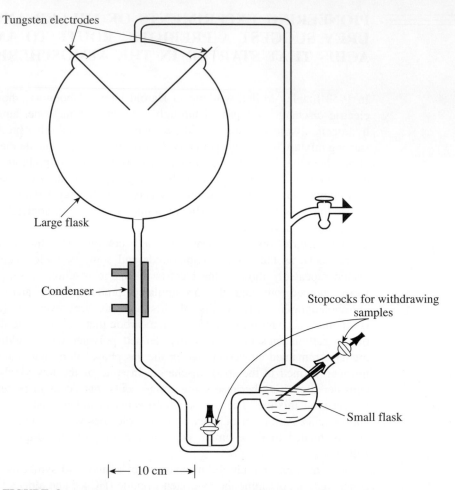

FIGURE 8   Apparatus used by Miller for the synthesis of amino acids. (From Miller, S. L. and Orgel, L. E. *The Origins of Life on Earth*. Englewood Cliffs, NJ: Prentice-Hall, 1974. Reprinted by permission of the author and Prentice-Hall, Inc., Upper Saddle River, NJ.)

comparable with the corresponding amino acids (Table 3), glycine and alanine, respectively, suggested that cyanide addition to the original aldehyde [Eq. (2′)] and its imino derivative [Eq. (2)] occurred about equally.

The Miller experiment was repeated using a mixture of methane, nitrogen, and water, with only traces of ammonia. The modified conditions and improved analytic procedures resulted in detection of several additional amino acids and a broader array of other products (see Table 3). Glycine and alanine were obtained in considerably greater yield than valine, leucine, and isoleucine were. In addition, some amino acids

TABLE 2

Yields from Sparking a Mixture of CH_4, NH_3, H_2O, and H_2[a,b]

Compound	Yield (μM)	Yield (%)
Glycine	630	2.1
Glycolic acid	560	1.9
Sarcosine	50	0.25
Alanine	340	1.7
Lactic acid	310	1.6
N-Methylalanine	10	0.07
α-Amino-n-butyric acid	50	0.34
α-Aminoisobutyric acid	1	0.007
α-Hydroxybutyric acid	50	0.34
β-Alanine	150	0.76
Succinic acid	40	0.27
Aspartic acid	4	0.024
Glutamic acid	6	0.051
Iminodiacetic acid	55	0.37
Iminoaceticpropionic acid	15	0.13
Formic acid	2330	4.0
Acetic acid	150	0.51
Propionic acid	130	0.66
Urea	20	0.034
N-Methyl urea	15	0.051

[a]Added as CH_4 was 59 mmol (710 mg) of carbon. The percentage yields are based on the carbon.

[b]From Miller, S. L. and Orgel, L. E. *The Origins of Life on Earth*. Englewood Cliffs, NJ: Prentice-Hall, 1974. Reprinted by permission of the author and Prentice-Hall, Inc., Upper Saddle River, NJ.

that do not occur in proteins (e.g., aminobutyric acid, valine, and norleucine) also were obtained. Certain nonprotein amino acids (e.g., N-ethyl glycine, γ-aminoisobutyric acid, and isovaline) that were detected in the product mixture are noteworthy because some of them also have been found in carbonaceous meteorites (Table 4).

The findings of Miller underscore the importance of knowing organic chemistry. It seems highly likely that the early steps in many if not most prebiotic reactions are to be found to strongly resemble certain well-established organochemical pathways.

The composition of gases used in the pioneering Miller experiments was based on the predictions of Oparin and Urey that Earth had a strongly reducing atmosphere around the time of the origin of life. As indicated in Chapter 5, there has been a shift toward the notion that the early atmosphere was not such a strongly reducing one. This notion starts with the belief that most of the gases in the early atmosphere arose from outgassing of Earth's mantle (see Chapter 5).

In response to the concern that the atmosphere might not have been as strongly reducing as Oparin and Urey thought, additional prebiotic simulation experiments

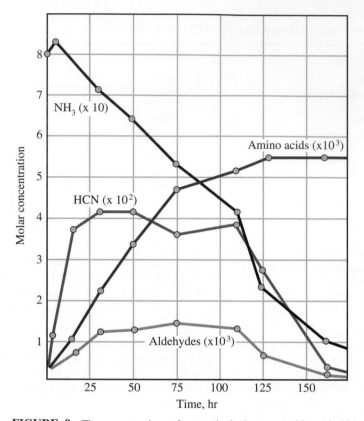

FIGURE 9 The concentrations of ammonia, hydrogen cyanide, and aldehyde in the U-tube; and the concentration of amino acids in the small flask while sparking a mixture of methane, ammonia, water, and hydrogen in the apparatus shown in Fig. 8. (From Miller, S. L. and Orgel, L. E. *The Origins of Life on Earth.* Englewood Cliffs, NJ: Prentice-Hall, 1974. Reprinted by permission of the author and Prentice-Hall, Inc., Upper Saddle River, NJ.)

were done with less reduced starting materials. The general findings were that as the oxidation state of the gases is increased (by stepwise substitutions for methane—at first carbon monoxide and then carbon dioxide—and of nitrogen for ammonia), both yields and product diversity decrease. In the early 1980s Schlesinger and Miller reproted a series of synthesis experiments in which different carbon sources were used in a mixture containing water, nitrogen, and varying amounts of hydrogen. The experiments were performed in the presence and absence of ammonia. The presence of ammonia made little difference as long as sufficient hydrogen was present. The carbon source was much more important. In the absence of methane, glycine was virtually the only amino acid synthesized, regardless of the level of hydrogen. This result led Miller to the conclusion that methane was absolutely required for the synthesis of significant concentrations of amino acids other than glycine.

$$R{-}\overset{\overset{\textstyle O}{\|}}{C}{-}H + NH_3 \ \rightleftharpoons\ R{-}\overset{\overset{\textstyle OH}{|\ *}}{\underset{\underset{\textstyle NH_2}{|}}{C}}{-}H \ \rightleftharpoons\ R{-}\overset{\underset{\textstyle NH}{\|}}{CH} + H_2O \tag{1}$$

$$R{-}\overset{\overset{\textstyle O}{\|}}{\underset{\underset{\textstyle NH}{\|}}{C}}{-}H + HCN \ \rightleftharpoons\ R{-}\overset{*}{\underset{\underset{\textstyle NH_2}{\|}}{CH}}{-}C{\equiv}N \tag{2}$$

$$R{-}\overset{*}{\underset{\underset{\textstyle NH_2}{|}}{CH}}{-}C{\equiv}N \ \xrightarrow{H_2O}\ R{-}\overset{*}{\underset{\underset{\textstyle NH_2}{|}}{CH}}{-}\overset{\overset{\textstyle O}{\|}}{C}{-}NH_2 \ \xrightarrow{H_2O}\ R{-}\overset{*}{\underset{\underset{\textstyle NH_2}{|}}{CH}}{-}COOH + NH_3 \tag{3}$$

$$R{-}\overset{\overset{\textstyle}{\|}}{\underset{\underset{\textstyle O}{\|}}{C}}{-}H + HCN \ \rightleftharpoons\ R{-}\overset{*}{\underset{\underset{\textstyle OH}{|}}{CH}}{-}C{\equiv}N \tag{2'}$$

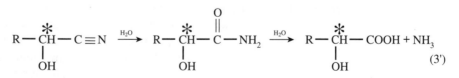

FIGURE 10 Explanation given for amino acid and α-hydroxyacid synthesis in the Miller experiment.

TABLE 3

Yields and Mole Ratios of Amino Acids from Sparking 336 M of CH_4[a,b]

Compound	Yield (μM)	Compound	Yield (μM)
Glycine	440	α,γ-Diaminobutyric acid	33
Alanine	790	α-Hydroxy-γ-aminobutyric acid	74
α-Amino-n-butyric acid	270	Sarcosine	55
α-Aminoisobutyric acid	~30	N-Ethylglycine	30
Valine	19.5	N-Propylglycine	~2
Norvaline	61	N-Isopropyglycine	~2
Isovaline	~5	N-Methylalanine	~15
Leucine	11.3	N-Ethylalanine	<0.2
Isoleucine	4.8	β-Alanine	18.8
Alloisoleucine	5.1	β-Amino-n-butyric acid	~0.3
Norleucine	6.0	β-Amino-isobutyric acid	~0.3
tert-Leucine	<0.02	γ-Aminobutyric acid	2.4
Proline	1.5	N-Ethyl-β-alanine	~5
Aspartic acid	34	N-Ethyl-β-alanine	~2
Glutamic acid	7.7	Pipecolic acid	~0.05
Serine	5.0	α,β-Diaminoproponic	6.4
Theonine	~0.8	Isoserine	5.5
Allothreonine	~0.8		

[a]The yields of glycine and alanine, based on the carbon, are 0.26 and 0.71%, respectively. The total yield of amino acids in the table is 1.90%

[b]From Miller, S. L. and Orgel, L. E. *The Origins of Life on Earth.* Englewood Cliffs, NJ: Prentice-Hall, 1974. Reprinted by permission of the author and Prentice-Hall, Inc., Upper Saddle River, NJ.

Carbohydrates Could Have Provided the Carbon Skeletons for Many of the Amino Acids Synthesized on Primitive Earth

An extended series of experiments conducted by Miller and coworkers have shown that mixtures of gases such as CH_4, CO, CO_2, H_2O, N_2, and H_2S subject to electric discharge or ionizing radiation can be used for the synthesis of certain amino acids. However, except for alanine and glycine most commonly occurring amino acids are produced in very small amounts if at all. Indeed it is difficult to see how larger amino acids could be produced by a gas-phase procedure. Because carbohydrate chains up to 5 and 6 carbons are relatively easy to make in prebiotic simulation experiments (see Chapter 12), it occurred to us that carbohydrates might supply the carbon skeletons for amino acids as they do in biochemical systems. Aldoses are the most likely carbohydrate precursors for amino acids because they are immediately receptive to forming α-amino acids by the Strecker procedure (see Fig. 10). Recall that the Strecker synthesis procedure results in a reaction of the aldehydic functional group with ammonia to form an imino derivative, which then gets converted to an iminonitrile by reaction with HCN. Finally, the iminonitrile is converted to an α-amino acid by hydrolysis.

TABLE 4
Amino Acids in the Murchison Meteorite[a]

1 Isovaline	27 L-Pipecolic acid
2 α-Aminoisobutyric acid	28 Glycine
3 D-Valine	29 Neutral cyclic
4 Linear neutral (C_6)	30 Neutral cyclic
5 L-Valine	31 β-Alanine
6 N-Methylalanine	32 Neutral cyclic
7 D-α-Amino-n-butyric acid	33 Polyfunctional linear aliphatic
8 D-Alanine	34 D-Proline
9 Linear neutral (C_5)	35 L-Proline
10 L-α-Amino-n-butyric acid	36 Linear neutral (C_5)
11 L-Alanine	37 Unknown[c]
12 Linear neutral (C_5)	38 Unknown[c]
13 NH_3[b]	39 Linear neutral (C_5)
14 Linear neutral (C_5)	40 Unknown[c]
15 Linear neutral (C_6)	41 γ-Aminobutyric acid
16 N-Methylglycine	42 D-Aspartic acid
17 N-Ethylglycine	43 L-Aspartic acid
18 D-Norvaline	44 Polyfunctional linear aliphatic
19 L-Norvaline	45 Polyfunctional linear aliphatic
20 Linear neutral (C_5)	46 Polyfunctional linear aliphatic
21 Linear neutral (C_6)	47 Polyfunctional linear aliphatic
22 D-β-Aminoisobutyric acid	48 Polyfunctional linear aliphatic
23 L-β-Aminoisobutyric acid	49 Unknown[c]
24 β-Amino-a-butyric acid	50 D-Glutamic acid
25 Unknown[c]	51 L-Glutamic acid
26 D-Pipecolic acid	52 Unknown[c]

[a]With kind permission from Kluwer Academic Publishers "Amino Acids in the Murchison Meteorite", from Lawless, J. G. and Peterson, E. Amino Acids in Carbonaceous Chondrites, *Origins of Life* 6:3–8, 1975.

[b]Present in blank.

[c]Peaks labeled "unknown" do not appear to be amino acids. Their identification awaits the use of high-resolution mass spectrometry.

Because one carbon is added to the carbohydrate skeleton in the Strecker synthesis, the appropriate carbohydrate precursor for any particular amino acid should contain one less carbon than the amino acid for which it is the precursor. We have considered the appropriate carbohydrate precursor for 17 of the 20 commonly encoded amino acids (Table 5); the three aromatics (tyrosine, phenylalanine, and tryptophan) have been excluded from consideration because their side chains do not fit into this simple scheme. As indicated in Table 5, 14 of these amino acids contain from 3 to 6 carbons in a straight chain. These all could have been derived from straight-chain carbohydrates. The remaining three amino acids (valine, leucine, and isoleucine) contain branched carbon chains that would have required branched-chain carbohydrate precursors. Branched-chain carbohydrates are also well-known products of the formose reaction.

TABLE 5
Possible Prebiotic Route to Amino Acids

Amino acid	Length of carbon chain in amino acid	Likely carbohydrate precursor in the prebiotic world	Changes needed to produce amino acid after Strecker synthesis modification
Alanine	3	Glycolaldehyde	Removal of terminal —OH
Valine	5 (branched)	Branched-chain aldotetrose	Removal of three —OH groups
Leucine	6 (branched)	Branched-chain ketopentose	Removal of four —OH groups
Isoleucine	6 (branched)	Branched-chain aldopentose	Removal of four —OH groups
Proline	5	Aldotetrose	Cyclization and removal of —OH
Methionine	4	Glyceraldehyde	Removal of two —OH groups and addition of an —SCH$_3$ group
Glycine	2	Formaldehyde	No change
Serine	3	Glycolaldehyde	No change
Threonine	4	Glyceraldehyde	Removal of an —OH group
Cysteine	3	Glycolaldehyde	Replacement of an —OH group by an —SH group
Asparagine	4	Glyceraldehyde	Amide formation from aspartate
Glutamine	5	Aldotetrose	Amide formation from glutamate
Aspartate	4	Glyceraldehyde	Oxidation of —OH group to acid and removal of internal —OH groups
Glutamate	5	Aldotetrose	Removal of internal —OH groups
Lysine	6	Aldopentose	Removal of —OH group and addition of terminal amino group
Arginine	5	Aldotetrose	Removal of three —OH groups and addition of guanidinium group
Histidine	5	Aldotetrose	Reaction with formamidine and further modifications

In the last column of Table 5 the modifications that would have to be undertaken to create the commonly occurring amino acids are listed. Only glycine and serine require no further modification. Considerable modifications would be necessary to remove the hydroxyl groups if the goal was to make exact replicas of the amino acids found in proteins. However, as far as amino acid function was concerned in the prebiotic world, it is not clear if the extra hydroxyl groups would have been an asset or a liability. At a later stage in the evolutionary process these hydroxyl groups would have seriously impaired the polypeptide chain-folding process; however, this is not directly relevant

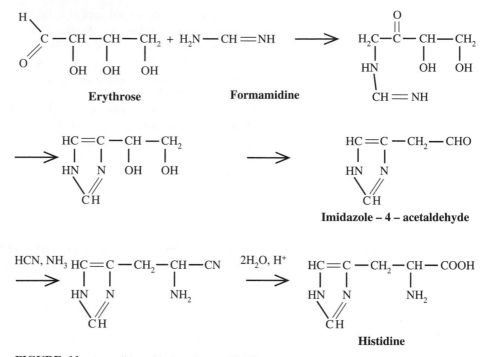

FIGURE 11   A possible prebiotic pathway to histidine.

to the discussion here because polypeptide chains with defined sequences of amino acids belong to a different era, and the source of carbon skeletons for amino acids by that time and the pathways would have shifted to ones more like the contemporary amino acid biosynthetic pathways.

Finally, it seems appropriate to mention a particularly elegant pathway that has been proposed by Oro and Miller for the prebiotic synthesis of histidine. This pathway starts with erythrose, a tetrose that should be accessible from the formose reaction, and formamidine, which is formed by the ammonolysis of HCN. Reaction between erythrose and formamidine gives rise to imidazole-4-acetaldehyde (Fig. 11). This reaction bears a striking resemblance to the formation of imidazole glycerol phosphate in the biosynthetic pathway (see Fig. 7). From imidazole-4-acetaldehyde the amino acid histidine results from a standard Strecker synthesis.

EXTRATERRESTRIAL SOURCES OF ORGANIC MATERIAL

Throughout most of this text we describe the basic organics used in the origin of life as having originated on Earth or in its atmosphere. There is little doubt that some organic compounds came to Earth from extraterrestrial sources. The question is, how much?

The major sources of exogenous organics would appear to be comets and dust, with asteroids and meteorites being minor contributors. Asteroids would have impacted Earth frequently in prebiotic times, but the amount of organic material brought in was probably small. Comets are the most promising source of exogenous organics as first pointed out by Oro. Comets usually contain mostly water and small amounts of HCN and formaldehyde, as well as CO_2 and CO.

There are considerable differences in my opinion on the relative importance of extraterrestrial sources of organic compounds that led to the origin of life. I choose to leave this as an open question. Like panspermia the evidence is too thin to come to firm conclusions one way or the other. So we must move ahead in areas that are amenable to experimentation.

SUMMARY

In this chapter we have discussed the biosyntheis of amino acids and have considered possible prebiotic pathways to some of the amino acids that may have been important to the origin of life.

1. The pathways to amino acids arise as branchpoints from a few key intermediates in the central metabolic pathways that include the glycolytic pathway and the tricarboxylic acid cycle. Amino acids derived from a common carbohydrate are said to be in the same family.

2. The glutamate family of amino acids includes glutamate, glutamine, proline, lysine, and arginine. The synthesis of glutamine is the main route for bringing nitrogen into the biological world.

3. The serine family includes serine, glycine, and cysteine. Serine is made first and serves as a common intermediate for glycine and cysteine. The synthesis of cysteine is the main route for bringing sulfur into the bioligical world.

4. The aspartate family of amino acids includes aspartate, asparagine, methionine, lysine, threonine, and isoleucine. Aspartate is made first and serves as a common intermediate for the other amino acids.

5. Histidine biosynthesis is unusual in two respects: it is in a family by itself; and both its structure and its pathway show a strong interplay with the purine pathway, which suggests it may have been an important amino acid early in the so-called RNA-only world.

6. Two compounds, hydrogen cyanide and formaldehyde, are likely to have been key compounds in the prebiotic synthesis of amino acids. Both of these compounds may well have been produced in abundance in prebiotic times with the help of light and electric discharge in the atmosphere. Organochemical routes to amino acid synthesis have been known since the turn of the century, and a variety of products of importance have been demonstrated in prebiotic simulation experiments. The list of products includes several amino acids, simple organic acids, and urea. If hydrogen sulfide is included in the prebiotic mix, the yields of organic compounds are considerably heightened; and cysteine is synthesized in addition to the amino acids glycine, alanine,

aspartic acid, glutamic acid, and serine. Because the basic amino acids are quite difficult to synthesize, there is some doubt about basic amino acids other than histidine
being made in significant amounts in prebiotic times.

7. Carbohydrates could have provided the carbon skeletons for many of the amino
acids synthesized on primitive Earth.

Problems

1. How do you suppose the glutamate used in transamination reactions is
   regenerated?
2. Why do you suppose the precursor for D-amino acids in biomolecules always is
   formed from the comparable L-amino acid?
3. If the cell has all the free tryptophan it needs, this is likely to inhibit further
   tryptophan synthesis. Indicate a mechanism for this inhibition.
4. How is tRNA implicated in porphyrin synthesis, and in bacterial cell wall
   synthesis? *Hint*: It might be useful to refer to Zubay's biochemistry text to
   answer this one.
5. During the biosynthesis of histidine, the α-amino group is added before the formation of the α-carboxylate group. Is this a typical pattern for the biosynthesis
   of amino acids?
6. How would you synthesize serine in the prebiotic world? How would you
   synthesize it in the biochemical world? Clearly illustrate all the steps and
   compare.
7. One of the reasons we believe that histidine was present in very early evolution
   is that it contains an imidazole side chain. Name at least three cases where
   imidazole groups have been implicated in prebiotic systems in previous
   chapters.
8. Why might extra hydroxyl groups on certain amino acid side chains be likely to
   interfere with polypeptide chain folding in aqueous solution?
9. Indicate five amino acids that are likely to have played important roles in
   pretranslation times. Defend your choices.

References

Keller, M., Boechl, E., Wachtershauser, G., and Stetter, K. O. Formation of Amide Bonds without a Condensation Agent and Implications for the Origin of Life. *Nature* (*London*) 368:836, 1994.

Miller, S. L. Production of Amino Acids under Possible Primmitive Earth Conditions. *Science* 117:528,
1953. (This reports one of the first important experimental studies related to the origin of life.)

Schlesinger, G. and Miller, S. L. Prebiotic Synthesis in Atmospheres Containing CH_4, CO_2 and CO. I.
Amino Acids, *J. Mol. Evol.* 19:376, 1983. (To make amino acids more complex than glycine you need
methane in the primitive atmosphere.)

Schlesinger, G. and Miller, S. L. Prebiotic Synthesis in Atmospheres Containing CH_4, CO_2 and CO. II. Hydrogen Cyanide, Formaldehyde and Ammonia. *J. Mol. Evol.* 19:383, 1983.

Summers, D. P. and Change, S. Prebiotic Ammonia from Reduction of Nitrite by Iron(II) on the Early Earth. *Nature (London)* 365:630, 1993.

Zubay, G. *Biochemistry*, 4th ed. New York: McGraw-Hill, 1998, Chapters 24 and 25.

Zubay, G. Did Carbohydrates Provide Carbon Skeletons for the First Amino Acids to be Synthesized on Planet Earth? *Chemtracts-Biochem. Mol. Biol.* 10:704, 1997.

CHAPTER 16

Chemistry of Translation

In Chapter 9 we describe the cellular machinery of protein synthesis. We see that there are three classes of RNA involved in translation: messenger RNA (mRNA), which serves as the template for translation; transfer RNA (tRNA), which transports amino acids to the template; and ribosomal RNA (rRNA), which together with ribosomal protein serves as the site where tRNA, mRNA, and various enzymes meet to carry out the process of translation. We also briefly describe the three stages in protein synthesis and the relationship between the nucleotide sequence in the mRNA and the amino acid sequence in the related polypeptide chain.

In this chapter we elaborate on the chemistry of the translation process and some aspects of the genetic code.

STEPS IN TRANSLATION

Translation starts with the attachment of amino acids to specific tRNA molecules. Subsequent steps take place on the ribosome; amino acids are transported to the ribosome by their tRNA carriers and the amino acids do not leave the ribosome until they have become an integral part of a polypeptide chain.

Amino Acids Are Activated before Becoming Linked to Transfer RNA

The incorporation of free amino acids into proteins requires their intermediate covalent attachment of tRNAs. This attachment serves two functions: (1) the tRNA directs the amino acid to a designated place on the mRNA template so that the amino acid is incorporated at the designated location in the polypeptide chain; and (2) the linkage between the amino acid and the tRNA activates the amino acid, making the subsequent formation of a peptide linkage energetically favorable.

The attachment of an amino acid to a tRNA is a two-step process (Fig. 1). In the first step an aminoacyl–AMP complex is formed from an amino acid and ATP. This step activates the amino acid so that it is energetically capable of forming a complex with the tRNA. In the next step the amino acid exchanges its AMP for an ester linkage with tRNA. Both of these steps are catalyzed by a specific enzyme that recognizes the amino acid and its cognate tRNA.

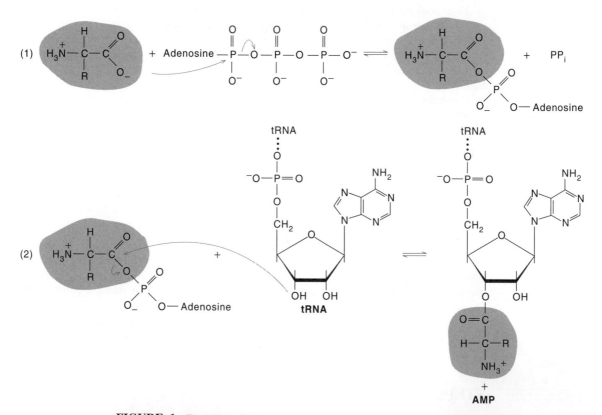

FIGURE 1   Formation of aminoacyl tRNA. This is a two-step process involving a single enzyme that links a specific amino acid to a specific tRNA molecule. In the first step (1) the amino acid is activated by the formation of an aminoacyl–AMP complex. This complex then reacts with a tRNA molecule to form an aminoacyl–tRNA complex (2).

Polypeptide Synthesis Is Initiated on the Ribosome

The small ribosomal subunit forms a ternary complex with the mRNA and the first aminoacyl tRNA, which is invariably a methionyl tRNA (Fig. 2). In this complex the mRNA is oriented so that the codon bases in the mRNA and the anticodon bases in the tRNA are oriented in an antiparallel way. Following this, the large ribosomal subunit adds to the initiation complex. This addition is accompanied by the hydrolysis of a guanosine 5'-triphosphate (GTP) to guanosine diphosphate (GDP) and P_i. Energetically speaking, the hydrolysis of a GTP to GDP is equivalent to the hydrolysis of an adenosine triphosphate (ATP) to adenosine diphosphate (ADP) and P_i. Thus the binding reaction involving the two ribosomal subunits has special energy requirements. This initiation reaction also requires a family of soluble proteins, which are collectively referred to as *initiation factors (IFs)*.

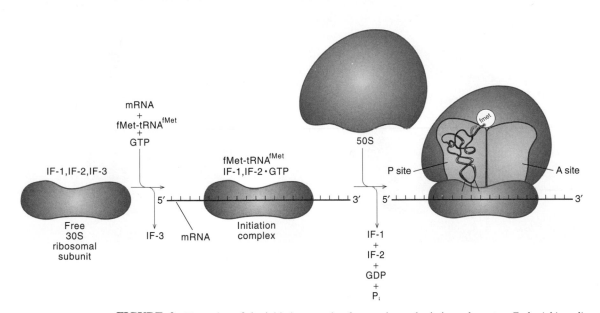

FIGURE 2   Formation of the initiation complex for protein synthesis in prokaryotes. *Escherichia coli* has three initiation factors bound to a pool of 30S ribosomal subunits. One of these factors, IF-3, holds the 30S and 50S subunits apart after termination of a previous round of protein synthesis. The other two factors, IF-1 and IF-2, promote the binding of both fMet-tRNA^{fMet} and mRNA to the 30S subunit. The binding of mRNA occurs so that its Shine–Dalgarno sequence pairs with 16S rRNA and the initiating AUG sequence with the anticodon of the initiator tRNA. The 30S subunit and its associated factors can bind fMet-tRNA^{fMet} and mRNA in either order. Once these ligands are bound, IF-3 dissociates from the 30S subunit, permitting the 50S subunit to join the complex. This releases the remaining initiation factors and hydrolyzes the GTP, which is bound to IF-2.

Elongation Reactions Involve Peptide Bond Formation and Translocation

Elongation involves all the reactions related to peptide synthesis, from the first peptide linkage to the last. A new group of dissociable protein factors, called *elongation factors (EFs)*, comes into play at the time of elongation.

Elongation begins with the binding of a second aminoacyl tRNA at a site adjacent to the methionyl tRNA (Fig. 3). The methionyl tRNA binds to the *P site* on the larger ribosomal subunit, while the second aminoacyl tRNA binds to the *A site.* The initiator tRNA is the only aminoacyl tRNA that binds directly to the P site. All the remaining aminoacyl tRNAs bind first to the A site and later become translocated to the P site as described later. A GTP hydrolysis occurs in conjunction with the ribosome binding of each aminoacyl tRNA.

With two aminoacyl tRNAs complexed to the ribosome, the system is ready for peptide bond formation. The actual formation of the peptide bond, called *transpeptidation,* is the only subreaction of protein synthesis that does not require participation of either a nonribosomal protein or GTP. This reaction is catalyzed by

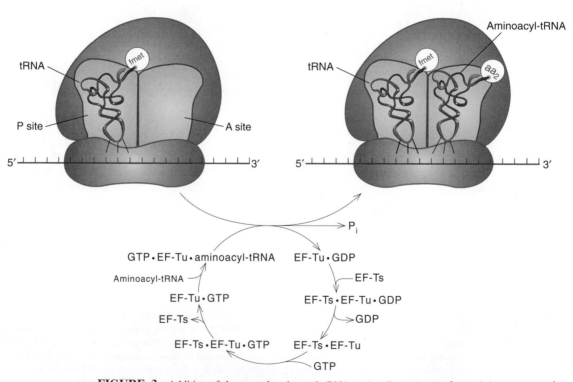

FIGURE 3   Addition of the second aminoacyl tRNA to the ribosome complex and the accompanying EF-Tu, EF-Ts cycle in *E. coli.* The purpose of the cycle is to regenerate another protein aminoacyl–tRNA complex suitable for transferring further aminoacyl tRNAs to the A site on the ribosome.

peptidyl transferase, a catalytic activity that is located in the RNA of the large ribosomal subunit.

The free amino group of the newly bound aminoacyl tRNA attacks the carbonyl group of the adjacently bound methionyl tRNA in a nucleophilic displacement reaction entailing the replacement of the ester bond with a peptide bond. The net result of this reaction is the transfer of the methionine from one tRNA to the next, simultaneous with peptide bond formation (Fig. 4).

Following transpeptidation, the ribosomal P site is occupied by a deacylated tRNA, and a peptidyl tRNA occupies the A site. Further peptide synthesis cannot occur until the peptidyl–tRNA complex is moved from the A site to the P site. At the same time, the mRNA must move precisely three bases along the ribosome so that interaction between the peptidyl tRNA and the mRNA is preserved. The combined movement of the peptidyl tRNA and the mRNA is known as the *translocation reaction* (Fig. 5). Translocation entails a reordering of the binding so as to expel the vacant tRNA, reposition the peptidyl tRNA at the P site, and place the next adjacent codon at the A site for its subsequent decoding. These collective reactions require the hydrolysis of yet another GTP.

The three reactions just described—absorption of the aminoacyl tRNA to the ribosome A site, transpeptidation, and translocation—are successively repeated until each codon on the mRNA has been translated to produce the completed polypeptide chain.

Termination of Translation Requires Special Termination Codons

The translation of natural messages yields protein products that are detached from tRNA and released from the ribosome. The overall reaction, the release reaction, comes about through the reading of termination codons by a special group of proteins, the *release factors*. When a termination codon is bound to the decoding site on the ribosome, then the release factors, in turn, bind. This codon-directed binding causes some as yet unknown change in the peptidyl transferase so that it transfers the completed polypeptide chain to water instead of to an amino group of an aminoacyl tRNA. Subsequently, the mRNA dissociates from its complex with the ribosome and the two ribosomal subunits separate. The termination step also requires a GTP hydrolysis.

All the steps in protein synthesis are summarized in Chapter 9, Fig. 14. In this figure it should be noted that proteins undergo posttranslational modification and are targeted to the locations of the cell where they function.

RULES FOR BASE PAIRING BETWEEN TRANSFER RNA AND MESSENGER RNA

In Chapter 9 we state that base pairing occurs between tRNA and mRNA in such a manner that the correct aminoacyl tRNA is properly juxtaposed for addition to the growing polypeptide chain. As we have indicated, this pairing involves an antiparallel orientation between the interacting bases in the tRNA and the mRNA. At first it was

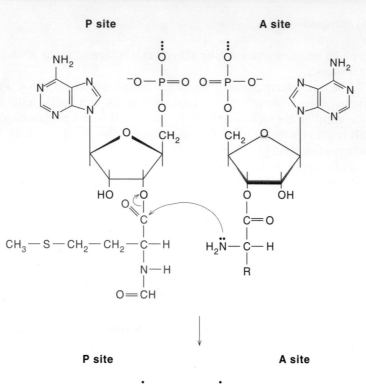

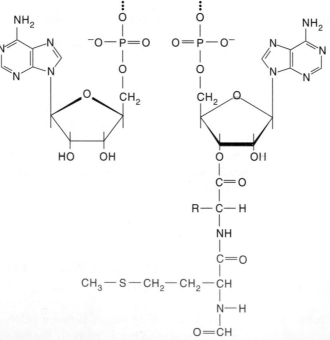

FIGURE 4 Formation of the first peptide linkage. The formylmethionine group is transferred from its tRNA at the P site to the amino group of the second aminoacyl tRNA at the A site of the ribosome. This involves nucleophilic attack by the amino group of the second amino acid on the carboxyl carbon of the methionine. The resulting bond formation attaches both amino acids to the tRNA at the A site.

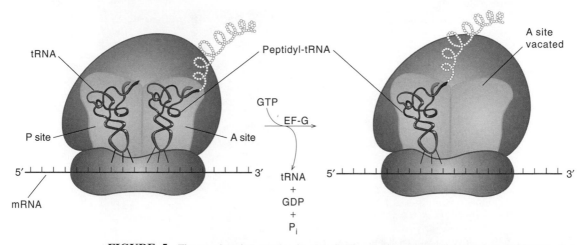

FIGURE 5 The translocation reaction in *E. coli*. The translocation reaction occurs immediately after peptide synthesis. It involves displacement of the discharged tRNA from the P site and concerted movement of the peptidyl tRNA and mRNA so that the peptidyl tRNA is bound to the P site and the same three nucleotides in the mRNA. The A site is vacated and ready for the addition of another aminoacyl tRNA. Translocation in eukaryotes is similar except that the EF-2 factor is involved instead of the EF-G factor.

thought that only Watson–Crick base pairs were used in this pairing but the situation is more complicated than this.

Code Is Highly Degenerate

A triplet code, one made from four nucleotides taken three at a time, generates a total of 64 different triplet sequences or codons. Three of these codons are utilized to terminate translation and are not generally used to specify amino acids. The remaining 61 codons and the 20 amino acids can be summarized neatly by grouping codons with the same first and second bases into a grid as we have seen (see Chapter 9, Table 2). The four horizontal sections are composed of codons with the same first base. The four vertical sections are composed of codons with the same second bases. The boxes representing the vertical–horizontal intersections contain codon families whose members differ only in their 3′-terminal base. For example, the codons UCU, UCC, UCA, and UCG comprise a family encoding serine. Thus the genetic code is degenerate (i.e., one amino acid is generally specified by multiple or synonymous codons).

The codon AUG is the only one that is generally used to specify methionine, but it serves a dual function in that it also is used to initiate translation. The UGA triplet also serves a dual function; it is usually recognized as a stop codon, but on rare occasion it serves as a codon for the unusual amino acid selenocysteine.

Wobble Introduces Ambiguity into Codon–Anticodon Interactions

For 61 triplets to act as codons, tRNAs must interact specifically with each triplet. Strict Watson–Crick base pairing between codon and anticodon would require 61

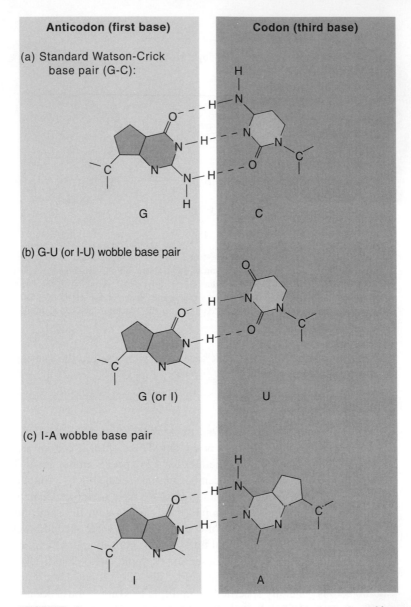

FIGURE 6   Examples of standard (a) and wobble (b and c) base pairs formed between the first base in the anticodon and the third base in the codon.

different anticodons and, correspondingly, 61 different tRNAs. As the characterization of tRNAs progressed it became clear that in many cases individual tRNAs could recognize more than one codon. In all cases the different codons recognized by the same tRNA were found to contain identical nucleotides in the first two positions and a

TABLE 1

The Wobble Rules of Codon–Anticodon Pairing

5′ Base of anticodon	3′ Base of codon
C	G
A	U
U	A or G
G	C or U
I	U, C, or A

different nucleotide in the third position (the 3′ position of the codon). This relationship fits with the neat arrangement of codon families that differ in the third base (see Chapter 9, Table 2).

The commonly observed 3′-terminal redundancy of the genetic code and its mechanistic basis were discovered by Francis Crick. In 1966 Crick proposed that codons and anticodons interact in an antiparallel manner on the ribosome in such a way as to require strict Watson–Crick pairing (i.e., A-U and G-C) in the first two positions of the codon but to allow other pairings in its 3′-terminal position. Nonstandard base pairing between the 3′-terminal position of the codon and the 5′-terminal position of the anticodon alters the geometry between the paired bases. Crick's proposal, labeled the *wobble hypothesis,* is now viewed as correctly describing the codon–anticodon interactions that underline the translation of the genetic code.

By inspecting the geometry that would result from different wobble pairings (Fig. 6) and recognizing that inosine, the deaminated form of adenine, frequently occurs in tRNAs in the 5′ position of the anticodon, Crick grasped the relationship between codon and anticodon. According to this relationship (Table 1), when C or A occurs in the 5′ position of an anticodon, it can pair only with G or U, respectively, in the 3′ position of a codon. Transfer RNAs containing either G or U in the 5′ or wobble position of the anticodon can each pair with two different codons, while an inosine (I) in this position produces a tRNA that can pair with three codons differing in the 3′ base. Subsequent sequence analysis of many tRNAs has proved this hypothesis as a feature of the translation of the "universal" genetic code.

SUMMARY

In this chapter we have discussed the chemistry of the translation process and some aspects of the genetic code.

1. Amino acids are activated by formation of aminoacyl–AMP complexes. In the next step the amino acid exchanges its AMP for an ester linkage to a specific tRNA The attachment of the amino acid to the tRNA is catalyzed by a specific aminoacyl tRNA synthase, which recognizes all the cognate tRNAs for a specific amino acid.

2. All the remaining reactions in protein synthesis take place on the ribosome. First, the small ribosomal subunit forms a ternary complex with the mRNA and the first aminoacyl tRNA. In this complex the mRNA is oriented so that the codon bases in the mRNA and the anticodon bases in the tRNA are oriented in an antiparallel way. Following this the large ribosomal subunit adds to the initiaton complex. The initiation reaction also requires a family of soluble proteins that are collectively referred to as initiation factors.

3. Elongation involves all the reactions related to peptide synthesis, from the first peptide linkage to the last. A new group of dissociable protein factors, called elongation factors, join the ribosome complex during the time of elongation.

4. Termination of translation is signaled by a special sequence at the end of the translatable message. A third set of dissociable proteins, the release factors, catalyzed the release of the nascent polypeptide chain from the ribosome and the dissociation of the tRNA and the ribosomal subunits.

5. The genetic code is the sequence relationship between nucleotides in the mRNA and amino acids in the proteins they encode. Triplet codons are arranged on the mRNA in a nonoverlapping manner without spacers.

6. The genetic code is highly degenerate, with most amino acids represented by more than one codon. In many cases the 3′ base in the codon may be altered without changing the amino acid that is encoded.

7. The codon–anticodon interaction is limited to Watson–Crick pairing for the first two bases in the codon but is considerably more flexible in the third position.

Problems

1. How many ATP equivalents are required to make a peptide linkage in the translation process? How does this compare with the bond energy of the peptide linkage. Explain any peculiarities if you can.

2. Assume that you have a copolymer with a random sequence containing equimolar amounts of A and U. What amino acids would be incorporated, and in what ratio, when this copolymer is used as an mRNA?

3. A single tRNA can insert serine in response to three different codons: UCC, UCU, or UCA. What is the anticodon sequence of this tRNA?

4. During protein synthesis, GTP hydrolysis is involved in translocation and in transport of aa-tRNA to the ribosome, but not in peptide bond formation. Where does the energy for peptide bond formation come from?

5. What are the possible amino acid changes that could result from a single nucleotide change in a GAA codon? If you know the structures of the amino acids, what would you predict the effects of the altered amino acids to be?

Reference

Zubay, G. *Biochemistry,* 4th Ed. New York: McGraw-Hill, 1998, Chapter 34.

CHAPTER 17

Early Developments
in Polypeptide Synthesis

In this chapter we discuss schemes that have been used to link amino acids into peptides and polypeptides with random sequences. Then we turn to cases where simple biochemical systems have been used for synthesis of peptides with defined sequences. Finally, we turn to a highly speculative scheme I present for the first time for the early evolutionary steps that may have been involved in translation systems. This scheme has not been tested experimentally.

NONINSTRUCTED SYNTHESIS OF POLYPEPTIDES

Noninstructed Peptide Bond Formation Results in Polypeptides with Random Sequences of Amino Acids

When a solution containing amino acids is heated in an open container, the water evaporates and peptides are formed. What could be more simple. Unfortunately, there are many problems with this approach for making peptides. Many amino acids do not react well when they are dried down. Excessive heating of amino acids usually results in a carbonized tar. A partial remedy to this problem was found by Sidney Fox who made an amino acid melt by using one or two low-melting amino acids as a solvent. The pair of amino acids that Fox found most useful for this purpose included aspartic and glutamic acids. A mixture of these two acidic amino acids added to other amino acids and heated to about 150°C gives a melt. The water distilled away, thereby encouraging the equilibrium to shift in favor of peptide bond formation. After a suitable heating period the reaction mixture was plunged into water to give rise to an aggregated structure that formed spherules called proteinoids.

Table 1 gives the composition of a crude proteinoid mixture. In this particular reaction the starting material was 33% aspartic acid, 17% glutamic acid, and 3% in each of the remaining amino acids. It is apparent that the composition of the amino acids incorporated into the proteinoid differs significantly from that of the starting mixture, a result showing that there is some selectivity involved in the uptake of amino acids. In particular, the aromatic amino acids, tyrosine and phenylalanine, appear to be overrepresented.

TABLE 1

Amino Acid Composition of Proteinoid Prepared with 200 PPA[a]

Amino acid	Percentage (%)	Amino acid	Percentage (%)
Thr	0.55	Lys	2.79
Ser	0.63	His	2.53
Pro	1.04	Arg	1.83
Gly	2.93	Total	7.15
Ala	1.31		
Val	1.33		
Met	0.86	Asp	51.9
Iso	0.71	Glu	13.3
Leu	3.44	Total	65.2
Tyr	3.87		
Phe	5.87		
NH$_2$	5.02		
Total	27.6		

[a]Asp: Glu: basic and neutral amino acid ratio = 2:1:3; 100°C, for 150 hr.

As an alternative to the Fox procedure for making peptide linkages, chemical condensing agents can be introduced to make oligopeptides from amino acids. In favorable cases condensing agents increase the yields of oligomers and permit lower temperatures to be used. For example, a dry-state reaction involving trimetaphosphate, imidazole, and Mg^{2+} resulted in small oligomers of glycine at temperatures as low as 65°C. Glycine oligomers also can be made in a dry-state reaction by heating in the presence of cyanamide, adenosine triphosphate (ATP), and aminoimidazole carboxamide (AICA). Incidentally, it should be remembered that AICA is a likely intermediate in the prebiotic synthesis of purines (see Chapter 13, Fig. 11).

Under favorable circumstances dry-state heating, with or without condensing agents, results in random-sequence oligomers from mixtures of amino acids. No prebiotic simulation experiments have resulted in ordered polypeptide synthesis. Therefore, it seems likely that ordered polypeptide synthesis was an evolutionary event that required more sophisticated conditions such as might result from specially designed ribozymes.

It also has been shown that limited peptide synthesis takes place in solution with appropriately activated amino acids. Thioester derivatives always have been a favorite for this and their activated state provides the necessary free energy for peptide bond formation.

Clays may have supplied a support facility for appropriately activated amino acids. In this connection it was reported several years ago by Mella Paecht-Horowitz that amino acids in the form of activated aminoacyladenylates are efficiently polymerized after being absorbed to the clay montmorillonite. Although this result is potentially exciting, little has been said about it in recent years so we must view it with some skepticism. It is believed that clay in this reaction functions as a binding surface for bringing activated amino acids in close proximity so that they are more likely to react with one another. The possibility that the clay also contributes specific catalytic functions to the reaction cannot be excluded. The linkage between amino acid and phosphorus in this activated complex is identical to that found by amino acids just prior to their attachment to transfer RNA molecules. Although aminoacyladenylates may or may not have been prebiotic compounds, their behavior is clearly very closely related to that of the activated amino acid normally found in the biochemical pathway to protein synthesis.

Finally, it should be mentioned that there is a most serious limitation to prebiotic systems for peptide and polypeptide synthesis. In a large pool of peptides or polypeptides with random sequences, it is highly likely that occasionally one is produced that is quite useful for structural or catalytic purposes. However, in a dilute sea of other useless peptides, its function can never have a meaningful impact. Without the amplification and selection provided by the Darwinian process there is no apparent way in which such a system could be useful.

For this reason we believe that the emergence of useful peptides had to await the development of ribozymes. Once ribozymes existed that could catalyze the synthesis of useful peptides, a Darwinian selection process could be exploited.

RIBOZYME-INSTRUCTED SYNTHESIS OF PEPTIDES AND POLYPEPTIDES

First Nonrandom Synthesis of Polypeptides with Defined Amino Acid Sequences Was Probably Ribozyme Directed

The process of translation is obviously very complex, so complex in fact that one feels compelled to seek a simpler strategy for the synthesis of the first functional polypeptides. The challenge is to find a simpler approach from which a translation system could have evolved. If we look at the three main biopolymers, polynucleotides, polypeptides, and polysaccharides, we find that all three have specific sequences. In the case of the first two, the sequences are controlled by a template-directed process. In contrast polysaccharides do not use a template mechanism. Instead each linkage made is catalyzed by an enzyme that recognizes the sugars it is linking. In the simplest cases one finds homopolymers of glucose as in glycogen or cellulose that only require a single enzyme to catalyze the linkages from an activated form of the glucose monomer. Another well-known homopolymer, chitin, contains N-acetyl-D-glucosamine monomers that are recognized by a specific polymerizing enzyme. At the heteropolymer level strict alternating copolymers abound in the family of linear polysaccharides known as glycosaminoglycans; two enzymes are employed for polymerizing the glycosaminoglycans. Hyaluronic acid, which contains D-glucuronic acid and N-acetyl-D-glucosamine monomers, is a well-known example of this type of heteropolymer; there are many others.

Oligosaccharides that become linked to the extracellular side of cell membranes contain as many as six different types of monomers in a specific sequence. This is remarkable in view of the fact that no template mechanism is involved in their synthesis. In all these cases the only general rule is that a distinct enzyme is required for the catalysis of each type of linkage; the linkages are all glycosidic linkages but the sugars that are linked are different.

Returning to polypeptide synthesis we are confident that from an evolutionary vantage point the type of situation we have described here for polysaccharides most likely existed for polypeptides at one time. The likelihood of such systems emerging should depend primarily on the availability of amino acids of different types, an activation system for the amino acid, and an enzyme (most likely a ribozyme) that is specific for the amino acids it links together.

We have suggested a way in which simple polypeptides with well-defined repeating sequences could be made. Of course, such systems would not be selected unless they had survival value. This is not hard to imagine. Homopolypeptides, such as polyglycine, polylysine, or polyglutamic acid, should require only one ribozyme each to polymerize an appropriately activated amino acid. Such ribozymes could have evolved from a phosphotransferase. These polymers might make stable complexes with nucleic acids while others might stabilize or provide devices for transport of metabolites across cellular membranes.

At the next level of complexity a heteropolypeptide with a strict alternating sequence of amino acids, such as an alternating copolymer of glycine and serine (or alanine), could make a sheetlike structure like that found in silk proteins with this sequence. To make a polymer of this type would require two ribozymes, one that catalyzes the addition of serine to glycine at the growth point and one that catalyzes the addition of glycine to serine. The next level of complexity would involve the synthesis of a heteropolymer with a triplet repeat such as one finds in collagen where the triplet repeat is glycine, proline, hydroxyproline. It should be obvious that the synthesis of such a polymer would require a minimum of three ribozymes. As one increases the number of amino acids in the repeat unit, a point ultimately would be reached where this approach would no longer be effective. This is one of the disadvantages in this strategy of making polypeptides. Another is the fact that only polypeptides with repeating sequences can be synthesized by this approach.

Alternative Routes Are Used for the Biosynthesis of Peptides

Before we turn to the evolution of template-directed systems, we discuss two of the biosystems that have evolved for the synthesis of well-defined peptides. Hopefully, this gives us more insight into the general question of what it takes to make specific linkages without the involvement of a nucleic acid template.

In the synthesis of the tripeptide glutathione the order of amino acids is determined by a specific enzyme for each peptide linkage made, two enzymes in all. By contrast, the ordering of amino acids in the larger peptide gramicidin S is mainly a function of the attachment points of amino acids on the enzyme surface. Quite remarkably one of the two enzymes involved in gramicidin S biosynthesis appears to be serving the dual role of enzyme catalyst and template.

Glutathione, γ-glutamylcysteinylglycine, is found in nearly all cells and plays a variety of roles. The tripeptide is formed in two steps catalyzed by ATP-requiring reactions. The first step is the condensation of glutamate with cysteine:

$$\text{glutamate + cysteine + ATP} \xrightarrow{\gamma\text{-glutamylcysteine synthase}} \gamma\text{-glutamylcysteine + ADP + P}_i. \qquad (1)$$

The second step is the condensation of the dipeptide with glycine:

$$\gamma\text{-glutamylcysteine + glycine + ATP} \xrightarrow{\text{glutathione synthase}} \text{glutathione + ADP + P}_i. \qquad (2)$$

The antibiotic gramicidin S produced by the bacterium *Bacillus brevis* is a cyclic decapeptide composed of a repeated sequence of five amino acid (-D-Phe-L-Pro-L-Val-Orn-L-Leu)$_2$. Although in glutathione biosynthesis the order of amino acids is determined by a specific enzyme for each peptide linkage made, in gramicidin the ordering is mainly a function of the attachment points of amino acids on the enzyme surface. Thus one of the two enzymes involved in gramicidin appears to be serving a dual role of enzyme and template.

Gramicidin synthesis is divided into five phases.

1. Activation, thioesterification, and racemization of L-phenylalanine. The light enzyme of the gramicidin-forming system activates L-phenylalanine at the aminoacyl-ladenylate and transfers the phenylalanine residue to a thiol group on the enzyme. Racemization of L-phenylalanine occurs at this thioester stage:

$$ESH + ATP + \text{L-Phe} \longrightarrow (\text{L-Phe} \sim AMP)ESH + PP_i \longrightarrow$$
$$ES \sim \text{L-Phe} + AMP \longrightarrow ES \sim \text{D-Phe}. \qquad (3)$$

2. Activation and thioesterification of L-proline, L-valine, L-ornithine, and L-leucine. The heavy enzyme of the gramicidin-forming system activates the other four amino acids found in gramicidin S and transfers each to a specific thiol-containing site on the protein:

$$\frac{E}{\underset{H\ \ H\ \ H\ \ H}{S\ \ S\ \ S\ \ S}} + Pro + Val + Orn + Leu + 4ATP \longrightarrow$$

$$\frac{E}{\underset{\substack{P\ \ V\ \ O\ \ L \\ r\ \ a\ \ r\ \ e \\ o\ \ l\ \ n\ \ u}}{S\ \ S\ \ S\ \ S}} + 4AMP + 4PP_1. \qquad (4)$$

3. Transfer of D-phenylalanine to the heavy enzyme and initiation of peptide formation. The heavy enzyme contains a covalently linked 2-nm-long pantetheine arm that is thought to serve as a carrier of the growing peptide chain. It is to the—SH group of this pantetheine arm on the heavy enzyme that the D-phenylalanyl group is probably transferred from the light enzyme (transthiolation). The first peptide bond would then be formed by a transpeptidation reaction that liberates the pantetheine thiol group.

4. Elongation. The liberated pantetheine arm is now free to undergo transthiolation with the newly formed phenylalanylpropyl residue and to move to the valyl–thiol site to repeat the transpeptidation step. This step is repeated at the ornithinyl and leucyl sites.

5. Cyclization. After the pentapeptide is formed, it is cyclized with an identical peptide in a head-to-tail fashion. One possible mechanism, implied in Fig. 1, involves the transfer of the first pentapeptide to a thiol "waiting" site; and when a second pentapeptide has been completed, cyclization occurs by two additional transpeptidation reactions (see Fig. 1). Another possibility is that the cyclization occurs by an intermolecular transpeptidation involving two heavy enzymes, each containing one completed pentapeptidyl residue at the terminal thiol site (not shown).

STEPS IN THE EMERGENCE OF A TRANSLATION SYSTEM

A ribozyme-directed system for polypeptide synthesis, like the protein enzyme-directed system currently used by carbohydrates should have had the capacity for producing long chains with repetitive sequences (as in the case of glycosaminoglycans) or

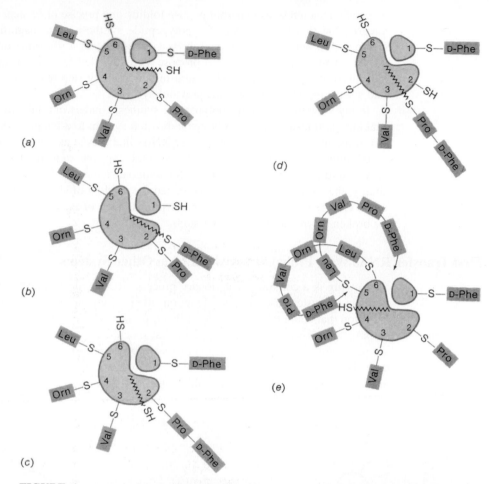

FIGURE 1 The formation of gramicidin S on a protein template. (a) The activated amino acids are held in thioester linkage on the light and heavy enzymes. The phenylalanyl residue has undergone racemization and is being transferred to the pantetheine arm on the heavy enzyme. (b) The first peptide bond is about to be formed by transfer of the phenylalanyl group from the pantetheine group to the prolyl residue. (c) The phenylalanylprolyl residue is about to be transferred to the free thiol group of the pantetheine arm. The light enzyme has accepted another phenylalanyl residue that already has undergone racemization. (d) The phenylalanylprolyl residue is about to be transferred to the valyl residue. (e) The first pentapeptidyl group after being made was transferred to waiting site 6. The second pentapeptidyl group just has been completed and is about to be condensed with the first to yield the decapeptide gramicidin S. The pantetheine arm is now free to repeat the process.

much shorter chains with nonrepetitive but unique sequences (as in the case of membrane-bound oligosaccharides). In this pretranslation period, polypeptides probably served structural functions but it is unlikely that they served enzymatic functions because the demands on enzyme systems are so much greater. To function as an enzyme a polypeptide chain must be long and each amino acid side chain must occupy a specific location so as to permit precise folding and precise placement of active sites. Clearly, the translation system for polypeptide synthesis is a magnificient piece of craftmanship but we need to know how it evolved from a ribozyme-directed system. The challenge we face here is to propose steps that led from the ribozyme-directed system to the template-directed system in which every step resulted in a system that had a selective advantage over the previous system.

In my view the system for polypeptide synthesis underwent three major changes in making the transition from a ribozyme-directed system to a template-directed system: (1) in the first step, a family of transfer RNAs that served as carriers for specific activated amino acids appeared; (2) in the second step, the large ribosomal subunit that served initially as a binding surface for aminoacyl transfer RNAs emerged; and (3) finally, the template strand that encouraged the binding of specific tRNAs to the ribosome emerged. These are the major transitions; each of these major transitions could be broken down into many smaller steps.

First Transfer RNAs Were Probably Borrowed from Other Systems

Borrowing is a common evolutionary process. Biochemical systems are very opportunistic. A system used for one function often is borrowed because there is another function that could take advantage of its properties. This seems to be the case with the tRNAs. A tRNA is a complex molecule that specifically forms a linkage to one type of amino acid (Fig. 2). This linkage is mediated by an enzyme that recognizes the tRNA and the amino acid. It seems likely that there were tRNAs long before there were translation systems. In support of this contention I can point to two systems in which tRNAs are used for other purposes. In the case of porphyrin synthesis, a specific

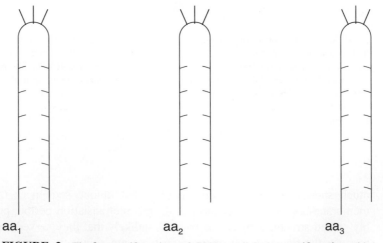

FIGURE 2   The first specific aminoacyl tRNAs are linked to specific amino acids by specific ribozymes. These complexes containing activated amino acids were initially used in specific biochemical processes other than translation.

transfer RNA for glutamine is used; and in the case of bacterial cell wall synthesis, a specific tRNA for glycine is used. In both of these cases the tRNA serves as the specific carrier for the activated amino acid in a biosynthetic process. It is easy to imagine that a series of tRNAs evolved in this way, each one serving as a carrier for an activated amino acid in a process other than polypeptide synthesis. Thus the evolution of a small family of tRNAs and enzymes (ribozymes) can be explained in terms of the selective advantage such molecules would have offered for specific biochemical processes. Ultimately, these aminoacyl tRNAs were recogized by the ribozyme system as a convenient source of activated amino acids for polypeptide synthesis.

Large Ribosomal Subunit Could Have Served as a Binding Surface for Aminoacyl tRNAs and as a Catalyst for Polypeptide Synthesis

Once some tRNAs had evolved for purposes other than polypeptide synthesis, it became possible for these tRNAs to explore additional functions. It seems likely that tRNAs carrying activated amino acids could have improved the efficiency of polypeptide synthesis as they carry activated amino acids and if there were some way of bringing them close to one another that would increase the efficiency of polypeptide synthesis even further. Naturally occurring clays may have provided the first surfaces for binding aminoacyl tRNAs together. Ultimately, a primitive large ribosomal subunit is likely to have evolved as a superior binding surface for aminoacyl tRNAs. Such a situation is likely to have facilitated polypeptide synthesis but the specificity for making different linkages remained with the ribozymes for some time.

Transition from a Ribozyme-Directed System to a Template-Directed System Was Probably a Gradual Process

The tRNAs that bound to a large ribosomal subunit could have overhung the ribosomal subunit on their ends that contain unpaired bases (Fig. 3). The first messenger RNAs

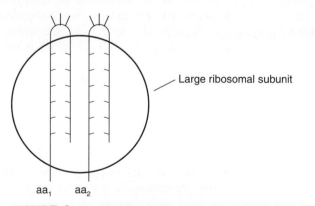

FIGURE 3   The first large ribosomal subunits were probably very crude. Their main value in polypeptide synthesis was to provide a surface on which aminoacyl tRNAs would bind. This brings them in close proximity, which increases the chances that a ribozyme will link the carboxyl group of amino acid 1 to the amino group of amino acid 2 and so on.

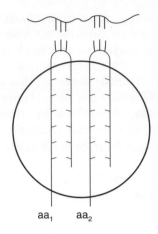

aa₁   aa₂

FIGURE 4 Small oligonucleotides that bind to the unpaired bases at the amino acid distal ends of the aminoacyl tRNAs were probably the first messengers. An oligonucleotide that makes a strong complementary complex with the tRNA is likely to encourage the binding to the large ribosomal subunit and thereby increases the frequency with which that amino acid becomes incorporated into peptidyl linkages.

(mRNAs) could have been short oligonucleotides that stabilized the binding of the tRNAs to the ribosome by hydrogen bonding to them (Fig. 4). An oligonucleotide that actually carried a sequence of nucleotides complementary to the exposed trinucleotide cluster at the ends of the tRNAs is likely to have facilitated the binding of specific tRNAs next to one another and thereby increased the frequency with which certain combinations of amino acids would become linked to one another. This clearly would have had a selective advantage in cases where the resulting amino acid sequence produced a useful polypeptide, one that gave the system a selective advantage. This is the beginning of a template-directed system. The ribozyme is still in charge of linking the amino acids together, but the template is increasing the number of times that specific combinations of amino acids are brought together on the ribosome by their tRNA carriers.

Following further evolution during which time the effectiveness and specificity of the template mRNA–tRNA interaction was increased, the ribozyme system became redundant as far as directing specific types of linkages was concerned. Thus there was no need to have ribozyme specificity any longer because the template was doing the same job. Hence the specific ribozymes gradually disappeared and were replaced by a nonspecific ribozyme for catalyzing peptide bond formation. This general ribozyme became permanently embedded in the large ribosomal subunit where it resides to the present day.

SUMMARY

In this chapter we have discussed the evolution of polypeptide synthesis. The complex process of translation at first thought seems like a process for which the evolution is almost too difficult to imagine. Here more than anywhere else it is important that we keep in mind basic principles of Darwinian evolution that evolution proceeds one step at a time and each step that flourishes must have a selective advantage.

1. In prebiotic times and possibly beyond that there must have been some polypeptide synthesis in which the amino acid sequences were random. There may have been some value in polypeptide chains with random sequences but we shall not attempt to assign any.

2. Useful polypeptides probably had to have defined sequences. It seems highly likely that such polypeptides were not synthesized until ribozymes had evolved that could catalyze specific peptide linkages. This strategy has been used in polysaccharide and oligosaccharide synthesis where each glycosidic linkage is catalyzed by a specific enzyme in a nontemplate catalyzed reaction.

3. One can imagine a large number of polypeptides with repetitive sequences that could have been very useful in an RNA-only world.

4. The value of being able to place amino acids in a nonrepetitive sequence was the driving force for further evolution of the polypeptide synthesizing machinery.

5. The first tRNAs were probably borrowed from other systems. We know of some cases where specific tRNAs carrying select amino acids function in other metabolic processes, one for glycine and one for glutamine are well known. Once a few of these tRNAs with specific recognition sites for ribozymes and amino acids were accessible in the living system it seems likely that the ribozymes would have adapted to catalyzing the linkage between specific aminoacyl tRNAs. Prior to ribosomes clays may have served as surfaces for bringing different tRNAs together in a nonspecific way. However, if living systems were already membrane-enclosed, it is difficult to see how clays could have helped. Some sort of a surface for bringing aminoacyl tRNAs together would have served as the first ribosomes. The first mRNAs were probably not very specific but merely served as a stabilizing force for bringing tRNAs together on the ribosome surface. The specificity of linkage was still the primary function of the ribozyme. Evolution of messages with complementary nucleotide sequences to the anticodon hairpin regions on specific tRNAs would have helped to bring specific tRNAs together. Eventually the interaction between anticodons on the tRNAs and codons on the mRNAs became so strong that the ribozyme specificity properties became redundant and gradually gave way to this new way of determining specificity of sequence. Template-directed translation systems were on the way at this point.

Problems

1. Keeping in mind that evolution occurs in single steps, each one conferring a selective advantage, can you suggest the evolutionary steps that led to protein synthesis as we know it today?

2. Why do proteins make superior enzymes to nucleic acids? See Problem 3 in Chapter 9.

3. It has been said that most biochemical pathways could be improved over the more or less accidental pathways that have evolved over the course of evolution. Could you suggest a superior assortment of amino acids to be used in protein synthesis over those that are used currently?

4. What possible functions might polypeptides with random amino acid sequences have served in those days before mechanisms existed for synthesis of polypeptides with ordered amino acid sequences?

5. The first polypeptide chains that had specific amino acid sequences probably did not contain all 20 amino acids found in modern-day proteins. Name three amino acids that you consider are most likely to have been present in the first proteins and explain your choices.

References

Eigen, M., Biebricher, C. K., and Gebinoga, M. The hypercycle. Coupling of RNA and Protein Biosynthesis in the Infection Cycle of an RNA Bacteriophage, *Biochemistry* 30: 11007, 1991.

Ferris, J. P., Hill, A. R., Liu, R., and Orgel, L. E. Synthesis of Long Prebiotic Oligomers on Mineral Surfaces, *Nature (London)* 381: 59, 1996.

Noller, H. F., Hoffarth, V., and Zimniak, L. Unusual Resistance of Peptidyl Transferase to Protein Extraction Procedures, *Science* 256: 1416, 1994.

Orgel, L. E. The Origin of Polynucleotide-Directed Protein Synthesis, *J. Mol. Evol.* 29: 465, 1989.

Paecht-Horowitz, M., Berger, J., and Katchalsky, A. Prebiotic Synthesis of Polypeptides by Heterogenous Polycondensation of Amino Acid Adenylates, *Nature (London)* 228: 636, 1970.

Prusiner, S. B. The Prion Diseases, *Sci. Am.* 272: 48, 1995. (Diseases in which proteins may be the infectious agents are discussed. Put the emphasis on may be.)

Shen, C., Mills, M., and Oro, J. Prebiotic Synthesis of Histidyl-histidine, *J. Mol. Evol.* 31: 175, 1990.

CHAPTER 18

Lipid Metabolism and Prebiotic Synthesis of Lipids

Fatty acids, as components of phospholipids, are important structural elements of biological membranes and, like carbohydrates, are important sources of energy. In this chapter we focus on the synthesis and breakdown of fatty acids, reactions leading from fatty acids to phospholipids, and possible pathways to lipids in the prebiotic world.

We see that fatty acids are both assembled and degraded in blocks of two-carbon atoms (Fig. 1). Despite this obvious similarity, the processes of synthesis and breakdown differ in major respects: they use completely different sets of enzymes, and they

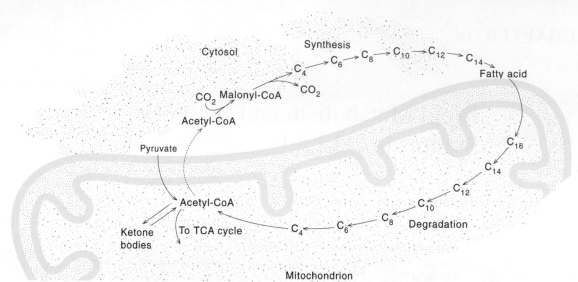

FIGURE 1   Synthesis and degradation of fatty acids. Both synthesis and degradation of fatty acids occur by multistep pathways involving totally different enzymes located in different parts of the cell. Both pathways are similar in that two-carbon atoms are either added (in synthesis) or deleted (in degradation) in each round of a repetitious, cyclic process.

The acetyl-CoA used as substrate for fatty acid synthases in the cytosol originates from the mitochondrion. This acetyl-CoA condenses with CO_2 to form malonyl-CoA, which eliminates CO_2 after an initial condensation reaction. After three more steps a two-carbon unit is added to a growing fatty acid chain. This cycle repeats itself many times until a fatty acid chain with 16 carbon atoms has been synthesized. Variations of this pathway give rise to unsaturated and polyunsaturated fatty acids.

The degradation of fatty acids also involves a repetitious process yielding a two-carbon unit (acetyl-CoA) in every cycle. This acetyl-CoA can be fed into the TCA cycle; alternatively, it can be used to make ketone bodies.

Elaborate controls prevent synthesis and degradation of fatty acids from occurring simultaneously.

occur in different cellular compartments. These differences facilitate regulatory mechanisms that ensure that biosynthesis and catabolism do not occur simultaneously.

In the first section of this chapter we consider reactions associated with fatty acid breakdown. The second section covers the pathway for fatty acid biosynthesis. The third section deals with the conversion of fatty acids into phospholipids, the immediate precursors of lipids in membranes. Finally, we take a stab at the difficult question of how membrane lipids were synthesized in the prebiotic world as precursors for the first cells.

FATTY ACID DEGRADATION

Fatty Acid Breakdown Occurs in Blocks of Two-Carbon Atoms

Investigations into the mechanisms of fatty acid catabolism began around the turn of the century when Fritz Knoop reported experiments that indicated fatty acids were degraded by removal of two carbons at a time. The data indicated that C-3 of a fatty acid was oxidized with subsequent cleavage between C-2 and C-3.

$$\underset{\beta\ \ \alpha}{\overset{3\quad 2\quad 1}{RCH_2CH_2COOH}} \longrightarrow \underset{\beta\ \alpha}{\overset{O\\\parallel\ 2\ \ 1}{RCCH_2COOH}} \longrightarrow \overset{O\\\parallel}{RCOH} + \underset{\alpha}{\overset{2\ \ 1}{CH_3COOH.}} \qquad (1)$$

C-2 also is known as the alpha (α) carbon and C-3, as the beta (β) carbon. Hence the term β *oxidation* was coined.

Oxidation of Saturated Fatty Acids Occurs in the Mitochondria in Eukaryotes

We now explore the remarkable process by which a long-chain saturated fatty acid is converted into two-carbon units (acetate), which can be oxidized to CO_2 and H_2O via the tricarboxylic acid (TCA) cycle and the electron–transport chain. Fatty acids that enter cells are activated to their coenzyme A (CoA) derivatives by the enzyme acyl-CoA ligase and transported into mitochondrial intermembrane space where the CoA ligand is replaced by carnitine (($CH_3)_3\overset{+}{N}CH(OH)CH_2CO_2^-$). A specific carrier protein transports acyl-carnitine into the mitochondrial matrix where the acyl-carnitine is converted back into acyl-CoA and β oxidation follows.

The outline for the β oxidation of a saturated fatty acid is shown in Fig. 2. In the first reaction, the acyl-CoA is dehydrogenated by acyl-CoA dehydrogenase to yield the α,β(or 2,3)*trans*-enoyl-CoA and reduced flavin-adenine dinucleotide ($FADH_2$). The electrons from $FADH_2$ are channeled into the electron transport chain (see Chapter 11). Enoyl-CoA is subsequently hydrated stereospecifically by enoyl-CoA hydrase to yield the 3-L-hydroxyacyl-CoA. The hydroxyl group is oxidized by 3-hydroxyacyl-CoA dehydrogenase with nicotinamide adenine dinucleotide (NAD^+) as coenzyme to yield β-ketoacyl-CoA and reduced NAD (NADH). The final step in the sequence is catalyzed by thiolase to form acetyl-CoA and an acyl-CoA that is two carbons shorter than the initial substrate for β oxidation. This acyl-CoA can undergo another cycle of β oxidation to yield $FADH_2$, NADH, acetyl-CoA, and an acyl-CoA with two fewer carbons (Fig. 2b). The enzymatic steps are cycled until, in the last sequence of reactions, the four-carbon unit

$$\text{Butyryl-CoA } (CH_3CH_2CH_2\overset{O\\\parallel}{C}\!-\!CoA)$$

is degraded to two acetyl-CoAs.

Fatty Acid Oxidation Yields Large Amounts of Adenosine Triphosphate

Oxidation, in combination with the TCA cycle and respiratory chain, provides more energy per carbon atom than any other energy source, such as glucose or amino acids, provide. The equations for the complete oxidation of palmitoyl-CoA are

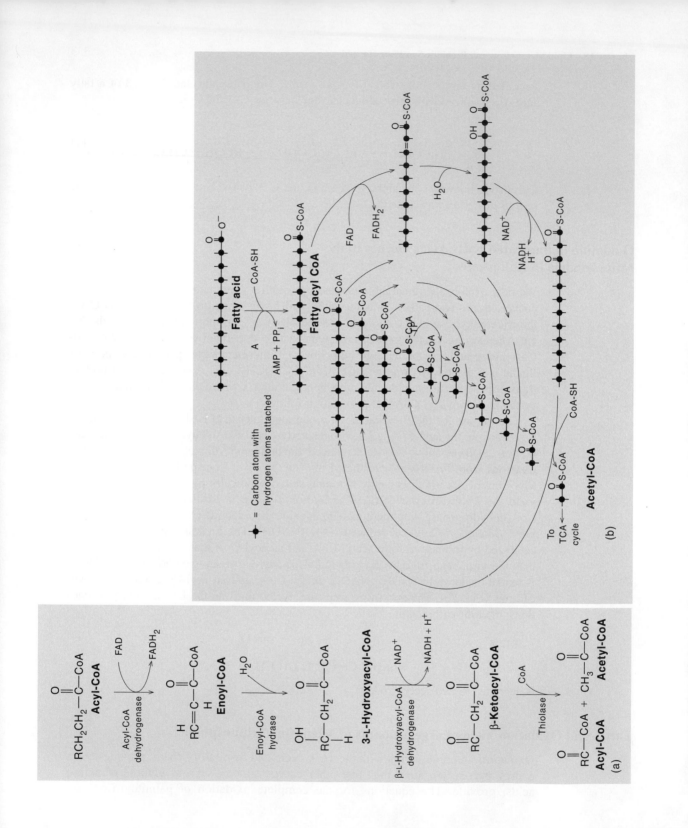

TABLE 1

Equations for the Complete Oxidation of Palmitoyl-CoA to CO_2 and H_2O

1. $CH_3(CH_2)_{14}CO\text{-}CoA + 7FAD + 7H_2O + 7CoA + 7NAD^+ \longrightarrow 8Ch_3CO\text{-}CoA + 7FADH_2 + 7NADH + 7H^+$
2. $7FADH_2^a + 10P_i^b + 10ADP + 3\frac{1}{2}O_2 + 10H^+ \longrightarrow 7FAD + 17H_2O + 10ATP$
3. $7NADH^a + 18P_i + 18ADP + 3\frac{1}{2}O_2 + 25H^+ \longrightarrow 7NAD^+ + 25H_2O + 18ATP$
4. $8CH_3CO\text{-}CoA + 16O_2 + 80P_i + 80ADP + 80H^+ \longrightarrow 8CoA + 88H_2O + 16CO_2 + 80ATP$
5. $CH_3(CH_2)_{14}CO\text{-}CoA^c + 108P_i + 108ADP + 23O_2 + 108H^+ \longrightarrow 108ATP + 16CO_2 + CoA + 123H_2O$

[a]The yield ATP per oxidation of $FADH_2$ and NADH is assumed to be 1.5 and 2.5, respectively.

[b]Remember that condensation of ADP and P_i yields a molecule of H_2O as well as ATP.

[c]See Chapter 13, Fig. 4 when accounting for the production of H_2O from the oxidation of acetyl-CoA because each TCA cycle uses two molecules of H_2O.

shown in Table 1. Equation (1) in the table shows the oxidation of palmitoyl-CoA by the enzymes of β oxidation. Each of the products of Eq. (1) to (4) are summed; the result is Eq. (5). Hence, complete oxidation of one molecule of palmitoyl-CoA yields $108ATP + 16CO_2 + 123H_2O$ and CoA. The water generated by β oxidation seems almost incidental to this process, but is crucial in several animal species. For example, the oxidation of fatty acids is used as a major source of H_2O by the killer whale. This animal lives in the sea but cannot drink the saltwater. Similarly, the fat stored in the hump of the camel serves as a source of energy and water in the desert.

The yield of ATP from the oxidation of palmitoyl-CoA can be compared with the yield from glucose. Both are major sources of body energy. Perhaps the most meaningful comparison is in terms of the number of ATPs produced per carbon atom. For glucose the yield is 30/6 = 5ATPs per carbon, and for palmitoyl-CoA it is 108/16 — almost 7ATPs per carbon. Thus, the oxidation of palmitoyl-CoA yields more ATP per carbon. The chemical reason for this difference is that 15 of the 16 carbons in palmitate are in the completely reduced state (i.e., no oxygen substitution) whereas all 6 carbons of glucose are paritally oxidized (each carbon has a hydroxyl group). The more a carbon atom is substituted with oxygen, the less energy can be harnessed by oxidation to CO_2, which is the derivative of carbon that is completely oxidized.

In this section we see that fatty acids are oxidized in units of two-carbon atoms. The immediate end products of this oxidation are $FADH_2$ and NADH, which supply energy through the respiratory chain; and acetyl-CoA, which has multiple pos-

FIGURE 2   Outline for β oxidation of a saturated fatty acid. (a) The β oxidation of fatty acids consists of four reactions in the matrix of the mitochondrion. Each cycle of reactions results in the formation of acetyl-CoA and an acyl-CoA with two fewer carbons. (b) The cyclic nature of the β-oxidation cycle. In some illustrations the —SH group of CoA is indicated for emphasis. Likewise in some illustrations the —S— group in an acyl-CoA structure is indicated.

sible uses in addition to the generation of energy via the TCA cycle and respiratory chain.

BIOSYNTHESIS OF FATTY ACIDS

Research on the mechanism of fatty acid biosynthesis goes back to the very early years of biochemical investigations. The finding that most fatty acids contain an even number of carbon atoms led Rapier to postulate in 1907 that fatty acids were produced by condensation of an activated two-carbon compound. In some of the first isotopic tracer experiments performed in the late 1930s and early 1940s, David Rittenberg and Konrad Bloch used the newly developed heavy isotopes of carbon (carbon-13) and hydrogen (deuterium) and implicated acetate as the two-carbon compound. Subsequently, Feodor Lynen showed a central role for acetyl-CoA in fatty acid biosynthesis. Precisely how acetyl-CoA was converted into fatty acids eluded workers until the late 1950s when Salih Wakil discovered the involvement of malonyl-CoA. Subsequent progress was rapid, and the scheme for fatty acid biosynthesis as we know it today was elucidated.

The synthesis of saturated fatty acids is very similar in all organisms. The overall reaction for the formation of palmitic acid is

$$
\underset{\text{Acetyl-CoA}}{CH_3\overset{\overset{\displaystyle O}{\|}}{C}CoA} + 7^-\underset{\text{Malonyl-CoA}}{\overset{\overset{\displaystyle O}{\|}}{O}CCH_2\overset{\overset{\displaystyle O}{\|}}{C}CoA} + 14NADPH + 14H^+ \longrightarrow
$$

$$
\underset{\text{Palmitic acid}}{CH_3(CH_2)_{14}COO^-} + 7CO_2 + 8CoA + 14NADP^+ + 7H_2O.
$$

(2)

First Step in Fatty Acid Synthesis Is Catalyzed by Acetyl-Coenzyme A Carboxylase

Eight enzyme-catalyzed reactions are involved in the conversion of acetyl-CoA into fatty acids. The first reaction is catalyzed by acetyl-CoA carboxylase and requires ATP. This is the reaction that supplies the energy that drives the biosynthesis of fatty acids. The properties of acetyl-CoA carboxylase are similar to those of pyruvate carboxylase, which is important in the gluconeogenesis pathway (see Chapter 11). Both enzymes contain the coenzyme biotin covalently linked to a lysine residue of the protein via its ϵ-amino group.

Acetyl-CoA carboxylase of *Escherichia coli* is a multienzyme complex that consists of three protein components that can be isolated individually: biotin carboxyl carrier protein (BCCP), biotin carboxylase, and carboxyltransferase (Fig. 3). The reaction sequence involves an initial carboxylation of BCCP, catalyzed by biotin carboxylase.

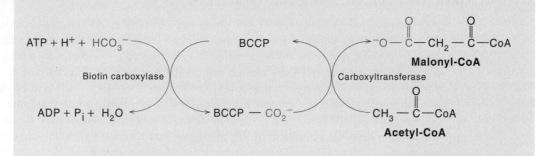

FIGURE 3   Reactions catalyzed by acetyl-CoA carboxylase. In *E. coli*, (BCCP) and the two enzymatic activities (biotin carboxylase and carboxyltransferase) can be separated from each other. In contrast, in the liver all three components exist on a single multifunctional polypeptide.

The CO_2 is covalently linked to one of the nitrogen atoms of biotin (Fig. 4). Subsequently, the CO_2 is transferred from BCCP to acetyl-CoA in a reaction catalyzed by carboxyltransferase, which yields malonyl-CoA.

A distinctly different form of acetyl-CoA carboxylase is found in the cytosol of animal tissues. The rat liver enzyme is a dimer composed of two identical subunits [relative molecular weight (M_r) of each = 265,000] with one biotin per subunit. In contrast to the multienzyme complex in *E. coli*, the three functional parts of acetyl-CoA carboxylase in rat liver occur in a single multifunctional polypeptide.

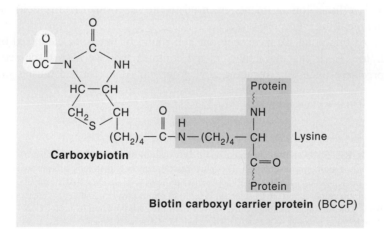

FIGURE 4   Structure of N'-carboxybiotin linked to biotin carboxyl carrier protein (BCCP). BCCP is one of the components of acetyl-CoA carboxylase isolated from *E. coli*.

Seven Reactions Are Catalyzed by the Fatty Acid Ligase

After malonyl-CoA synthesis, the remaining steps in fatty acid synthesis occur on the fatty acid ligase, which exists as a multienzyme complex. In the initial reactions acetyl-CoA and malonyl-CoA are transferred onto the protein complex by acetyl-CoA transacylase and malonyl-CoA transacylase (step 1 and step 2 in Fig. 5). The acceptor for the acetyl and malonyl groups is acyl carrier protein (ACP). ACP also carries all the intermediates during fatty acid biosynthesis. The prosthetic group that binds these intermediates is the coenzyme phosphopantetheine that is bound to ACP by an ester linkage between the phosphate of the coenzyme and a serine hydroxyl side chain of the protein (Fig. 6). The sulfhydryl group of the phosphopantetheine is the attachment site for the intermediates during fatty acid synthesis. Note that CoA also contains this phosphopantetheine moiety.

In the condensation reaction, the acetyl group is initially transferred from ACP on to an SH group of 3-ketoacyl-ACP. This acetyl moiety then reacts with malonyl-ACP (step 3 in Fig. 5) so that the acetyl component becomes the methyl terminal two-carbon unit of the acetoacetyl-ACP. The release of CO_2 in this condensation reaction provides the extra thermodynamic push to make the reaction highly favorable.

The object of the next three reactions (steps 4 to 6 in Fig. 5) is to reduce the 3-carbonyl group to a methylene group. The carbonyl group is first reduced to a hydroxyl by 3-ketoacyl-ACP reductase. Next, the hydroxyl is removed by a dehydration reaction catalyzed by 3-hydroxyacyl-ACP dehydrase with the formation of a *trans* double bond. This double bond is reduced by NADPH catalyzed by 2,3-*trans*-enoyl-ACP reductase. Chemically, these reactions are nearly the same as the reverse of three steps in the β-oxidation pathway except that the hydroxyl group is in the D-configuration for fatty acid synthesis and in the L-configuration for β oxidation (compare Figs. 2a and 5). Also remember that different cofactors, enzymes, and cellular compartments are used in the reactions of fatty acid biosynthesis and degradation.

At this point we have seen one full round of reactions catalyzed by the fatty acid ligase. Each enzyme activity of the complex that we have discussed has been used precisely once. The resulting acyl group on ACP (formed during step 6, Fig. 5) is

---

FIGURE 5 Outline of the reactions for fatty acid biosynthesis. Fatty acids grow in steps of two-carbon units on a multienzyme complex. The initial reactions of fatty acid biosynthesis are shown. In the first reaction, acetyl-CoA reacts with ACP (acyl carrier protein) to form acetyl-ACP (step 1). ACP is shown with its SH group emphasized (see Chapter 19, Fig. 13) to remind readers that the acyl derivatives are linked to ACP via a thioester bond. Malonyl-CoA, derived from the carboxylation of acetyl-CoA (see Fig. 3), reacts with ACP to yield malonyl-ACP (step 2). These two ACP derivatives then condense to form 3-ketoacyl-ACP with the release of ACP and CO_2 (step 3). Step 4 involves the reduction of the 3-keto group with NADPH and step 5, the dehydration of 3-hydroxyacyl-ACP to form 2,3-*trans*-enoyl-ACP. The final reaction in the cycle is reduction of the double bond with NADPH (step 6) to give an acyl-ACP with four carbons. Screens permit tracking of different molecular groupings.

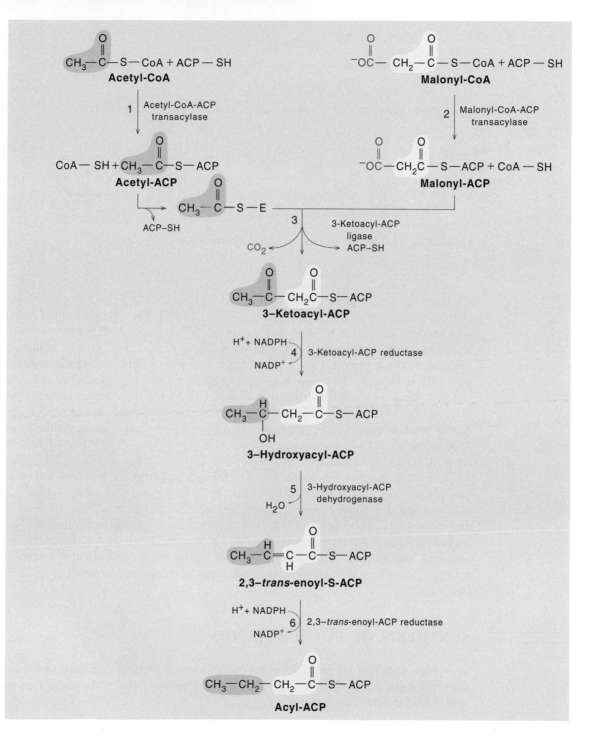

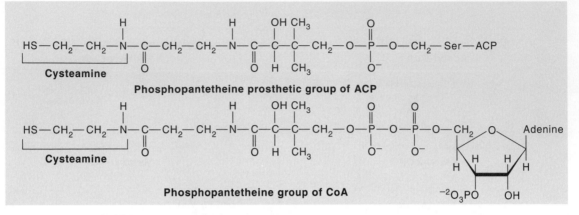

FIGURE 6   The phosphopantetheine group in acyl carrier protein (ACP) and in CoA.

transferred to the SH group of the active site of 3-ketoacyl-ACP ligase (the condensing enzyme activity) as was the acetyl group in the first cycle. The acyl group then reacts with another molecule of malonyl-ACP catalyzed by 3-ketoacyl-ACP synthase to yield a six-carbon intermediate with a ketone on the three-carbon (Fig. 7). The reduction reactions (steps 4 to 6 in Fig. 5) occur again and a six-carbon intermediate, hexanoyl-ACP is formed. The hexanoyl group is transferred to the 3-ketoacyl-ACP ligase and condenses with another malonyl-ACP followed by the three reduction reactions (Fig. 7). The biosynthetic process continues to recycle until palmitoyl-ACP is made. At this point, in animal cells, palmitate is hydrolyzed from the phosphopantetheine of ACP by the activity of a thioesterase whereas in *E. coli* palmitoyl-ACP is used directly for phospholipid biosynthesis.

Figure 8 illustrates the processes of synthesis and degradation in a way that emphasizes the parallels and differences. In both cases, two-carbon units are involved and the reactions involved are very similar. However, different enzymes and coenzymes are utilized in the biosynthetic and degradative processes. The second striking difference between the two pathways is that malonyl-CoA is the principal substrate in the anabolic pathway but plays no role in the catabolic pathway. Moreover, the processes take place in different compartments of the cell (see Fig. 1).

SYNTHESIS OF PHOSPHOLIPIDS

Most fatty acids that are not utilized for energy storage perform important structural roles as integral components of membrane lipids. Phospholipids are the major type of lipids found in membrane structures.

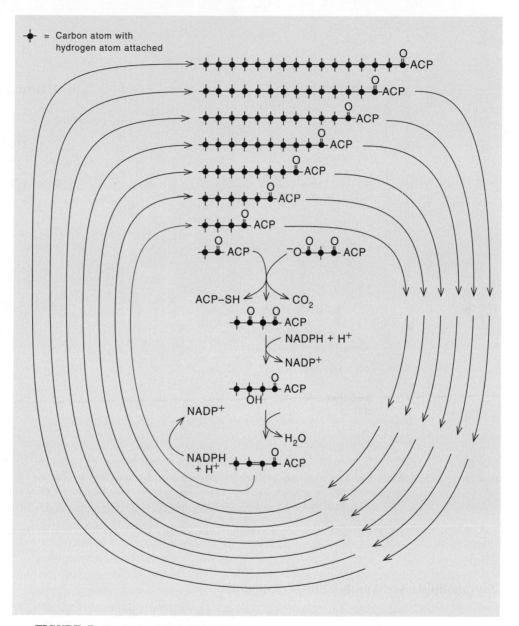

FIGURE 7 Synthesis of fatty acids. The initial and continuation reactions of fatty acid synthesis are depicted. The acyl-ACP generated from the first cycle of fatty acid synthesis is condensed with another molecule of malonyl-ACP to yield the 3-keto, six-carbon intermediate that is reduced, dehydrated, and reduced to give an acyl-ACP with six carbons. The cycle is then repeated five times with the final yield of an acyl-ACP that contains 16 carbons (palmitoyl-ACP).

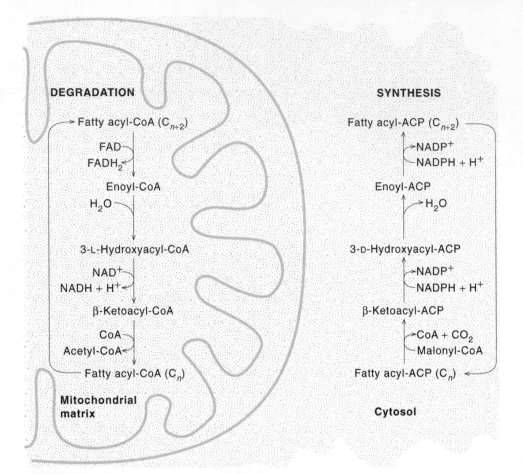

FIGURE 8   Comparison between synthesis and degradation of fatty acids in the liver. Both processes involve two carbons at a time and very similar intermediates, even though they go in opposite directions. CoA also is involved heavily in both processes. Here the similarities end. The enzymes used in the two processes are totally different, and the coenzymes are also different. In the degradative direction FAD and NAD^+ are used, whereas in the synthetic direction the coenzyme NADPH is used. Degradation occurs in the mitochondrial matrix, and synthesis occurs in the cytosol.

Phospholipids Are Amphipathic Compounds

Phospholipids are ideal compounds for making membranes because of their amphipathic nature. Thus the polar headgroups of phospholipids prefer an aqueous environment, whereas the nonpolar acyl substituents do not. As a result, phospholipids spontaneously form bilayer structures that prevent the unrestricted transport of most molecules other than water into the cell. Similarly, the phospholipid bilayer prevents leakage of metabolites from the cell. The amphipathic nature of phospholipids has a great influence on the mode of their biosynthesis. Thus, most of the reactions involved

TABLE 2
Polar Head Groups Associated with Major Classes of Phospholipids

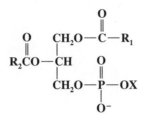

	X Substituent	
Name of X—OH	**Formula of X**	**Name of phospholipid**
Water	—H	Phosphatidic acid
Choline	—CH$_2$CH$_2$N$^+$(CH$_3$)$_3$	Phosphatidylcholine
Ethanolamine	—CH$_2$CH$_2$N$^+$H$_3$	Phosphatidylethanolamine
Serine	—CH$_2$—CH$\overset{\overset{+}{N}H_3}{\underset{COO^-}{|}}$	Phosphatidylserine
Glycerol	—CH$_2$CH(OH)CH$_2$OH	Phosphatidylglycerol
Phosphatidylglycerol	—CH$_2$CH(OH)—CH$_2$—O—P—O—CH$_2$	Diphosphatidylglycerol (cardiolipin)

in lipid synthesis occur on the surface of membrane structures catalyzed by enzymes that are themselves amphipathic.

In *Escherichia coli* Phospholipid Synthesis Generates Phosphatidyethanolamine, Phosphatidylglycerol, and Diphosphatidylglycerol

Escherichia coli contains three important classes of phospholipids: phosphatidylethanolamine (75–85%), phosphatidylglycerol (10–20%), and diphosphatidylglycerol (5–15%).

In eukaryotes phosphatidyl choline and phosphatidyl ethanolamine are quantitatively the most important phospholipids (Table 2). We focus on the synthesis of

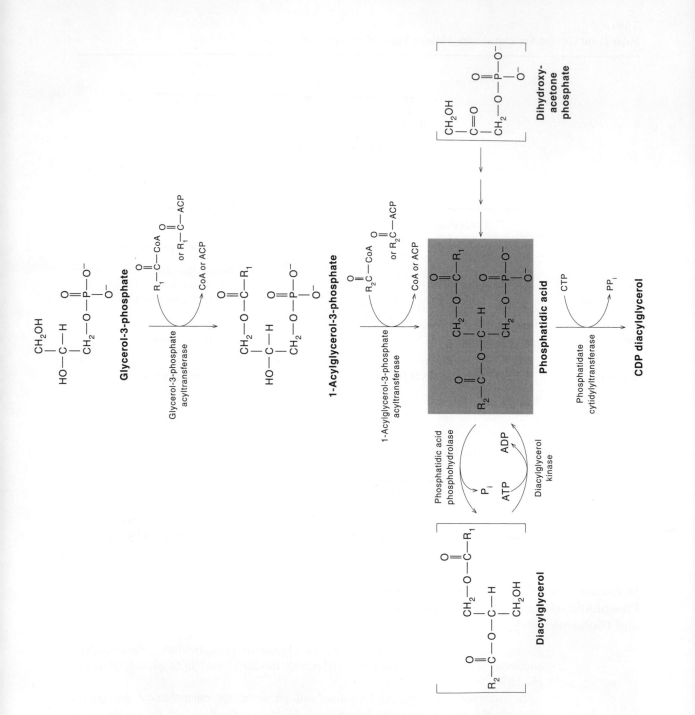

phospholipids in *E. coli*. All three of the major phospholipids found in *E. coli* share the same biosynthetic pathways up to the formation of cytidine 5′-diphosphate (CDP)-dia-cylglycerol (Fig. 9).

Most of the enzymes for phospholipid synthesis are located on the inner plasma membrane of *E. coli*. Glycerol-3-phosphate acyltransferase, the first enzyme in the pathway, preferentially utilizes saturated fatty acyl derivatives (palmitoyl-CoA or palmitoyl-ACP) for the initial acylation of glycerol-3-phosphate. The second enzyme (see Fig. 9) 1-acylglycerol-3-phosphate acyltransferase, catalyzes phosphatidic acid formation; this enzyme shows a preference for acyl residues with a double bond. The substrate specificities of these two acyltransferases account for saturated fatty acids usually being in the SN-1 position and unsaturated fatty acids, in the SN-2 position of phospholipids. A third transferase reaction converts phosphatidic acid to CDP-diacylglycerol with cytidine 5′-triphosphate (CTP) as a cosubstrate. Interest-ingly, CTP is the only high-energy nucleotide used for the synthesis of phospholipids. CDP-diacylglerol is a branchpoint intermediate. It is converted either to phos-phatidylserine enroute to phosphatidyl-ethanolamine or to phosphatidylglycerol phos-phate enroute to phosphatidylglycerol and diphosphatidylglycerol (Fig. 10). The hydroxyl group of glycerol-3-phosphate or serine reacts with the high-energy pyrophosphate bond of CDP-diacylglycerol.

Phosphatidylglycerol and diphosphatidylglycerol also are synthesized in the mito-chondria of eukaryotes by a pathway similar to that in prokaryotes. The only difference is that diphosphatidylglycerol in eukaryotes is made by the reaction of CDP-diacylglycerol with phosphatidylglycerol instead of the condensation of two molecules of phosphatidylglycerol that occurs in *E. coli*.

PREBIOTIC SYNTHESIS OF LIPIDS

Most likely membrane lipids were of great importance either before or shortly after the origin of life, because membranes provide a uniquely effective way of containing a complex mixture of water-soluble organic compounds that must function together. For this reason it behooves us to explore possible prebiotic routes for the synthesis of lipids from which the first membranes were formed.

The principal component in all biological membranes is a cluster of (two or more) hydrocarbon chains containing carboxyls at one terminus that are usually esterified to glycerol (see Fig. 9). The formation of these membrane lipids is biphasic: the first and most demanding phase involves synthesis of the carboxylated hydrogen carbon chains (the fatty acids); the second phase involves their esterification to suitable derivatives such as glycerol or a derivative thereof.

As we have seen, fatty acids are biosynthesized in a series of cyclically repeating

FIGURE 9   The first phase of phospholipid synthesis in *E. coli* and eukaryotes. Additional routes to and from phosphatidic acid, found predominantly in eukaryotes, are shown in brackets.

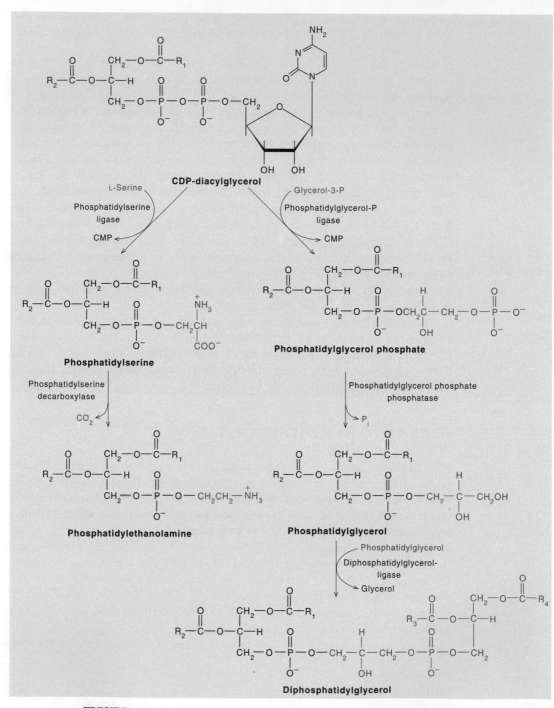

FIGURE 10   The second phase of phospholipid synthesis in *E. coli*, from CDP-diacylglycerol to the end products.

steps, each step entailing the addition of two carbons to the growing chain. Although this process is quite complex, it is possible that simplifying modifications could be made that preserve some of the main features of the biosynthetic pathway. It also seems worthwhile to consider cruder conditions under which the very first lipids could have been synthesized.

Strenuous Conditions Have Been Used by Organic Chemists to Synthesize Fatty Acid Chains

The first pathway to fatty acids was derived from a major industrial process for manufacturing hydrocarbons. This process, developed by Fischer and Tropsch in 1933, starts by treating coke with steam at high temperatures

$$C + H_2O \longrightarrow CO + H_2. \tag{3}$$

The resutling carbon monoxide can be hydrogenated at high temperatures to methane in the presence of nickel or iron catalysts. Conditions of hydrogenation can be modified to produce a complex mixture of aliphatic hydrocarbons

$$nCO + 2nH_2 \longrightarrow C_nH_{2n} + nH_2O \tag{4}$$

$$nCO + (2n + 1)H_2 \longrightarrow C_nH_{2n+2} + H_2O. \tag{5}$$

This process has been used since its discovery for the commercial production of saturated hydrocarbons.

In 1979 Nooner showed that the Fischer–Tropsch process could be modified to produce fatty acids up to 21 carbons in length, instead of saturated hydrocarbons, by introducing carbonates such as potassium carbonate as promoters and meteoritic iron as a catalyst. This reaction is conducted at high temperatures and high-hydrogen pressures. Clearly, such conditions are far removed from the average conditions believed to have existed on primitive Earth. However, hydrothermal vents found under the sea might have provided conditions compatible with the Fischer–Tropsch process. There are more than a hundred hydrothermal vents dispersed on tectonically active areas of Earth. The temperatures and pressures reached in these vents often rise above the critical point for water, creating a bizarre but poorly understood aqueous environment conducive to extremes of temperature and pressure. In some of these vents there are huge amounts of catalytic clays and minerals interspersed in a reducing environment rich in hydrogen and carbon monoxide.

Saturated Hydrocarbons Can Be Synthesized under Mild Conditions with the Help of Iron Sulfide

Observations made by Gunter Wachterhauser and colleagues indicate that the iron–sulfur mineral pyrite may have presented an environment in which fatty acid synthesis could be carried out under considerably milder conditions. They found that iron sulfide is a powerful reducing agent that offers a possible solution to the problem of synthesizing a wide range of reduced organic compounds. In particular, it provides a route for the synthesis of saturated hydrocarbons that make up the bulk of fatty acid chains. A combination of iron sulfide and hydrogen sulfide can be used to carry out the following reactions:

$$NO_3^- \longrightarrow NH_3 \tag{6}$$

$$HC\equiv C \longrightarrow H_2C=CH_2,\ H_3C-CH_3 \tag{7}$$

$$-CH_2-CO- \longrightarrow -CH=CH-,\ -CH_2-CH_2- \tag{8}$$

$$HS-CH_2-COOH \longrightarrow CH_3-COOH. \tag{9}$$

Iron sulfide alone can carry out these additional reactions:

$$HS-CH_2-CH_2-X \longrightarrow CH_2=CH_2, \tag{10}$$

where $X = OH$, SH, and NH_2.

The experimental conditions required for these reductions are geochemically plausible: 100°C, and an aqueous environment having nearly neutral pH and being fastidiously anaerobic — conditions that would have prevailed on prebiotic Earth. Wachterhauser's findings represent the first proposal for synthesizing highly reduced carbon chains under readily attainable prebiotic conditions.

Because of the very high stability of fatty acids, they could have accumulated over a period of millions of years before they were condensed to form lipids and became incorporated as parts of cellular membranes. The unusual requirements for their synthesis may have led to their being synthesized in an environment incompatible with the first living systems. Once formed, cohesive lipid molecules could have floated as bilayer vesicles through the oceans until they encountered the first living systems and formed the first membrane-bounded living systems.

A Prebiotic Pathway That Resembles the Biosynthetic Pathway

The two pathways we have just described for fatty acid synthesis bear no resemblance to the biosynthetic pathway. In this section we describe a pathway that bears a striking resemblance to the contemporary biochemical pathway and that may have been its direct precursor.

First, let us review the strong and the weak points of the biosynthetic pathway insofar as it concerns a potential prebiotic pathway. The use of CoA-thioester intermediates assists fatty acid biosynthesis in two ways: first, the thioester bond irreversibly attaches the intermediates to a common carrier (CoA or acyl carrier protein) that standardizes the synthetic process; second, the electrophilic carbonyl group of the CoA-thioester intermediates gives them some of the reactive characteristics of aldehydes (or ketones). However, carbon–carbon bond synthesis by condensation of thioesters like acetyl-CoA is energetically unfavorable by about 7 kcal/mol and therefore it must be coupled to the energy-yielding hydrolysis of ATP. This coupling process, which proceeds via malonyl-CoA as an energized intermediate, depends on selective catalysis by enzymes. Because of the degree of catalytic sophistication required, this process seems beyond the range of rudimentary catalysts one would expect to find in the prebiotic world or in the first living systems.

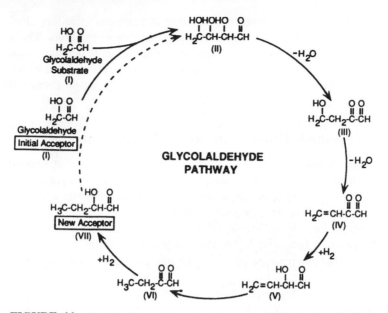

FIGURE 11   Primitive fatty acid synthesis from glycolaldehyde. I + I $\rightleftharpoons$ II, the $\Delta G^{0'} \approx -5$ kcal/mol was estimated from the average of related aldol condensations. (From Weber, A. L. Origin of Fatty Acid Synthesis: Thermodynamics and Kinetics of Reaction Pathways, *J. Mol. Evol.* 32: 93–100, 1991.)

To overcome these problems and yet preserve some of the main features of the biosynthetic pathway, Arthur Weber proposed an alternative pathway for primitive fatty synthesis with glycolaldehyde as the main substrate (Fig. 11). It should be recalled that glycolaldehyde is believed to play a pivotal role in the prebiotic synthesis of ribose and also has been invoked as a potential intermediate in primordial amino acid synthesis. In contrast to the contemporary biosynthetic pathway, Weber's proposed pathway is not dependent on an exogenous source of phosphoanhydride energy (ATP). Furthermore, the chemical spontaneity of its reactions suggest that it could have been readily catalyzed by the rudimentary catalysts available at an early stage in the origin of life.

Figure 11 depicts a plausible reaction cycle for the primitive synthesis of fatty acids from glycolaldehyde. All reactions of the pathway have been shown to occur nonenzymatically in the presence of water and are energetically downhill. Moreover, the condensation, dehydration, isomerization, and reduction reactions of the pathway are susceptible to general acid–base and metal ion catalysis.

The glycolaldehyde pathway differs fundamentally from the present-day pathway in its use of intermediates at the aldehyde (ketone) level of oxidation instead of CoA-thioesters at the carboxylic acid level of oxidation. This gives reactions of the primitive pathway an enhanced reactivity, more favorable energetics, and immunity to hydrolytic interference. The reactivity of aldehydes (ketones) can be attributed to the strong electrophilicity of their carbonyl group that gives them a relatively acidic α-carbon, enabling them to undergo a wide range of reactions including aldol condensations, β eliminations (dehydration), enolizations, and isomerizations. The spontaneous transformations of the aldehydes (ketones) of the glycolaldehyde pathway as depicted in Fig. 11 could have provided a core of chemical reactions from which fatty acid synthesis developed as rudimentary biocatalysts began to act on chemically generated intermediates.

Final Step in Lipid Synthesis Could Be Driven by Dehydration

David Deamer has done some pioneering experiments that suggest a simple way of connecting fatty acids to a phosphorylated derivative of glycerol. He mixed fatty acids, glycerol, and inorganic phosphate together; heated the mixture at 65°C to dryness, and analyzed the product. Deamer found that some phosphatidic acids were produced. In solution this reaction would be energetically unfavorable, but because water is one of the reaction products, the removal of water by drying converts this into an energetically favorable reaction for which we suggest the following stoichiometry:

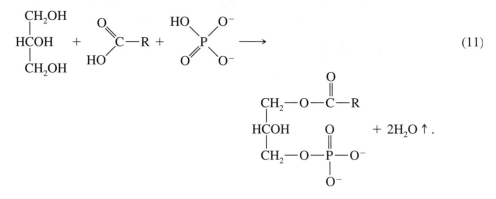

$$(11)$$

▤ SUMMARY

In this chapter we have discussed fatty acid breakdown, synthesis membrane biosynthesis, and possible prebiotic pathways for the synthesis of lipids.

   1. The degradation of fatty acids occurs by an oxidation process. The active substrate is the acyl-CoA derivative of the fatty acid. Breakdown of a long-chain fatty acid occurs in blocks of two-carbon atoms by a cyclic process. Each cycle involves four discrete enzymatic steps. In the process of oxidation the energy is sequestered in the form of reduced coenzymes of FAD and NAD$^+$. These reduced coenzymes lead to ATP production.

   2. Fatty acid biosynthesis takes place in eight steps. All except the first step take place on a multienzyme complex. Fatty acid biosynthesis also occurs in steps of two-carbon atoms. Energy is consumed instead of produced in fatty acid synthesis.

3. Phospholipids are biosynthesized by acylation of either glycerol-3-phosphate or dihydroxyacetone phosphate to form phosphatidic acid. Phosphatidic acid reacts with CTP to yield CDP-diacylglycerol, which in bacteria is converted to phosphatidylserine, phosphatidylglycerol, or diphosphatidylglycerol.

4. Organic chemists can synthesize fatty acids by a modification of the process used industrially to make hydrocarbons. This process involves starting from carbon and water to make carbon monoxide. The carbon monoxide is hydrogenated at high temperatures and high hydrogen pressures to make complex mixtures of aliphatic hydrocarbons. By introducing potassium carbonate into the system, one can modify this process to make fatty acids.

5. Saturated hydrocarbons can be synthesized under considerably milder conditions with the help of iron sulfide catalysts, which must have been abundant in the prebiotic world.

6. Perhaps the most exciting proposal for fatty acid synthesis is that of Arthur Weber, who has proposed a pathway for primitive fatty synthesis with glycolaldehyde as the main substrate. It should be recalled that glycolaldehyde has been strongly implicated in ribose synthesis and also has been suggested as an intermediate in the fabrication of the carbon skeletons for prebiotic amino acid biosynthesis.

7. The final step in lipid synthesis could be driven by dehydration. David Deamer has demonstrated that lipids can be made by heating mixtures of fatty acids, glycerol, and inorganic phosphate at 65°C to dryness.

Problems

1. Examine the chemistry in Fig. 5. Where else in metabolism have you seen a similar sequence of chemical events? Which feature or features are identical and which are different?
2. Is β oxidation best described as a spiral or a cylic process? Why?
3. Which catalytic activity of the mammalian fatty acid synthase determines the chain length of the fatty acid product?
4. Order the following substances from the least to the most oxidized carbon: formaldehyde, carbon dioxide, methane, formic acid, and methanol.
5. Fatty acids are unlikely to have played a significant role as energy suppliers in the early evolution of living organisms. Why?
6. Which reactions involving fatty acids or lipids should we focus on if we are concerned with the most primitive functions of lipids?
7. Biochemical reactions could be said to be driven by the chemical energy produced by ATP hydrolysis. Where does the energy come from that drives prebiotic reactions?
8. Could you calculate the $\Delta G°$ for the following reaction:

$$CH_2OHCHOHCHOHCHO \rightleftharpoons 2CH_2OHCHO?$$

See Fig. 11. Also see Problem 5 in Chapter 12.

References

Bochl, E., Keller, M., Wachterhauser, G., and Stetter, K. O. Reactions Depending on Iron Sulfide and Linking Geochemistry and Biochemistry, *Proc. Natl. Acad. Sci. USA* 89: 8117, 1992.

Huber, C. and Wachterhauser, G. Activated Acetic Acid by Carbon Fixation on (Fe, Ni) S Under Primordial Conditions, *Science* 276: 245, 1997.

Keller, M., Blochl, E., Wachterhauser, G., and Stetter, K. O. Formation of Amide Bonds without a Condensation Agent and Implications for the Origin of Life, *Nature* (*London*) 368: 836, 1994.

Weber, A. L. Nonenzymatic Formation of "Energy-Rich" Lactoyl and Glyceroyl Thioesters from Glyceraldehyde and a Thiol, *J. Mol. Evol.* 20: 157, 1984.

Weber, A. L. Origin of Fatty Acid Synthesis: Thermodynamics and Kinetics of Reaction Pathways, *J. Mol. Evol.* 32: 93, 1991. (Weber proposes an alternative pathway for fatty acid synthesis using glycolaldehyde.)

Zubay, G. *Biochemistry.* 4th ed. New York: McGraw-Hill, 1998, Chapters 21 and 22. (This discusses the biochemistry of fatty acids and lipids.)

CHAPTER 19

Properties of Membranes and Their Evolution

Every cell is surrounded by a plasma membrane that creates a compartment where the functions of life can proceed. The plasma membrane keeps proteins and other essential materials inside the cell. However, the plasma membrane is much more than an inert barrier; it serves to bring nutrients into the cell and extrude waste products. Other membrane-bound proteins sense the cell-surrounding, communicate with other cells, or act to move the cell to a new location. In aerobic bacteria, plasma membranes house the electron-transfer reactions that provide the cell with energy.

In addition to their plasma membrane, eukaryotic cells also contain internal membranes that define a variety of organelles. Each of these organelles is specialized for particular functions. Although membranes from different organelles or cells may have very different activities, they share several basic properties:

1. As a rule, biological membranes are impermeable to polar molecules or ions. An ion or polar molecule can cross a biological membrane only if the membrane has a protein that is a specific transporter for that molecule.

2. Membranes are flexible structures that can change their shape and size.

3. Membranes are durable. The plasma membrane of an erythrocyte, for example, experiences constant buffeting as the blood courses through the capillaries, and yet it survives for the lifetime of the cells. The membrane either must be remarkably resistant to damage or must reseal very quickly if it is breached.

4. When viewed with an electron microscope after thin-sectioning and staining, membranes typically have a trilaminar appearance of two dark lines separated by a lighter space, with a total thickness on the order of 40 Å.

5. Membranes contain proteins that are not simply structural in nature, but have a variety of enzymatic activities. The identities and amounts of these proteins vary in different cells and may change with time in response to changing conditions.

Our first goal in this chapter is to describe the structures found in biological membranes. This is followed by a description of some of the main functions of these membranes. Finally, we consider how membranes came into being, how membranes became integrated into living systems to form the first cells, and how the functions of membranes multiplied with time.

STRUCTURES OF BIOLOGICAL MEMBRANES

Biological membranes consist primarily of lipids and proteins. The relative amounts of these two components vary considerably, depending on the source of the membrane. At one extreme, the inner mitochondrial membrane is about 80% protein and 20% lipid by weight; at the other, the myelin sheath membrane that surrounds nerve fibers is about 80% lipid and 20% protein. Many membranes also contain small amounts of carbohydrates. These carbohydrates are almost always covalently attached to either proteins (as glycoproteins) or lipids (as glycolipids or lipopolysaccharides).

Membranes Contain Complex Mixtures of Lipids

The predominant phospholipids in most membranes are phosphoglycerides, which are phosphate esters of the three-carbon alcohol, glycerol. A typical structure is that of phosphatidylcholine (lecithin):

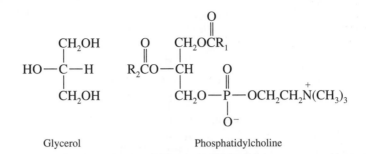

Glycerol   Phosphatidylcholine

Here R_1 and R_2 are long, fatty acid side chains. The parent fatty acids R_1CO_2H and R_2CO_2H usually have an even number of carbon atoms; 16- and 18-carbon acids are the most common. The acid esterified to the hydroxyl group on C-1 of the glycerol (that at the top of phosphatidylcholine as shown in the preceding structure) usually has a fully saturated chain, whereas the acid attached at C-2 often has one or more double bonds, which are almost always *cis* double bonds. Table 1 lists some the fatty acids commonly found in these positions. A phosphatidylcholine that has palmitic acid esterified at both the C-1 and C-2 positions of the glycerol is known by the name dipalmitoyl phosphatidylcholine. One with palmitic acid at C-1 and oleic acid at C-2 is called 1-palmitoyl-2-oleoylphosphatidylcholine.

The phosphate group in phosphatidylcholine forms ester linkages both with the hydroxyl group on C-3 of glycerol and also with a second alcohol, choline ($HOCH_2CH_2N^+(CH_3)_3$). In other phospholipids, a variety of other alcohols occupy the second position, and like choline, these all contain polar or electrically charged substituents. Table 2 shows the most common of these alcohols and indicates how the phosphoglycerides containing them are named by appending the name of the alcohol to the prefix "phosphatidyl." Free phosphatidic acid, in which the phosphate group is not esterified in this position (see Fig. 9) is an intermediate in phospholipid biosynthesis and metabolism but is not a major constituent of most biological membranes.

Phospholipids Spontaneously Form Ordered Structures in Water

Phospholipid molecules are said to be *amphipathic*, a term derived from the Greek, meaning having ambivalent feelings. The polar head group of the molecule is intrinsically soluble in water; the fatty acid tails are hydrophobic. In this regard, phospholipids resemble detergents such as sodium dodecylsulfate (SDS), which has an ionic sulfate head and a single, long hydrocarbon tail:

TABLE 1

Fatty Acids Frequently Found in Membrane Phospholipids

Name	Structure	Abbreviation[a]
Saturated fatty acids[b]		
Myristic acid	$CH_3(CH_2)_{12}CO_2H$	14:0
Palmitic acid	$CH_3(CH_2)_{14}CO_2H$	16:0
Stearic acid	$CH_3(CH_2)_{16}CO_2H$	18:0
Unsaturated fatty acids[c]		
Palmitoleic acid	$CH_3(CH_2)_5\overset{\displaystyle H}{C}=\overset{\displaystyle H}{C}(CH_2)_7CO_2H$	$16:1^{\Delta9}$
Oleic acid	$CH_3(CH_2)_7\overset{\displaystyle H}{C}=\overset{\displaystyle H}{C}(CH_2)_7CO_2H$	$18:1^{\Delta9}$
Linoleic acid	$CH_3(CH_2)_4\overset{\displaystyle H}{C}=\overset{\displaystyle H}{C}CH_2\overset{\displaystyle H}{C}=\overset{\displaystyle H}{C}(CH_2)_7CO_2H$	$18:2^{\Delta9,12}$
Linolenic acid	$CH_3CH_2\overset{\displaystyle H}{C}=\overset{\displaystyle H}{C}-CH_2-\overset{\displaystyle H}{C}=\overset{\displaystyle H}{C}-CH_2-\overset{\displaystyle H}{C}=\overset{\displaystyle H}{C}(CH_2)_7COOH$	$18:3^{\Delta9,12,15}$
Arachidonic acid	$CH_3(CH_2)_3(CH_2-\overset{\displaystyle H}{C}=\overset{\displaystyle H}{C})_4(CH_2)_3CO_2H$	$20:4^{\Delta5,8,11,14}$

[a]The abbreviation indicates the total number of carbon atoms and, following the colon, the number of double bonds and their positions relative to the carboxyl group. The Δ superscript indicates the start location of the double bond or bonds. Numbering of carbons starts from the carboxyl carbon. Thus $18:2^{\Delta9,12}$ indicates two double bonds, one between carbons 9 and 10 and one between carbons 12 and 13.

[b]A saturated fatty acid usually is esterified at C-1 of the glycerol.

[c]An unsaturated fatty acid usually is present at the C-2 position. Some bacterial phospholipids have a fatty acid with an internal cyclopropane ring here.

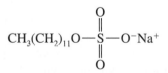

Sodium dodecylsulfate

When a detergent is mixed with water, it aggregates spontaneously into spherical or ellipsoidal micelles, in which the polar headgroups of the molecules are exposed to water, but the hydrophobic tails are sequestered (Fig. 1 a). Surprisingly, this process is driven, not by a decrease in energy, but instead by an increase in entropy associated with removing the hydrocarbon chains from water. If a hydrocarbon is dissolved in water, the water molecules surrounding it adopt a netlike structure that is more highly ordered than the structure of pure liquid water. Burying the hydrocarbon tails of the detergent molecules in the center of a *micelle* frees many water molecules from these nets and increases the overall amount of disorder in the system.

Phospholipids themselves generally do not form micelles, because, instead of having a single hydrophobic tail, they have two. The two tails are too bulky to pack

TABLE 2
Major Phosphoglycerides

Alcohol (HO—X)	Formula	Phospholipid
Choline	$HO-CH_2CH_2N(CH_3)_3^+$	phosphatidylcholine (lecithin)
Ethanolamine	$HO-CH_2CH_2NH_3^+$	phosphatidylethanolamine
Serine	$HO-CH_2CHNH_3^+$ $\mid$ CO_2^-	phosphatidylserine
Glycerol	$HO-CH_2CHCH_2OH$ $\mid$ OH	phosphatidylglycerol
Phosphatidylglycerol	$HO-CH_2CHCH_2$ $\mid$ OH	diphosphatidylglycerol (cardiolipin)
myo-Inositol		phosphatidylinositol

together in a spherical or ellipsoidal micelle. They hold the individual molecules apart enough so that portions of the tails remain in contact with water. There is, however, another solution to the problem: the phospholipids can form a *bilayer*. In this type of aggregate, the hydrocarbon tails of two monolayers of phospholipids pack together while the polar headgroups face outward and remain in contact with water on either side (see Fig. 1 b). A sheetlike planar bilayer removes the hydrophobic tails from water everywhere except around the edges of the sheet; and by curling up into a spherical vesicle, the bilayer can get rid of its edges. Such a vesicle is called a *liposome*.

Although other types of aggregates can form under special conditions, when phospholipids are agitated in the presence of excess water, they tend to aggregate spontaneously to form bilayers. Electron micrographs of bilayers immediately call to mind the trilaminar appearance of biological membranes. Furthermore, the idea that biological membranes might be built of phospholipid bilayers suggests an explanation for the observation that membranes are largely impermeable to ions. To cross a phospholipid bilayer, an ion would have to traverse the apolar region of hydrocarbon tails, where it would not be well solvated. Small, lipid-soluble molecules, on the other hand, are expected to pass across such a membrane relatively easily, and this is indeed

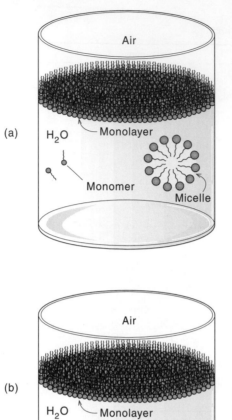

FIGURE 1 Structures formed by (a) detergents and (b) phospholipids in aqueous solution. Each molecule is depicted schematically as a polar headgroup (●) attached to one or two long, nonpolar chains. Most detergents have one nonpolar chain; phospholipids have two. At very low concentrations, detergents or phospholipids form monolayers at the air–water interface. At higher concentrations, when this interface is saturated, further molecules form micelles or bilayer vesicles (liposomes).

observed to occur. Finally, the presence of a phospholipid bilayer explains the flexibility of natural membranes and the ability of membranes to reseal if they are punctured. A bilayer is held together, not by bonds or electrostatic attractions between individual phospholipid molecules, but instead by the entropy increase that results when the hydrocarbon side chains come together and shed their coats of water.

Membranes Have Both Integral and Peripheral Proteins

In spite of the arguments presented in the previous section, it is reasonable to ask whether biological membranes are held together by phospholipid bilayers alone.

Considering that all membranes contain a substantial amount of proteins, any model for membrane structure must confront the questions of how these proteins are positioned and what role, if any, proteins play in stabilizing the membrane structure.

Washing preparations of biological membranes with salt solutions usually only removes a minor fraction of the total protein. Such observations have led to the realization that biological membranes contain two classes of proteins: *peripheral* and *integral*. The peripheral proteins can be removed by washing with salts. In addition to high salt concentrations, ethylenediaminetetraacetic acid (EDTA), a chelator of divalent cations, and urea often are used to solubilize these proteins. As a group, peripheral membrane proteins have amino acid compositions similar to those of soluble proteins. On the order of 30% of their residues may be hydrophobic and 70%, hydrophilic or neutral. They exhibit the full range of secondary structures and, again, are not remarkable in this regard.

Integral membrane proteins are much more difficult to extract. To solubilize them, it usually is necessary to resort to the use of detergents like SDS. Detergents disrupt the phospholipid bilayer of the membrane and incorporate lipids and the hydrophobic portions of proteins into their micelles. After a membrane has been disrupted, integral membrane proteins can be purified, but they almost always require the continued presence of a detergent to remain in solution.

Integral Membrane Proteins Contain Transmembrane α Helices

As we might expect from the fact that they are soluble only in the presence of detergents, integral membrane proteins tend to have comparatively high contents of hydrophobic amino acid residues. Between 40 and 60% of their side chains may be hydrophobic. In addition, hydrophobic residues often show a curious distribution in the amino acid sequence. They frequently appear in strings of approximately 20 residues, separated by stretches of hydrophilic residues.

An α helix of 20 amino acid residues is approximately 30 Å long (1.5 Å per residue). This is about the right length to reach across the hydrocarbon region of a bilayer of phospholipids containing 16- and 18-carbon fatty acids. Physical measurements of secondary structure in integral membrane proteins support the view that integral membrane proteins typically have substantial amounts of α-helical structure oriented perpendicular to the plane of the membrane.

One of the best studied integral membrane proteins is bacteriorhodopsin, an integral membrane protein of the halophilic bacterium *Halobacterium halobium*. This protein forms well-ordered arrays in two-dimensional sheets that can be studied by electron diffraction. Measurements of the diffraction patterns show clearly that bacteriorhodopsin has seven transmembrane helices (Fig. 2). This cluster of helices has a cavity down the center that allows for the selective passage of protons.

Proteins and Lipids Have Considerable Lateral Mobility within Membranes

In 1972 Jon Singer and Garth Nicolson incorporated all the available structural information about membranes into a model they called the *fluid-mosaic model*. This model,

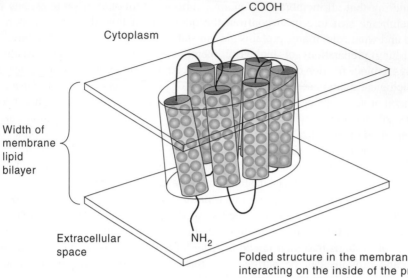

Folded structure in the membrane with charged R-groups interacting on the inside of the protein and hydrophobic surface exposed to membrane lipid side chains.

FIGURE 2   A model for the structure of bacteriorhodopsin, a membrane protein from *Halobacterium halobium*. The protein has seven membrane-spanning segments connected by shorter stretches of hydrophilic amino acid residues.

which is generally accepted, pictures the phospholipid bilayer as the primary structural element of biological membranes. Integral membrane proteins are embedded in the bilayer, in some cases only partially, but in other cases extending all the way across (Fig. 3). Peripheral proteins are attached more loosely by ionic interactions with protruding portions of integral proteins or with phospholipid headgroups. A key feature of the fluid-mosaic model is that the integral proteins are, in most cases, not linked together by protein–protein interactions. They are free to diffuse laterally in the bilayer or to rotate about an axis perpendicular to the plane of the membrane. The entire structure thus has the potential of being dynamic instead of static. It should be mentioned that the lateral mobility of a minor number of integral membrane proteins is restricted because they are attached to networks that emanate from the cytoplasm. The lipid components of membranes also have considerable lateral mobility like most of the proteins.

Biological Membranes Are Asymmetrical

Although phospholipids diffuse laterally in the plane of the bilayer and rotate more or less freely about an axis perpendicular to this plane, movements from one side of the bilayer to the other are a different matter. Diffusion across the membrane, a transverse, or flip-flop motion, requires getting the polar headgroup of the phospholipid through the hydrocarbon region in the center of the bilayer. Flip-flop motions of phospholipids do occur, and they can be catalyzed enzymatically, but they are much slower than the

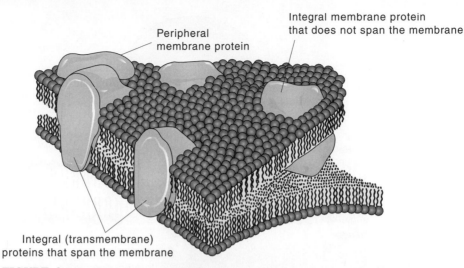

Peripheral
membrane protein

Integral membrane protein
that does not span the membrane

Integral (transmembrane)
proteins that span the membrane

FIGURE 3   The fluid-mosaic model for biological membranes as envisioned by Singer and Nicolson. Integral membrane proteins are embedded in the lipid bilayer; peripheral proteins are attached more loosely to protruding regions of the integral proteins. The proteins are free to diffuse laterally or to rotate about an axis perpendicular to the plane of the membrane. (For further information, see Singer S. J. and Nicolson, G. L. The Fluid Mosaic Model of the Structure of Cell Membranes, *Science* 175: 720, 1972.)

other types of motions we have described. The same arguments apply even more forcefully to membrane proteins. For a protein to invert its orientation in the membrane, the hydrophilic domain that sticks out into the solution on one side of the membrane has to pass through the center of the bilayer. This rarely occurs. Once a protein has been inserted into a membrane in a particular orientation, it usually retains that orientation indefinitely. An important consequence of these considerations is that biological membranes are both structurally and functionally asymmetrical. An enzyme or receptor embedded in the membrane usually provides a binding site for its substrate or effector only on one side of the membrane.

Glycophorin is one of the numerous glycoproteins found in the plasma membranes of erythrocytes and other cells. These proteins provide good illustrations of membrane structural asymmetry, because the oligosaccharides attached to them are almost always on the outside of the cell. Glycophorin has about 100 carbohydrate residues attached to the protein at 16 sites between residues 2 and 50; these are all on the outside of the erythrocyte (Fig. 4).

Lipids also show asymmetrical distributions between the inner and outer leaflets of the bilayer. In the erythrocyte plasma membrane, most of the phosphatidylethanolamine and phosphatidylserine are in the inner leaflet, whereas the phosphatidylcholine is located mainly in the outer leaflet. A similar asymmetry is seen even in artificial liposomes prepared from mixtures of phospholipids. In liposomes containing a mixture of phosphatodylethanolamine and phosphatidylcholine, phosphatidylethanolamine localizes preferentially in the inner leaflet, and phosphatidylcholine

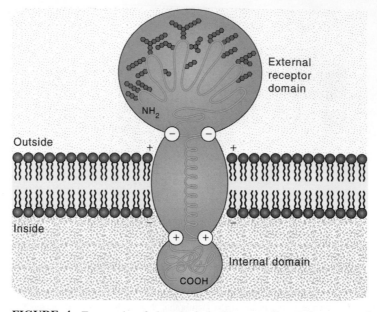

FIGURE 4   Topography of glycophorin in the mammalian erythrocyte membrane. Carbohydrate residues (small hexagons) are attached to the hydroxyl groups of threonine and serine residues in the N-terminal domain of the protein. The N terminus and all the carbohydrates are outside the cell; the C-terminal domain of the protein is inside. The hydrophobic, membrane-spanning domain is flanked by charged amino acid residues that may interact electrostatically with the polar headgroups of the phospholipids.

localizes in the outer leaflet. For the most part, the asymmetrical distributions of lipids probably reflect packing forces determined by the different curvatures of the inner and outer surfaces of the bilayer. By contrast, the disposition of membrane proteins reflects the mechanism of protein synthesis and insertion into the membrane. We return to this point later in this chapter.

Newly Synthesized Lipid Components Are Formed on Preexisting Membranes

The subject of membrane biogenesis is concerned with the lipid components and the protein components that follow independent pathways.

The basic structure of membranes is determined by the lipid bilayer structure. Newly synthesized lipid components are formed on the scaffolding of preexisting membranes instead of the creation of new ones. In eukaryotes this synthesis is confined to the cytoplasmic face of the endoplasmic reticulum (ER). Distribution of these lipids to other membranes involves the budding off of vesicles from the ER and their subsequent fusion with other membranes.

The location of specific phospholipids on the two leaflets of a membrane is a function of where they are made and the opposing tendencies of two classes of enzymes:

the ATP-dependent *translocases* and the *flipases*. As suggested by their name, flipases catalyze the exchange of phospholipids between the leaflets of a common bilayer. Flipases tend to equilibrate the distribution of different phospholipids across a bilayer. Thus the net transport of a phospholipid by a flipase is from the side of the bilayer with the higher concentration of the specific phospholipid to the opposite side. By contrast, translocases catalyze the transport of specific phospholipids from the side of a bilayer that has the lower concentration of the phospholipid to the opposite side, thereby augmenting the asymmetry of the membrane.

Locations and Orientation of Membrane Proteins Are Determined by Specific Targeting Mechanisms

All polypeptide chains are synthesized on ribosomes. Free ribosomes synthesize cytosolic proteins, mitochondrial proteins, and nuclear proteins. Membrane-bound ribosomes are involved in the synthesis of membrane-bound proteins, and proteins destined for secretion as well as some proteins that go to other locations in the cytoplasm. There is a specific targeting mechanism for every type of protein; we focus our remarks on the targeting mechanism for some of the proteins made on membrane-bound ribosomes.

The targeting of proteins that are synthesized on the endoplasmic reticulum starts with a special sequence at or near the N-terminal end of the polypeptide chain. All proteins made on the ER contain an N-terminal *signal peptide*. Because the N-terminal portion of the polypeptide is synthesized first, the signal peptide protrudes beyond the ribosomal surface early in the synthesis of a polypeptide. A protein known as the *signal recognition particle* (SRP) binds to this sequence that temporarily halts polypeptide synthesis. Peptide synthesis does not resume until the nascent polypeptide N terminus has passed through the membrane into the lumen (interior) of the ER.

Shortly after the signal peptide has entered the lumen, it is cleaved from the growing polypeptide by a membrane-bound signal peptidase. As the nascent polypeptide chain begins to fold into its native conformation, other enzymes commence the process of adding carbohydrates to the nascent protein to form glycoproteins. On completion of peptide synthesis, the ribosome dissociates from the ER and the newly formed protein. Secretory proteins, and proteins that remain in the ER pass completely through the ER membrane into the lumen. In contrast, integral membrane proteins contain a hydrophobic stop–transfer sequence near the C terminus that arrests the passage of the growing polypeptide chain through the membrane (Fig. 5). This ensures that the integral membrane protein remains embedded in the ER membrane with its C terminus on the cytoplasmic side. Sometime after completion of polypeptide synthesis, the partially processed transmembrane proteins and secretory proteins appear in the *Golgi apparatus*. As proteins transit the compartments of the Golgi, they undego further modification. Proteins are carried from one compartment to the next by specialized membranous transport vesicles. In the final Golgi compartment, fully processed proteins aggregate into homogeneous clusters and become transported to their final cellular destinations. Membranes and secretory proteins are transported in protein-coated vesicles. A typical vesicle buds off from its membrane of origin and

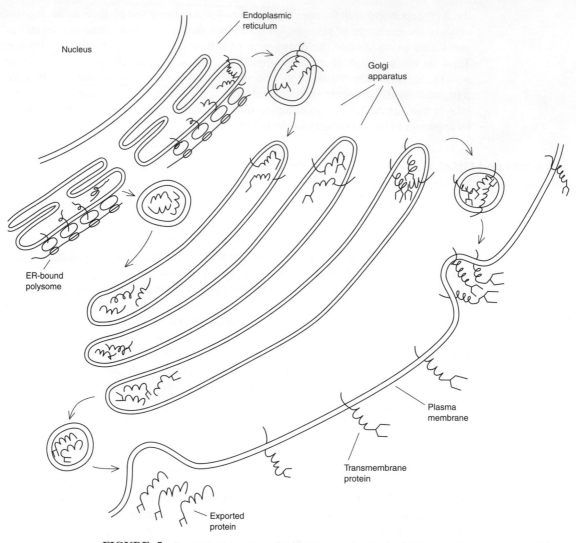

FIGURE 5   Synthesis and pathway followed by proteins destined to become transmembrane proteins or proteins for export. The synthesis of these two classes is very similar. It starts on ribosomes that become bound to the endoplasmic reticulum early during synthesis. In the ER, synthesis of the proteins continues, and processing and modification of the proteins begin. The proteins destined to become transmembrane proteins uniquely carry a stop transfer sequence near their C termini so that they remain attached to the membrane. By budding and fusion the proteins are transferred to the various compartments of the Golgi apparatus where further modification of the proteins takes place. Finally, by budding and fusion the vesicles carrying the proteins become part of the plasma membrane.

subsequently fuses to a target membrane. This process preserves the orientation of the transmembrane protein so that the lumens of the ER and the Golgi apparatus are topologically equivalent to the outside of the cell. As a result the carbohydrate of integral membrane glycoproteins occurs only on the external surfaces of plasma membranes.

FUNCTIONS OF BIOLOGICAL MEMBRANES

Whereas a major function of biological membranes is to maintain the status quo by preventing loss of vital materials and entry of harmful substances, membranes also must engage in selective transport processes. Living cells depend on an influx of phosphate and other ions, and of nutrients such as carbohydrates and amino acids. They extrude certain ions, such as Na^+, and rid themselves of metabolic end products. How do these ionic or polar species traverse the phospholipid layer of the plasma membrane? How do pyruvate, malate, tricarboxylic acid (TCA) citrate, and even adenosine triphosphate (ATP) move between the cytosol and the mitochondrial matrix? The answer is that biological membranes contain proteins that act as specific transporters, or *permeases*. These proteins behave much like conventional enzymes: they bind substrates and they release products. Their primary function, however, is not to catalyze chemical reactions but to move materials from one side of a membrane to the other. In this section we discuss the general features of membrane transport.

Most Solutes Are Transported across Membranes by Specific Carriers

For purposes of comparison, consider the movement of glucose across a porous membrane such as a piece of dialysis tubing (Fig. 6 a). The pores in dialysis tubing are large enough so that glucose and water molecules can diffuse from either side of the membrane to the other with little hindrance. If the solutions on the two sides initially contain different concentrations of glucose, molecules diffuse from the more concentrated solution to the more dilute solution. The net rate of this diffusion is proportional to the difference between the two concentrations (ΔC), as shown in Fig. 6 b and c.

By contrast, the kinetics of transport across a biological membrane usually does not exhibit such a linear dependence on ΔC. Instead, the rate approaches an asymptote at high concentrations (Fig. 6 d). This behavior suggests that the transport protein has a specific binding site for the material that it transports. The overall rate of transport is limited by the number of these sites in the membrane. It can be shown that the overall rate is given by the following expression:

$$\nu = \frac{V_{max}[\Delta C]}{K_m + [\Delta C]},$$

where K_m is an algebraic function of the microscopic rate constants for binding, dissociation, and translocation of the substrate in either direction.

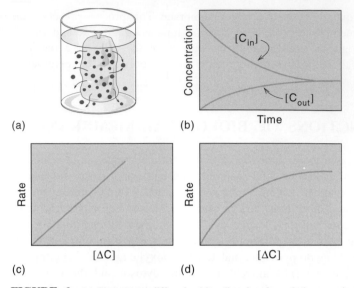

(a)

(b)

(c)

(d)

FIGURE 6 (a) Glucose can diffuse in either direction through the pores in dialysis tubing. (b) If the glucose concentration inside the dialysis bag is initially higher than the concentration outside, glucose diffuses out spontaneously until the two concentrations are equal. (c) At any given time, the net rate of diffusion is proportional to the difference between the two concentrations. $v_{in \to out} = k[C_{in}] - k[C_{out}] = k[\Delta C]$. (d) The measured rate of transport across a biological membrane usually is not simply proportional to the concentration difference across the membrane but instead approaches a maximum at high values of $[\Delta C]$.

The high specificity of biological transport systems provides compelling support for this interpretation. For example, although dialysis tubing is equally permeable to D- and L-glucose, most eukaryotic cells take up D-glucose rapidly, but not L-glucose. In bacteria, the synthesis of a specific transport system for a particular nutrient often can be induced by adding a small amount of the nutrient to the growth medium. Thus, if *Escherichia coli* is grown on medium without lactose, transporters for lactose usually are not found in the plasma membrane. When lactose is added to the culture medium, however, synthesis of a protein that tranports lactose is switched on, and cells soon begin to demonstrate this activity.

Some Transporters Facilitate Diffusion of a Solute down a Concentration Gradient

The presence of a specific transport system in the plasma membrane does not necessarily imply that a cell can pump the substrate in a direction opposite to the way the substrate spontaneously diffuses. In some cases, the net flow always proceeds down the concentration gradient, just as glucose always diffuses out of the more concentrated of two solutions separated by a dialysis membrane. The glucose transporter in the plasma membranes of most animal cells behaves in this way. Muscle cells continue to take up glucose from the blood only because the molecules that enter the cell are quickly modified by metabolic reactions that keep the cytosolic glucose concentration low. Yet, for a given concentration gradient, the uptake of D-glucose is much faster than the uptake of L-glucose or other materials that the cell is not equipped to transport. Transport that proceeds in the same net direction as simple diffusion but that is catalyzed by a specific transport system is termed *facilitated diffusion.*

For a neutral solute the free energy change associated with moving 1 mol of the material from a solution with concentration $[C_1]$ to a solution with concentration $[C_2]$ is

$$\Delta G_{1\to 2} = 2.3 \text{ RT log} \frac{[C_2]}{[C_1]}.$$

This expression, which reflects the entropy increase resulting from distributing the molecules more randomly between the two solutions, holds for both free and facilitated diffusion. If $[C_1] > [C_2]$, $\Delta G_{1\to 2}$ is negative and molecules diffuse spontaneously from solution 1 to solution 2. Net flux across the membrane ceases when $[C_1] = [C_2]$ and $\Delta G_{1\to 2}$ is zero.

If the solute the charged, there is an additional thermodynamic effect of moving the charge across any difference in electric potential that exists between the solutions on the two sides of the membrane.

Active Transport against a Concentration Gradient Requires Energy

Cells can transport some materials against gradients of concentration. For example, eukaryotic cells pump out Na^+ ions even though the external Na^+ concentration is greater than the cytosolic concentration. Many cells take up amino acids even though the cytosolic concentration exceeds the external concentration, and cells that line the stomach pump out protons against a concentration gradient of more than a million to one. Transport against such concentration and electric gradients is called *active transport.*

To drive active transport, a cell must couple transport to another process that is thermodynamically favorable, so the total ΔG is negative. Cells have developed a variety of schemes for this coupling (Fig. 7). Some transport systems, including the $Na^+ - K^+$ pump of animal cells, drive active transport by the hydrolysis of ATP (see Fig. 7 a). The $Na^+ - K^+$ pump may consume as much as 70% of the ATP that nerve cells synthesize, and even other types of cells spend a substantial portion of their energy resources in this way.

Active transport of a solute against a concentration gradient also can be driven by a flow of an ion down its concentration gradient. In some cases, the ion moves across the membrane in the opposite direction to the primary substrate (*antiport*); in others, the two species move in the same direction (*symport*). Many eukaryotic cells take up neutral amino acids by coupling this uptake to the inward movement of Na^+ (see Fig. 7 c). As we discussed previously, Na^+ influx is downhill thermodynamically because the $Na^+ - K^+$ pump keeps the intracellular concentration of Na^+ lower than the extracellular concentration.

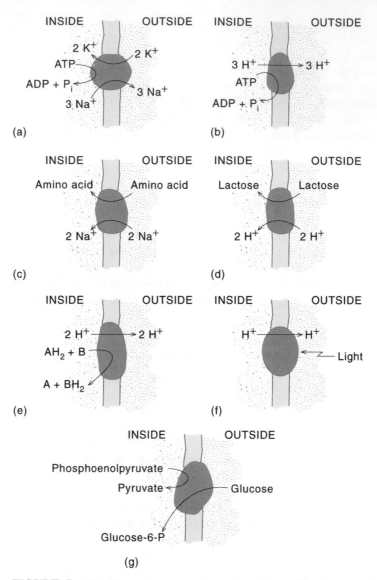

FIGURE 7 Cells drive active transport in a variety of ways. The plasma membrane Na^+–K^+ pump of animal cells (a) and the plasma membrane H^+ pump of anaerobic bacteria (b) are driven by the hydrolysis of ATP. Eukaryotic cells couple the uptake of neutral amino acids to the inward flow of Na^+ (c). Uptake of β-galactosides by some bacteria is coupled to inward flow of protons (d). Electron-transfer reactions drive proton extrusion from mitochondria and aerobic bacteria (e). In halophilic bacteria, bacteriorhodopsin uses the energy of sunlight to pump protons (f). *Escherichia coli and some other* bacteria phosphorylate glucose as it moves into the cell and thus couple the transport to hydrolysis of phosphoenolpyruvate (g).

Escherichia coli couples the uptake of lactose to the inward flow of protons. Proton influx is downhill because electron-transfer reactions set up gradients of pH across the membrane.

Proton pumps driven by electron-transfer reactions exist in mitochondria, bacteria, and chloroplasts (Fig. 7 e). Bacteriorhodopsin, which was mentioned earlier in connection with the structure of membrane proteins, uses light energy to pump protons out of *Halobacterium halobium* (see Fig. 7 f). Some bacterial cells can energize sugar transport by still another strategy, *group translocation*. They modify the sugar by phosphorylation as they transport it (see Fig. 7 g). The overall free energy change thus includes the negative ΔG of the phosphorylation reaction, which in this case is provided ultimately by hydrolysis of phosphoenolpyruvate.

Asymmetrical Orientation of Membrane Transport Proteins Is the Key to Vectorial Transport

Transport proteins typically contain multiple α helices stretching across the membrane in an asymmetrical manner. The sequence of the lactose transporter (lactose permease) of *E. coli* illustrates this point. This protein probably has 12 transmembrane α helices linked by short runs of more hydrophilic amino acid residues. Amino acid residues that are implicated in binding of lactose are distributed among five of the helices. Although the three-dimensional structure of the protein is not known, it seems likely that these helices pack together in a parallel bundle. Such a bundle could enclose a tubular channel that stretches across the membrane. However, the channel probably is partially blocked by amino acid side chains, because it does not render the membrane freely permeable to ions. Even protons can negotiate the channel only in the presence of the substrate.

Figure 8 shows a schematic model of how lactose permease could couple the movement of protons and lactose across the membrane. In this model, the transmembrane channel is always plugged at one end or the other. The lactose binding site is centrally located and is accessible to either the extracellular solution or the intracellular solution, depending where the channel is open. Transitions between these two states might occur by relatively minor conformational changes in the protein. The lactose binding site itself also is presumed to switch between two states with different dissociation constants for lactose, depending on whether or not a nearby amino acid residue is protonated. Binding of a proton from the solution on either side favors the binding of lactose from the same side. The transporter thus would tend to pick up both a proton and a lactose from the solution with the lower pH and to release them on the other side. A critical feature of the model is that, if the proton-binding site is occupied, conformational changes that expose this site to the opposite side also are occupied. This restriction prevents the protein from catalyzing proton transport in the absence of lactose transport.

Signals Can Be Transported without the Actual Flow of Substance

Hormones are agents for cellular communication between cells in multicellular organisms. Typically a hormone is released from an endocrine gland in response to a meta-

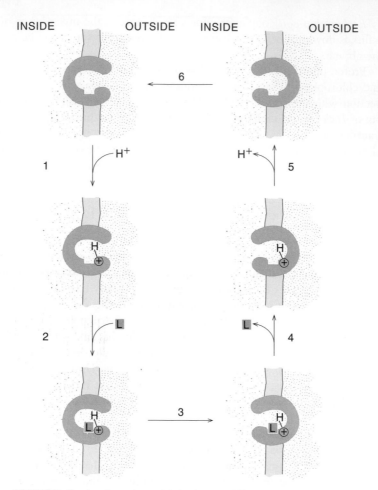

FIGURE 8 A schematic model for lactose–H^+ symport catalyzed by the lactose permease. In the model, the protein forms a channel with gates that can open to expose lactose- and proton-binding sites to the solution on one side of the membrane or the other. In step 1, binding of a proton from the extracellular solution increases the affinity of the lactose-binding site for lactose. Binding of lactose (L) from the extracellular solution (step 2) results in a conformational change that switches the states of the gates (step 3). Lactose and a proton dissociate to the intracellular solution (steps 4 and 5), and the protein relaxes to a conformation that has a low affinity for both lactose and protons. In step 6, the gates switch back to their original state, exposing the unloaded binding sites to the external solution. In living cells, the cycle is driven in the indicated direction by the electrochemical potential gradient for protons (the intracellular solution has a higher pH and a negative electric potential relative to the extracellular solution). The protein can transport a proton into the cell only if a galactoside moves along with it. (From Collins, J. C., Permuth, S. F., and Brooker, R. J. Isolation and Characterization of Lactose Permease Mutants with an Enhanced Recognition of Maltose and Diminished Recognition of Cellobiose, *J. Biol. Chem.* 264:14698, 1989.)

bolic need. The hormone targets to specific cells that are triggered to respond in an appropriate manner. For example, when the blood sugar is low, the pancreas releases the hormone glucagon that binds to specific receptor sites on the outer surface of the plasma membrane of liver cells. This triggers metabolic changes on the cytosolic side of the membrane that ultimately results in the secretion of glucose into the bloodstream. This "signal transduction" process like ordinary transport is completely dependent on the asymmetrical orientation of the membrane-bound receptor and the other proteins linked to it on the cytosolic side that bring about the response.

EVOLUTION OF MEMBRANE STRUCTURE AND FUNCTION

In Chapter 10 we suggest that membrane-enclosed cells came into being some time after the first ribozymes and definitely before the advent of translation systems. It is highly likely that these primitive living systems were sequestered in some way, possibly by adhering to clay surfaces (see Chapter 20). It is also likely that the first fatty acids used to make cellular membranes were made under conditions that would have been too harsh to share with living systems that are far more delicate. In view of this we must ask how the first membranes made contact with the early membrane-free living systems. Then we must consider how the early living systems became enclosed by these membranes and how the membranes of these most primitive cells evolved.

Macromolecules Can Be Encapsulated in Bilayer Vesicles under Simulated Prebiotic Conditions

If the first fatty acids and membranes were made in a location some distance from the early living systems, we have to ask three questions. How were the first membranes assembled? How did these membranes make contact with the living systems? Finally, how did the living systems become encapsulated? We have simple answers to all three of these questions. First, it is a simple step from fatty acids to liposomes because of the tendency for lipids to spontaneously organize into bilayer vesicles or liposomes. Liposomes are very stable structures and should be able to move about freely in the oceans. Second, it seems likely that the liposomes traveled across the ocean surfaces until they made chance encounters with living systems that may have existed in tidal regions where they were subject to alternate dry–wet cycles. Third, the encapsulation of the living systems into the liposomes was probably a simple process that required no more than one or two dry–wet cycles. This is based on the experiments of David Deamer and colleagues who have shown that if one starts with a mixture of liposomes and RNA, the liposomes break into linear structures on dehydration. On rehydration the liposomes reform and in the process about half of the RNA becomes encapsulated (Fig. 9).

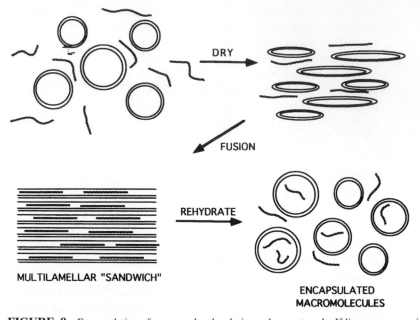

FIGURE 9   Encapsulation of macromolecules during a dry–wet cycle. If liposomes are mixed with soluble RNA and dried, the liposomes fuse to produce a multilamellar structure in which the macromolecules are sandwiched between the lipid layer. On rehydration, vesicles form that encapsulate a significant fraction of the RNA. (From Deamer, D. W. The First Living Systems: A Bioenergetic Perspective, *Microbiol. Mol. Biol. Rev.* 61:239–261, 1997.)

Primary Functions of the First Membranes Were to Sequester the Living Systems and to Give Them Greater Mobility

It seems likely that the first membranes, had two selective advantages. First, they prevented the components of the living systems from becoming dispersed. Second, they gave the living systems great mobility. Once comfortably lodged in its lipid enclosure, the RNA system could travel on the seas without fear of dispersal. This would facilitate the exploration of different environments that might be more favorable for survival and growth.

First Membranes Could Have Impeded the Flow of Substrates to the Interior of the Living Cells

Cell membranes in the modern world are complex lipoprotein structures with sophisticated mechanisms for regulating the transfer of substrates and other substances. If the first membrane structures contained lipids but no proteins, then there is concern that the hydrophilic substrates would not be able to pass through the cell membrane.

To explore this problem David Deamer and colleagues conducted a series of experiments where they created a situation in which polyadenylic acid and an enzyme for polynucleotide synthesis, polynucleotide phosphorylase, were encapsulated in liposomes made from lipids with different carbon chain lengths. Keep in mind that contemporary membranes contain mostly C_{16} and C_{18} fatty acids. In Deamer's experiments nucleotide diphosphate substrates were added to the solution containing the loaded liposomes and measurements were made to see if any polynucleotide synthesis occurred inside the liposomes. It was found that if the lipid chains were relatively short (C_{10} to C_{12}), the bilayers were highly permeable due to the large numbers of fluctuating transient defects in the bilayer. Substrate could potentially reach the encapsulated polymerase enzyme but the high permeability also allowed product polynucleotide to escape. On the other hand, permeation of substrate across bilayers with 16 to 18 carbons was too slow for the substrate to be useful to the polymerizing enzyme. However, with an appropriate choice of lipid with 14 carbon chains, the enzyme could be supplied with substrate at a sufficient rate while still being protected against loss.

First Transmembrane "Proteins" Probably Functioned as Simple Permeases for Facilitated Transport

Although pure liposomes of the right chain length would appear to allow limited access to external substrate, it clearly would be a major advantage to have a more reliable mechanism for permeation. In view of this problem it seems likely that if polypeptides were not part of the first cellular membranes, they were very early and were welcomed additions.

In Chapter 17 we describe the possibility for the synthesis of polypeptides with repetitive sequences and how they are likely to have preceded translation systems. In our view membranes could not have evolved very far without membrane polypeptides for facilitating transport of small molecules. Thus it seems likely that membranes in a very early stage of their evolution were dependent on polypeptides that were amenable to serving as transmembrane polypeptides. What would be the requirements for a transmembrane polypeptide? First, it should be able to form an α helix. Second, it should be able to form polypeptides greater than the width of the membranes. This would require segments of α helix 20 or more residues long. Third, the amino acid composition of these polypeptides should have a high percentage of hydrophobic amino acids; otherwise they would not be able to form the necessary complexes with the membrane lipids. Finally, the polypeptides should have a heptapeptide repeating sequence because this makes it possible for helical segments to cluster in a way that leads to effective pores down the center of the polypeptide complexes.

There are probably numerous solutions to this problem. One possibility might be a combination of alanine (ala), serine (ser), and methionine (met) with the heptapeptide repeating sequence

-ala-ser-met-ala-ser-met-ala-.

Most structure experts would agree that a polypeptide with this repeating unit is very likely to form an α helix and make a stable complex with lipid. Incidentally, this polypeptide should only require four ribozymes for its synthesis: one to make the ala-ala bond between oligopeptide repeats, one to make the two ala-ser bonds, one to make the two ser-met bonds, and one to make the two met-ala bonds. This does not seem like an excessive demand to make on the ribozyme system.

Although a cluster of polypeptides with this composition is highly likely to make a stable transmembrane complex with the lipid bilayer, it should still be sufficiently hydrophilic to allow for the passage of water-soluble small molecules, like amino acids and nucleotides, down its core.

The first permeases must have been very limited in number and specificity. Additional specificities would have required considerable evolution. Active transport and signal transduction (discussed earlier) would have introduced considerable complexity and the need to orient the transmembrane polypeptides asymmetrically in the membrane.

SUMMARY

In this chapter we have discussed the structures and functions of biological membranes and the evolution of membrane structure and function.

1. Biological membranes consist primarily of proteins and lipids whose relative amounts vary considerably.

2. Because of their amphipathic nature, phospholipids spontaneously form ordered structures in water. When phospholipids are agitated in the presence of excess water, they tend to aggregate spontaneously to form bilayers, which strongly resemble the types of structures they form in biological membranes.

3. Membranes contain proteins that merely bind to their surface (peripheral proteins) and those that are embedded in the lipid matrix (integral proteins). Integral membrane proteins contain transmembrane α helices.

4. Proteins and lipids have considerable lateral mobility within membranes.

5. Biological membranes are asymmetrical. Consistent with this asymmetry, a protein that has been inserted into a membrane in a particular orientation usually retains that orientation indefinitely.

6. Biological membranes contain proteins that act as specific transporters of small molecules into and out of the cell. Most solutes are transported by specific carriers that are invariably proteins.

7. Some transporters facilitate diffusion of a solute from a region of relatively high concentration or down a favorable electrochemical potential gradient. Such transporters do not require energy, because the transport is in the thermodynamically favored direction.

8. Other transporters move solutes against an electrochemical potential gradient and require an energy-producing process to make them functional. Cells drive such active transport processes in a variety of ways. The transport can be coupled to the

hydrolysis of a high-energy phosphate, to the cotransport of another molecule down an electrochemical potential gradient, or to the modification of the transported molecule soon after it crosses the membrane.

9. Signals can be transported without the actual flow of substance; most hormones function this way.

10. Macromolecules can be encapsulated in bilayer vesicles under simulated prebiotic conditions.

11. The primary functions of the first membranes were to enclose living systems and to give them greater mobility.

12. The first transmembrane proteins probably functioned as simple permeases for facilitated transport.

Problems

1. Both triacylglycerols and phospholipids have fatty acid–ester components, but only one can be considered amphipathic. Indicate which is amphipathic, and explain why.

2. The relative orientation of polar and nonpolar amino acid side chains in integral membrane proteins is "inside-out" relative to that of the amino acid side chains of water-soluble globular proteins. Explain.

3. Differentiate between peripheral and integral membrane proteins with respect to location, orientation, and interactions that bind the protein to the membrane. What are some strategies used to differentiate between peripheral and integral proteins by means of detergents or chelating agents?

4. Elementary portrayals of transport systems often have utilized "revolving doors" as analogies. After reading this chapter what objection or objections do you have to the revolving door idea?

5. Although highly regular membranes should make strong bilayer vesicles, somewhat irregular membranes that permitted the in and out passage of small organic molecules would have some advantages in the early evolution of living systems. Explain.

6. Membranes must have become an important component of living systems prior to the evolution of translation systems. Why?

7. Membranes must have undergone considerable evolution prior to their serving useful functions in living systems. Explain.

References

Chakrabarti, A. C., Breaker, R. R., Joyce, G. F., and Deamer, D. W. Production of RNA by a Polymerase Protein Encapsulated within a Phospholipid Vesicle, *J. Mol. Evol.* 39:555–559, 1994. (This is as provocative as it is perplexing.)

Deamer, D. W. The First Living Systems: A Bioenergetic Perspective, *Microbiol. Mol. Biol. Rev.* 61:239–261, 1997. (This is a good general source with excellent references. Special attention should be paid to the sections on membranes.)

Hargreaves, W. R., Mulvihill, S. J., and Deamer, D. W. Synthesis of Phospholipids and Membranes in Prebiotic Conditions, 276: 78–80, 1997. (Membrane-forming lipids require only three precursors: glerol, fatty acid or aldehyde, and phosphate.)

Pohorillle, A. and Wilson, M. A. Molecular Dynamics Studies of Simple Membrane-Water Interfaces: Structure and Functions in the Beginnings of Cellular life, *Origins Life Evol. Biosphere* 25: 21–46, 1995. (A computer model leads to a proposal for how simple ions are transport across a simple lipid membrane.)

CHAPTER 20

Possible Roles of Clays and Minerals in the Origin of Life

Two properties are invariably associated with a living system: the capacity for self-reproduction and the capacity to undergo change. Nucleic acids possess both of these properties, but they require an elaborate support system for maintenance and growth. How did the surface chemistry on Earth evolve to the stage where this support system could be sustained? Finding possible answers to this question represents a major goal for prebiotic chemists.

The stringent needs of delicate organic polymers such as polynucleotides or polypeptides stand in sharp contrast to the simpler requirements of the more robust polymers from which clays are constructed. Because evolution usually proceeds from the simple to the complex, this makes us wonder if clays could have comprised the first living systems.

To pursue this possibility we turn to a consideration of clays and closely related minerals, their structures and chemistry, and their potential roles in the chemical evolution leading to the origin of life.

DID "LIVING CLAYS" PRECEDE NUCLEIC ACIDS?

By comparison with nucleic acids, inorganic polymers use much simpler, more accessible building blocks whose polymerization requirements are far less sophisticated. In particular, they do not require special activating groups for polymerization as nucleic acids do. The most visible inorganic polymers are clays and nonclay minerals. The existence of vast quantities of clays and nonclay minerals of many varieties offers numerous possibilities for a simple form of organized chemical activity. Clays do not require a sophisticated environment for growth or maintenance and they are stable for long periods of time. Thus a favorable combination of clays could have constituted a support system for prebiotic evolution. Alternatively, a clay or combination of clays could have evolved into a living system that predated the nucleic acid-living system.

CLAYS ARE COMPLEXES OF CATIONIC AND ANIONIC POLYMERS

Clays are covalently linked complexes of cationic and anionic polymers. In this section we consider the components of clays and then the complexes they can make.

Oxycations Form Polycationic Polymers

Metallic cations become hydrated in aqueous solution. A common arrangement of water molecules, the octahedral complex, has six water molecules arranged on mutually orthogonal axes with the oxygens facing the cation

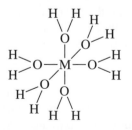

These hydrated cations can function as Lewis acids that donate protons to water molecules

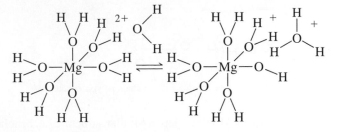

The conjugate base so formed can undergo further reactions to form dimers joined by one or two oxygen bridges, with the elimination of one or two water molecules, respectively. Further deprotonations and dehydrations lead to higher oligomers and finally to polymers

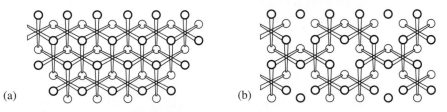

As the pH of a solution containing a given cation increases, so does the tendency to polymerize.

Polymers of Al^{3+}, Mg^{2+}, Ca^{2+}, Mn^{2+}, Fe^{2+}, Co^{2+}, and Ni^{2+} all form hydroxide crystal structures, as exemplified by brucite, a magnesium-containing polymer. In brucite two-dimensional polycationic polymers are stacked on top of each other (Fig. 1); each hydroxyl oxygen is coordinated with three magnesium cations. The hydroxyl bonds are normal to the plane of the layers. Hydrogen atoms in any layer are nested into a surface niche created by a triplet of hydrogen atoms on an adjacent layer. We can think of a layer of brucite as two close-packed sheets of hydroxyl groups, with the much smaller magnesium cations occupying the octahedral sites thereby created (see Fig. 1).

FIGURE 1   Two examples of polycationic layer structures formed by oxycations. (a) Brucite. Spheres in the upper (large) and lower (small) planes represent hydroxyl groups. Magnesium ions (Mg^{2+}) are located at spoke intersections. (b) Gibbsite. Gibbsite is very similar in structure to brucite, but because the cation (Al^{3+}) is trivalent instead of divalent as in brucite, only two out of every three of the octahedral positions are occupied by a cation.

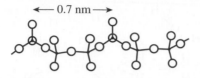

FIGURE 2   An extended straight-chain polyphosphate.

Gibbsite, $Al(OH)_3$, has a similar structure to brucite, but because of the greater charge on the metal cation ($+3$ instead of $+2$), only two out of three of the octahedral sites are occupied (see Fig. 1b). In gibbsite the oxygen atoms at the base of one layer are directly above those at the top of the underlying layer. Only half of the hydrogen atoms point toward adjacent layers, an arrangement permitting complete interlayer hydrogen bonding.

Metal sulfides also form cationic polymers. Colloidal iron sulfides were probably the precursors of present-day iron–sulfur proteins in oxidation–reduction reactions.

Oxyanions Form Polyanionic Polymers

Whereas aqueous cations tend to polymerize when the pH is raised, many oxyanions form polynuclear species when the pH is lowered. Molybdenum and tungsten behave this way in their higher oxidation states. High-molecular-weight polyphosphates are formed under dehydrating conditions. Branched-chain polyphosphates are rapidly degraded in water, while straight chains remain metastable for a considerable time. This metastability is exploited by all contemporary organisms that use adenosine triphosphate (ATP) as the main source of their high-energy phosphate. Many bacteria also use straight-chain polyphosphates as a source of high-energy phosphate (Fig. 2).

Like phosphates, sodium polysilicate $(Na_3SiO_3)_n$ forms under dehydrating conditions. Sodium polysilicates dissolve in water to give viscous concentrated alkaline solutions. On dilution, smaller oligomers are found, and below 1 mM the monomeric species predominates. On acidification, a solution of sodium silicate gives the hydrate oxide. This process can be reversed by raising the pH.

Although the extent of sodium silicate polymerization increases as the pH is reduced, the rate of polymerization has a maximum between pH 8 and 9. Because this is close to the pH (9.5) at which silicic acid and its monoanion are present in equal concentrations, a biomolecular nucleophilic reaction may be the rate-limiting step in polymerization:

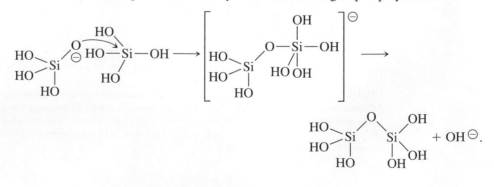

Unlike polyphosphates, branched-chain polysilicates are stable. This property, along with the tetravalence of the silicic acid unit, allows for the formation of densely cross-linked polymers.

Cationic and Anionic Polymers Form Clays

Most of the minerals that make up the rocks in Earth's crust are silicates. As a group they combine the structural features of both cationic and anionic polymers, although there are also many new features. Silicates are classified according to the state of polymerization of the formally anionic silicate component. This component may extend in one, two, or three dimensions. We focus on the class called layer silicates, in which the covalence is restricted to two-dimensional sheets. Layer silicates contain separable two-dimensional macromolecules that tend to replicate in kind.

The main types of layer silicates are composed of an extended siloxane net fused on either one or both sides of a gibbsite sheet (Fig. 3). This would give the ideal (1:1) kaolinite or (2:1) pyrophyllite structures, respectively. These clays are described as dioctahedral layer silicates because only two of the three octahedral sites are occupied. The corresponding trioctahedral layer silicates are those in which the siloxane net is fused to brucite on one or both sides. The four major structural classes of layer silicates are

$1:1$ Dioctahedral with $M_2Si_2O_5(OH)_4$ as monomeric unit

$1:1$ Trioctahedral with $M_3Si_2O_5(OH)_4$ as monomeric unit

$2:1$ Dioctahedral with $M_2Si_4O_{10}(OH)_2$ as monomeric unit

$2:1$ Trioctahedral with $M_3Si_4O_{10}(OH)_2$ as monomeric unit.

A typical layer silicate crystal consists of a stack of such units either directly in contact with each other (e.g., kaolinite), or with intervening water molecules (e.g., halloysite), or cations (e.g., micas), or both (e.g., montmorillonite). Real layers always are distorted from the ideals. Add to this the possibilities inherent in cation substitutions within the layers as well as between them, and one gets the picture of what a tremendous variety of structures is possible with these mixed inorganic polymers.

We focus on the structures of two clays, kaolinite and montmorillonite, because they have attracted the greatest interest of prebiotic chemists. The detailed structure of a layer of kaolinite, together with the manner of stacking of successive layers with a crystal, is shown in Fig. 4. Each layer occupies a thickness of 0.72 nm. The layer is depicted as electrically neutral, but in reality it carries a small negative charge, which is due to a small amount of isomorphous replacement of trivalent Al^{3+} cations by divalent cations (e.g., Mg^{2+}). This built-in negative charge gives kaolinite a small cation exchange capacity.

The superposition of oxygen and hydroxyl planes of successive layers within a single kaolinite crystal gives rise to interlayer hydrogen bond formation. The hydrogen bonds, together with the forces due to nonspecific van der Waals interactions, hold adjacent layers together so tightly that penetration of the kaolinite interlayers by organic or inorganic compounds is difficult. As a consequence adsorption is most often

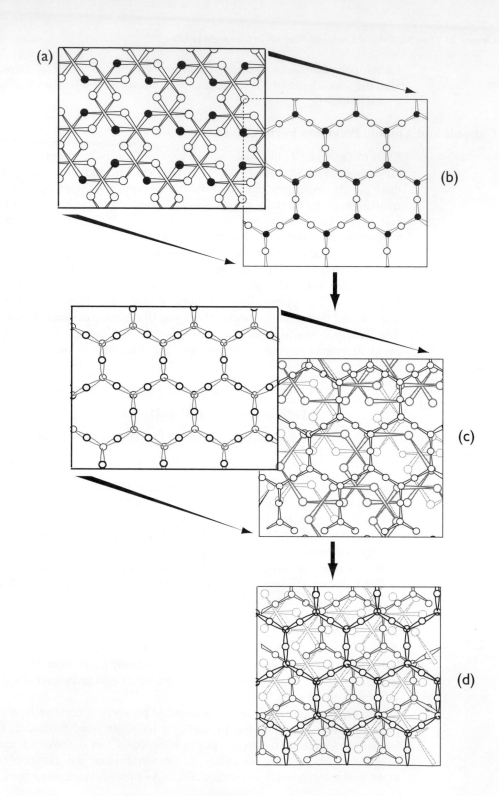

confined to the external crystal surfaces. In this connection the edges of kaolinite crystals are of particular importance. This is so because the edges containing unsatisfied valencies (broken bonds) occupy an appreciable portion (10 to 20%) of the accessible crystal surface. On the other hand, in montmorillonite only about 2% of the total external crystal and interlayer area is apportioned to the crystal edges. Hence the influence of crystal edges on the pH-dependent charge and sorption of anions, of electron-transfer reactions involving organic compounds, and of initiation and/or inhibition of polymerization of organic monomers is much more in evidence with kaolinite than with montmorillonite.

Montmorillonite is a 2:1 layer silicate. It is an excellent sorbent for organic compounds because of its large surface area, its high cation exchange capacity, and the relative ease with which it forms interlayer complexes with a variety of organic molecules.

The structure of montmorillonite is similar to that of the dioctahedral clay pyrophyllite (Fig. 4 b). It differs from pyrophyllite by occasional substitutions of trivalent Al^{3+} with divalent cations (e.g., Mg^{2+}, Fe^{2+}) in the octahedral position, and less frequently of Si^{4+} by Al^{3+} in the tetrahedral sheet. The net result of isomorphous replacement in the pyrophyllite structure is a layer that carries a permanent negative charge. This positive charge deficiency is balanced by absorption of exchangeable cations, which—apart from those associated with external crystal surfaces—are situated between the randomly superposed layers within a crystal. Water also is readily adsorbed in the interlayer space.

Formation of Clays Is a Slow Process Attended by Covalent Bond Formation

Making a clay involves more than just an ordered emplacement of units, as does the crystallization of sodium chloride or sucrose. Covalent bonds are made and broken in the process, and several molecules must be eliminated as each unit adds to the growing polymer. To make and break covalent bonds, precise orientations of groups are required and activation energy barriers have to be overcome. Under such circumstances, metastable polymers can form and persist. The orderly unit-by-unit processes of crystal growth can become blocked by tangles of polymer attached to the crystals. Only at concentrations low enough to prevent spontaneous net polymerization in

---

FIGURE 3   One way of looking at layer silicate structures. Starting with a gibbsite sheet (a), remove two-thirds of the hydroxyls (shaded) from the lower side and insert instead the apical oxygens of an extended siloxane network. Such a network is shown in (b) with the apical oxygens shaded. This would give the 1:1 dioctahedral (kaolinite) layer type. (c) Performing a similar operation also on the upper surface of the gibbsite sheet would lead to the 2:1 dioctahedral (pyrophyllite) structure. (d) Repeating these operations, but starting from brucite instead of gibbsite, would lead to the corresponding trioctahedral archetypes (serpentine and talc). In these models the balls are either oxygen or hydroxyl atoms. Cations are located at spoke intersections. (From Cairns-Smith, A. G. *Genetic Takeover and the Mineral Origins of Life.* New York: Cambridge University Press, 1982, p. 196. © Cambridge University Press 1982. Reprinted with the permission of Cambridge University Press.)

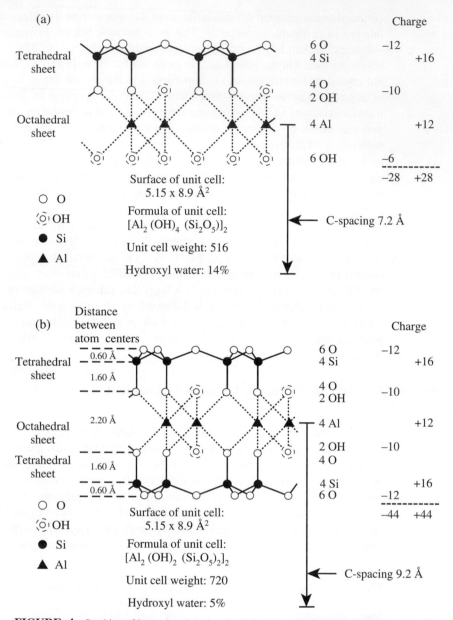

FIGURE 4 Stacking of layers in a 1:1 clay; kaolinite (a), and a 2:1 pyrophyllite clay (b). (From H. van Olphen, *An Introduction to Clay Colloid Chemistry*. New York: John Wiley-Liss, Inc., a division of John Wiley & Sons, 1977, p. 62. Reprinted by permission.)

solution would you expect such tangles to be unstable so that the crystal surfaces are cleared to expose sites at which orderly growth can proceed. As a rule, more regular crystalline clays result when the concentrations of precursors are just above saturation. Conversely, amorphous clays are more likely to form when the concentration of precursors are substantially above saturation.

The polymerization process itself is spontaneous, requiring no special functional groups for activation as in nucleic acid polymerization. The precursors for clays come from the weathering of rocks. The broad abundance of clays testifies to the availability of sufficient supplies of precursors for growing clays and the stability of clays once formed.

ARE CLAYS CAPABLE OF INHERITABLE CHANGE?

Clays clearly satisfy one of the requirements for living systems. They are capable of growing in a reproducible manner. As a rule the composition and structure of each layer are similar to those of the previous layer, features suggesting that any particular clay surface serves as the template for continuous growth of the same type of clay. The products of clay growth do not, however, separate regularly like the products from nucleic acid duplication. Instead, at ill-defined stages, clays fragment so that they maintain a colloidal size.

The second requirement of living systems is that they be capable of inheritable change. This is a difficult requirement to test for clays because of the lack of documentation. Most of the data are on crystalline clays that produce well-defined X ray diffraction patterns. Clays that are irregular or that have imperfections do not produce characteristic diffraction spots; their patterns tend to be more blurred, weaker in some cases, and lacking certain diffraction spots found in regular crystals.

Most clay minerals possess appreciable disorder. Small crystal size and structural disorder together produce X-ray powder patterns that are difficult to evaluate because diffraction procedures are usually limited to the analysis of regular structure features.

Structural disorders in layer silicates are of many kinds:

1. Disorder in the distribution of cations is common in silicate minerals. When Al^{3+} replaces Si^{4+} in tetrahedral sites, and when Mg^{3+}, Fe^{2+}, Fe^{3+}, and other ions replace Al^{3+} in octahedral sites, disordered arrangements occur.

2. Short-range order is compatible with long-range disorder. For example, in silica glass every silicon atom is surrounded by a tetrahedral group of four oxygen atoms, but little or no order occurs over a long range. Cation replacements in crystals may possibly be ordered within small domains, but may change from one domain to another so that in a crystal composed of many domains there is apparent disorder.

3. Disorder in layer stacking is common in clay mineral structures.

4. Order–disorder occurs in mixed-layer systems, which can arise in two ways. First, a major characteristic of many clay mineral structures is their ability to swell by incorporating inorganic and organic material between the layers. Water is the most

important of these materials. Generally the swelling does not take place equally between all layers at the same time. Second, the geometric similarity of layer silicate structures facilitates the formation of structures containing different kinds of layers. Both ordered and, more generally, disordered sequences of layers of many kinds are found.

5. Structures with nonplanar layers also occur. In layer silicates with markedly polar layer structures, the layers curl into rolls or spiral forms. Successive curved layers do not mesh together in the normal crystalline manner.

Because we are interested in clays as possible genetic substances, we are primarily concerned with the types of disorder that can be propagated from one layer to the next. A disordered arrangement originates at the surface of a growing clay, but can this disorder be propagated from one layer to the next? That is the crucial question as far as evolutionary matters are concerned. A priori we might expect that disorders due to cation distribution and to layer-stacking-related effects could be propagated in favorable cases. However, as far as cation distribution is concerned, there is no evidence for the propagation of disorders. This is disappointing because it is changes in the monomeric units that would be most likely to give rise to new chemical information that could generate new types of catalytic sites on the clay surfaces.

COULD CLAYS OF CLEAR NONCLAY MINERALS HAVE AIDED IN THE DEVELOPMENT OF THE FIRST LIVING SYSTEMS?

Although clays may not qualify as living systems in their own right, the possibility remains that clays may have been an essential component of the first living systems. There are several reasons why we look to clays and nonclay minerals in this regard. One of the main ones is that it is difficult to see how chemical evolution preceding the emergence of nucleic acids could have occurred without some way of maintaining the chemical components of the system as cohesive units. Membranes do this job in living cells, but to carry out their function they require mechanisms to discriminate between molecules important to the system and molecules that are detrimental to the system (see Chapter 19). The readily available surfaces of clays and nonclay minerals provide a possible alternative for retaining the molecular components of an evolving prebiotic system. Containment or retention is dictated simply by the affinity of the molecules of the system for the clay surface. We gave a simple example of how evolution might occur on such surfaces in Chapter 10, see Fig. 5. Molecules that bind to such surfaces are less likely to be swept away by the surrounding oceans and more likely to undergo further reactions with one another.

Another potential advantage of clay surfaces is that they supply some catalytic groups of their own. Examples are discussed later in this chapter.

Certain nonclay minerals offer additional advantages to a developing prebiotic system. For example, the mineral hydroxylapatite (HAP) is a potential source for the all-important anion phosphate, while iron pyrite is a potent source of reducing power (see Chapter 18).

In the following sections we consider the binding properties of clays and nonclay minerals, the types of reactions that they might catalyze, and how they may have participated in the evolution of the first living system.

Associations between Montmorillonite and Organic Molecules

We confine our remarks on clay binding of organic molecules to the 2:1 clay montmorillonite. The 2:1 clays have a high capacity for binding small organic molecules because the 2:1 sheets are capable of indefinite expansion in water. By contrast, in a 1:1 clay like kaolinite the individual layers tend to remain close-packed because of strong hydrogen bonding between them. Recall that montmorillonite carries a net negative charge. This charge is partially neutralized by the binding of cations. Bound monovalent cations such as Na^+ and K^+ are readily displaced by divalent cations such as Ca^{2+} or Zn^{2+}. A homionic clay, in which all the dissociable cations are of one type, may be prepared by treating the clay with an excess of the appropriate salt and then removing the nonbonded salt by repeated water rinsing. Such clays show modified binding properties. For example, studies on binding nucleotides by homionic clays of montmorillonite have shown that adenosine nucleotides bind much more to copper montmorillonite than to sodium or magnesium montmorillonite.

We prefer to focus on the properties of sodium montmorillonite in the presence of Mg^{2+}, because these conditions seem more likely to represent typical prebiotic conditions. In Table 1 some binding results for sodium montmorillonite in a Mg^{2+}-containing solution are presented. The most important factor for small-molecule binding to montmorillonite appears to be the possession of one or more planar organic ring structures. It is presumed that these ring structures lie flat on the sheetlike interlayer surfaces of the clay. Because of the net negative charge of the clays, positively charged organic molecules generally bind more strongly than neutral molecules bind, while molecules with negative charges frequently show diminished binding. Small organic molecules that bind strongly to montmorillonite generally bind until the sheetlike surfaces are almost covered. Despite their net negative charge, polynucleotides in limited amounts bind strongly to montmorillonite. The fact that polynucleotides can bind strongly, while mononucleotides bind very weakly, is believed to be due to the accumulative effect of many weak electrostatic interactions.

Binding to Hydroxylapatite Has a Strong Electrostatic Component

HAP is a nonclay mineral with the approximate composition $Ca_{10}(PO_4)_2 (OH)_2$. HAP may have been a major source of phosphate for bioorganic molecules, because most of the inorganic phosphate in Earth's crust is found in this mineral. A great deal is known about HAP because it has been used for more than a quarter century to analyze nucleic acids in the laboratory. Although all nucleic acids have a high affinity for HAP, double helices bind to it more strongly than single-stranded nucleic acids. This difference in

TABLE 1

Binding of Select Bioorganic Molecules to Montmorillonite and Hydroxylapatite (HAP)[a]

Compound	Montmorillonite	HAP
Adenine	Strong	Very weak
Adenosine	Moderate	Very weak
AMP	Very weak	Weak
ADP	Very weak	Moderate
ATP	Very weak	Moderate
Hypoxanthine	Strong	Very weak
Uracil	Weak	Very weak
DAMN	Weak	—
AICA	Strong	—
FMN[b]	Strong	—
FAD[c]	Strong	Moderate
tRNA	Strong (in small amounts)	Strong

[a]Results of Winter, D. and Zubay, G. Binding of Adenine and Adenine-Related Compounds to the Clay Montmorillonite and the Mineral Hydroxylapatite, *Origins Life Evol. Biosphere* 25: 61–81, 1995. Most compounds were tested at a starting concentration of 1 mM in 0.1 M NaCl, 0.02 M MgCl$_2$, pH 8.0. Concentration dependencies were usually studied over three orders of magnitude. Strong, >80%; moderate, 20–80%; weak, 5–20%; very weak, 0.5%.

[b]Flavin mononucleolide.

[c]Flavin-adenine dinucleotide.

binding is quite remarkable, because most absorbents bind single-stranded nucleic acids more strongly than duplex nucleic acids, probably as a result of the greater exposure of the purine and pyrimidine bases in single-stranded structures. It has been conjectured that HAP once may have served the dual function of harboring nucleic acid duplexes, including duplexes in the process of growth, as well as providing a source of phosphate for making nucleotides.

The most important factor for small-molecule binding to HAP is negative charge. Without negative charge little or no binding to HAP is detectable. The higher the negative charge is, the stronger the binding. From these findings it is concluded that the main affinity between organic molecules and HAP is electrostatic, involving the negative charges on the small molecules and the positive charges originating from the Ca^{2+} ions in the HAP. Similar conclusions have been reached by others in studies on HAP and other nonclay minerals such as gypsum, a calcium sulfate mineral, and iron pyrite, an iron sulfide mineral.

Comparative Observations on Binding to Montmorillonite and Hydroxylapatite

Collectively, montmorillonite and HAP show a significant affinity for all the small molecules listed in Table 1. Most frequently the binding to these two inorganic polymers is

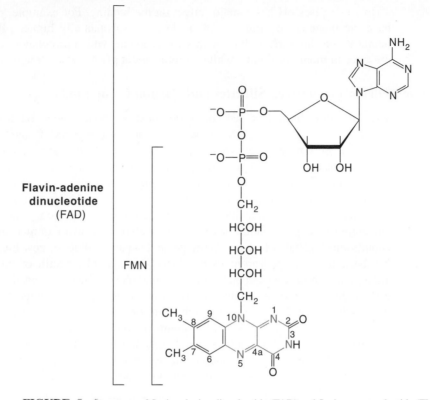

FIGURE 5   Structures of flavin-adenine dinucleotide (FAD) and flavin mononucleotide (FMN).

complementary. For example, in the series adenine, adenosine, adenosine 5' monophosphate (AMP), adenosine diphosphate (ADP), and adenosine triphosphate (ATP), the first two molecules bind to montmorillonite, while the last three bind much more strongly to HAP. Flavin-adenine dinucleotide (FAD) is exceptional in that it binds strongly to montmorillonite and moderately to HAP. Most likely the negative charges originating from the pyrophosphate group on FAD account for the affinity to HAP (Fig. 5); the affinity for montmorillonite is almost certainly due to the three-ring structure of the flavin.

In a number of cases the ratio of bound to free small molecules increases as the total amount of small molecules increases. This suggests that *cooperative binding*, such as would result from favorable contacts between the small molecules, supplements the affinity for the clay and mineral surfaces. Examples of molecules that bind cooperatively include adenine, adenosine, hypoxanthine, uracil, and aminoimidazole carboxamide (AICA) on montmorillonite; and FAD and AMP on HAP. Very few studies of cooperative binding have been performed in which mixtures of two different small molecules are tested simultaneously. Cooperative binding could be important because when two or more molecules bind cooperatively, the chances that they will react further with one another is increased.

In some cases pH has a major effect on the binding. For example, molecules that have the potential to protonate often do so at substantially higher pH levels when bound to the clay surface. We see this with adenine, which protonates at pH 7.5 in the presence of montmorillonite, while it protonates at pH 4.1 when free in solution.

Covalent Bond Formation between Silicates and Carbon Compounds

Thus far our discussion has been confined to noncovalent binding of organic molecules to clays. In this section we focus on some aspects of covalent binding, in particular, on covalent bond associations between silicates and carbon compounds. These associations are of special significance because they could have provided an important bridge between the inorganic and organic worlds.

Many acidic mucopolysaccharides (protein–polysaccharide complexes) found in biological systems contain high levels of tightly bound silicon. The list of silicon-containing mucopolysaccharides includes hyaluronic acid of human umbilical cord, chondroitin-6-sulfate of human cartilage, and heparin sulfate of cow lung. Silicon also has been found in polyuronides of plants, for example, pectin of citrus fruits and alginic acid from horsetail kelp. In these complexes silicon is covalently bonded to carbon through oxygen atoms (Fig. 6). Silicon can act as a bridge between two or more polysaccharide chains and thereby contribute to the maintenance of the structure and integrity of connective tissues. Proteins in connective tissues, especially collagen, also have been found to contain bound silicon. The fact that pectin and alginic acid contain high levels of tightly bound silicon suggests that the same principle of cross-linking by silicon is employed in the plant kingdom.

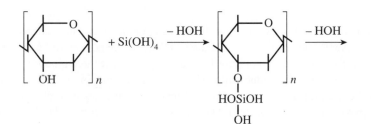

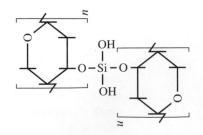

FIGURE 6   Possible covalent structure formed between carbohydrates and silicic acid.

The cell walls of diatoms, a group of small marine organisms, consist of siliceous frustules encased in an organic coating (Fig. 7). A major portion of this organic coating is composed of proteins. The silica of diatoms is similar in chemical composition to a silica gel, SiO_2nH_2O. For several species in which the amino acid composition of the organic coating has been determined, the proteins contain more serine and threonine than the protein within the cell contains. It has been suggested that a silicic acid molecule can undergo condensation reactions with hydroxyl groups on adjacent serine residues:

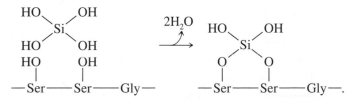

Silica or silica gel also is used as a skeletal material in a large number of organisms. For example, an enormous number of planktonic protozoa called radiolarians have considerable silica in their skeletons; the variation and elegance of their shapes are extraordinary. Amorphous silica is the most abundant inorganic skeletal compound in the phylum *Porifera*. The siliceous spicules are often bound together by a secondary deposit of silica or by a collagenous cement.

Silica is often deposited as amorphous silica gel in the shoots of vascular plants, especially horsetails and grasses. Silicon is considered to be essential for the normal growth of these plants. For example, *Equisetum arvense* (a horsetail) has been shown to collapse when grown on silicon-free nutrient solutions but to be normally erect when sodium metasilicate is included in the growth medium.

The findings of covalent bond formation between silica and carbohydrates or proteins suggest a unique function for silicate-based clays. Such associations in prebiotic times could have been important in establishing a cooperative interaction between clays and organics. The *cis* hydroxyl group on ribose may constitute a good attacking group for making a covalent bond between the silica and the sugar in an appropriate clay.

Organic Reactions Catalyzed by Clays

In addition to binding organic molecules, some clays and minerals have been found to catalyze various organic reactions. The catalytic activities of clays have been attributed to several properties: (1) highly specific surfaces with a specific distribution of surface charge, (2) exchangeable metal ions bound to the surface, and (3) variable adsorptive capacity for different organic materials. Given these properties, clays may be considered as crude inorganic analogs of enzymes. They may help to bring the reactants into juxtaposition, catalyze surface reactions, stabilize intermediates, and catalyze subsequent reactions.

Despite the exciting potential that clays offer to prebiotic chemists, few impressive reactions have been carried out on clays. Conditions have been found where a small

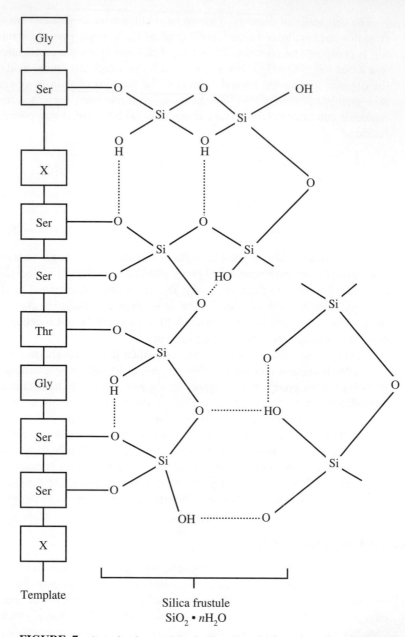

FIGURE 7 A molecular model of silica frustule formation of a diatom. (From Hecky, R. E., Mopper, K., Kilham, P., and Degens, E. T. The Amino Acid and Sugar Composition of Diatom Cell-Walls, *Mar. Biol.* 19:323, 1973.)

amount of synthesis of carbohydrates, lipids, amino acids, and purines and pyrimidines is clay dependent, but alternative, nonclay-dependent reactions of equal efficiency usually have been found. Two notable exceptions were discussed in Chapters 14 and 17: the polymerization of polypeptides from carbodiimide-activated amino acids on illite and HAP, and the formation of oligonucleotides up to 55 monomers long from nucleoside 5′-phosphorimidazolides on montmorillonite. Ferris and colleagues who made these observations have suggested that the minerals on primitive Earth provided a "library" of surfaces for the exploration of molecular evolution.

Clays can function as proton-donating Brønsted acids. We have already noted that in the presence of montmorillonite, adenine can protonate at a much higher pH than expected (7.5 observed versus 4.1 expected). The proton-donating potential of clays should be manifest in acid-catalyzed reactions. In this connection we have found that FAD bound to montmorillonite at pH 7.5 is completely hydrolyzed in 16 hr at 25°C. The products of this breakdown are flavin mononucleotide (FMN) and AMP, a result indicating that the pyrophosphate linkage has been cleaved.

Even in cases where the clays themselves do not possess particular catalytic activities, they may absorb the catalyst in question so that it can act on other absorbed molecules. The absorption properties of clays also could have served as a selective agent for concentrating certain organic molecules by surface binding.

Not to be overlooked is the potential role of clays for catalyzing asymmetrical synthesis. However, no generally confirmed examples of clay-catalyzed asymmetrical synthesis are known.

SUMMARY

In this chapter we have described inorganic polymers, especially clays, and the possible roles they may have played in the origin of life.

1. The complex problems associated with the origin and maintenance of a nucleic acid system have led to the suggestion that they may have been preceded by clay–polymer systems in the origin of life.

2. Clays are complexes of cationic and anionic polymers that form spontaneously from simple monomeric units.

3. Most clays are silicates that combine the structural features of cationic and anionic polymers.

4. In layer silicates, covalence is restricted to two dimensions. Layer silicates are separable into two-dimensional macromolecules that tend to replicate in kind. The main structural types of layer silicates are composed of an extended siloxane net on either one side or both sides of a polycationic sheet. A typical layer silicate crystal consists of a stack of such units, directly in contact with each other, with intervening water molecules, cations, or with both. The different ways of stacking and compositions lead to a tremendous variety of structures.

5. Although insufficient information is available, it is doubtful that clays can propagate discrete alterations in their structure. Thus clays satisfy only one of the two main

criteria for living systems: they are capable of replication, but they do not appear to be capable of inheritable change.

6. Montmorillonite is an ideal sorbent because of its swelling properties, which expose the siliceous sheets of either side of the individual layers. Organic compounds, particularly those containing ring structures that can lie flat, bind moderately or strongly to montmorillonite. This binding is enhanced by a negative charge on the clay.

7. Nonsiliceous minerals must not be overlooked as potentially important contributors to the origin of life. The calcium phosphate mineral HAP is a particularly interesting mineral, because it may have been the major source of phosphate on primitive Earth. It also shows strong binding to compounds with multiple negative charges.

8. Siliceous materials can make covalent bonds with organic compounds. A number of living systems show such associations, which usually involve covalent bond formation between silicon and carbon through oxygen atoms. Serine side chains in polypeptides and hydroxyl groups from polysaccharides make covalent bonds of this type.

9. Although it does not appear that inorganic polymers ever could have constituted living systems on their own, a very plausible first living system might be one consisting of clays complexed with organic compounds organized on the clay and/or mineral surfaces.

Problems

1. What is the difference between a 1:1 and a 2:1 clay? Which is more suitable for binding large amounts of simple organic molecules?

2. How would you expect RNA to bind to a 2:1 clay? Once bound would it be able to move from one location to another?

3. Describe three roles that clays are likely to have played in the origin of life.

4. If you found clays in a meteorite, what would this tell you about the origin of the meteorite?

5. Do you think you could make a clay in the laboratory? Explain your answer.

6. What type of molecules are likely to bind to

(a) Kaolinite?

(b) Hydroxylapatite?

(c) Montmorillonite?

References

Acevedo, O. L. and Orgel, L. E. Template-Directed Oligonucleotide Ligation on Hydroxylapatite, *Nature* (*London*) 321:790, 1986.

Cairns-Smith, A. G. *Genetic Takeover and the Mineral Origins of Life*. Cambridge: Cambridge University Press, 1982.

Cairns-Smith, A. G., Hall, A. J., and Russell, M. J. Mineral Theories of the Origin of Life and an Iron-Sulfide Example, *Origins Life Evol. Biosphere* 22:161, 1992.

Ferris, J. P. and Ertem, G. Oligomerization of Ribonucleotides on Montmorillonite: Reaction of the 5'-Phosphorimidazolide of Adenosine, *Science* 257:1387, 1992.

Ferris, J. P., Hill, A. R., Liu, R., and Orgel, L. E. Synthesis of Long Prebiotic Oligomers on Mineral Surfaces, *Nature* (*London*) 381:59, 1996. (This refers to minerals as presenting a library of surfaces for different reactions and is very persuasive.)

Ferris, J. P., Huang, C.-H., and Hagen, W. J. Montmorilonite, a Multifunctional Mineral Catalyst for the Prebiological Formation of Phosphate Esters, *Origins of Life Evol. Biosphere* 18:121, 1988.

Gibbs, D., Lohrmann, R., and Orgel, L. E. Template-Directed Synthesis and Selective Adsorption of Oligoadenylates on Hydroxylapatite, *J. Mol. Evol.* 15:347, 1980.

Lazard, D., Lahav, N., and Orenberg, J. B. The Biogeochemical Cycle of the Adsorbed Template. II. Selective Adsorption of Mononucleotides on Adsorbed Polynucleotide Templates, *Origins Life Evol. Biosphere* 18:347, 1988.

Rao, M., Odom, G., and Oro, J. Clays in Prebiological Chemistry, *J. Mol. Evol.* 15:317, 1980.

Schwartz, A. W. and Orgel, L. E. Template-Directed Polynucleotide Synthesis on Mineral Surfaces, *J. Mol. Evol.* 21:299, 1985.

Wachtershauser, G. Before Enzymes and Templates: Theory of Surface Metabolism, *Microbiol. Rev.* 52:454, 1988.

Wachtershauser, G. Groundworks for an Evolutionary Biochemistry: The Iron-Sulphur World, *Prog. Biophys. Molec. Biol.* 15:58–85, 1992.

Winter, D. and Zubay, G. Binding of Adenine and Adenine-Related Compounds to the Clay Montmorillonite and the Mineral Hydroxylapatite, *Origins Life Evol. Biosphere* 25:61, 1995. (Many small organic molecules that bind to minerals bind cooperatively.)

PART IV

Evolution of Living Systems

Part IV follows chronologically on Part III as we deal with the development of living systems in that long interval between the first living organisms and the present. This material is fascinating not only for the light it sheds on today's living systems, but also for the clues it may provide to many obscure puzzles relating to the origin of life. Chapter 21 is devoted to a general description of the evolution of living systems, with emphasis on the genetic and environmental factors that are the agents of evolution. The remaining three chapters focus on specific biochemical processes.

In Chapter 22 we examine the evolution of the tricarboxylic acid (TCA) cycle that was introduced in Chapter 11. The main function of the TCA cycle is to oxidize carbon compounds from the three-carbon state to carbon dioxide. The TCA cycle requires molecular oxygen and thus could not function in the normal way in organisms before there was a reasonable supply of molecular oxygen. However, a variety of present-day organisms either cannot use oxygen or can function without it. From studies of such organisms, it appears that most of the reactions of the TCA cycle existed before the cycle itself existed. This supposition indicates that these reactions must have functioned in other capacities in different species.

In Chapter 23 we discuss the evolution of photosynthesis. This chapter begins with a detailed account of how photosynthesis works in existing organisms. It then deals with specific aspects of the evolution of oxygenic photosynthetic systems.

Finally, in Chapter 24 we consider the evolution of the genetic code. The first part of this chapter deals with the origin of the triplet code; the second part examines the nature and possible causes of differences in the code in different organisms.

CHAPTER 21

Evolution of Organisms

Contemporary organisms are the end products of about 3.9 billion years of evolution. Scientists from many different disciplines have made important contributions to the science of evolution; included in this list should be geologists, chemists, and biologists. Whereas our understanding of the various factors that influence the course of evolution is vast, it is difficult to assess the relative contributions of the different factors involved. As a consequence, the field of evolution looms with controversy. We have spent a good deal of time talking about prebiotic evolution and protocellular evolution in Part III. This chapter starts with "conventional" living cells that have DNA, RNA, and protein all in their usual roles. I attempt to introduce the reader to the significant parameters that have influenced cellular and organismic evolution and some of the tools that have been used to investigate the process.

EXISTING ORGANISMS HAVE A COMMON ORIGIN

Of the species that have become extinct, we have records for only about 250,000. Because this is only a small fraction of the total species that have existed, it should be clear that our records of the evolutionary process are incomplete. How do we fit the pieces of this puzzle together when most of them are missing? We must make many guesses and to do this we must consider evidence from as many different sources as we can find.

Fortunately, in recent years the information that has been derived from classical morphological comparisons has been considerably bolstered by biochemical information. Such information has permitted relatedness studies even in cases where morphological differences are skimpy. Thus, for example, we can make quantitative comparisons of amino acid sequences or of nucleotide sequences of functionally related genes between any two species. Such comparisons have led us to the conviction that all species are related in having a common origin. It is comforting to find that for plants and animals the closer two organisms are morphologically, the closer they are in their genomic nucleic acid sequences.

CLASSICAL EVOLUTIONARY TREE IS BASED ON MORPHOLOGY

The classical depiction of evolutionary relationships resembles a branching tree in which all existing organisms are shown at the tips of the branches (Fig. 1). At the base of this tree is the simple ancestral prokaryote, from which there arise principal branches representing the various kingdoms. Each of these branches gives rise to further and further branches.

Prokaryotes for the most part have remained as relatively undifferentiated single-celled organisms with a single chromosome. By contrast, *eukaryotes* have changed dramatically. They all contain at least two and usually more than two chromosomes, which are separated from the cytoplasm by a nuclear membrane. Frequently, eukaryotes contain other membrane-bounded organelles such as chloroplasts or mitochondria. Although the organisms on many branches of the eukaryotic part of the evolutionary tree have remained as relatively undifferentiated, single-celled organisms, those on many other branches have evolved into multicellular organisms in

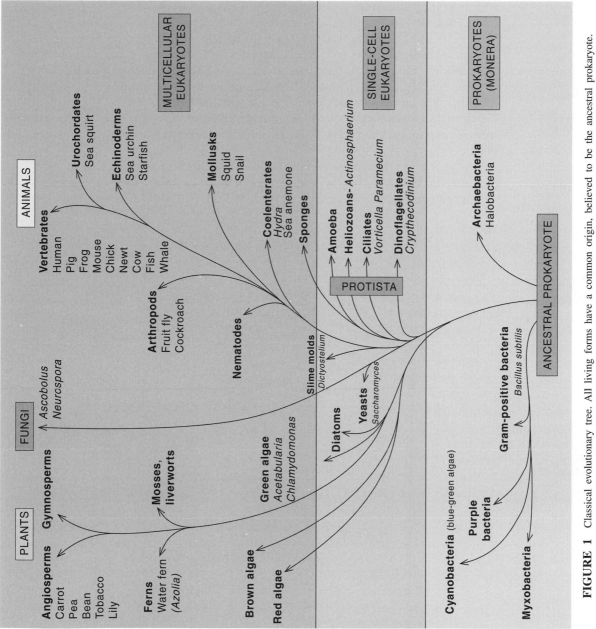

FIGURE 1 Classical evolutionary tree. All living forms have a common origin, believed to be the ancestral prokaryote. Through a process of evolution, some of these prokaryotes changed into other organisms with different characteristics. The evolutionary tree indicates the main pathways of evolution.

which the individual cells of the organisms differentiate to serve different functions. Multicellular eukaryotes are divided into three groups: plants, animals, and fungi.

It is important to realize that *species* are the only tangible entities in the evolutionary tree. Although there is individual variation within a species, all members of the same species share the same gene pool, and are reproductively isolated from members of other species. That is, if mating between species occurs at all, the offspring are infertile. Thus the species has objective existence. The twigs, branches, and limbs of the classical evolutionary tree represent the perceived relationships or groupings that we superimpose on the species. These relationships, called the higher taxa, are listed in Fig. 2.

The gaps in our knowledge of evolution become apparent when making morphological comparisons between well-known species. For example, among the primates we find various species of monkeys, gibbons, and apes, as well as humans. All these

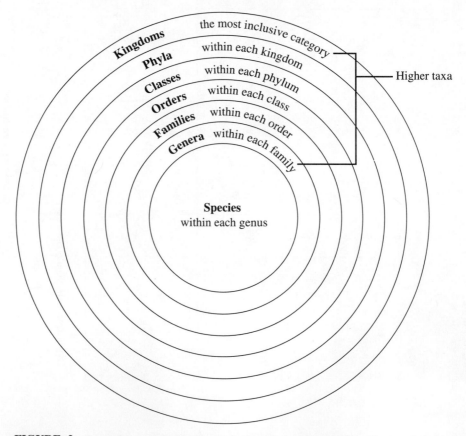

FIGURE 2 Categories used to describe the large-scale patterns, trends, and rates of change among species.

species have certain features that we identify as primatelike. Nevertheless, a close comparison of any two members from different species of primates would reveal numerous morphological differences between them (Fig. 3). Certainly these differences did not occur in one step. Instead, small genetic changes probably accumulated in descendants of a given ancestor until those differences made it impossible for mating to take place between them and other descendants of that ancestor. Only then could we say that a species change had resulted. After a species change occurs, further genetic changes can occur within each new species, but there is little or no genetic exchange between species.

Fossil Record Supplies Some of the Missing Pieces of the Evolutionary Tree Puzzle

To understand the relationship between existing species of primates or any other organisms, we try to find direct evidence for intermediates leading to common branchpoints. Evidence concerning extinct species comes from fossils. To be preserved as a fossil, an organism must be buried before it deteriorates, and the resulting sedimentary rock layer in which it becomes trapped must remain relatively undisturbed by geologic forces. More recent fossils are found in ice or tar pits. Some types of organisms and environments are more likely than others to yield fossils. For this reason animals with shells or skeletons are better represented in the fossil records than soft-bodied animals such as jellyfish or worms. Fossilization is favored in certain locales. For example, tar pits favor fossilization, whereas land surfaces undergoing erosion do not.

Ancient fossils can be dated with precision according to the sedimentary layer in which they are located. Sedimentary layers are formed by the gradual deposition of erosion products. As a rule, the sedimentary layers that lie deepest beneath Earth's surface were formed first, followed by those overlying them. Sometimes, however, layers are missing as a result of erosion or nondeposition in some places. Layers also can be deformed, or even reversed in sequence, by geologic upheavals. Thus a comprehensive understanding of sedimentary layers is required to make sound interpretations of the overall pattern.

The most reliable estimates for the age of different sedimentary layers come from radioactive dating methods. This process consists of measuring the amount of a particular radioactive substance and its decay products in a given sample. The age of the sample can be estimated from the relative amounts of the unstable isotope and its decay products, given the known half-life of the unstable isotope. This technique was discussed in Chapter 4. Different isotopes are best suited for measuring different spans of time; the most accurate measurements are obtained by using an isotope with a half-life comparable with that of the sample.

Information obtained from the isotopic dating of fossil-containing sedimentary rocks has been used to construct a geologic timescale like the one shown in Fig. 4. This scale is divided into two *eons*, the Precambrian and the Phanerozoic. Eons are divided into *eras*, eras are divided into *periods*, and some periods are divided into *epochs*. Through the combined efforts of geologists and paleontologists, a great deal

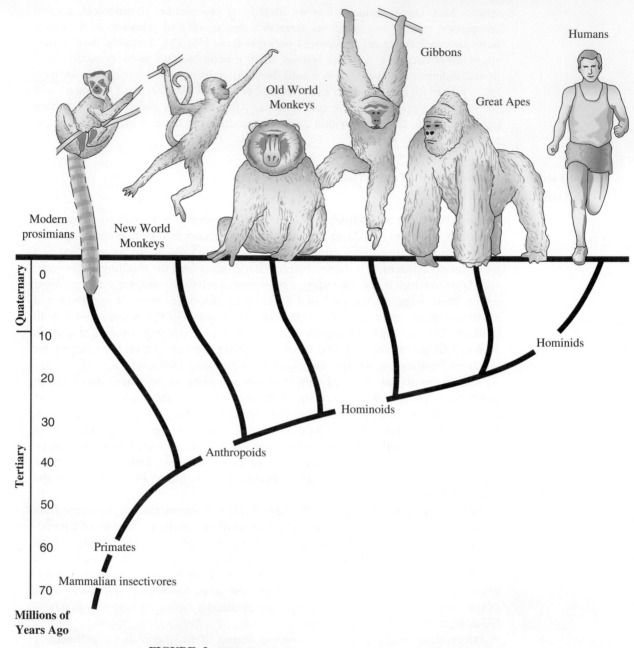

FIGURE 3 Evolution of primates that took place during the Cenozoic era.

can be learned about Precambrian times. Fossils that are found in various sedimentary rock layers are assumed to have a similar age to the sediments. Prior to isotope dating, the main divisions of geologic time were recognized by abrupt transitions in the fossil sequence. We now know that these transitions resulted from mass extinctions, a topic that is taken up in the next section. A summary of some of the main events based on the geologic timescale and the fossil evidence for Precambrian times is presented in Fig. 4.

Biological Evolution Has Been Punctuated by Mass Extinction

During the course of evolution new species are constantly appearing while established species tend to become extinct. The average lifetime of a species has been estimated at about 4 million years. In addition to a steady disappearance of existing species with time, there have been periods when the number of species being extinguished has risen sharply. In fact we are probably in the midst of just such a period caused by human overpopulation. Never before has one species been such a threat to so many other species. In the past, most mass extinctions were the result of geologic or meteorological upheavals.

In the last 600 million years, there have been five mass extinctions, the time and severity of which are reflected in Fig. 5 by sharp drops in the number of families. For example, it can be seen that during the Permian extinction, which occurred 220 million years ago, about half of the existing families were eliminated over a period of a few million years. This included about 90% of all marine species. The less severe Cretaceous extinction, some 65 million years ago, had a greater effect on terrestrial organisms. Later in this chapter we consider explanations for these mass extinctions.

BIOCHEMICAL RECORD

Although mostly limited to existing organisms, the biochemical record has provided a wealth of quantitative information on the relatedness of species and has led to a somewhat different picture from that provided by morphological comparisons.

Biochemical methods for determining the relatedness of organisms started with comparisons of specific proteins from different species. Such comparisons were first made by measuring immunologic cross reactivities or electrophoretic mobilities of proteins with comparable enzymatic activities. The most valuable measurements on proteins have been made by examining the amino acid sequences of comparable proteins. Although such comparisons are more work and require the prepurification of the proteins to be compared, the rewards in terms of information content more than justify the extra effort.

Zuckerkandle and Pauling pioneered in attempts to draw conclusions from the amino acid sequences of related proteins in different species. They argued that the accumulation of sequence differences is approximately proportional to the geologic time elapsed since separation of two species. Therefore, the number of differences

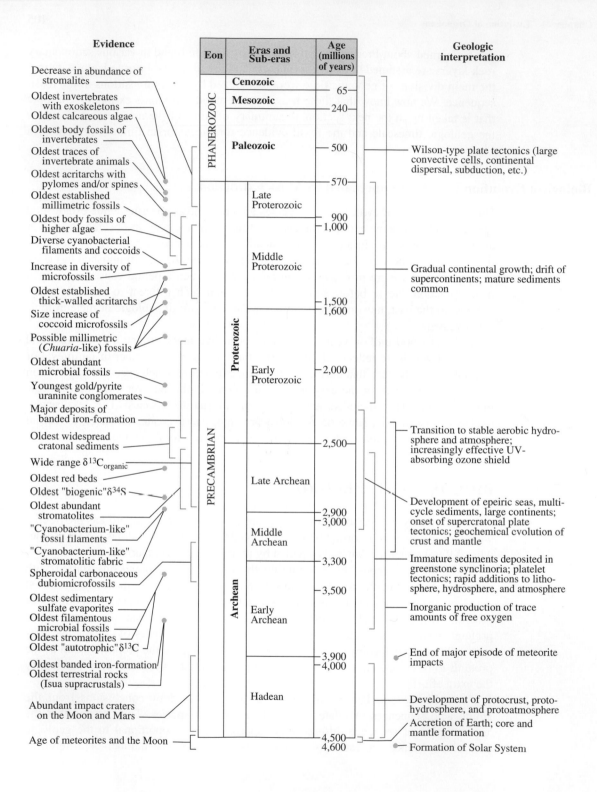

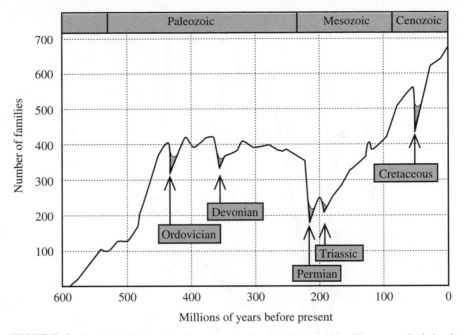

FIGURE 5 Family level diversity of marine animals over the past 600 million years. Periods of mass extinction are labeled.

observed should lead to an estimate of the time of evolutionary separation, when this is unknown for two species, provided that at least for another pair of species the time actually elapsed is known. The linearity of the number of differences versus time is shown for three proteins in Fig. 6. The relationships in this figure indicate that there has been an appreciable difference in the rate of change for different proteins, even though closely related proteins may show a linear rate of change with time. Of the three protein types shown in Fig. 6, fibrinopeptides have the fastest rate of change and cytochrome c has the slowest.

Linear relationships such as the one observed for the three classes of proteins discussed here have implications for the kinds of mutations involved. Mathematically it can be shown that linearity would be expected for *neutral mutations*, those that have no significant beneficial or deterimental effects on the organism. According to what is known as the neutralist theory, most of the amino acid sequence changes would result from neutral mutations. We return to this point in a later section.

A surprising fact suggested by the linearity relationship is that the mutation rate as a function of time is the same in quite different organisms. Because the generation

FIGURE 4 A geologic timescale showing some of the evidence geologists and paleontologists have used to reconstruct the Precambrian history of life. From Schopf, J. W. *Earth's Earliest Biosphere*. 1983, Princeton, NJ: Copyright 1983 by Princeton University Press, p. 368. Reprinted by permission of Princeton University Press.

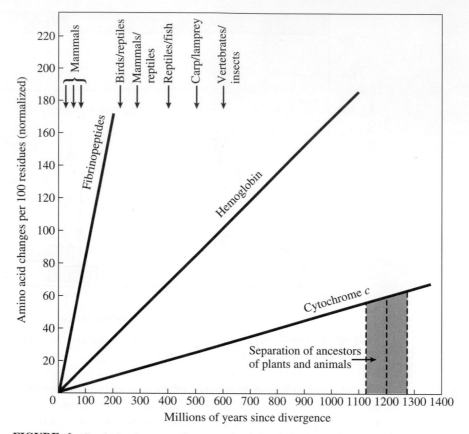

FIGURE 6   Graph showing rates of macromolecular evolution in the fibrinopeptides, hemoglobin, and cytochrome *c*. (After Dickerson, R. E. *J. Mol. Evol.* 1:26–45, 1971.)

time is quite different for different organisms, one also might expect the number of mutations per generation to vary in different organisms, but it appears not to do so. There is no simple explanation for this; indeed, it is possible that numerous exceptions to the linearity relationship are to be found as more and more species are examined.

Hemoglobin Family of Proteins Provides a Database for Evolutionary Investigations

No family of proteins has been studied more intensively than the hemoglobins. Hemoglobin and hemoglobin-related proteins comprise a family of closely related proteins involved in the transport of oxygen from the atmosphere to the tissues where it is consumed. In humans hemoglobin is a tetramer containing very similar protein subunits, the α-(141 amino acids) and the β-(146 amino acid) polypeptide chains. Closely related to hemoglobin is myoglobin (153 amino acids), a monomer whose major function is to store oxygen in the tissues where it is consumed.

Amino acid comparisons between the polypeptide chains of hemoglobin and myoglobin indicate a strong evolutionary relatedness. It seems likely that in early evolution there was only one gene for a primitive hemoglobin-like molecule, which probably in the course of time underwent different point mutations in different species. At later stages of evolution, successive gene duplications must have taken place and these genes also must have undergone considerable change. By using the assumption that evolutionary relatedness is proportional to the number of amino acid changes, a composite tree of the globin family can be constructed (Fig. 7). In this tree the lengths of the branches are proportional to the number of amino acid changes between the proteins in different species. Those branches with numbered circles represent duplications that gave rise to different genes. In general the evolutionary relationships of the different species derived from this biochemical tree are similar to those on the evolutionary tree developed from morphological comparisons.

Comparison of Ribosomal RNA Sequences Leads to a Universal Phylogenetic Tree

With the emergence of rapid methods for nucleic acid sequencing, phylogeny (the history of evolutionary development) has become a tractable problem even with bacteria, whose other attributes made such studies untenable. We are beginning to get a comprehensive phylogenetic picture of organisms that is based on comparative analyses of the rRNA sequences of the small ribosomal RNA subunit. The 16S rRNA molecule nucleotide differences in regions of the alignment (about 950 nucleotides) for which there are primary and/or secondary structural similarities were unambiguous. The values for sequences were corrected for multiple nucleotide substitutions per site, and the resulting estimates of sequence divergence (mutations fixed per sequence position) were used to infer the phylogenetic tree. The lineages must have diverged from a common ancestral sequence in the interior of the network, and the contemporary sequences are at the tips of the branches. The length of each segment in the tree reflects the amount of sequence change along that segment. Major groups are identified by designations proposed by C. R. Woese and colleagues.

In Fig. 8 organisms fall into three primary lines of evolutionary development: the eukaryotes (Eucarya), the eubacteria (Bacteria), and the archaebacteria (Archaea). The line segments connecting the various contemporary species to their common ancestors are not the same length. This means that members of the three lines have not evolved at the same rate, a fact evident from comparisons of protein sequences as well. Perhaps the most striking characteristic of this tree is the distinctness of the three primary kingdoms, as evidenced by the large sequence distances that separate each kingdom from the others.

The root of the universal tree in Fig. 8 cannot be identified by using sequences of 16S rRNA only. However, recent phylogenetic studies (e.g., see N. R. Pace) of gene families that originated before the last common ancestor of the three domains have positioned the root of the universal tree deep on the bacterial line. Therefore eukaryotes and archaebacteria had a common history that excluded the descendants of the eubacterial line. This period of evolutionary history shared by eukaryotes and

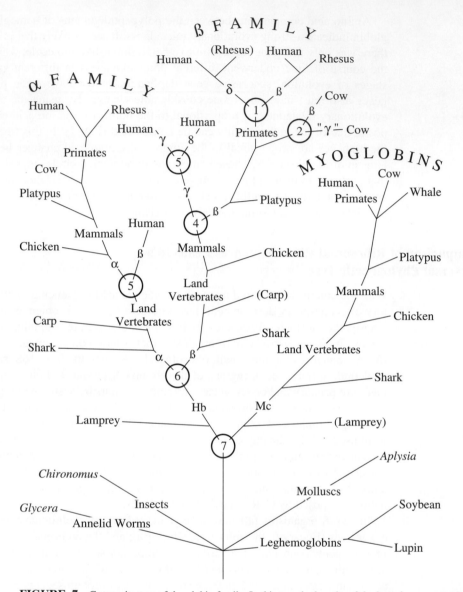

FIGURE 7   Composite tree of the globin family. In this tree the lengths of the branches are proportional to the number of amino acid changes between the globins in different species. Branchpoints with numbered circles represent gene duplications, and simple unmarked branches are species divergences. (From Dickerson, R. E. and Geis, I. *Hemoglobin: Structure, Function, Evolution, and Pathology*. Menlo Park, CA: Benjamin/Cummings, 1983, p. 82. Reprinted by permission of the author.)

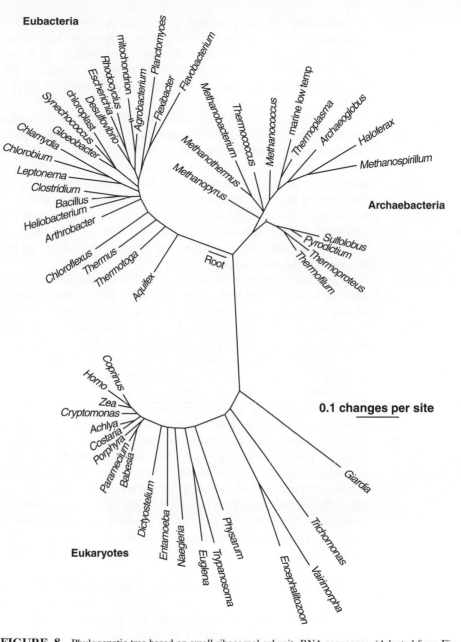

FIGURE 8   Phylogenetic tree based on small ribosomal subunit rRNA sequences. (Adapted from Fig. 1 in Pace, N. R. A Molecular View of Microbial Diversity and the Biosphere, *Science* 276:734–740, 1997. Reprinted with permission from *Science* © 1997 American Association for the Advancement of Science.)

archaebacteria was an important time in the evolution of cells, during which the refinement of the primordial information processing mechanisms occurred. Thus modern representatives of eukaryotes and archaebacteria share many properties that differ from bacterial cells in fundamental ways. This includes the basic structure of the RNA polymerase and the way in which it identifies start sites on the DNA template.

It is striking that in both bacterial kingdoms close to the root most of the known species are hyperthermophiles. There are two possible explanations for this. Perhaps a late great impact heated the oceans, so that although the very first life may have lived in cooler water, only those few organisms that could live in very hot water survived to become the ancestors of modern life. Alternatively, prephotosynthetic cells may all have lived in hot conditions around early volcanoes. In volcanic hydrothermal systems, heat drives water circulation through the rock: the water becomes charged with a wide variety of chemical species. Where the water emerges from the rock, a diverse biological community can be sustained by the chemical disequilibrium between the hot fluid and the external environment.

Before these observations on ribosomal RNA sequences and other sequences, it was believed largely on the basis of fossil remains that the first eukaryotes arose about 1.4 billion years ago. In the light of this sequence information it is highly likely that eukaryotes arose long before this. However, it must always be kept in mind that there are no absolute time coordinates on the Woese evolutionary tree.

In the light of the phylogenetic relationships that emerge from rRNA sequence comparisons, it is of interest to consider the five kingdoms defined by the classical evolutionary tree; animals, plants, fungi, protists (unicellular eukaryotes), and prokaryotes (monera). The classical tree focuses attention on morphological diversity, especially of the higher eukaryotes, without recognizing the molecular diversity now known to exist within the kingdoms of monera and protista. The five kingdoms of the classical evolutionary tree were considered fundamental units; no interkingdom affiliations credibly could be made. With the current phylogenetic perspective, however, the relationships between different eukaryotes in different kingdoms can be more precisely defined. The four eukaryotic kingdoms of the classic tree in fact form a phylogenetically coherent unit that as a whole ranks with monera. Monera is now known to comprise two separate kingdoms, the eubacteria and the archaebacteria. Within the eukaryotes, the protists do not appear to form a phylogenetically distinct unit.

Two other features illustrated in Fig. 8 are particularly noteworthy. First, there is relatively little sequence divergence between animals (e.g., *Homo*) and plants (e.g., *Zea*). The eukaryotic sequence diversity is dominated by the protists, which are represented by *Dictyostellium* (cellular slime mold), *Trypanosoma* (trypanasoid flagellate), and *Euglena* (euglenoid flagellate).

Second, note that there are some surprising relationships found between organellar rRNAs and certain bacterial species. Among the eubacterial sequences the green plant chloroplast is specifically related to the photosynthetic eubacteria. Similarly, the plant mitochondrial 16S rRNA is affilitated with a purple bacterium *Agrobacterium tumefaciens.* An explanation for these similarities requires that we consider an atypical form of genetic information transfer.

Vertical and Horizontal Transfers of Genetic Information Both Have Had Major Influences on the Phylogenetic Tree

Since the time of Charles Darwin it has been conventional to think in terms of genetic transfer being confined within a species. Furthermore, all evolutionary diagrams suggest that evolution occurs by the *vertical transfer* of genetic information from one generation to the next. However, in the past 30 years if not longer there have been numerous indications that gene transfer could occur between species. Sequence studies leave little doubt that interspecies transfer of genetic information called *horizontal transfer* is a reality. In eukaryotes the notion of interspecies transfer, championed by Lynn Margulis in the 1960s, became accepted once it was found that the nucleic acid sequences of the mitochondrial and chloroplastic rRNAs were unmistakably related to those found in certain species of eubacteria. Mitochondrial transfer may have been a unique event, but chloroplast transfer is believed to have occurred several times. How these transfers occurred is uncertain but it has been speculated that phagocytosis of the prokaryotic organism by the larger eukaryotic organism was probably involved. Subsequently, the adopted prokaryotic systems were trimmed to create a specialized organelle for oxidative phosphorylation, on the one hand, and a specialized organelle for photosynthesis, on the other. Both of these organelles still carry many organelle-specific genes. Christian deDuve has speculated that peroxisomes were the first organelles to be adopted, but this is hard to prove because peroxisomes carry no genetic component to verify their origin.

There is a great deal of suggestive evidence to indicate that smaller gene segments have been horizontally transferred. A. Jeltsch and A. Pingoud have made a strong case for the horizontal transfer of DNA restriction modification enzymes between bacterial species. These genes in fact protect bacteria from invasion by foreign DNA. Bacteria usually do not phagocytose one another. For horizontal gene transfer there is a large family of small circular DNAs called plasmids that can be promiscuously transferred from one bacterium to another of a different species. In view of the existence of plasmids we can anticipate that horizontal gene transfer occurs with a relatively high frequency between bacteria.

The existence of the horizontal transfer of genetic information greatly complicates relatedness studies and always must be taken into account in considering evolutionary origins.

RECOMBINATION AND MUTATION, THE ULTIMATE SOURCES OF GENETIC VARIABILITY

Recombination and mutation are the agents of genetic change. In this section we consider the different types of recombination and mutation and the differences in the selection processes between *haploid* organisms that carry one gene for each characteristic and *diploid* organisms that carry two genes for each characteristic.

Reproduction in Diploids Leads to Recombination of Genetic Determinants

A *gene* is a hereditary determinant that specifies a certain trait or traits and that behaves as a unit of inheritance occupying one (usually) contiguous region of a chromosome. A gene usually carries the DNA reading frame encoding the information for the synthesis of a single polypeptide chain; it also carries whatever else is necessary for the gene to be expressed. In writing, genes are specified by one or more small letters in italic. Frequently, a superscript is appended to the genetic symbol. The two most common superscripts are a plus sign, indicating a normal gene, also called a *wild-type* gene; and a minus sign, indicating a *mutant* (often nonfunctioning) gene. Different representations of the same gene are referred to as *alleles*.

Bacteria usually carry a single chromosome with a unique representation of each gene. Eukaryotes can be either diploid or haploid. When they are diploid, they carry pairs of homologous chromosomes. Each chromosome of a homologous pair carries the same pattern of genes, but the genes may be represented by different alleles on the two chromosomes. If both representations are the same, the individual is said to be *homozygous* for the trait in question; if they are different, the individual is said to be *heterozygous*. In both haploids and diploids, the total gene content of the chromosomal DNA (disregarding copies on homologous chromosomes) is referred to as the *genome* of the organism.

In diploids, alleles interact in different ways with one another. A *dominant* allele is expressed in both homozygous and heterozygous individuals. A *recessive* allele is expressed only in homozygous individuals. Usually, but not always, a normal or wild-type allele is dominant to a mutant or defective allele. Put another way, the defective allele is recessive to the wild-type allele. The situation also arises in which two alleles are *codominant*, meaning that both are expressed. From this discussion, it should be clear that the pattern of gene expression does not necessarily reflect all the alleles present in the chromosomes of the individual. For this reason, we often need to distinguish between the *genotype*, the collection of gene representations carried by that individual and the *phenotype,* the observable traits displayed.

In sexual reproduction, which occurs only in diploids, the chromosomes of the parent organisms undergo an intricate process of replication and reshuffling, with an end product in the offspring being a new combination of genes, half of them inherited from each parent. For a given trait, it is a matter of chance which alleles an individual inherits. The genotype is a composite of the genotypes of the parents; the phenotype produced by this new combination depends on the interaction between the particular alleles that have been inherited. Because different offspring of the same parents can inherit different combinations of alleles, we see genetic variability even at the level of a single generation. That variability is not the result of any change in the genes themselves, and in a whole population of individuals the proportion of alleles present from generation to generation does not undergo radical change. To bring about changes in genes, there must be inheritable changes in the DNA sequences, that is, mutations.

Mutations Are Retained or Eliminated from the Population by the Process of Natural Selection

From the standpoint of evolution, the variation that arises by recombination is a device for testing different combinations of alleles. Recombination in this case occurs between genes. There are also types of recombination that occur within genes and that are difficult to distinguish from mutation because they often result in new genes. Intragenic recombination occurs between homologous genes much more frequently than between nonhomologous genes. In both cases the new alleles are tested in the evolutionary process of selection, but in the latter case it is difficult to distinguish recombination from mutation.

Mutations can involve single base changes, such as insertions or deletions of bases; or multiple changes, in the form of translocations, inversions, or duplications. All such changes contribute new alleles that can be tested for their survival value.

In this section we discuss the fate of new mutations and the way in which they are either adopted or eliminated by the process of natural selection. *Fitness* is a major factor in determining the future of a newly introduced mutation. By fitness we mean not merely the capacity to survive, but also the ability to reproduce.

In organisms that are haploid, it is easiest to predict the fate of an organism that bears a new mutation. If the mutation is beneficial, it is adopted within the species population at an exponential rate (Fig. 9). The steepness of the exponential curve is proportional to the relative fitness of the new mutant organism over preexisting organisms in the population. If the new mutation is lethal or merely detrimental, it is eliminated with great rapidity without a buildup in its frequency. If the new mutation is neither beneficial nor detrimental, that is, *neutral*, one cannot predict whether it is to become integrated into the species or eliminated; this is a matter of chance. Many analyses suggest that most mutations that become fixed in the population of both haploid and diploid organisms may be of the neutral type. This is probably because many more of the mutations that occur are of the neutral type than of the beneficial type.

The fate of new mutations in a diploid population is more complex. New alleles that are introduced as a result of mutation are usually adopted if they are favorable, whereas if they are unfavorable they are usually dropped. The extent to which an unfavorable mutant allele lingers in a population at equilibrium depends on whether the allele is dominant, recessive, or codominant. A recessive lethal at equilibrium is present at a concentration equal to the square root of the mutation rate. As a consequence, the homozygous frequency for a recessive lethal is equal to the mutation rate (Box 21A). This is usually a very small number.

If the new mutation is beneficial, the rate of adoption depends on whether the new mutation is dominant, codominant, or recessive (Fig. 10). The rate of adoption in all

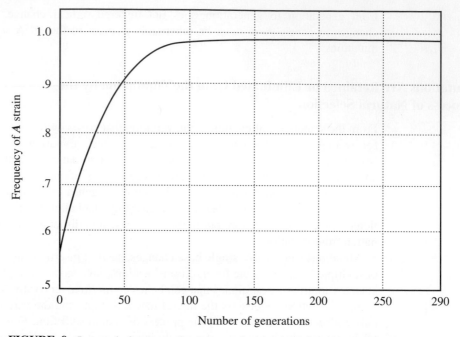

FIGURE 9 Increase in frequency of a mutant strain A bearing a favorable mutation in a haploid population.

cases is represented by a sigmoidal curve instead of an exponential curve as was seen for haploids. Moreover, the rate for dominant and codominant cases is roughly the same, but the time of adoption for a favorable recessive is greatly prolonged. This is hardly surprising, because at a low frequency the recessive mutation has very little

---

BOX 21A   Frequency of Homozygous Recessives

If q is the probability of a recessive allele in a diploid population, then the probability of getting a homozygous recessive is q^2. If the homozygous recessive is lethal, there is a simple relationship between the frequency of homozygous recessives and the mutation rate in a population that has reached genetic equilibrium. At equilibrium the rate of loss of recessive alleles must be equal to the rate of introduction of new recessive alleles. Hence at equilibrium

$$q^2 = \mu \quad \text{or} \quad q = \sqrt{\mu},$$

where μ is the mutation rate.

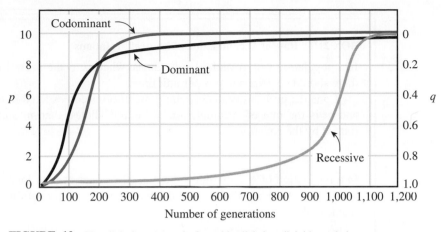

FIGURE 10 The allele frequency of a favorable allele in a diploid population.

impact on survival, as a result it is expressed only in the homozygous state, which is very infrequent. However, if the recessive mutation survives the crisis period at very low frequency, its concentration may rise to the point where homozygous recessives constitute a significant portion of the population. From this point on, its frequency should rise quite rapidly. It seems likely that if a mutation resulting in a beneficial recessive allele occurs at a reasonable frequency, eventually it becomes a major allele in the gene pool, unless there are other beneficial mutations arising simultaneously that are competitive with it.

The situation with neutral mutations, which are neither beneficial nor detrimental, is especially interesting because, as noted earlier, the vast majority of mutations that become fixed in the population are believed to be neutral mutations. It can be shown that the rate at which a particular gene mutates is equal to the mutation rate for a neutral mutation (Box 21B). Thus, in spite of the fact that many neutral mutations are discarded by chance, many others survive to become significant alleles. It seems likely that the majority of point mutations, those involving single nucleotide changes, are neutral. Indeed, that is the rationale for using such changes as a yardstick for biochemical time, as described earlier in this chapter.

Mutations That Arise Naturally Are Called Spontaneous Mutations

In the 19th century, Darwin proposed that populations within a given species include members with varying characteristics and that environmental conditions favor the survival and propagation of those variants most suited to the environment. An alternative explanation of evolution, proposed earlier by Lamarck, was that individuals within a population adapt to the environment and can pass on these adaptations to their progeny.

The vast majority of individuals carrying conspicuous mutations seem to be worse off than their more normal brethren are. This type of observation and Darwin's

---

BOX 21B   Rate of Accumulation of Neutral Mutations

In a population with N diploid individuals, where μ is the neutral mutation rate, there should be $2N\mu$ mutants per unit time.

If p is the probability that a neutral mutation becomes fixed in the population and m is the rate of accumulation of neutral alleles, then m and p should be related by the equation

$$m = 2N\mu p. \tag{1}$$

In a population of N individuals there are $2N$ genes at each locus. If the alleles are neutral, then all alleles should have an identical probability of becoming fixed, so that

$$p = \frac{1}{2N}. \tag{2}$$

If we insert this value for p into Eq. (1), we find that

$$m = 2N\mu\left(\frac{1}{2N}\right) = \mu. \tag{3}$$

Thus the rate of accumulation of neutral mutations should be equal to the rate at which the neutral mutation arises by the mutation process.

---

observations on natural selection persuaded most biologists to believe that mutation was a random process; the environment was viewed as favoring survival of rare mutant organisms that were more fit but as playing no role in the occurrence of such organisms. Ironically, the bacteriologists were the last to accept the randomness concept of mutation but the first to "prove it."

Following pioneering investigations by Luria and Delbruck, a most convincing demonstration of the Darwinian doctrine was made by the Lederbergs in 1952. They plated a large number of bacteria (10^7) on a nutrient agar plate. So many bacteria were spread on the plate that it was not possible to see bacterial colonies arising from individual bacterial cells. The seeded plate appeared as a uniform "lawn" of bacteria, but in reality it consisted of very small colonies that merged to give a uniform appearance. In the next step of the investigation, a piece of velvet was lightly pressed against the surface of this lawn, and some cells from each colony stuck to the velvet. Several identical impressions of the lawn were transferred to other plates, which contained the virus, by lightly pressing the velvet against the agar surface. Some virus-resistant colonies appeared on each of these transferred plates. Most revealing was the finding that the pattern of colonies that developed on all the impressed plates was very similar (Fig. 11). This result indicates that small virus-resistant colonies had been present

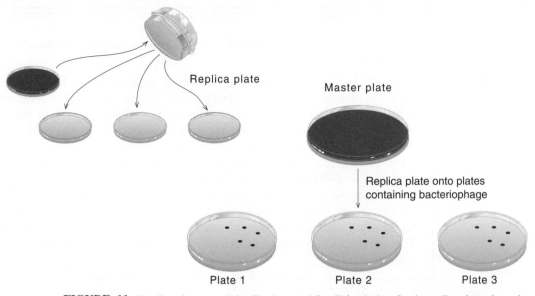

FIGURE 11 Replica plates containing T1 phage and five T1′ colonies. On the replica plates the resistant colonies are the same in number and the same in location. The master plate contains normal *E. coli* cells grown on medium free of phage. The replica plates contain the same medium but with phage. Insert (upper left) illustrates replica platting technique.

from the outset in the population of cells on the original plate that never had been exposed to the virus. Because the pattern of virus-resistant colonies was the same before and after exposure to the virus, the virus could not have been a causal agent in producing variants (however, see Box 21C). Instead, addition of the virus served as a selective agent, eliminating all other cells and permitting replication of the virus-resistant cells, which then developed into large, visible colonies.

Other selective agents or selective growth conditions have been used to demonstrate the general usefulness of this replica plating technique. A large population of seemingly homogeneous cells invariably contains variants that are similar in the vast majority of their genes but contain randomly distributed mutations in many if not most of the cells. Thus a large population contains all sorts of variants that can be selectively amplified by the appropriate environmental circumstances. The Lederbergs' procedure was useful both in a theoretical sense, for verifying the existence of mutants before exposure to the environmental stress factor, and in a practical sense, for the isolation of variants.

Spontaneous mutation rates, measured in mutations per genome per generation vary over several orders of magnitude, from about 2×10^{-4} to 4×10^{-10} (Table 1). The wide variation in frequency appears to be a function of both the organism involved and the type of mutation. For example, in *Escherichia coli* the mutation to streptomycin resistance occurs at a very low frequency (4×10^{-10}). The low mutation frequency is

---

BOX 21C   Reservations Concerning the Significance of the Luria–Delbruck–Lederberg Experiment

From the Luria and Delbruck experiment as modified by Lederberg, one is given the impression that the original bacteria used in the experiment are naive in that they have never been exposed to the T phage. However, since *E. coli* bacteria as a species have had to do battle with T phage throughout much of their history, they must have evolved a strategy for protecting themselves. One possibility is that they have a sequence that mutates with a moderately high frequency to a virus-resistant configuration so that some members of the population, be it ever so small, survive to repopulate the threatened colony. As far as I know this possibility has not been tested. An affirmative to this explanation would entail finding a sequence that always is changed in the same way when threatened and that reverts to the original sequence when the threat is not present.

---

partially due to the fact that the phenotype can result only from specific base pair changes in either of two ribosomal protein genes. On the other hand, mutations that cause a deficiency in the ability to ferment lactose occur at a much higher frequency (2×10^{-7}), because a change at many points within the gene for the lactose-degrading enzyme produces this deficiency. Mutations in corn appear to occur with a higher average frequency than those in *E. coli*. There is no simple explanation for such differences, because it is probably a function of many factors. However, it is generally true that mutation rates in prokaryotes are lower by about an order of magnitude than those in eukaryotes. This is true even if we confine the comparison to mutations per nucleotide pair per generation. In these units, prokaryotes measure from about 10^{-8} to 10^{-10}, whereas eukaryotes measure from about 10^{-7} to 10^{-9}.

Spontaneous mutations result from a variety of causes. There are mutagenic agents within the environment, such as cosmic radiation and mutation-causing chemicals. The cosmic rays that cause chromosomal alterations are mainly γ-rays. Most of the chemicals that are responsible for mutations are industrial pollutants in the air and water; some are intentional additives in foodstuffs. The chlorine deliberately added to drinking water may be the most serious mutagen to which the average individual is exposed.

Even in an environment free of mutagens there would be factors associated with the DNA metabolism that would result in mutations. Thus there is a small probability that the DNA replication enzyme adds an incorrect base to a growing chain. Even if the DNA repair system recognizes this mispaired region after replication, there is no assurance that the mispaired bases are to be repaired so that the original base pair is restored.

TABLE 1
Mutation Rates of Specific Genes in Various Organisms

Organism	Trait	Mutations per genome per generation
Bacteriophage T2 (virus)	Host range	3×10^{-9}
	Lysis inhibition	1×10^{-8}
Escherichia coli (bacterium)	Streptomycin resistance	4×10^{-10}
	Streptomycin dependence	1×10^{-9}
	Resistance to phage Tl	3×10^{-9}
	Lactose fermentation	2×10^{-7}
Salmonella typhimurium (bacterium)	Tryptophan independence	5×10^{-8}
Chlamydomonas reinhardi (alga)	Streptomycin resistance	1×10^{-6}
Neurospora crassa (fungus)	Adenine independence	4×10^{-8}
	Inositol independence	8×10^{-8}
Zea mays (corn)	Shrunken seeds	1×10^{-6}
	Purple seeds	1×10^{-5}
Drosophila melanogaster (fruit fly)	Electrophoretic variants	4×10^{-6}
	White eye	4×10^{-5}
	Yellow body	1×10^{-4}
Mus musculus (mouse)	Brown coat	8×10^{-6}
	Piebald coat	3×10^{-5}
Homo sapiens (human)	Huntington's chorea	1×10^{-6}
	Aniridia (absence of iris)	5×10^{-6}
	Retinoblastoma (tumor of retina)	1×10^{-5}
	Hemophilia A	3×10^{-5}
	Achondroplasia (dwarfness)	$4–8 \times 10^{-5}$
	Neurofibromatosis (tumor of nerve tissue)	2×10^{-4}

Perhaps the most common source of spontaneous mutations results from special DNA segments, sometimes called jumping genes of movable genetic elements, but better known as *transposable elements*. These genetic elements vary in size from a few hundred to several thousand base pairs. Each organism carries several different types of elements. The involvement of such elements in mutagenesis can be directly assessed by the presence of the element at the mutated site in the genome. Through limited studies of this sort, it has been estimated that in fruit flies about 80% of the so-called spontaneous mutations are the result of the insertion of transposable genetic elements. As we see in a later section, it seems highly likely that these elements have played a major role in the evolution of most species.

Induced Mutations Result from Deliberate Exposure to Mutagenic Agents

There are numerous mutagenic agents, some chemical, some physical, that can augment the mutation frequency by up to several orders of magnitude. The types of mutations produced and even the location of the mutation in many cases depend on the mutagenic agent. The use of mutagens to increase mutation frequency has been an invaluable tool for obtaining mutants useful in gene mapping and gene characterization.

Chemical mutagens can be classified according to the types of genetic changes they encourage. Some mutagens, such as chemically modified bases, encourage base replacements; others, such as the dye acridine orange, induce small deletions or additions. Mutagens with known properties have been very useful in basic genetic investigations, not just for increasing the frequency of mutations, but for obtaining mutants of the desired type.

When the goal is to produce a defined change at a specific genetic locus, it is usually most efficient to modify the gene *in vitro* and reinsert it into the organism. This approach to making mutagens is called *directed mutagenesis*. Discrete alterations can be made in a variety of ways on any DNA *in vitro* and the effect of such alterations can be subsequently tested *in vivo*. The techniques of DNA recombinant methodology are becoming increasingly sophisticated and are being applied to more and more species. We may be approaching the time when directed mutagenesis is to be a major factor in determining the course of evolution.

High Frequency of Transposable Genetic Elements Challenges Classical Evolutionary Theories

Thus far we have discussed evolution under the assumptions that mutation rates are fixed and that natural selection as Darwin proposed operates on variants already present in the population. The experiments of Luria and Delbruck and the Lederbergs, starting in the early 1940s, suggested that phage resistance of bacteria could be explained in this way. However, exposure to phage leads to sudden death for sensitive bacteria; this test of evolutionary mechanisms did not leave any opportunity for the bacteria to adapt to the environmental stress. Most stresses that encourage natural selection, such as nutritional deprivation, permit more time for the organisms to adapt or to mutate to another form.

For many test situations in *E. coli*, the gene for β-galactosidase has been used. The investigator starts with a variant gene encoding nonfunctional β-galactosidase and looks for revertants, that is, mutants that make functional enzymes. The tests are usually conducted in a selective medium. As a rule, the selective medium contains a small amount of glucose and a large amount of lactose as the sole carbon sources. The β-galactosidase enzyme is required only for the utilization of lactose. A bacterial strain that carries defective β-galactosidase grows for a short time, just until it uses up the glucose. If this strain is grown on plates containing the mixed carbon source, small colonies develop but stop growing after a day or so. Nevertheless, in time (2 days to 2 weeks), outgrowths of cells able to utilize the lactose appear. The time lag before the appearance of these lactose-utilizing subcolonies indicates that they did not exist in the original culture. Thus these cells must have arisen after exposure to the selective growth medium.

A refined analysis of this situation indicates that the condition of extreme nutritional deprivation greatly increases the mutation rate. The rate is stimulated in a random way so that mutants that can utilize the lactose are no more likely to appear than any other types of mutants. However, in the presence of lactose the mutants that can utilize the lactose start to grow very rapidly because they are the only ones that can utilize this carbon source. This finding leads to a completely different conclusion from that indicated by the work of Luria and Delbruck and the Lederbergs. It indicates that nutritional deprivation stimulates the rate of mutagenesis and that some of the mutants so formed are favored for growth on the selective medium.

What is the mechanism involved in this increased rate of mutagenesis? A complete answer is not possible at the present time. However, the discovery of *transposable genetic elements* in the 1970s suggested a mechanism for genetic change that had previously not been widely considered. Moreover, it was subsequently found that transposable genetic elements are hyperactive under conditions of metabolic stress, and thus may be the means for producing variants at a higher frequency after exposure to selective agents. It is possible that transposable genetic elements are major agents of evolution in many species. If this is the case, then we can see why mutation rates rise during times of stress and how a new mutant could arise in a stressful situation even in the absence of cell division. Such a possibility exists because transposable genetic elements do not require gross chromosomal replication to alter the genome.

EARTH: AN EVER-CHANGING ENVIRONMENT

Recombination and mutation supply the genetic variability; the environment determines which variants flourish and which do not survive. There is no such thing as equilibrium of the environment. The environment is constantly changing. Sometimes these changes are fast, as when an asteroid hits Earth; sometimes they are slow, as in the case of oxygen buildup or continental movements. Changes in the environment have an appreciable effect on the distribution of alleles in a steady-state population. A mutation that is neutral or even detrimental under one set of circumstances can become beneficial if the environment changes in a certain direction. Similarly, a mutation that is beneficial can have a detrimental effect if the environment changes.

Earth has provided an ever-changing environment for the evolution of life. Some environmental changes are the result of geologic or extraterrestrial factors, while others, such as alterations in the composition of the atmosphere, have been caused by living organisms. Figure 12 surveys some of the major environmental changes that have had a major impact on biological evolution.

Continental Movements Slowly Change the Local Environment

The crust of Earth consists of 13 large plates that move in different directions over its surface (see Chapter 5, Fig. 4). These plates move at a rate of 1 to 6 cm/year, drifting apart, coming together, or sliding past one another.

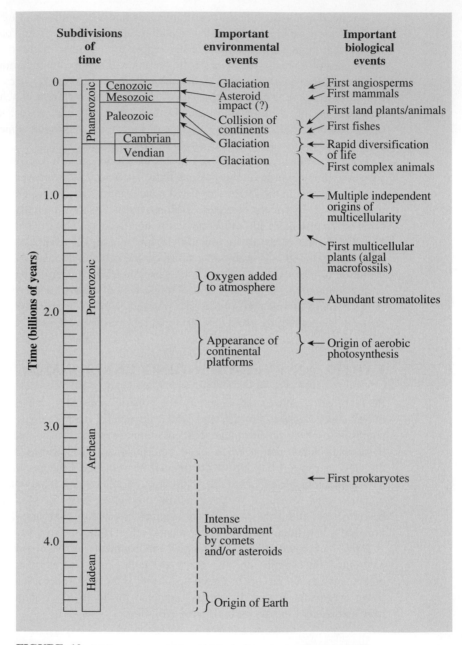

FIGURE 12   Major environmental and biological events over the past 4.6 billion years.

When plates drift apart, molten rock from below the crust flows up to fill the space between them, producing a ridge of new crust. Such ridges are present only in ocean basins and are called midocean ridges.

When two plates come together, one of them slides beneath the other, and a trench is formed on the bottom where this descent occurs. The plate that slides down into Earth's mantle melts, along with its load of sediment. Some of this material rises again as molten rock and erupts in volcanoes at the surface. Along the western coast of South America, where the Eastern Pacific plate is sliding under the American plate, there are volcanoes just inland, and a deep trench just offshore. If the plates moving toward each other carry continents, the continents eventually collide and form a larger continent. The sediment that has accumulated near the margins of each plate is squeezed and pushed up into a mountain range. The Himalayas were formed this way when the Indian subcontinent collided with Asia.

When two plates slide past each other, a faulting results in earthquakes and volcanism; this has occurred along the western coast of North America, where the Pacific plate is moving northward past the American plate. The relative positions of the continents have been changing since the early Precambrian period.

About 500 million years ago continents were isolated blocks scattered around the equator (Fig. 13), and the major ocean basins were centered at the poles. During the next 250 million years these isolated continents came into contact to form a single supercontinent, Pangaea. Pangaea persisted for approximately 100 million years and then began to fragment. Our modern geography, characterized by three north–south-oriented continental masses (the Americas, Europe–Africa, and Pacific and Indian Oceans), has developed in the past 160 million years.

Geographic changes resulting from plate movements have played a major role in evolution. Distinctive species such as the eucalyptus tree and the marsupials of Australia resulted. In general, dispersion leads to an increase in the number of species. Conversely, when isolation is reduced, the number of species is usually lowered. Under such conditions organisms find themselves in competition with species that they had never met before, and competitive elimination occurs between species in close proximity. This happened when the Panamanian land bridge developed between North and South America in Pliocene time, approximately 7 million years ago; the total family level of diversity for both continents combined is estimated to have dropped by about 20%.

In the latter half of the Palezoic era, when widely separated continents began to converge to form Pangaea, the sea levels fell drastically. This produced a drainage of shallow continental shelves, loss of coastlines due to the clumping on continents, and substantial climatic changes resulting from altered oceanic current patterns. Such dramatic changes in the environment were probably the main cause of the Permian mass extinction (see Fig. 5). The loss of shallow marine habitats essential to marine organisms resulted in the extinction of a large number of groups.

The breakup of Pangaea is well documented by fossil remains. Both animals and plants with very little ability to migrate by air or water have been found on continents that are presently separated by large bodies of water (Fig. 14). For example, fossil

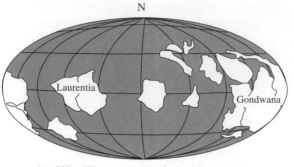

(a) 500 million years ago (Ordovician period)

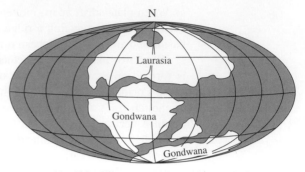

(d) 180 million years ago (Jurassic period)

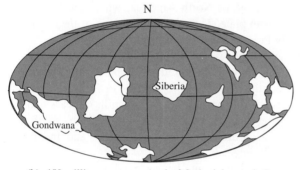

(b) 450 million years ago (end of Ordovician period)

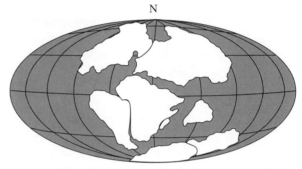

(e) 135 million years ago (end of Jurassic period)

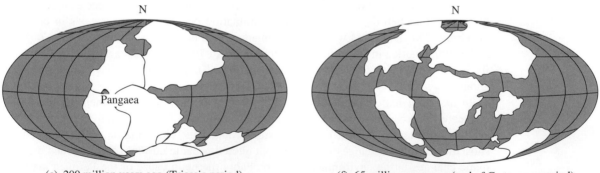

(c) 200 million years ago (Triassic period)

(f) 65 million years ago (end of Cretaceous period)

FIGURE 13 Continental drift over the past 500 million years (myr): (a) 500, (b) 450, (c) 200, (d) 180, (e) 135, and (f) 65 myr ago.

remains of Cynognathus, a Triassic land reptile, have been found both in South America and in Africa. Similarly, fossil remains of the Triassic land reptile Lystrosaurus have been found in Africa, Antarctica, and India; and fossils of the seed fern *Glossopteris* have been found in South America, Africa, Australia, India, and Antarctica.

Continental drift has had a major impact on the climate of different regions. Many sediments deposited in earlier times reflect climates that are inappropriate for the latitudes in which these sediments are located today. For example, late Ordovician glacial deposits occur in North Africa, tropical reefs occur in Silurian and Devonian deposits of the American midcontinent, great tropical coal swamps of the Carboniferous period occur in North America and Europe, Permian desert deposits are found in Germany, and Jurassic desert dune sands are found in Utah. These deposits were formed when the tectonic plates containing the land masses moved through latitudinal zones where the appropriate climatic conditions prevailed. The regional climatic changes, together with the past positions of the continents, affected oceanic circulation patterns and controlled animal and plant distributions as well.

Asteroid Bombardments Can Result in Sudden Traumatic Changes in the Environment

Bombardment of Earth by asteroids was far more common in Earth's early history than it is today. Only a few asteroids now remain in orbits crossing that of Earth, and the probability of further collisions in the future is lower but still finite. Walter Alvarez proposed that a sizable asteroid collided with Earth about 65 million years ago. He based this proposal on the observation that a thin stratum of rock formed precisely at the boundary between the Cretaceous and Tertiary periods. This stratum contains high concentrations of the element iridium, which is rare on Earth's surface but is far more common in asteroids. The thin worldwide layer of iridium-rich rock may have formed from the dust that settled following the collision. A collision of sufficient magnitude could have created a dust cloud that would have blocked the Sun for a period of months or years and drastically reduced temperatures. It is speculated that this asteroid bombardment was the event that triggered the Cretaceous extinction.

Increasing Oxygen Levels Created a Hospitable Surface Environment for Aerobic Organisms

At present, oxygen constitutes about 21% of the atmosphere. Additions to atmospheric oxygen occur primarily from the photosynthetic activities of plants. Photosynthesis effectively splits water into two hydrogen atoms, which enter the biosphere as newly produced organic matter, and an oxygen atom is combined with another and released to the hydrosphere or atmosphere as molecular oxygen. In the process, carbon dioxide also is incorporated into organic matter. These oxygen additions to the atmospheric, hydrospheric, and organic reservoirs are balanced by oxygen consumption resulting from respiration of organisms, combustion of organic matter, and rock weathering.

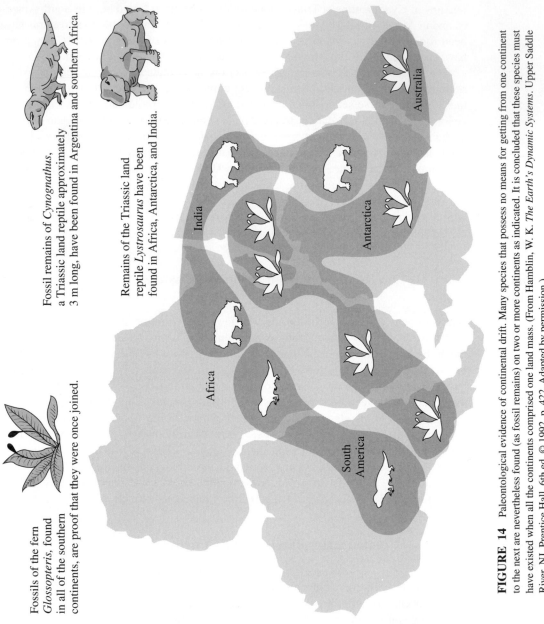

Fossils of the fern *Glossopteris*, found in all of the southern continents, are proof that they were once joined.

Fossil remains of *Cynognathus*, a Triassic land reptile approximately 3 m long, have been found in Argentina and southern Africa.

Remains of the Triassic land reptile *Lystrosaurus* have been found in Africa, Antarctica, and India.

India

Africa

South America

Antarctica

Australia

FIGURE 14 Paleontological evidence of continental drift. Many species that possess no means for getting from one continent to the next are nevertheless found (as fossil remains) on two or more continents as indicated. It is concluded that these species must have existed when all the continents comprised one land mass. (From Hamblin, W. K. *The Earth's Dynamic Systems*. Upper Saddle River, NJ. Prentice Hall. 6th ed. © 1992, p. 422. Adapted by permission.)

Banded iron formations, found in discrete sedimentary layers (see Fig. 4), give us the best measure of the periods when the oxygen content of the atmosphere was on the rapid rise. This is because banded iron formations most likely resulted from the reaction of molecular oxygen with water-soluble ferrous iron to form water-insoluble ferric iron compounds, which precipitated to produce massive, distinctive bands. Geologists have dated the major banded iron formations to the period from about 2.5 to 1.8 billion years ago. Prior to this time the atmospheric content of oxygen was very low. Most of the oxygen produced during the initial periods after photosynthesis got started was taken up in oxidation reactions, mainly with atmospheric hydrogen, or ferrous ion in the ocean or upper mantle. By 1.8 billion years ago the atmospheric content of oxygen had reached about 20% of its current level.

Most of the organisms that thrived in the absence of oxygen were killed by the oxygen. Some, however, evolved to oxygen-resistant species, or they withdrew to ecological niches that were relatively free of atmospheric oxygen. Finally, some evolved the ability to take advantage of oxygen in aerobic metabolism. Some prokaryotes appear to have developed the capacity for aerobic oxidations and passed it on to eukaryotes by horizontal evolution as indicated earlier in this chapter.

The impact of oxygen on evolution also was felt in another way. As the oxygen content of the atmosphere grew, so did the ozone content of the upper atmosphere. This ozone layer shields Earth's surface from the damaging effects of ultraviolet (UV) radiation. Were it not for the ozone layer, no life as we know it would be possible on the land surfaces, so this indirect effect of atmospheric oxygen had a major effect on increasing the number of land-based organisms.

Changes in Atmospheric Carbon Dioxide Levels Could Have a Significant Environmental Impact

Carbon dioxide serves as the source of carbon for all organisms, even though the direct acquisition of carbon from carbon dioxide is restricted to a limited number of producer organisms. Carbon dioxide is transferred from Earth's mantle and crust to the atmosphere and hydrosphere by outgassing through volcanoes and steam vents. It returns to solid Earth, first by conversion to bicarbonate in solution and then by precipitation as calcium carbonate in the hydrosphere. Within this slow cycle runs a much faster one, in which carbon is incorporated into the living biomass by photosynthesis and is returned to the atmosphere–hydrosphere system by the natural decomposition of organisms, by respiration of organisms, and by fires. Some of the reduced carbon is returned to the solid crust in the form of organic matter. Changes in the rates of global volcanism and weathering must be reflected by changes in carbon dioxide levels in the atmosphere and the hydrosphere.

Most plants operate optimally at much higher carbon dioxide pressures than we have in the present atmosphere. They would cease to function, however, if the carbon dioxide pressure were to drop to a quarter of its current value. The present terrestrial plants are therefore in the lower part of what is considered to be their "carbon dioxide stability field."

Analyses of gases from ice formed in the Wisconsin glacial interval suggest that carbon dioxide levels were substantially below the present levels in the last ice age. This lower level approaches the lower limit of tolerance of higher plants. It is not known whether life on Earth has been limited at times by low availability of atmospheric carbon dioxide.

Higher carbon dioxide levels can influence climates by enhancing the greenhouse effect (see Chapter 5). It has been estimated that a doubling of the atmospheric carbon dioxide content would raise the global mean annual temperature by several degrees Celsius. This effect would not be globally uniform, but would be much smaller in the tropics (where heat is largely converted into water vapor) and much larger at higher latitudes (where increased rainfall would release much latent energy).

SUMMARY

In this chapter we have discussed the main ways in which we measure evolutionary change, and the major environmental factors that influence extinction and emergence of new species.

1. The classical evolutionary tree is based on morphology. It relies heavily on the fossil record and on being able to date fossils by dating the sedimentary layers in which they are found. Radioactive isotopes with well-defined half-lives have provided a most reliable, objective way of dating such sedimentary layers.

2. Evolution does not appear to have been a steady gradual process. Instead, it appears that major steps in evolution have been coincident with periods of mass extinctions. During these periods abrupt environmental changes appear to have resulted in the disappearance of many species and the emergence of new ones.

3. The biochemical record gives us a new way of determining the relatedness of different organisms. Sequence comparisons between related proteins and related RNA molecules have led to a somewhat different view of the evolutionary tree. Although biochemical comparisons are mostly limited to the study of existing organisms, they provide an objective criterion for relatedness. They are especially useful in studying microorganisms, where morphological comparisons are virtually impossible.

4. Mutations provide examples of genetic change on which evolutionary pressures can act. The fate of a mutation depends on the type of organism, diploid or haploid; and on whether, in terms of fitness, the mutation is beneficial, detrimental, or neutral. In diploid organisms, it also depends on whether the mutant allele is dominant or recessive. Detrimental mutations are rapidly eliminated in haploid organisms, whereas beneficial mutations are usually rapidly adopted and neutral mutations may or may not be adopted. In diploid organisms, the situation is somewhat different. Dominant detrimental mutations are virtually eliminated. Dominant beneficial mutations usually are adopted. Recessive detrimental mutations never are eliminated totally. Instead, they are kept at a low level. Recessive beneficial mutations frequently are adopted, although it may take considerable time because of the small impact they have when they are present at a low frequency.

5. Numerous environmental factors have influenced selection during the history of Earth. Among the most important factors we have considered are continental movements, sea level fluctuations and changes in habitat area, regional climatic changes, topographic changes, asteroid bombardments, and changes in atmospheric oxygen and carbon dioxide levels.

Problems

1. What is the difference between vertical and horizontal evolution?
2. If two isolated islands suddenly were connected by an isthmus, would the number of species be expected to increase or decrease? Explain.
3. When did oxygen become abundant in the atmosphere? What is the evidence?
4. What is the evidence for the first prokaryotes? What is the evidence for the first eukaryotes?
5. How does directed mutagenesis relate to the usual Darwinian selection process as a vehicle for evolution?

References

Darwin, C. *On the Origin of Species by Means of Natural Selection.* London: John Murray, 1859.

deDuve, C. The Birth of Complex Cells, *Sci. Am.* April: 50–57, 1996. (Humans, together with all other animals, plants, and fungi, owe their existence to the momentous transformation of tiny, primitive bacteria into large, intricately organized cells.)

Gupta, R. S. and Golding, G. B. The Origin of the Eukaryotic Cell, *TIBS* 21:166, 1996. (The eukaryotic cell nucleus is a chimera that received major contributions from a Gram-negative eubacteria and an archaebacterium.)

Hamblin, W. K. *The Earth's Dynamic Systems,* 5th ed. New York: Macmillan, 1985.

Jeltsch, A. and Pingoud, A. Horizontal Gene Transfer Contributed to the Wide Distribution and Evolution of Type II Restriction-Modification Systems, *J. Mol. Evol.* 42:91–96, 1996.

Kimura, M. *The Neutral Theory of Molecular Evolution.* Cambridge: Cambridge University Press, 1983.

Knoll, A. H., Bambach, R. K., Canfield, D. E., and Grotzinger, J. P. *Science* 273:452–457, 1996.

Lang, B. F. *et al.* An Ancestral Mitochondrial DNA Resembling a Eubacterial Genome in Miniature, *Nature (London)* 387:493–4997, 1997.

Martin, W. and Muller, M. *Nature (London)* 392:37–41, 1998. The hydrogen hypothesis for the first eukaryote.

Pace, N. R. A Molecular View of Microbial Diversity and the Biosphere, *Science* 276:734–739, 1997.

Schidlowsky, M. A 3,800-Million Year Isotopic Record of Life from Carbon in Sedimentary Rocks, *Nature (London)* 333:313, 1996.

Sueoka, N. Directional Mutation Pressure: Mutator mutations and dynamics, *J. Mol. Evol.* 37:137–153, 1993.

Tattersall, I. Out of Africa Again...and Again? Africa Is the Birthplace of Humanity, *Sci. Am.* April: 60–67, 1997.

Woese, C. R. Bacterial evolution, *Microbiol. Rev.* 51:221–271, 1987.

Woese, C. R. and Pace, N. R. Probing RNA Structure, Function and History by Comparative Analysis. In R. F. Gesteland and J. R. Atkins (Eds.) *The RNA World.* Cold Spring Harbor, NY: Cold Spring Harbor Laboratory Press, 1993.

CHAPTER 22

Evolution of the Main Energy-Producing Pathway for Aerobic Metabolism

The Tricarboxylic Acid Cycle

In its infancy Earth had almost no molecular oxygen in its atmosphere. About 2.5 billion years ago photosynthesis began to utilize water as a source of electrons; molecular oxygen was a by-product of this process (see Chapter 23). When it first appeared, atmospheric oxygen was harmful to most living organisms because of its oxidizing power. Many organisms succumbed to the toxic effects of oxygen. Others adjusted by developing a protective system or by retreating to subterranean locations. A few not only adjusted, but learned to use the oxygen in reactions that supply chemical energy to cells. In this chapter we consider the ways in which some species learned to take advantage of the oxygen. Consistent with the typical stepwise advancement of the evolutionary process, the learning process did not require an entirely new system of enzymes, but instead the addition of one or two enzymes to an ongoing system. Clearly, with time the oxygen-energy-producing system underwent many additional changes that led to an improved efficiency.

CARBOHYDRATE METABOLISM IN ANAEROBES AND AEROBES

Whatever the source of energy of materials, the immediate chemical agent used to supply energy for specific chemical reactions in all organisms is adenosine triphosphate (ATP). In the ATP–adenosine diphosphate (ADP) cycle (see Chapter 7), an input of chemical energy drives the linkage of ADP and a phosphate group into ATP; then the ATP donates the phosphate group elsewhere, leaving ADP to be recharged (Fig. 1). Carbohydrates are directly linked to energy-producing pathways in all heterotrophs and many autotrophs (see Chapter 7, Table 1). In this chapter we focus on the evolution of carbohydrate utilization in energy-producing pathways; however, before we discuss the evolutionary aspects of these pathways we review how the pathways work in contemporary organisms. For a fuller discussion of this subject Chapter 11 should be consulted.

Glycolytic Pathway Is Used by Both Anaerobes and Aerobes

The process of glucose catabolism involves a reaction sequence that is the most ubiquitous pathway in all energy metabolism (see Chapter 11, Fig. 1). This pathway, known as the glucolytic pathway, occurs in the majority of living cells. Glycolysis is generally regarded as a primitive process because it is thought to have arisen early in biological history, before oxygen was a prominent component in the atmosphere.

The initial steps in glycolysis are energy-requiring, necessitating the input of ATP. First, glucose receives a phosphate group from ATP and undergoes internal rearrangements; then another phosphate group from another ATP is donated to the process. The resulting intermediate is split into two different phosphorylated three-carbon molecules. Each of these molecules becomes the substrate of an enzyme that transfers some

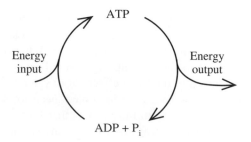

FIGURE 1 The biochemical energy cycle involving ATP. Energy input drives the formation of ATP from ADP + P_i. The ATP donates the phosphate to an energy-requiring process, leaving ADP to be recharged.

of its electrons to the coenzyme nicotinamide adenine dinucleotide (NAD^+). The enzyme-bound intermediate now combines with inorganic phosphate (P_i) to form glycerate-1,3-bisphosphate, which donates one of its phosphate groups to ADP. The resulting ATP is thereby formed by the direct, enzyme-mediated transfer of a phosphate group from a substrate to ADP.

Following replacement of the three-phosphate group by a two-phosphate group, a hydrogen atom and a hydroxyl group are stripped from the intermediate, leaving an internally unstable molecule called phosphoenolpyruvate (PEP). Its bonds shift and its electrons become redistributed in ways that increase the amount of energy available for the next reaction. That energy is enough to drive the transfer of a phosphate group from PEP to ADP. After this second substrate-level phosphorylation, the molecule remaining is pyruvate.

Glycolysis takes place in both anaerobes and aerobes. However, the fate of pyruvate produced from glycolysis is very different in the two types of organisms. Whatever route is traversed by the pyruvate, the reduced NAD (NADH) produced in glycolysis must be converted back to NAD^+ if glycolysis is to continue. In some organisms, the final electron acceptor is actually an intermediate or product of the glucose molecule. For example, in yeast, the pyruvate resulting from glycolysis is broken down to acetaldehyde, which by accepting electrons from NADH, converts it to ethanol. In some animal cells that normally rely on aerobic respiration, the pyruvate can be converted into the three-carbon compound lactate under anoxic conditions (see Chapter 11, Fig. 1). In both of these cases the NADH is converted back to NAD^+, permitting glycolysis to continue.

Tricarboxylic Acid Cycle Is Used Only by Aerobes

In aerobes, oxygen is involved in carbohydrate metabolism. Aerobic respiration is an oxygen-dependent pathway; in eukaryotes, it takes place in mitochondria, while in prokaryotes, it occurs on or in the vicinity of the cytoplasmic membrane. The pyruvate resulting from glycolysis is completely dismantled in this pathway. Its carbon and oxygen atoms are ultimately converted to carbon dioxide and water molecules, respectively. Its hydrogen atoms and electrons are transferred to NAD^+ and flavin-adenine dinucleotide (FAD), which in turn transfer them to an electron-transport system. The transport system finally surrenders the electrons to oxygen atoms (which combine with hydrogen to form water). Operation of the transport system leads to the formation of many ATP molecules. From start to finish the aerobic pathway typically has a net yield of around 30 molecules of ATP per molecule of glucose degraded. An overview of the energy-producing pathways of carbohydrate metabolism is shown in Fig. 2.

The cyclic process whereby pyruvate is broken down begins with acetyl-coenzyme A (acetyl-CoA), which is obtained either by oxidative decarboxylation of the pyruvate made available by glycolysis or by the oxidative cleavage of fatty acids (see Chapter 11, Fig. 4). Regardless of its source, the acetyl-CoA transfers its acetyl group to a four-carbon acceptor, oxaloacetate, thereby generating citrate, a six-carbon compound. The citrate is subjected to two successive decarboxylations and several oxidative

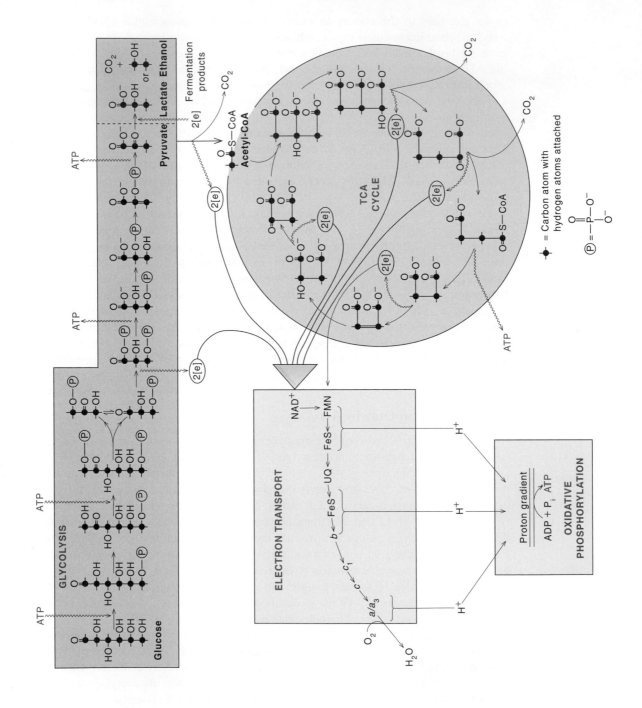

events, leaving a four-carbon compound from which the starting oxaloacetate moiety is eventually regenerated. This cyclic process is known as the tricarboxylic acid (TCA) cycle, or as the Krebs cycle (after its discoverer). Each turn of the cycle involves the entry of two carbons from acetyl-CoA and the release of two carbons as carbon dioxide, thus providing for regeneration of the oxaloacetate with which the cycle begins. As a result, the cycle is balanced with respect to carbon flow and functions without the net consumption or buildup of oxaloacetate or any other intermediate, unless side reactions occur that either feed carbon into the cycle or drain it off into alternative pathways.

Almost all the ATP-yielding reactions in aerobic metabolism result from reactions that are coupled with the electron-transport reactions. The electron-transport phase is catalyzed by enzymes and other proteins embedded in the inner membrane of the mitochondrion in eukaryotes or in the cytoplasmic membrane in bacteria (see Chapter 11, Fig. 6). These protein enzymes are organized into two types of reaction sites: (1) an electron-transport system that establishes concentration and electrical gradients across the membrane; (2) an ATP–synthase system that phosphorylates ADP to form ATP. The reactions begin in the inner compartment of the mitochondrion, where NADH and reduced FAD ($FADH_2$) formed during the action of the TCA cycle transfer hydrogen and electrons to an electron-transport system. At some transfer points in this system, the electrons are accepted and passed on, but the hydrogen (in the form of H^+) is left behind in the outer compartment. The protons accumulate in the outer compartment, and this establishes concentration and electrical gradients across the membrane. The ions follow the gradients and flow inward, through the ATP–synthase system that serves as a channel across the membrane. The energy associated with this flow drives ATP formation.

MANY ORGANISMS USE PARTS OF THE KREBS CYCLE

Life existed long before there was more than a trace oxygen in Earth's atmosphere. Most of the molecular oxygen in Earth's atmosphere arose from photosynthesis. Oxygenic photosynthesis began about 2.8 billion years ago, but molecular oxygen did not appear in the atmosphere in significant quantities until about 2.2 billion years ago (Fig. 3). The 0.6-billion-year lag between oxygen production by photosynthetic organisms and its significant accumulation in the atmosphere is attributed to scavenging of the oxygen by an enormous excess of reduced compounds in Earth's crust, most

FIGURE 2   The components of respiratory metabolism include glycolysis, tricarboxylic acid (TCA) cycle, electron-transport chain, and oxidative phosphorylation of ADP to ATP. Glycolysis converts glucose to pyruvate; the TCA cycle fully oxidizes the pyruvate (by means of acetyl-coenzyme A) to carbon dioxide by transferring electrons stepwise to coenzymes; the electron-transport chain reoxidizes the coenzymes at the expense of molecular oxygen; and the energy of coenzyme oxidation is conserved in the form of a transmembrane proton gradient that is then used to generate ATP.

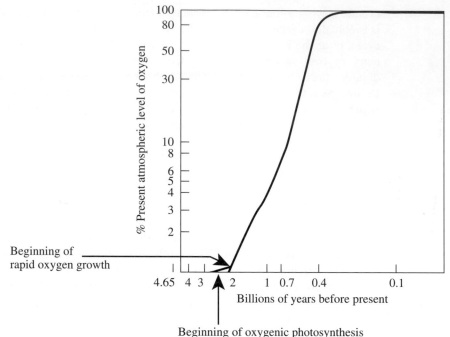

FIGURE 3   History of atmospheric oxygen level on Earth.

notably ferrous iron. When large-scale scavenging came to an end, the oxygen level began to rise. Around 0.8 billion years ago the oxygen content of the atmosphere had reached 5% of the present level. Because oxygen is not required for glycolysis but is required for aerobic respiration, it is hardly surprising that glycolysis is believed to have been the more primitive form of carbohydrate metabolism. Not only the reactions of the TCA cycle and coupled electron transport are far more complex, but also they have an absolute requirement for atmospheric oxygen as the ultimate electron acceptor (see Fig. 2; and Chapter 11, Fig. 6). Once oxygen was abundant, aerobic respiration became prominent because it is much more efficient in producing chemical energy than anaerobic metabolism is.

There are strong indications that the TCA cycle did not spring forth suddenly as oxygen became available and aerobic respiration became possible. Instead, most of the intermediates and most, if not all, of the enzymes were already in use, and relatively small changes were necessary to adapt the collection of reactions of function in a cyclic manner in aerobic metabolism. This becomes apparent when we examine organisms that do not use the TCA cycle in the normal way or do not use it at all. The development of the TCA cycle reflects the conservative and opportunistic nature of biochemical evolution. If a particular metabolite or chemical reaction step is capable of performing an essential function, organisms are likely to retain it, adaptive and refining it to suit many diverse situations, instead of developing a new solution from scratch for each novel situation. A consequence of this is that the anatomic and physiological features of present life forms are not necessarily optimum solutions for the present situation.

In general, biochemical evolution does not provide optimum solutions. The systems it provides are very efficient, but they have been strongly influenced by evolutionary events starting from the origin of life. For this reason studying the history of biochemical evolution can give us important leads, not only to the evolution of living organisms but also to the origin of life itself.

A most effective way of studying the history of biochemical evolution is by a close examination of comparative biochemical pathways of organisms that occupy different branches of the evolutionary tree. We use this approach to argue that most elements of the aerobic pathway of energy production existed before there was such a pathway.

Some Anaerobes Make Extensive Use of the Sequence between Oxaloacetate and Succinate

From the oxidation–reduction history of Earth and the microfossil record, we can confidently assert that the earliest successful cell lines must have obtained energy by anaerobic mechanisms. The comparative biochemistry of prokaryotes and eukaryotes, as well as the central roles of sugars in the structures, and functions of all cells, strongly suggests that sugar fermentation was the process at the heart of the primeval bioenergetic system. Why then are so many complicated fermentation patterns observed in contemporary prokaryotic anaerobes? They apparently represent alternative evolutionary "solutions" for achieving oxidation–reduction (redox) balance.

In simple anaerobic glycolysis, oxidation and reduction are balanced internally, that is, by reduction of the pyruvate generated from sugar cleavage (see Chapter 11, Fig. 1). The disposal of electrons from NADH can be seen as a crucial feature in the sugar metabolism of growing fermentative anaerobes. If this is obligatorily coupled to pyruvate reduction, the pyruvate becomes unavailable for biosynthesis. Howard Gest proposed that before the TCA cycle existed, an important step in metabolic evolution consisted of alterations that improved efficiency by minimizing the use of pyruvate simply as an electron sink.

It is reasonable to assume that biological energy conversion mechanisms evolved from relatively inefficient anaerobic fermentations to more efficient processes, and that this occurred through a lengthy sequence of gradual improvements that were integrated with the evolution of essential biosynthetic processes. This notion is the basis for the suggestion that the earliest fermentations were internally balanced with respect to oxidation and reduction, and became altered through the employment of "accessory" oxidants, thus sparing pyruvate (or PEP) for biosynthetic use.

Carbon dioxide was probably present in the early atmosphere at a considerably higher concentration than it currently is; this and other considerations make it a prime candidate for exploitation as an accessory oxidant. Thus the carboxylation of pyruvate (or PEP) can supply four-carbon dicarboxylic acids that function as electron sinks as follows:

$$CO_2 + \text{pyruvate (or PEP)} \rightarrow \text{oxaloacetate} \xrightarrow{2\,e^-} \text{malate} \rightarrow \text{fumarate} \xrightarrow{2\,e^-} \text{succinate}.$$

This sequence can fulfill the redox requirements of glucose fermentation, using only one of the pyruvates derived from cleavage of a six-carbon sugar while leaving the other pyruvate available for synthesis of additional compounds. Use of the four-carbon sequence as an electron sink in fermentative metabolism is found in many prokayotes and in many invertebrates that can survive long periods in anaerobic habitats. Examples of bacterial fermentations in which succinate is formed as a major end product are given in Table 1.

The abundance of fermentations in which succinate accumulates suggests that the reductive four-carbon sequence was employed as an early redox balancing device. When Krebs formulated the TCA cycle, and for some time thereafter, the reductive conversion of fumarate to succinate was believed to be catalyzed by succinate dehydrogenase acting in reverse. In 1957, however, the existence of an essentially unidirectional "fumarate reductase" was recognized in studies with the strict anaerobe *Veillonella alcalescens*. Extracts of this bacterium were shown to contain a soluble, flavin-linked fumarate reductase activity that clearly could not be ascribed to conventional succinate dehydrogenase acting in the "reverse" (reductive) direction. Accordingly, it was suggested that the reductase was similar to a succinate dehydrogenase, but represented a catalyst modified to meet the physiological requirements of anaerobic growth.

In early fermentation systems in which fumarate was used as an accessory oxidant, the electron-transfer catalysts were soluble (cytoplasmic) and served only to ensure

TABLE 1
Examples of Glucose Fermentations in Which Succinate is a Major Product[a]

Products	mmol/100 mmol glucose fermented			
	E. coli	*Aerobacter*[b] *acrogenes*	*Serratia marcescens*	*Erwinia carotovora*
2,3-Butanediol	0.3	19.2	52.7	15.0
Acetoin	0.059	—	7.99	—
Glycerol	1.42	4.2	1.29	—
Ethanol	49.8	51.5	29.6	66.2
Formic acid	2.43	68.4	3.32	134.0
Acetic acid	36.5	51.9	8.69	64.3
Lactic acid	79.5	10.1	20.9	23.1
Succinic acid	10.7	13.1	8.84	11.2
Carbon dioxide	88.0	79.6	158.9	13.2
Hydrogen (H_3)	75.0	—	0.22	—

[a]From Gest, H. Evolutionary Roots of the Citric Acid Cycle in Prokaryotes, *Biochem. Soc. Symp.* 53:3–16, 1987. Reprinted by permission.

[b]Now known as *Enterobacter aerogenes*.

redox balance (certain inorganic compounds could have functioned in similar fashion). A number of prokaryotic systems show the capacity to produce one ATP per NADH by a complex electron-transfer sequence that starts with the oxidation of NADH and ends with the reduction of fumarate. Fumarate reductase and the other catalysts of such sequences are usually membrane-bound. This suggests that the original soluble cytoplasmic fumarate reduction system may have evolved by addition of intermediary electron carriers and incorporation of most of the system into the cytoplasmic membrane. The idea that reduction of oxaloacetate to succinate was first used as an improved means of achieving redox balance in sugar fermentation has been elaborated by Wilson and Lin into a more detailed scheme for the role of fumarate reductase in the evolution of a membrane-bound redox–proton pump. In any event, various organisms that obtain the bulk of their ATP by fermentation (substrate-level phosphorylation) also can generate additional ATP by anaerobic electrophosphorylation associated with operation of a TCA cycle sequence that proceeds in the reductive direction. The reductive four-carbon pathway also can fulfill a biosynthetic purpose in furnishing succinate for conversion to succinyl-coenzyme A (CoA); the latter is a precursor of metalloporphyrins (see Chapter 23). Thus it seems likely that an early root of the TCA cycle was the reductive four-carbon electron sink pathway that emerged as an adjunct to hexose fermentation and also provided a useful biosynthetic intermediate.

Alternative Routes to Glutamate Involve Reactions of the Tricarboxylic Acid Cycle

The starting metabolites for the synthesis of many amino acids are carbohydrates that are intermediates in the glycolytic pathway and the TCA cycle. The TCA cycle intermediate α-ketoglutarate is the starting metabolite for glutamate synthesis (see Chapter 15, Fig. 1). In fact, glutamate is made in one step from α-ketoglutarate by a transamination reaction. Because α-ketoglutarate is an intermediate in the normal oxidative TCA cycle, the question arises as to how glutamate was synthesized by anaerobes before the availability of oxygen in the atmopshere. A logical design feature in the blueprint of such organisms would be to make α-ketoglutarate by the reductive carboxylation of succinyl-CoA. In this connection, it is instructive to consider two strict anaerobes that depend on TCA cycle reactions for biosynthetic functions, the archaebacterium *Methanobacterium thermoautotrophicum* and the green sulfur photosynthetic bacterium *Chlorobium limicola*.

Methanobacterium thermoautotrophicum is an autotroph that derives energy in some fashion from the reaction

$$CO_2 + 4H_2 \rightarrow CH_4 + 2H_2O.$$

Biosynthesis is initiated by the acetyl-CoA pathway of autotrophic growth, in which two carbon dioxide molecules are converted to actyl-CoA. The latter provides other biosynthetic intermediates as follows:

$$\text{Acetyl-CoA} \rightarrow \text{pyruvate} \rightarrow \text{PEP} \xrightarrow{CO_2} \text{oxaloacetate} \xrightarrow{a}$$

$$\text{succinyl-CoA} \xrightarrow{b} \text{ketoglutarate,}$$

where a it the four-step reductive four-carbon sequence and b is the reductive carboxylation catalyzed by α-ketoglutarate synthase.

The related methanogen, *Methanosarchina barkeri*, generates α-ketoglutarate from oxaloacetate by the oxidative six-carbon segment of the classical TCA cycle,

$$\text{Oxaloacetate} \rightarrow \text{citrate} \rightarrow \text{isocitrate} \rightarrow \alpha\text{-ketoglutarate.}$$

Methanosarcina barkeri lacks fumarate reductase and α-ketoglutarate synthase, two of the enzymes required to make α-ketoglutarate by the reductive TCA cycle. It also appears to lack α-ketoglutarate dehydrogenase, which might prevent funneling of the α-ketoglutarate that is needed for other biosynthetic purposes. It is not clear how *M. barkeri* makes the succinate or the succinyl-CoA it needs for biosynthetic purposes.

Carbon Dioxide Is Fixed by *Chlorobium limicola* by Running the Tricarboxylic Acid Cycle Backward

Evans and his coworkers (1996) proposed that the carbon dioxide fixation pathway of *Chlorobium limicola* is a reductive cycle that, in overall effect, is a reversal of the entire TCA cycle, leading to the net conversion of carbon dioxide to acetyl-CoA. The suggested cycle, shown in Fig. 4, failed to gain general acceptance at first because the activity of several of the required enzymes appeared to be too low to account for carbon dioxide fixation during growth. Moreover, the presence of citrate lyase was questionable. Subsequent studies, however, demonstrated an ATP-dependent lyase, and provided other evidence in support of the conclusion that the reductive cycle is the sole mechanism of carbon dioxide fixation in *Chlorobium*. The reductive α-ketoglutarate synthase reaction is essential for operation of the cycle, and α-ketoglutarate dehydrogenase is not detectable in extracts of *Chlorobium*.

The fact that the enzymes used to carry out the conversion of α-ketoglutarate to succinate are different from those used to carry out the reverse conversion is an example of unidirectionality in metabolism. Another example is the fumarate-to-succinate interconversion discussed earlier. Daniel Atkinson has pointed out that nearly every metabolic sequence is paired with a sequence that caries out the same conversion in the opposite direction. In many cases both sequences are present in the same cell, but they always function under quite different circumstances.

One other anaerobic permutation of TCA cycle activity in bacteria is especially worthy of note. Beghardt and his coworkers have provided convincing evidence that the strict anaerobe *Desulphuromonas acetoxidans* oxidizes acetate to carbon dioxide with elemental sulfur (instead of oxygen) as the electron acceptor by means of the carbon conversions of the conventional TCA cycle.

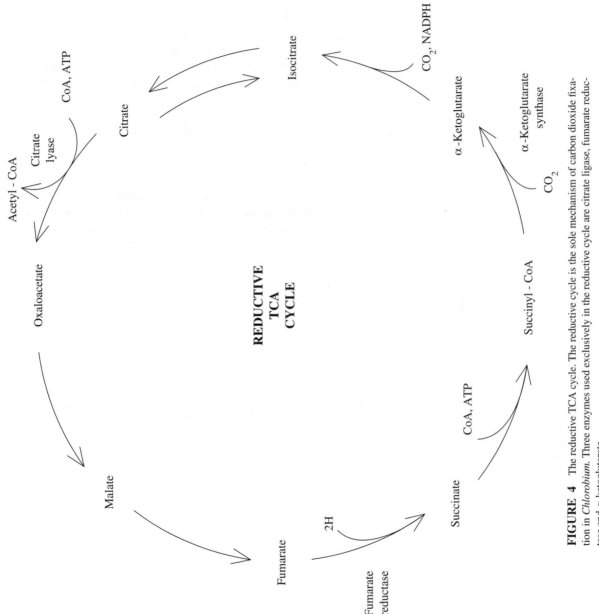

FIGURE 4 The reductive TCA cycle. The reductive cycle is the sole mechanism of carbon dioxide fixation in *Chlorobium*. Three enzymes used exclusively in the reductive cycle are citrate ligase, fumarate reductase and α-ketoglutarate.

Aerobic "Specialists" Use Parts of the Tricarboxylic Acid Cycle

It is generally supposed that an anaerobic (nonoxygenic) green or purple bacterium gave rise to the first oxygenic photosynthetic organisms; that bacterium was probably the ancestor of today's *Cyanobacterium* (see Chapter 23). The nature of TCA cycle sequences in cyanobacteria is consequently of considerable interest. Most cyanobacteria are incapable of growing to an appreciable extent as heterotrophic aerobes (in darkness). A Similar phenotype is shown by various aerobes that obtain energy from oxidation of inorganic compounds (aerobic lithotrophs) and organisms that obtain energy from aerobic oxidation of methyl groups attached to atoms other than carbon (methylotrophs). These "aerobic specialists," listed in Table 2 assimilate [^{14}C] acetate carbon when appropriate energy sources and oxygen are provided, but in a restricted fashion. The isotopic label is found only in lipids and protein, with significant amounts in only four amino acids: glutamate, proline, leucine, and arginine. This, by itself, indicates that the TCA cycle is not being used in the usual manner. They also lack demonstrable α-ketoglutarate dehydrogenase activity, which is required for normal operation of the TCA cycle. These aerobic specialists, capable of using unconventional carbon and energy sources, probably existed before the development of a functional TCA cycle.

TABLE 2
Activities of TCA Cycle Enzymes in "Specialist" Organisms[a]

Organism	Enzyme activity (nmol/mg of protein per min)[b]					
	Aconitase	Isocitrate dehydrogenase	α-Ketoglutarate dehydrogenase	Succinate dehydrogenase	Fumarase	Malate dehydrogenase
Phototrophs						
Anacystis nidulans	—	36	0	6.9	—	7.8
Coccochloris peniocystis	—	36	0	2.1	—	7.1
Gloeocapsa alpicola	—	36	0	3.4	—	17
Anabaena variabilis	2.1	4.7	0	0.2	3.7	2.0
Lithotrophs						
Thiobacillus thiooxidans		80	0	7.9	—	11.2
T. thioparus	26	204	0	—	80	223
T. denitrificans	56	61	0	0.5	148	16
Methylotrophs						
Methylomonas methanica	10	15	0	12	34	430
M. albus	3	19	0	12	37	340
Methylococcus minimus	6	22	0	10	52	440
M. capsulatus	8	22	0	2	32	74
Methylotroph W1	—	5	0	0.8	—	69
Methylotroph C2A1	28	52	0	0	0	6
Methylotroph 4B6	10	75	0	0.6	9	0

[a] From Smith, A. J. and Hoare, D. S. Specialist Phototrophs, Lithotrophs, and Methylotrophs: A Unity among a Diversity of Prokaryotes? *Bact. Rev.* 41: 431, June 1977. Copyright © 1977 American Soceity for Microbiology, Washington, D.C. Reprinted by permission.

[b] Dashes mean not determined.

Facultative Aerobic Bacteria Use the Tricarboxylic Acid Cycle in Different Ways in the Presence and Absence of Oxygen

Starting about 2 billion years ago, molecular oxygen produced by photosynthetic prokaryotes began to accumulate in the atmosphere. This change led to widespread extinctions of numerous anaerobic cell lines because of the toxic effects of oxygen. Studies on the oxygen tolerance of resting cells of various species of bacteria show that strict anaerobes frequently die on exposure to air. The harmful effects of oxygen excess stem from the production of superoxide (O_2^-) and hydrogen peroxide (H_2O_2). Such compounds are the direct cause of DNA, protein, and membrane damage. Some bacteria, however, can tolerate exposure to oxygen. Prominent. in the survival group are facultative aerobes such as *Escherichia coli*, which can grow either as aerobes or as anaerobes. In *E. coli* there is an adaptive response leading to the induction of peroxide-destroying enzymes. These enzymes are normally repressed by a regulatory protein that becomes ineffective as a repressor when it is in the oxidized form.

As the partial pressure of oxygen in the atmosphere increased, some cell lines learned not only how to protect themselves against the toxic effects of oxygen but also how to utilize it as an efficient means of generating biochemical energy. The regulation of TCA cycle reactions and sequences in such organisms is of great value for obtaining a glimpse of microbial life before the onset of the aerobic age. The most extensive knowledge of this aspect of bacterial physiology comes from studies on *E. coli* and the purple photosynthetic bacterium *Rhodobacter capsulatus*. *Escherichia. coli* and *R. capsulatus* both have aerobic and anaerobic energy conversion systems subject to a variety of regulatory controls, and oxygen is particularly prominent as an environmental regulator in the metabolism patterns.

In *E. coli,* α-ketoglutarate dehydrogenase and the enzymes concerned with fumarate and succinate interconversion are the primary targets. Under anaerobic conditions with glucose as the energy source, α-ketoglutarate dehydrogenase is not formed; and the cycle becomes a noncyclic branched pathway for production of succinate and succinyl-CoA, on the one hand, and α-ketoglutarate, on the other. This change involves derepression of fumarate reductase synthesis, which is repressed under aerobic conditions. Synthesis also is derepressed under anaerobic conditions when lactate or glycerol are available as carbon and energy sources. With either of these nonfermentable substrates, growth occurs if fumarate is provided as a terminal hydrogen acceptor. On the other hand, when oxygen is present, fumarate reductase synthesis is repressed and synthesis of enzymes of the classical TCA cycle are derepressed.

In a general sense, the regulatory effects of oxygen on *R. capsulatus* are analogous to those on *E. coli*. Physiological and enzymatic studies with both wild-type and mutant strains of *R. capsulatus* provide evidence consistent with operation of the TCA cycle as the major bioenergetic mechanism for dark aerobic growth. When oxygen is supplied, biosynthesis of the alternative (membrane-bound) energy-converting system (anaerobic photophosphorylation) is repressed. Removal of oxygen leads to derepression of the photopigment system; and in cells growing anaerobically in the photoheterotrophic mode, the α-ketoglutarate dehydrogenase level decreases markedly.

TABLE 3

Fumarate Reductase (FR) and Succinate Dehydrogenase (SD) Activities in Extracts of Photosynthetic Bacteria[a]

Organism	Specific activity[b]	
	FR[c]	SD
Chlorobium limicola	38	4
Rhodopseudomonas gelatinosa	228	36
R. viridis	25	11
R. capsulata	<1	196
R. palustris	<1	49
R. sphaeroides	<1	24
Rhodospirillum rubrum	2	80
R. tenue	16	57
Chromatium vinosum	<1	26

[a] From Beatty, J. T. and Gest, H. Generation of Succinyl-Coenzyme A in Photosynthetic Bacteria, *Arch. Microbiol.* 129: 338, 1981. Copyright © 1981 Springer-Verlag. Reprinted by permission

[b] Values are expressed as nmol fumarate reduced or succinate oxidized per min per mg protein.

[c] Reduced methyl viologen used as electron donor.

In contrast to *E. coli, R. capsulatus* does not produce fumarate reductase, and the supply of succinyl-CoA for biosynthesis consequently depends on activity of α-ketoglutarate dehydrogenase. The level of activity needed for the biosynthetic function, however, is much lower than that required for the energy conversion function of the cycle. Thus mutants of *R. capsulatus* deficient in α-ketoglutarate dehydrogenase retain the capacity to grow photosynthetically (anaerobically) on malate or on carbon dioxide and hydrogen, but lose the ability to grow as aerobic heterotrophs. The low rate of succinyl-CoA production required for photosynthetic growth of such mutants is probably due to the residual α-ketoglutarate dehydrogenase activity of the deficient enzyme in the mutants that were studied.

The evolutionary phylogeny of photosynthetic bacteria indicates a significant divergence of the green species from the purples (see Chapter 23). This divergence seems to be reflected in the directionality of enzymes concerned with interconversions of fumarate and succinate. Thus when the ratio of fumarate reductase to succinate dehydrogenase activities is low (much less than 1), the physiological role of the protein in question is most likely that of a succinate dehydrogenase. Conversely, when the ratio is high (much greater than 1) in an organism that grows well anaerobically, it is reasonable to consider the protein as a fumarate reductase that functions *in vivo* either in a bioenergetic role (activation of fumarate as the terminal electron acceptor for energy-yielding electron flow) or in the capacity of an anabolic enzyme for the synthesis of tetrapyrroles, and so on. The results of Beatty and Gest, summarized in Table 3, show that high fumarate reductase to succinate dehydrogenase ratios were observed in *C. limicola* and *R. gelatinosa*. It is noteworthy that in these particular organisms,

α-ketoglutarate dehydrogenase either was undetectable or was present at relatively low specific activity.

From Table 3 it can be seen that most purple bateria do not produce fumarate reductase, whereas *C. limicola* depends on the enzyme for biosynthetic functions. These results reinforce the conclusion that the reductive formation of succinate is an important feature of *Chlorobium* metabolism. By contrast, *R. gelatinosa* may prove to be an interesting exception among the purple bacteria in that it may generate succinyl-CoA for biosynthesis either from reductively produced succinate or from α-ketoglutarate.

SCHEME FOR EVOLUTION OF THE TRICARBOXYLIC ACID CYCLE IN PROKARYOTES

All the bits of information about the use of the TCA cycle enzymes in different microorganisms have been used by Gest to advance an idea for connecting the evolution of fermentations with the beginnings of the TCA cycle: that evolution involved exploiting the highly oxidized carbon dioxide molecule for generating hydrogen acceptors that could spare utilization of pyruvate simply as a terminal electron sink for balancing oxidation and reduction. Thus a segment of the TCA cycle (oxaloacetate $\rightarrow$ malate $\rightarrow$ fumarate $\rightarrow$ succinate), running in reverse, could be viewed as an early innovation in the evolution of more efficient fermentation (Fig. 5).

The TCA cycle evolved from prokaryotic fermentative systems via a number of stages, each representing improved bioenergetic and biosynthetic efficiency. One evolutionary root of the cycle was an adjunct of anaerobic glycolysis employed for effecting redox balance without obligatory use of pyruvate as the electron acceptor, namely, the reductive sequence

$$\text{Oxaloacetate} \xrightarrow{2e^-} \text{malate} \rightarrow \text{fumarate} \xrightarrow{2e^-} \text{succinate.}$$

Accordingly, if oxaloacetate was generated from pyruvate (or PEP) by carboxylation with carbon dioxide, more three-carbon units derived from hexose were potentially available for biosynthesis. Later association of the fumarate $\rightarrow$ succinate reduction with cellular membranes, accompanied by gradual addition of intermediary electron carriers, including cytochrome *b*, added to the capacity for simultaneous electrophosphorylation.

Evolution of the reaction pyruvate $\rightarrow$ acetyl-CoA $\rightarrow$ CO_2 + H_2 provided an additional source of ATP (acetyl-CoA $\rightarrow$ acetyl phosphate $\rightarrow$ acetate + ATP) and paved the way for construction of the oxidative branch of the TCA cycle:

$$\text{Oxaloacetate} + \text{acetyl-CoA} \rightarrow \text{citrate} \rightarrow \alpha\text{-ketoglutarate.}$$

This sequence is likely to have functioned in some early anaerobic prokaryotes to provide the α-ketoglutarate necessary for glutamate formation. After oxygen accumulated in Earth's atmosphere, a bioenergetic TCA cycle became feasible. The sequence from oxaloacetate to succinate could now be reversed to operate in the oxidative

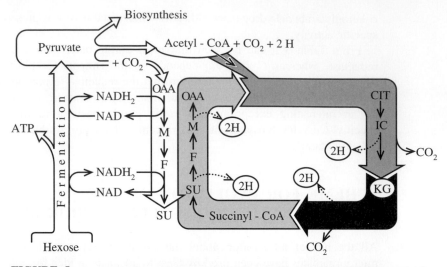

FIGURE 5   A scenario for evolution of the TCA cycle. It is suggested that the reductive sequence oxaloacetate–succinate was first used as a mechanism for regenerating NAD^+ required for hexose fermentation, thus minimizing utilization of pyruvate as a terminal electron acceptor. Conversion of succinate to succinyl-CoA provided an essential intermediate in the synthesis of tetrapyrroles. The production of citrate from acetyl-CoA and oxaloacetate, and subsequent conversions. presumably originated as a pathway for biosynthesis of α-ketoglutarate or glutamate. It appears that α-ketoglutarate dehydrogenase and succinate thiokinase were the final catalysts required for linking biosynthetic reaction sequences so as to yield a cyclic mechanism for providing a good flux of reducing equivalents to the electron-transport machinery. (OAA = oxaloacetate, M = malate, F = fumarate, SU = succinate, CIT = citrate, IC = isocitrate, and KG = α-ketoglutarate.)

direction; and with the final addition of an α-ketoglutarate dehydrogenase, subject to regulation by oxygen, the linear biosynthetic segments thereby became a cycle capable of supporting more efficient energy conversion. Early versions of the aerobic TCA cycle were no doubt less effective than the modern cycle is, partly because the electron-transport system used for oxidation of NADH with oxygen still was undergoing evolutionary refinement (cytochrome c and its oxidase were late additions).

SUMMARY

In this chapter we have analyzed the TCA cycle for oxidative carbohydrate metabolism in different microorganisms with the goal of determining how this pathway evolved.

1. Carbohydrate catabolism supplies the energy and the carbohydrate skeletons for biosynthesis. Carbohydrate catabolism is handled differently in typical anaerobes and in typical aerobes. Anaerobes catabolize glucose and other carbon compounds that can be converted into the three-carbon compound pyruvate. This results in the net production of two ATPs for every glucose molecule that is catabolized. The pyruvate is

usually reduced, not to supply further energy, but instead so that the NAD^+ coenzyme needed for glycolysis can be regenerated. Aerobes also convert glucose to pyruvate, but they go on to catabolize pyruvate all the way to carbon dioxide. This exhaustive degradation of glucose is coupled to an electron-transport system that consumes oxygen and produces large quantities of ATP, about 30 molecules of ATP per molecule of glucose.

2. Pyruvate is degraded to carbon dioxide by a cyclic process that begins with acetyl-coenzyme A, which is obtained either by oxidative decarboxylation of the pyruvate made available by glycolysis or by the oxidative cleavage of fatty acids. The acetyl-CoA transfers its acetyl group to a four-carbon acceptor (oxaloacetate), thereby generating citrate, the six-carbon compound that gave rise to one name for the cycle, which also is called the Krebs or TCA cycle. The citrate is next subjected to two successive decarboxylations and several oxidative events, leaving a four-carbon compound from which the starting oxaloacetate is eventually regenerated. Each turn of the cycle involves the entry of two carbons from acetyl-CoA and the release of two carbons as carbon dioxide, thus providing for regeneration of the oxaloacetate with which the cycle begins.

3. There are basic reasons for believing that most of the reactions of the TCA cycle existed before there was aerobic metabolism. Evolution is a process that proceeds in simple steps, with each step serving to improve the system in some way. Because the TCA cycle is very complex, it must have required many evolutionary steps to become functional. In the process, most, if not all, of the TCA cycle reactions must have existed for other reasons.

4. Although we cannot recreate the past, we can seek other ways to substantiate the working hypothesis that most of the reactions of the TCA cycle served some other purpose before there was a TCA cycle. The best approach may be to analyze the way in which reactions of the TCA cycle are used in organisms that do not use the TCA cycle in the normal way.

5. While pyruvate oxidation can regenerate the NAD^+ necessary to keep the glycolytic pathway in operation, it seems likely that a catabolic system that could preserve at least some of the pyruvate for other purposes might have a selective advantage. If carbon dioxide were used to carboxylate pyruvate to oxaloacetate, then the reactions between oxaloacetate and succinate, operated in the reductive direction, could supply an electron sink to keep glycolysis going while consuming only half of the pyruvate so produced. Such a pathway not only functions as an electron sink but also results in the synthesis of a useful biosynthetic intermediate, succinate.

6. In an aerobic organism, glutamate is synthesized in one step from α-ketoglutarate. Because α-ketoglutarate is an intermediate in the TCA cycle, we must ask how α-ketoglutarate is produced in systems that do not have a complete TCA cycle. In some anaerobes, it appears that α-ketoglutarate is produced by the reductive TCA cycle from oxaloacetate. In others, it appears that it can be produced oxidatively from oxaloacetate by the operation of the first three reactions of the cycle, operating oxidatively.

7. In certain photosynthetic microorganisms, carbon fixation occurs by reactions that entail the entire TCA cycle, operating in the reductive direction.

8. Some aerobes use only parts of the TCA cycle. Such organisms probably existed before development of a functional TCA cycle.

9. Facultative aerobes like *E. coli* can grow either like aerobes in the presence of molecular oxygen or like anaerobes in its absence. In the presence of oxygen, *E. coli* utilizes the TCA cycle in the normal way. In the absence, *E. coli* uses the first half of the TCA cycle in the normal oxidative direction and the second half of the cycle in the reductive direction.

Problems

1. When an "aerobic specialist" is fed [^{14}C]acetate, the ^{14}C label is ultimately found in only four amino acids: glutamate, glutamine, proline, and arginine. What does this say about the use of the TCA cycle?
2. Were you not surprised to see that *Chlorobium limicola* used the entire TCA cycle in the reductive direction? How do you suppose that this organism meets its energy needs?
3. Name three intermediates in the TCA cycle that are used in biosynthesis and indicate what important biomolecules are made from them.
4. Is pyruvate an intermediate in the TCA cycle?
5. The conversion of malate to oxaloacetate is highly endergonic. In what way is this useful to facultative aerobes like *E. coli*?
6. Why did Howard Gest consider the conversion of pyruvate to lactate a wasteful reaction?

Reference

Gest, H. Evolutionary Roots of the Citric Acid Cycle in Prokaryotes, *Biochem. Soc. Symp.* 54: 3–16, 1987.

CHAPTER 23

Evolution of Photosynthesis

Most organisms live by degrading complex molecules provided by other organisms. Life on Earth could not continue for very long in this manner. The energy that sustains the *de novo* synthesis of these complex molecules comes from the Sun, and is captured in the process of photosynthesis. Plants and other photosynthetic organisms convert about 10^{11} tons of carbon from carbon dioxide into organic compounds annually. About half of this carbon dioxide fixation takes place in the oceans and is carried out

TABLE 1

Properties of Photosynthetic Organisms

Group	Evolve O_2	Chloroplasts	Type of chlorophyll[a]	Number of photosystems
Prokaryotes				
Purple bacteria	No	No	BChl a or b	1
Green sulfur bacteria	No	No	BChl a and c	1
Cyanobacteria[b]	Yes	No	Chl a	2
Eukaryotes				
Green algae and higher plants	Yes	Yes	Chl a and b	2

[a] BCl = bacteriochlorophyll: Chl = chlorophyll.

[b] Prochlorophytes, a small subgroup of the cyanobacteria, contain Chl b in addition to Chl a.

by microorganisms. In addition to algae and higher plants, numerous types of bacteria are capable of photosynthesis.

An overall equation for carbon dioxide fixation as it occurs in plants is

$$CO_2 + H_2O + light \longrightarrow (CH_2O) + O_2, \tag{1}$$

where (CH_2O) denotes part of a carbohydrate. Electrons and protons are removed from water, oxygen is evolved, and carbon dioxide is reduced to the level of a carbohydrate. One group of photosynthetic bacteria, the cyanobacteria, carry out the same process. Other types of photosynthetic bacteria carry out similar processes except that they do not evolve oxygen because they use substances other than water as a source of electrons. Table 1 summarizes the distinctions between some of the major groups of photosynthetic organisms. The characteristics of the different groups are discussed in more detail later.

In plants, the reactions of photosynthesis take place in specialized organelles known as *chloroplasts* (Fig. 1). Chloroplasts are bounded by an envelope of two membranes and have a highly folded internal membrane called the thylakoid membrane. The *chlorophyll* found in chloroplasts is bound to proteins that are integral constituents of the thylakoid membrane, and it is here that the initial conversion of light into chemical energy occurs. The enzymes responsible for the actual fixation of carbon dioxide and the synthesis of carbohydrates reside in the stroma that surrounds the thylakoid membrane.

Algae contain chloroplasts similar to those of higher plants. Prokaryotic photosynthetic organisms, which include cyanobacteria and several groups of purple or green bacteria, do not have chloroplasts. In prokaryotes, the photochemical reactions of photosynthesis take place in the membrane that encloses the cell.

Our main interest in this chapter is to discuss how photosynthesis might have arisen on primitive Earth and how the various forms of photosynthesis found in present-day

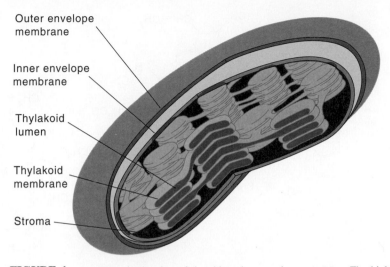

Outer envelope
membrane

Inner envelope
membrane

Thylakoid
lumen

Thylakoid
membrane

Stroma

FIGURE 1   A schematic drawing of the chloroplast membrane systems. The highly folded thylakoid membrane separates the thylakoid lumen from the stroma.

organisms could have evolved. We begin by considering the main features of photosynthesis as we know them today.

PHOTOSYNTHESIS DEPENDS ON THE PHOTOCHEMICAL REACTIVITY OF CHLOROPHYLL

All photosynthetic organisms that belong to the eubacterial and eukaryotic kingdoms take advantage of the photochemical reactivity of chlorophyll or bacteriochlorophyll. Figure 2 shows the structures of several of the different types of chlorophyll that occur in nature. The distribution of these molecules among some of the different groups of photosynthetic organisms is indicated in Table 1. Chlorophylls resemble hemes, the prosthetic groups of the cytochromes and hemoglobin, and are derived biosynthetically from protoporphyrin IX. Photosynthetic organisms also contain small amounts of pheophytins or bacteriopheophytins, which are the same as the corresponding chlorophylls or bacteriochlorophylls except that the magnesium atom is replaced by two hydrogen atoms (see Fig. 2). We see that the pheophytins and bacteriopheophytins play special roles as electron carriers in photosynthesis.

Whereas the long-wavelength absorption bands of hemes are relatively weak, chlorophyll *a* has an intense absorption band at 676 nm (Fig. 3). Chlorophyll *b* has a similar band at 642 nm. Bacteriochlorophylls *a* and *b* have strong absorption bands in

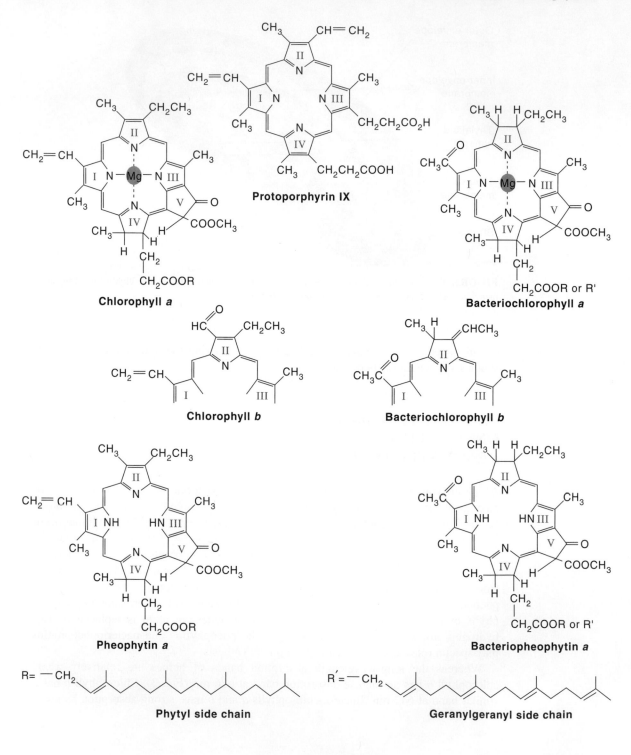

Protoporphyrin IX

Chlorophyll *a*

Bacteriochlorophyll *a*

Chlorophyll *b*

Bacteriochlorophyll *b*

Pheophytin *a*

Bacteriopheophytin *a*

R=   —CH₂

Phytyl side chain

R′=   —CH₂

Geranylgeranyl side chain

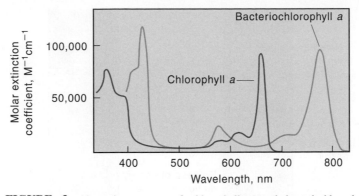

FIGURE 3 Absorption spectra of chlorophyll *a* and bacteriochlorophyll *a* in ether. In the chlorophyll–protein complexes found in photosynthetic organisms, the long-wavelength absorption band generally is shifted to even longer wavelengths. This probably reflects interactions between neighboring chlorophylls, which are bound to the proteins as dimers or larger oligomers.

the region of 780 to 790 nm (see Fig. 3). All the chlorophylls thus absorb light very well at long wavelengths.

When Chlorophyll Absorbs a Photon, the Energy of the Molecule Is Increased

The electrons in a molecule are held in a set of molecular orbitals, each of which is associated with a definite energy. In an isolated molecule, the energies of the orbitals depend mainly on the interactions of the electrons with each other and with the nuclei, and are more or less independent of time. A molecule thus can have a variety of energies, depending on how its electrons are distributed among the available orbitals. For an organic molecule with 2n electrons, the lowest overall energy usually is obtained when there are two electrons in each of the first n orbitals, leaving all the orbitals with higher energies empty (Fig. 4). This is the ground state of the molecule.

When light interacts with a molecule, it is possible for an electron to move from one of the occupied molecular orbitals to an unoccupied orbital with a higher energy, and for a photon to be absorbed in the process (see Fig. 4). Two requirements must be met for this transition to occur. First, the difference between the energies of the two

FIGURE 2 Structures of protoporphyrin IX (the prosthetic group of hemoglobin, myoglobin, and *c*-type cytochromes), and several types of chlorophyll and bacteriochlorophyll. Chlorophyll *a* and chlorophyll *b* are the main types of chlorophyll in plants and cyanobacteria. Purple photosynthetic bacteria contain either bacteriochlorophyll *a* or bacteriochlorophyll *b*, depending on the bacterial species. Chlorophyll *b* and bacteriochlorophyll *b* are the same as chlorophyll *a* and bacteriochlorophyll *a* except for the substituents on ring II. Pheophytins and bacteriopheophytins are the same as the corresponding chlorophylls or bacteriochlorophylls except that two hydrogen atoms replace the magnesium atom.

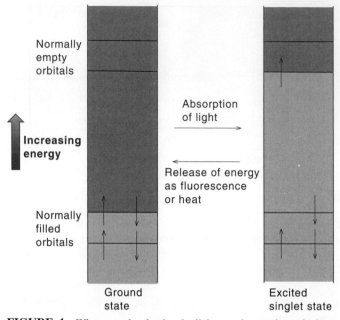

FIGURE 4   When a molecule absorbs light, an electron in excited to a molecular orbital with higher energy. The horizontal bars in this diagram represent molecular orbitals for electrons. Each orbital can hold two electrons with antiparallel spins (arrows pointing upward or downward). Only the top few of the many filled orbitals are shown here. The absorption of light can raise an electron from one of these orbitals to an orbital that is normally unoccupied. For this to occur, the energy of the photon must match the difference between the energies of the two orbitals. An excited molecule can return directly to the ground state by giving off energy as fluorescence or heat, or by transferring energy to another nearby molecule; or it can transfer an electron to another molecule (see Fig. 5).

orbitals must be the same as the photon's energy. Second, the two orbitals must be oriented in an appropriate way with respect to the oscillating light field.

If the chlorophyll (or any other molecule) absorbs a photon, it is excited to a state that lies above the ground state in energy. A molecule in such an excited state can decay back to the ground state by releasing energy in several different ways (see Fig 4). One possibility is to emit a photon; this is the phenomenon of *fluorescence*. Another decay mechanism is to transfer the excitation energy to a neighboring molecule by a process called *resonance energy transfer*. An excited molecule also can *transfer an electron* to a neighboring molecule.

Light Drives the Formation of Oxidants and Reductants by Electron Transfer

Electron transfer is a common path for the decay of an excited chlorophyll molecule, because an electron in the upper, normally unoccupied orbital is bound less tightly than one in a lower, normally filled orbital. This makes the excited molecule a stronger

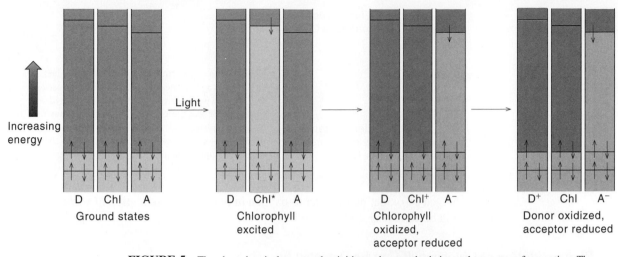

Increasing
energy

| D | Chl | A | D | Chl* | A | D | Chl⁺ | A⁻ | D⁺ | Chl | A⁻ |

Ground states   Chlorophyll   Chlorophyll   Donor oxidized,
                         excited   oxidized,   acceptor reduced
                                                acceptor reduced

FIGURE 5   The photochemical process that initiates photosynthesis is an electron-transfer reaction. The horizontal bars and vertical arrows represent molecular orbitals and electrons, as in Fig. 4. Absorption of light increases the energy of a chlorophyll complex (Chl) by $h\nu$, making electron transfer to an acceptor (A) thermodynamically favorable. The oxidized chlorophyll complex (Chl^+) extracts an electron from a donor (D).

reductant. Excitation facilitates the release of an electron from a molecule of chlorophyll. In the process the chlorophyll molecule is oxidized and another molecule becomes reduced (Fig. 5). The oxidized chlorophyll species that is formed is a relatively strong oxidant and can extract an electron from a third molecule.

The idea that light drives the formation of oxidants and reductants was first advanced by C. B. van Niel in the 1920s. It was known at the time that purple photosynthetic bacteria thrive only if they are provided with a reduced substrate. Some species grow well on reduced inorganic materials such as hydrogen sulfide; others prefer organic reductants. Van Niel noticed that although the bacteria did not evolve oxygen, the reactions they carried out had a formal resemblance to the process of photosynthesis that occurs in plants. If we use H_2B to represent a reduced substrate and B for an oxidized product, the process that occurs in the purple bacteria can be written

$$CO_2 + 2H_2B \xrightarrow{\text{light}} (CH_2O) + H_2O + 2B.$$

The fixation of carbon dioxide by plants and cyanobacteria [Eq. (1)] can be described similarly by inserting H_2O and O_2 in place of H_2B and 2B

$$CO_2 + 2H_2O \longrightarrow (CH_2O) + H_2O + O_2. \tag{2}$$

To van Niel, this suggested that the essence of photosynthesis is the separation of oxidizing and reducing power. The substance that is reduced in the photochemical reaction could be used to reduce carbon dioxide to carbohydrates in enzyme-catalyzed reactions that do not require light. The material that is oxidized photochemically could

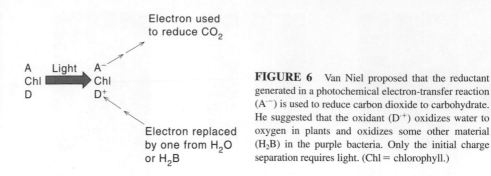

FIGURE 6   Van Niel proposed that the reductant generated in a photochemical electron-transfer reaction (A^{-}) is used to reduce carbon dioxide to carbohydrate. He suggested that the oxidant (D^{+}) oxidizes water to oxygen in plants and oxidizes some other material (H_2B) in the purple bacteria. Only the initial charge separation requires light. (Chl = chlorophyll.)

be discharged by oxidation of water to oxygen in plants, or by oxidation of some other material in bacteria (Fig. 6).

Photooxidation of Chlorophyll Can Be Detected by Optical Absorption Spectrum

The photooxidation of chlorophyll in photosynthetic systems can be detected by measuring changes in the optical absorption spectrum. Oxidation results in the loss of the molecule's characteristic absorption bands. In the 1950s Bessel Kok found that illumination of chloroplasts caused an absorbance decrease at 700 nm (Fig. 7). Kok suggested that this reflected the photooxidation of a reactive chlorophyll complex, which he called P700. A second type of reactive complex in chloroplasts, P680, was discovered subsequently. We see that P700 and P680 are parts of two distinct photochemical systems, photosystem I and photosystem II found in cyanobacteria and plants. Duysens found a similar observation on *Rhodospirillum rubrum,* a purple photosynthetic bacterium. Here the reactive bacteriochlorophyll complex absorbed at 870 nm.

Reactive Chlorophyll Is Found in Reaction Centers

The chlorophyll that undergoes photooxidation is bound to a protein in a complex called a *reaction center.* Reaction centers of purple bacteria generally contain three

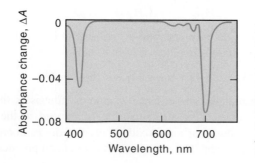

FIGURE 7   When a suspension of chloroplasts is illuminated, there are decreases in the suspension's optical absorbance in the regions around 430 and 700 nm. The absorbance changes reflect the oxidation of a special chlorophyll complex (P700). (After Kok, B. Partial Purification and Determination of Oxidation Reduction Potential of the Photosynthetic Chlorophyll Complex Absorbing at 700 μm, *Biochim. Biophys. Acta* 48:527, 1961, with permission from Elsevier Science.)

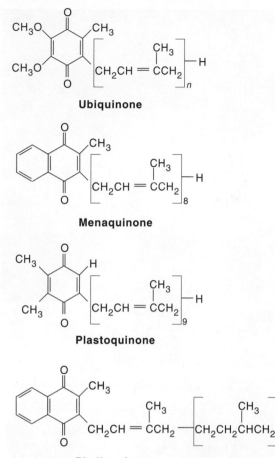

FIGURE 8   The structures of quinones found in photosynthetic reaction centers. Purple photosynthetic bacteria contain ubiquinone, menaquinone, or both, depending on the bacterial species; photosystem II contains plastoquinone; and photosystem I contains phylloquinone.

polypeptides with molecular weights of approximately 28,000, 31,000, and 34,000. These are referred to as the H, L, and M subunits, respectively. Bound noncovalently to the protein are four molecules of bacteriochlorophyll, two molecules of bacterio-pheophytin, two quinones, and one nonheme iron atom. In some bacterial species, both of the quinones are ubiquinone; in others, one of the quinones is menaquinone (Fig. 8).

The crystal structure of reaction centers from the purple bacterium *Rhodopseudomonas viridis* is shown in Fig. 9. In this reaction center the bacteriochlorophyll and bacteriopheophytin molecules, the nonheme iron atom, and the quinones are all bound to the L and M polypeptides, which are folded into a series of α helices that pass back

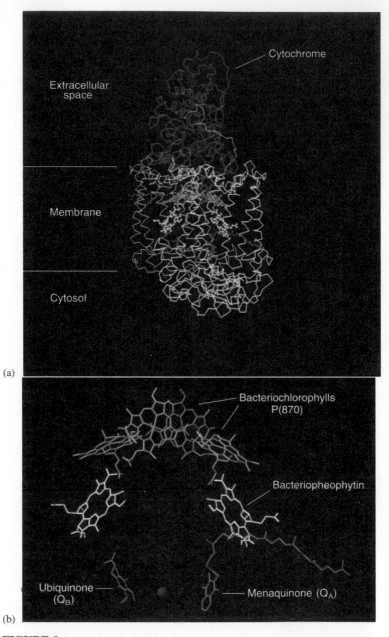

FIGURE 9   (a) Structure of the reaction center of *Rhodopseudomonas viridis*. (b) An expanded view of some of the components of the *R. viridis* reaction center. The bacteriochlorophyll dimer that undergoes photooxidation is at the top.

and forth across the cell membrane (Fig. 9 a). These two polypeptides have homologous amino acid sequences and similar folding patterns. The H polypeptide is located largely on the cytoplasmic side of the membrane, but it also has one transmembrane α helix. Reaction centers from *R. viridis* also have a bound cytochrome that sits on the external (periplasmic) surface of the membrane.

Figure 9 b shows the arrangement of the pigments in greater detail. Two of the four bacteriochlorophylls are packed closely together. Spectroscopic studies have identified this special pair as P870, the reactive complex that releases an electron when the reaction center is excited with light. One of the striking features of the crystal structure is the axis of approximate symmetry that runs from the center of P870 through the iron atom. A rotation of 180° about this axis interchanges the pigments on either side of the reaction center and interchanges homologous amino acid residues of the L and M subunits.

Photosystem II reaction centers found in eukaryotes resemble the reaction centers of the purple photosynthetic bacteria in a number of ways. The amino acid sequences of their two principal polypeptides (D1 and D2) are homologous to those of the L and M polypeptides of the bacterial reaction center (about 25% of the corresponding amino acid residues are identical). Also, photosystem II reaction centers contain a non-heme iron atom and two molecules of plastoquinone, a quinone that is closely related to ubiquinone (see Fig. 8), and they contain two molecules pheophytin *a*.

Reaction centers of photosystem I contain two major polypeptides with molecular weights of approximately 83,000, along with one or more small iron–sulfur proteins. The amino acid sequences of the two major polypeptides are homologous to each other, but appear to be unrelated to those of photosystem II or the purple bacterial reaction center. Photosystem I reaction centers differ in containing an iron–sulfur cluster with four iron atoms (see later discussion) in place of the single nonheme iron atom. They also have about 80 molecules of chlorophyll *a*, most of which do not enter into electron-transfer reactions but serve instead as an antenna that transfers energy to P700. However, recent crystallographic structures suggest that photosystem I resembles the purple bacteria and photosystem II in having an axis of approximate rotational symmetry stretching from P700 to the iron–sulfur cluster. Another point in common is the presence of two quinones. Reaction centers of the green sulfur bacteria are very similar to those of plant photosystem I.

Electrons Move from P870 to Bacteriopheophytin and Then to Quinone

Let us now consider the acceptors that remove an electron from P870 in the purple bacterial reaction center. The earliest detectable acceptor is a bacteriopheophytin. Note that the reaction center contains two bacteriopheophytins and two additional bacteriochlorophylls, in addition to the special pair of bacteriochlorophylls that make up P870 (see Fig. 9 b). When reaction centers are excited with a short flash of light, a transient excited state of P870 (P870*) is formed very rapidly. This state decays in about 3×10^{-12} s, and as it does a P870$^{\cdot+}$ BPh$^{\cdot-}$ radical pair is created. Here P870$^{\cdot+}$ is the radical formed by removing an electron from P870 and BPh$^{\cdot-}$ is the radical formed by adding an electron to a bacteriopheophytin (Fig. 10). Formation of the radical-pair

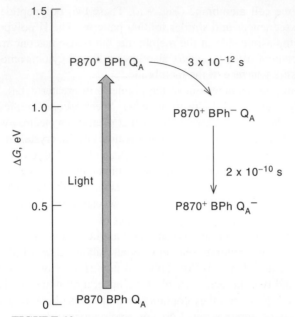

FIGURE 10   The initial electron-transfer sequence in reaction centers of purple photosynthetic bacteria. (P870* = the excited state of P870: BPh = bacteriopheophytin: Q_A = menaquinone or ubiquinone depending on the bacterial species.) In this scheme, various states of the photosynthetic apparatus are positioned vertically according to their free energies, with the states that have the highest free energies at the top. (Reprinted with permission from Woodbury, N. W., Becker, M., Middendorf, D., and Parson, W.W. Picosecond Kinetics of the Initial Photochemical Electron Transfer Reaction in Bacterial Photosynthetic Reaction Centers, *Biochemistry* 24:7516, 1985. Copyright 1985, American Chemical Society.)

state can be detected spectrophotometrically by the disappearance of absorption bands of P870 and of bacteriopheophytin, and the formation of new absorption bands attributable to the two radicals. The bacteriopheophytin that undergoes reduction is the one shown on the right in Fig. 9 a and b. The two bacteriopheophytins are bound to the L and M polypeptides in slightly different ways, and are distinguishable by their different absorption spectra. The $P870^{\cdot+} BPH^{\cdot-}$ state lasts for about 2×10^{-10}s, decaying by the movement of an electron from $BPh^{\cdot-}$ to one of the quinones (O_A). This leaves the reaction center in the state $P870^{\cdot+} Q_A^{\cdot-}$ where $Q_A^{\cdot-}$ is the semiquinone radical (see Fig. 10).

The electron carriers that participate in these first few steps are clustered in the reaction center with little freedom of motion. One indication of this is that the electron-transfer reactions, in addition to being phenomenally fast are almost independent of temperature. Thus little molecular motion is needed for electron transfer from one molecule to another in the complex. Further, the reactions are amazingly efficient, so that every time the reaction center is excited, an electron moves from P870 to Q_A.

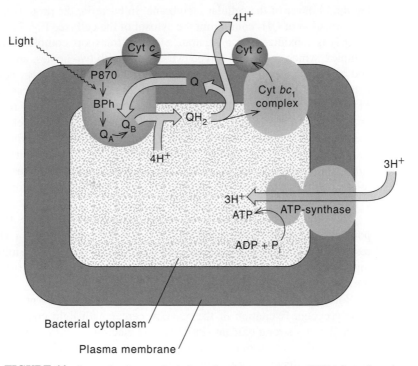

FIGURE 11   In purple photosynthetic bacteria, electrons return to P870$^+$ from the quinones Q_A and Q_B via a cyclic pathway. When Q_B is reduced with two electrons, it picks up protons from the cytosol and diffuses to the cytochrome bc_1 complex. Then the quinol QH_2 transfers electrons to a c-type cytochrome by an indirect, cyclic pathway (not shown). The cytochrome diffuses to the reaction center and reduces P870$^+$ either in a direct manner or by way of another cytochrome. The electron-transfer steps catalyzed by the cytochrome bc_1 complex are essentially identical to those catalyzed by complex III of the mitochondrial respiratory chain. In the course of these reactions, four protons probably are pumped out of the cell for every two electrons that return to P870. This outward proton translocation creates an electrochemical potential gradient across the membrane. Protons move back into the cell through an ATP-synthase, driving the formation of ATP.

Cyclic Electron-Transport Chain Moves Protons Across the Membrane

From Q_A^-, an electron moves to the second quinone that is bound to the reaction center (Q_B in Fig. 11). In the meantime, a c-type cytochrome replaces the electron that was removed from P870, preparing the reaction center to operate again. In *R. viridis*, the electron donor is the bound cytochrome with four hemes shown at the top of Fig. 9 a. In other species it often is a soluble c-type cytochrome with a single heme.

When the reaction center is excited a second time, a second electron is pumped from P870 to the bacteriopheophytin and on to Q_A and Q_B. This places Q_B in the fully reduced form, Q_B^{2-}. The uptake of two protons transforms the reduced quinone to the uncharged quinol, QH_2, which dissociates from the reaction center into the phospholipid bilayer of the cellular membrane. In bacteria, the protons that are taken up in the formation of QH_2 come from the cytosol of the cell (see Fig. 11).

QH_2 is reoxidized by a separate electron transport complex, the cytochrome bc_1 complex, which contains a b-type cytochrome with two hemes, an iron–sulfur protein, and cytochrome c_1. Electrons move through the cytochrome bc_1 complex to a c-type cytochrome, which then diffuses to the reaction center and reduces P870 (see Fig. 11). Movement of electrons through the cytochrome bc_1 complex results in uptake of protons from the cytosol and release of protons on the extracellular surface of the membrane. The flow of electrons from QH_2 back to P870 thus drives the movement of protons outward across the cell membrane, generating a transmembrane electrochemical potential gradient for protons. The inside of the cell becomes negatively charged relative to the external medium, and the pH of the cytosol becomes higher than the external pH.

The flow of the protons back into the bacterial cell, down the electrochemical potential gradient, is mediated by an adenosine triphosphate (ATP) synthase that resembles the proton-conducting ATP-synthase of eukaryotic mitochondria. As in mitochondria, movement of protons through a channel in the ATP-synthase is linked to the formation of ATP (see Fig. 11).

Note that the electron-transport system of purple photosynthetic bacteria is completely cyclic. Excitation of the reaction center with light creates a strong reductant ($BPh^{\cdot-}$) and a strong oxidant ($P870^{\cdot+}$); and electrons return from $BPh^{\cdot-}$ to $P870^{\cdot+}$ via quinones, the cytochrome bc_1 complex, and c-type cytochromes. The cyclic flow of electrons results in the formation of ATP, but no net oxidation or reduction. How then can we explain van Niel's observation that the bacteria carry out a net transfer of electrons from organic acids to carbohydrates? The biosynthesis of carbohydrates from carbon dioxide would seem to require a reductant such as the coenzyme reduced nicotinamide adenine dinucleotide phosphate (NADPH) in addition to ATP. The answer to this puzzle is that photosynthetic bacteria use ATP to support the transfer of electrons from succinate or other substrates to nicotinamide adenine dinucleotide phosphate ($NADP^+$). These reactions are carried out by dehydrogenases in the cell membrane.

Antenna System Transfers Energy to the Reaction Centers

The reactive chlorophyll or bacteriochlorophyll molecules of P700, P680, or P870 make up only a small fraction of the total pigment in photosynthetic membranes. Chloroplasts contain on the order of 400 chlorophyll molecules per P700 and P680, and the cell membranes of purple photosynthetic bacteria have from 25 to several hundred bacteriochlorophylls per P870, depending on the species. Most of the chlorophyll or bacteriochlorophyll serves as an antenna. When one of the molecules in the antenna system is excited with light, it can transfer its energy to a neighboring molecule by resonance energy transfer. Energy absorbed anywhere in the antenna migrates rapidly from molecule to molecule until it is trapped by an electron-transfer reaction in a reaction center (Fig. 12).

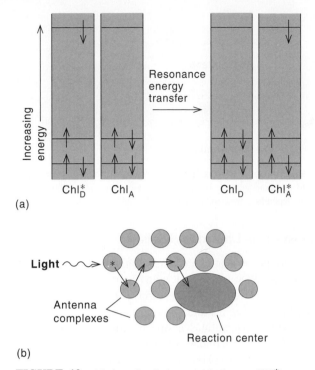

(a)

(b)

FIGURE 12   (a) A molecule in an excited state (Chl_D^*) can transfer its energy to another molecule (Chl_A). The donor molecule returns to its ground state (Chl_D) and the acceptor is elevated to an excited state (Chl_A^*). This requires that the difference in energy between Chl_D^* and Chl_D (the excitation energy of Chl_D) be the same as the difference between Chl_A^* and Chl_A (the excitation energy of Chl_A). When the energies match in this way, there is a resonance between the two states $\{Chl_D^*\ Chl_A\}$ and $\{Chl_D\ Chl_D^*\}$. (b) The energy of light absorbed by pigmen-protein complexes in the antenna system hops rapidly from complex to complex by resonance energy transfer, until it is trapped in an electron transfer reaction in a reaction center.

The chlorophyll or bacteriochlorophyll molecules that form the antenna are bound to small, integral membrane proteins, each of which typically carries only a few pigment molecules. These proteins aggregate into larger arrays, over which excitations can hop from complex to complex. Along with bacteriochlorophyll or chlorophyll, antenna systems contain a variety of other pigments such as carotenoids. These accessory pigments fill in the antenna's absorption spectrum in regions where the chlorophylls do not absorb well. As is shown in Fig. 3, chlorophyll *a* absorbs red or blue light well, but not green. Carotenoids, which are long, linear polyenes (Fig. 13), have absorption bands in the green region. The energy they absorb is transferred to chlorophyll molecules of the antenna, and from there to the reaction centers. Cyanobacteria and red algae have very large auxiliary antenna complexes called phycobilisomes, which contain open chain tetrapyrrole pigments covalently bound to proteins. Phycobilisomes typically line one side of the photosynthetic membrane and transfer energy to chlorophyll–protein complexes in the membrane.

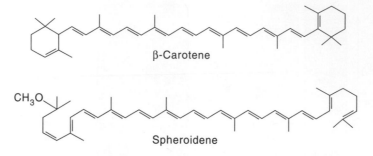

FIGURE 13   Structures of β-carotene, a major carotenoid in many types of plants, and spheroidene, a common carotenoid of photosynthetic bacteria. There are about 350 different natural carotenoids. These vary widely in color, depending principally on the number of conjugated double bonds.

Chloroplasts Have Two Photosystems Linked in Series

We have mentioned earlier that chloroplasts contain two types of reactive chlorophyll complexes, P700 and P680. Together with their antennae and their initial electron acceptors and donors, the reaction centers that contain P700 or P680 form two distinct assemblies called photosystem I and photosystem II, respectively (Fig. 14). Excitation of P700 (photosystem I) generates a strong reductant, which transfers electrons to NADP⁺ by way of several secondary electron carriers. Excitation of P680 (photosystem II) generates a strong oxidant, which oxidizes water to oxygen. The reductant formed in photosystem II injects electrons into a chain of carriers that connect the two photosystems. This scheme, first suggested by R. Hill and F. Bandall in 1960 is called the Z scheme. Note that the Z scheme differs from the bacterial electron-transport chain in being linear instead of cyclic.

The electron acceptors on the reducing side of photosystem II resemble those of purple bacterial reaction centers. The acceptor that removes an electron from P680 is a molecule of pheophytin *a*. The second and third acceptors are quinones (plastoquinone). As in the bacterial reaction center, electrons are transferred one at a time from the first quinone to the second. When the second quinone becomes doubly reduced, it picks up protons from the stromal side of the thylakoid membrane and dissociates from the reaction center.

The chain of carriers between the two photosystems includes the cytochrome *b/f* complex and a copper protein called plastocyanin. Like the bacterial cytochrome bc_1 complex, the cytochrome *b/f* complex contains two *b*-type hemes, an iron–sulfur protein, and a *c*-type cytochrome—cytochrome *f*. Electrons move through the cytochrome *b/f* complex from reduced plastoquinone to cytochrome *f*, and from there to plastocyanin, which transfers them to P700 in the reaction center of photosystem I.

Photosystem I Reduces Nicotinamide Adenine Dinucleotide Phosphate (NADP⁺) by Way of Iron–Sulfur Clusters

On the reducing side of photosystem I, the earliest electron acceptor appears to be a molecule of chlorophyll *a*. The next acceptor probably is a quinone, phylloquinone

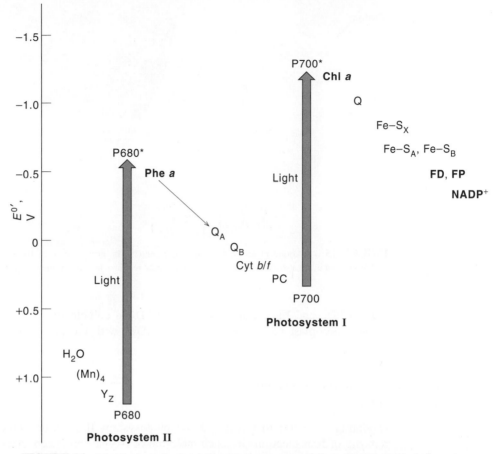

FIGURE 14   The Z scheme for the photosynthetic apparatus of plants. Two photochemical reactions are required to drive electrons from water to NADP. In this scheme, the solid arrows represent paths of electron flow. The electron carriers are positioned vertically according to their reducing strengths ($E^{0'}$ values), with the strongest reductants at the top. Electrons are pushed upward by light and flow downward spontaneously. $(Mn)_4$ = a complex of four manganese atoms bound to the reaction center of photosystem II; Yz = tyrosine side chain; Phe$_a$ = pehophytin a; Q_A and Q_B = two molecules of plastoquinone; Cyt b/f = cytochrome $b_6 f$: PC = plastocyanin; Chl$_a$ = chlorophyll a; Q = phylloquinone; Fe-S$_x$, Fe-S$_A$, and Fe-S$_B$ = iron–sulfur centers in the reaction center of photosytem I; FD = ferredoxin (a soluble iron-sulfur protein); FP = ferre-doxin-NADP oxidoreductase (a flavoprotsin).

(see Fig. 9), In these respects photosystem I resembles photosystem II and the purple photosynthetic bacteria, which use pheophytin a or bacteriopheophytin a followed by quinones as electron acceptors. From this point on, however, photosystem I is different. Its next set of electron carriers consists of four iron–sulfur clusters. These components are designated FD$_X$, FD$_A$, FD$_B$, and FD$_S$. FD$_X$, FD$_A$, and FD$_B$ each consists of four iron atoms and four inorganic sulfides held by four cysteines in the cubic struc-ture shown in Fig. 15. FD$_X$ resides on the two main polypeptides of the photosystem I reaction center; FD$_A$ and FD$_B$ are bound to a separate subunit. Electrons move from

Protein

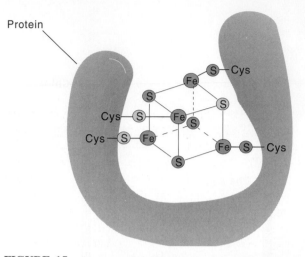

FIGURE 15   Structure of an Fe-4S iron–sulfur center. The arrangement of the sulfur ligands around each iron atom is approximately tetrahedral. The cysteine residues are part of the polypeptide chain.

FD_X to FD_A and FD_B, which then reduce FD_S, a soluble iron–sulfur protein found in the chloroplast stroma. FD_S reduces $NADP^+$ with the aid of a flavoprotein (see Fig. 14).

Oxygen Production Requires Accumulation of Four Oxidizing Equivalents

Oxidation of water to oxygen at the photosystem II reaction center requires the removal of four electrons for each molecule of oxygen produced. According to the Z scheme, each electron must traverse the photochemical reactions of both photosystems, so at least eight photons (two × four) must be absorbed for each oxygen molecule that is released.

The component that undergoes oxidation as the photosystem II reaction center progresses through its oxidation states is a complex of four atoms of manganese bound to the D1 polypeptide. The manganese atoms probably are oxidized sequentially from the Mn(III) level to Mn(IV). The manganese complex transfers electrons to P680 via the phenolic side chain of a tyrosine residue (Y in Fig. 14).

Flow of Electrons from Water to Nicotinamide Adenine Dinucleotide Phosphate ($NADP^+$) Is Linked to Proton Transport

As in mitochondria and purple photosynthetic bacteria, flow of electrons through the chloroplast's cytochrome b/f complex results in translocation of protons across the thylakoid membrane. When plastoquinone undergoes reduction in photosystem II and in the cytochrome b/f complex, protons are taken up from the stromal side of the membrane. When the quinone is reoxidized, protons are released into the thylakoid

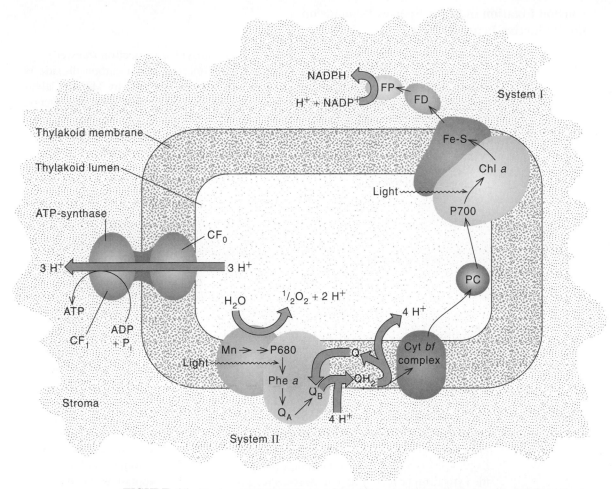

FIGURE 16   The transport of two electrons from photosystem II through the cytochrome *b/f* complex to photosystem I pumps four protons from the chloroplast stroma to the thylakoid lumen. Two more protons are released in the lumen for each molecule of water that is oxidized to oxygen, and one additional proton is taken up from the stroma for each molecule of $NADP^+$ that is reduced to NADPH. The movements of protons from the thylakoid lumen back to the stroma through an ATP-synthase drives the formation of ATP. The abbreviations used in this figure are the same as in Fig. 14.

lumen. This lowers the pH in the lumen and makes the inside positively charged with respect to the stroma. In current schemes of the reactions catalyzed by the cytochrome *b/f* complex, four protons move across the membrane for each pair of electrons that proceed to photosystem I. Two more protons are released in the lumen for each molecule of water that is oxidized to oxygen, and one proton is taken up from the stroma when $NADP^+$ is reduced to NADPH (Fig. 16). Proton movement back out through an ATP-synthase drives the formation of ATP.

Carbon Fixation in Many Species Depends on the Reductive Pentose Cycle

The ATP and NADPH that are generated by the photosynthetic electron-transfer reactions are used for fixation of carbon dioxide. Reactions in which carbon dioxide is incorporated into carbohydrates were discovered in the early 1950s by Melvin Calvin and coworkers in a series of studies that were among the first to use radioactive tracers. These experiments showed that 3-phosphoglycerate is formed from ribulose-1,5-bisphosphate. A molecule of carbon dioxide is incorporated in the process, so that one molecule of the five-carbon sugar ribulose-1,5-bisphosphate gives rise to two molecules of the three-carbon 3-phosphoglycerate. The reaction is catalyzed by ribulose bisphosphate carboxylase, which is found in the chloroplast stroma

$$CO_2 + \begin{array}{l} CH_2OPO_3^{2-} \\ | \\ C{=}O \\ | \\ CHOH \\ | \\ CHOH \\ | \\ CH_2OPO_3^{2-} \end{array} \xrightarrow{\text{ribulose bisphosphate carboxylase}} 2\begin{array}{l} CO_2^{-} \\ | \\ CHOH \\ | \\ CH_2OPO_3^{2-} \end{array} \quad . \tag{3}$$

Further studies revealed that other enzymes in the chloroplast stroma convert 3-phosphoglycerate back to ribulose-1,5-bisphosphate. The reactions by which this occurs are called the reductive pentose cycle. The reductive pentose cycle regenerates three molecules of ribulose-1,5-bisphosphate for every three that are carboxylated. In the process, three molecules of carbon dioxide are taken up, and there is a gain of one molecule of glyceraldehyde-3-phosphate. The expenses are nine molecules of ATP converted to ADP and six molecules of NADPH oxidized to $NADP^+$, or three molecules of ATP and two of NADPH consumed per molecule of carbon dioxide fixed. The glyceraldehyde-3-phosphate that is saved is exported from the chloroplast to the cytoplasm or is converted to hexoses for storage in the chloroplast as starch.

PHOTOSYNTHESIS AROSE EARLY IN EVOLUTION AND IS STILL WIDESPREAD AMONG EUBACTERIA

Photosynthesis began early in the evolution of living systems. There are fossil indications for the presence of prokaryotic photosynthetic organisms that date back 3.5 billion years. This fossil evidence centers on multilayered sedimentary formations termed stromatolites, which are found in South Africa and in western Australia. These ancient domed columns are similar to structures that have been deposited in shallow seawater by photosynthetic prokaryotes within the last 2000 years. In the same rock units from the period of 3.5 billion years ago, there are microscopic fossils of filamentous organisms resembling the present-day photosynthetic green nonsulfur bacterium *Chloroflexus aurantiacus*. However, it is unclear whether stromatolites were deposited by organisms of this nature or by cyanobacterial species that had the ability to evolve oxygen. Oxygen evolution must have been widespread 2.2 billion years ago because this is when O_2 began to accumulate in Earth's atmosphere.

Currently, photosynthetic organisms are found in all three kingdoms: archaebacteria, eubacteria, and eukaryotes. In eubacteria and eukaryotes, the photochemical reactions are based on the photooxidation of chlorophyll or bacteriochlorophyll. By contrast, several archaebacterial species (e.g., *Halobacterium halobium*) use a retinal-protein complex that undergoes isomerization on excitation with light. Although the photosynthetic archaebacteria resemble eubacteria and eukaryotes in using the energy of light to pump ions across a membrane, their photosynthetic apparatus is much less efficient than those of the systems that use chlorophyll or bacteriochlorophyll, and it does not provide the cells with enough energy to sustain true autotrophic growth. The two types of photosynthetic apparatus are totally unrelated structurally and evolutionarily. We focus on the more important eukaryotic and eubacterial systems.

Within the eubacteria kingdom, 5 of the 11 phyla include photosynthetic organisms: purple bacteria, green nonsulfur bacteria, green sulfur bacteria, Gram-positive bacteria, and cyanobacteria. Figure 17 shows the positions of these phyla in an evolutionary tree based on 16S-rRNA sequences (also see Fig. 8, chapter 21). The purple bacteria include the photosynthetic genera *Rhodopseudomonas, Rhodobacter, Chromatium*, and *Rhodospirillum*, along with many species of nonphotosynthetic bacteria such as *E. coli*. The green nonsulfur bacteria comprise the genera *Chlorobium* and *Prosthe-*

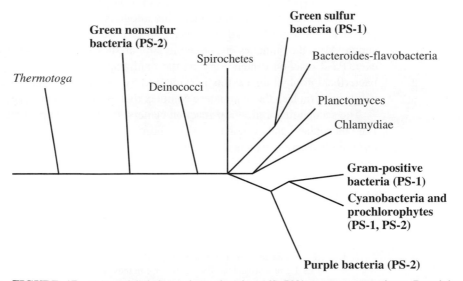

FIGURE 17 Eubacterial phylogenetic tree based on 16S-rRNA sequence comparisons. Branch lengths on the tree are approximately proportional to calculated evolutionary distances. Phyla that contain photosynthetic species are indicated with bold type. (Adapted from Woese, C. R. Bacterial Evolution, *Microbiol. Rev.* 51:221, 1987.)

cochloris. In the Gram-positive bacteria, there are three recently discovered photosynthetic genera (*Heliobacterium, Heliospirillum,* and *Heliobacillus*) in company with such nonphotosynthetic genera as *Bacillus.* Among the cyanobacteria there are numerous diverse photosynthetic genera, including *Anabaena, Synecococcus,* and *Synechocystis.* Also grouped with the cyanobacteria are two species that are called "prochlorophytes," *Prochloron didemna* and *Prochlorothrix hollandica.* Prochlorophytes resemble chloroplasts in having both chlorophyll *a* and chlorophyll *b,* and in lacking the phycobilisomes that are found in most other types of cyanobacteria. It is generally believed that chloroplasts of algae and higher plants arose by an endosymbiotic fusion of eukaryotic cells with organisms related to the prochlorophytes.

Reaction Centers Fall into Two Groups: PS-1 and PS-2

The 16S-rRNA sequences indicate that various photosynthetic species are related more closely to nonphotosynthetic species of the same phylum than they are to the photosynthetic organisms of other phyla that have similar photosynthetic systems (see Fig. 17). Moreover, photosynthetic reaction centers found in different eubacterial phyla and in eukaryotic chloroplasts have many features in common. All these reaction centers fall into one of two groups: PS-1 reaction centers and PS-2 reaction centers (Fig. 18).

In PS-2 reaction centers, the initial electron acceptor is a pheophytin or bacteriopheophytin and the second and third acceptors are quinones. This group includes the reaction centers of photosystem II in chloroplasts and cyanobacteria, and the reaction centers of purple bacteria and green nonsulfur bacteria. The quinones found in each of these reaction centers are functionally homologous in that Q_B always accepts electrons one at a time from Q_A and then picks up protons and dissociates from the reaction center. Also, the quinones are bound in similar sites near a nonheme iron atom. However, PS-2 reaction centers vary on the oxidizing side of the reactive chlorophyll or bacteriochlorophyll dimer; this is evidenced by the fact that the purple and green nonsulfur bacteria lack the manganese complex that is required for oxygen evolution.

The second major group of reaction centers, the PS-1 group, occurs in photosystem I of chloroplasts and cyanobacteria, in Gram-positive bacteria, and in green sulfur bacteria. PS-1 reaction centers are larger, more complex structures than PS-2 reaction centers contain many molecules of antenna pigments in addition to the few molecules that act as electron carriers. The initial electron acceptor is a chlorophyll or bacteriochlorophyll that transfers electrons to a set of iron–sulfur proteins by way of a quinone that is tightly bound to the reaction center.

FIGURE 18   The photosynthetic reaction centers found in plants and various types of bacteria can classified into the PS-1 group (a) and the PS-2 group (b). In these diagrams, the photochemical electron donors and acceptors are positioned vertically according to their reducing strengths ($E^{0'}$ values) A more negative $E^{0'}$ values means a stronger reductant. (For comparison, the $NADPH$–$NADP^+$ redox couple has an $E^{0'}$ of -0.32 V, and H_2O–O_2 has an $E^{0'}$ of $+0.82$ V.) Chl = chlorophyll *a*; Phe = pheophytin *a*, BChl = one of several different types of bacteriochlorophyll; BPh = bacteriopheophytin *a* or *b*.

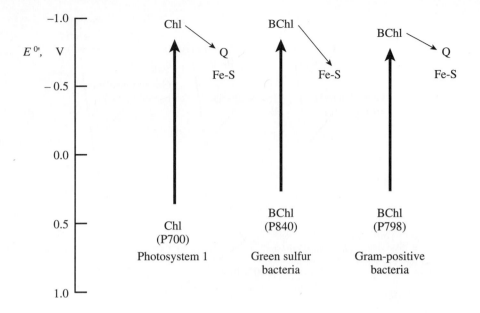

(a)   **PS -1 reaction centers**

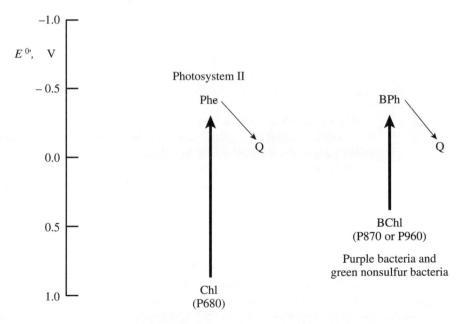

(b)   **PS -2 reaction centers**

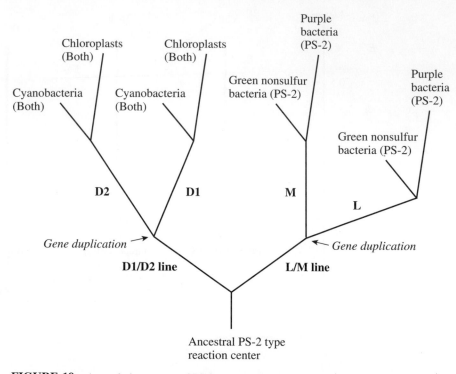

FIGURE 19   An evolutionary tree of PS-2-type reaction centers, based on "parsimony" analysis of the amino acid sequences of the main pair of polypeptides (subunits L and M or D1 and D2). (Adapted with kind permission from Kluwer Academic Publishers, from Blankenship, R. E. Origin and Early Evolution of Photosynthesis, *Photosynth. Res.* 33:91, 1992.)

Within the PS-2 reaction centers, comparisons of protein sequences indicate that the L subunit found in a given bacterial species is related more closely to the L subunits of other species than it is to the M subunit of the same species. Furthermore, the L and M subunits are related more closely to each other than to either the D1 or the D2 subunits of plants or cyanobacteria. These relationships are indicated in Fig. 19. Such analyses suggest that the L–M line of bacterial reactions diverged from the D1–D2 line of plant and cyanobacterial reaction centers at a relatively early stage of evolution (see Fig. 19). At this early stage of evolution of the reaction centers, both lines probably had only one subunit. Separate gene-duplication events in the two lines then led to the pairs of subunits that we know today. A similar gene duplication presumably gave rise to the pair of subunits found in PS-1 reaction centers.

The evolutionary trees in Figs. 17 and 19 show striking inconsistencies. For example, in the tree based on 16S-rRNA sequences (Fig. 17), green nonsulfur photosynthetic bacteria diverge from the other prokaryotes at an early stage, long before cyanobacteria split off from purple bacteria. By contrast, in the evolutionary tree of PS-2-type reaction centers, the D1–D2 line leading to cyanobacteria and the L–M line leading to green nonsulfur and purple bacteria diverge first. It has been suggested that this type of discrepancy is the result of horizontal gene transfer of some or all the reaction center components.

Reaction Centers Have Evolved from Water-Soluble to Membrane-Bound Systems

In search of a simplifying hypothesis Samuel Granick proposed that the biosynthetic pathway for photosynthetic porphyrins recapitulates the evolution of photosynthesis. In other words, we might expect the early members in the porphyrin pathway to have been active photosynthetic pigments in ancient times. The pathway for chlorophyll involves many steps and several branchpoints

$$Glycine + succinyl\text{-}CoA + \alpha\text{-}ketoglutarate$$
$$\downarrow$$
$$\downarrow$$
$$Uroporphyrinogen\ III \rightarrow \rightarrow vitamin\ B_{12}$$
$$\downarrow$$
$$Protoporphyrinogen$$
$$\downarrow$$
$$Protoporphyrin\ IX \rightarrow hemes$$
$$\downarrow Mg$$
$$Mg\text{-}Protoporphyrin\ IX$$
$$\downarrow$$
$$Protocholorophyllide\ a$$
$$\downarrow$$
$$Chlorophyllide\ a \rightarrow chlorophyll\ a \rightarrow chlorophyll\ b$$
$$\downarrow$$
$$Bacteriochlorophyllide\ a \rightarrow bacteriochlorophyll\ a.$$

The structures of the intermediates in the pathway are shown in Fig. 2 and later in Fig. 20. The porphyrinogens (uroporphyrinogen and protoporphyrinogen) are colorless cyclic compounds in which the four pyrrole rings are separated by nonconjugated methylene atoms. They can be formed nonenzymatically from simpler organic molecules under conditions designed to resemble those on prebiotic Earth. The first fully conjugated porphyrin in the pathway is protoporphyrin IX, which also serves as an intermediate for the biosynthesis of hemes. Because chlorophyll and bacteriochlorophyll are synthesized from protoporphyrin IX, we might expect that protoporphyrin IX functioned in primitive photosynthetic organisms. For similar reasons we might expect that chlorophyll was used before bacteriochlorophyll.

Although uroporphyrinogen and protoporphyrinogen have no absorption bands in the visible region of the spectrum and are not photochemically active, the corresponding porphyrins (uroporphyrin and protoporphyrin) have strong absorption bands in the deep-blue region near 400 nm. Protoporphyrin IX has a similar absorption band. When excited with light, these metal-free porphyrins undergo electron-transfer reactions. However, one ramification of Granick's hypothesis is that the photochemical reactions that occurred in the earliest photosynthetic organisms could have been different from the reactions that occur in contemporary organisms. In all present-day photosynthetic organisms, an excited molecule of chlorophyll or bacteriochlorophyll transfers an electron to a neighboring molecule, generating an oxidized chlorophyll that is reduced by a secondary donor. These three reactions are summarized by Eqs. (4) to (6).

$$P \xrightarrow{\text{light}} P^* \tag{4}$$

$$P^* + A \longrightarrow P^{\cdot +} + A^{\cdot -} \tag{5}$$

$$P^{\cdot +} + D \longrightarrow P + D^{\cdot +} \tag{6}$$

where P represents chlorophyll or bacteriochloroohyll, A is an electron acceptor, and D is a secondary electron donor. The third step [Eq. 6] is always much slower than the second step [Eq. 5]. Nonenzymatic reactions similar to these occur readily with chlorophyll, bacteriochlorophyll, and some other metalloporphyrins in solution. With metal-free porphyrins such as uroporphyrin or protoporphyrin, however, the excited molecule more generally reacts by first accepting an electron from a donor. The photo-reduced porphyrin can transfer an electron to another molecule (A), so that the final products are $A^{\cdot -}$ and $D^{\cdot +}$, just as they are in reactions (4) to (6), but the intermediates are different

$$P \xrightarrow{\text{light}} P^* \tag{7}$$

$$P^* + D \longrightarrow P^{\cdot -} + D^+ \tag{8}$$

$$P^{\cdot -} + A \longrightarrow P + A^{\cdot -}. \tag{9}$$

The distinction between these two reaction mechanisms hinges to a great extent on the metal atom. The magnesium atom in chlorophyll or bacteriochlorophyll makes the excited molecule easier to oxidize, but harder to reduce. This probably results mainly from the fact that the magnesium binds at least one additional electronegative ligand. In reaction centers and photosynthetic antenna complexes, this ligand usually is the imidazole side chain of a histidine residue; in solution it can be water or a polar organic molecule. The extra ligand stabilizes the positive charge on $P^{\cdot +}$, but it would destabilize the negative charge of $P^{\cdot -}$. An excited molecule of a metalloporphyrin is therefore a stronger reductant than is an excited molecule of the corresponding metal-free porphyrin.

These observations suggest that very early photosynthetic systems, based on metal-free porphyrins, such as protoporphyrin IX, might have served largely to generate oxidized organic molecules. These could have been either the oxidized form of the electron donor in Eq. (8) ($D^{\cdot +}$) or other molecules that were oxidized by $D^{\cdot +}$ in subsequent steps. Because the atmosphere of early Earth contained no oxygen and thus was more reducing than today's atmosphere is, and a variety of reduced organic mole-cules probably were formed by abiotic photochemical reactions, the generation of oxidized species could have been necessary for the synthesis of other biological molecules. The formation of the metalloporphyrin Mg-protoporphyrin IX would have changed the situation, making reaction sequences like Eqs. (4) to (6) more favorable and allowing the generation of stronger reductants [A$^{.-}$ in Eq. 5]. Initially, the new reductants could have been used simply to generate other new types of reduced organic molecules. Quinones evidently proved to be particularly useful electron carriers in these reactions, because they are still used by most, if not all, photosynthetic organisms. Eventually, the reoxidation of a reduced quinone also was used to drive protons across a cellular membrane.

There is another change in the chemistry of the porphyrins that occurs in parallel with the formation of metalloporphyrins from the metal-free molecules: a progressive decrease in the electrical charge on the molecule. Uroporphyrinogen III has a charge of -8 and is water-soluble molecule; coproporphyrinogen, an intermediate in the conversion of uroporphyrinogen to protoporphyrinogen, has a charge of -4 (Fig. 20). Protoporphyrinogen and protoporphyrin IX both have charges of -2. The charge on chlorophyllide a or bacteriochlorophyllide a is -1, and that on chlorophyll or bacteriochlorophyll is 0. The molecules thus become less and less soluble in water as the biosynthetic series proceeds, and increasingly likely to collect in a phospholipid membrane or to bind to a hydrophobic site on a protein. Localization of the photosynthetic apparatus in a cellular membrane was probably a key step in coupling it to a system that translocates protons.

As indicated in Fig. 16, present-day reaction centers of the PS-2-type drive proton translocation by reducing quinones on one side of the cellular or thylakoid membrane. Protons taken up on one side of the membrane are released on the opposite side when the quinone is reoxidized. An important feature of these systems is that they contain two quinones with distinctly different functions. The first quinone Q_A, takes electrons one at a time from the initial electron acceptor (pheophytin or bacteriopheophytin), and passes them quickly to the second quinone, Q_B (see Figs. 9 and 11). The second quinone, after collecting two electrons, picks up protons, dissociates from the reaction center, and diffuses to the cytochrome bc or b/f complex, where it is reoxidized. This specialization of the functions of the two quinones probably enhances the efficiency of the photochemical apparatus by allowing electron transfer from P870 to P680 to Q_A to occur rapidly without regard for the redox or protonation state of Q_B.

Judging from the strong homology of the two main polypeptides in the reaction center, the two-quinone system apparently arose by a gene-duplication process from a simpler one quinone system (see Fig. 19). The evolution of the two-quinone system for removing electrons from the PS-2 reaction center could have promoted comparable increases in the effectiveness of the components that returned electrons to the oxidized chlorophyll complex.

In the earliest reaction centers based on water-soluble, metal-free porphyrins, the electron donors [D in Eq. (8)] most likely included inorganic ferrous iron, which was abundant in the prebiotic oceans. As the reaction center moved into cellular membranes, adopted Mg porphyrins as the photochemically active molecules, and became more efficient at reducing quinones, simple inorganic reductants like ferrous iron were replaced by Fe porphyrins and ultimately by cytochromes.

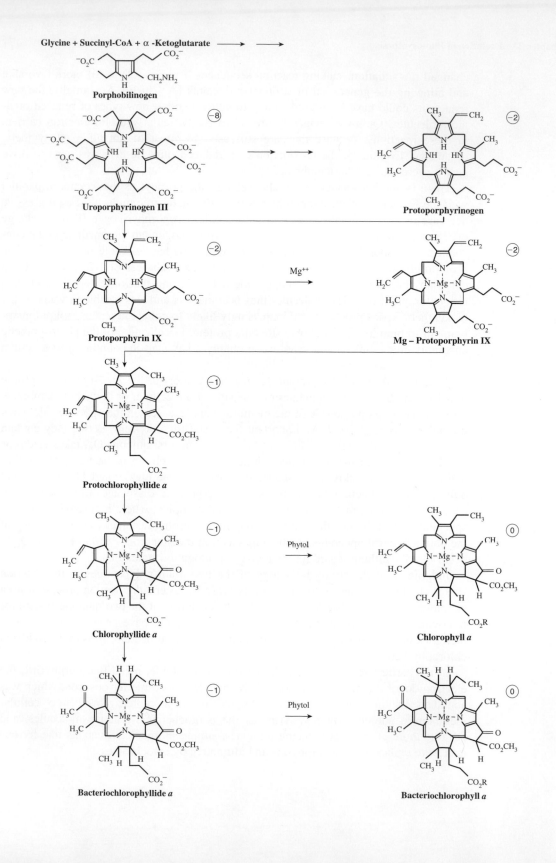

Glycine + Succinyl-CoA + α -Ketoglutarate ⟶ ⟶

Porphobilinogen

Uroporphyrinogen III

Protoporphyrinogen

Protoporphyrin IX

Mg^{++}

Mg – Protoporphyrin IX

Protochlorophyllide *a*

Chlorophyllide *a*

Phytol

Chlorophyll *a*

Bacteriochlorophyllide *a*

Phytol

Bacteriochlorophyll *a*

Evolution of Antenna Systems Favored Pigments with Long-Wavelength Absorption Bands

Antenna systems increase the reaction center's ability to absorb light. Pigments that comprise the antenna system must be efficient in transferring energy and they also must be efficient in absorbing a wide range of wavelengths from the solar source. Early developments in photosynthetic pigments benefited the former the most. To see this we must review the pigments starting with protoporphyrin IX that absorbs well near 400 nm but not well at longer wavelengths. The addition of a magnesium atom to form Mg protoporphyrin IX had only minor effects on the absorption spectrum. The subsequent evolution of protochlorophyllide a resulted in a pigment with a long-wavelength absorption band near 620 nm with an increased strength of about a factor of 3. Next, the formation of chlorophyllide a increased the absorption strength by about a factor of 4.5 and moved the absorption maximum farther to the red (662 nm). The ability of the reaction center to absorb and to use red light was increased by these changes, but even more important, the strengthening of the long-wavelength band made possible the development of antenna systems. This is because the rate constant for resonance energy transfer from one molecule to another is proportional to the product of the strengths of the long-wavelength absorption bands of the two molecules. Only the long-wavelength absorption band is important in this regard because the excited molecule almost always relaxes very quickly into the lowest excited state, which corresponds to this band. Thus the conversion of protophyrin IX of Mg protoporphyrin IX to chlorophyllide a or chlorophyll a, which increases the strength of the long-wavelength absorption band by a factor of about 17, would increase the rate constant for energy transfer by roughly the square of this factor, or almost 300-fold. The rate constant for energy transfer is critical if an antenna system is to be useful, because the movement of excitations to the reaction center has to compete with fluorescence and other nonproductive decay processes.

The rate of resonance energy transfer also is much higher for close-packed pigments. The close packing of pigments in the antenna was strongly aided by the decreased negative charge on pigments like chlorophyllide (-1) and chlorophyll a (0).

Much of the subsequent evolution of antenna systems represented a continuing search for ways to use all accessible wavelengths of light. By synthesizing chlorophyll b in addition to chlorophyll a, prochlorophytes and chloroplasts broadened the usefulness of their antennas in the red-orange region of the spectrum. Another advantage of using chlorophylls a and b for an antenna is that these molecules have absorption bands at both short wavelengths (between 350 and 450 nm), and long-wavelengths (630 to 700 nm). To capture light with wavelengths between 450 and 630 nm, contemporary photosynthesizers use a variety of accessory pigments, including many different types of carotenoids. Cyanobacteria and red algae specialized to the use

FIGURE 20   Intermediates in the biosynthesis of chlorophyll a and bacteriochlorophyll a (R = phytyl.) Encircled numbers indicate the electrical charge on the molecule.

of green and yellow light (from 500 to 600 nm) by developing phycobilisomes. Other types of bacteria by synthesizing bacteriochlorophyll a, moved the long-wavelength bands of their antenna systems to the near infrared, beyond the region of the spectrum that had been claimed by cyanobacteria and chloroplasts.

In addition to their role in improving the absorption spectrum of antenna, carotenoids had an even more important role in photosynthetic species as well as many nonphotosynthetic species. They protect living systems from the harmful effects of oxygen by discharging several highly reactive and toxic forms of oxygen, the superoxide radical ($O_2^{.-}$) and excited "singlet" O_2, which are generated when chlorophyll or other organic molecules are illuminated in the presence of molecular oxygen. The formation of superoxide and singlet O_2 would not have been a problem for the earliest photosynthesizers, when Earth's atmosphere has little free oxygen. It was only after the development of photosynthetic oxygen evolution and the eventual flooding of the atmosphere with oxygen that such a protective mechanism became essential.

Photosynthetic Systems Ultimately Turned Their Attention to Water as a Source of Electrons

Primitive photosynthesizers probably used reductants such as $Fe(OH_2)$, hydrogen, hydrogen sulfide, and other reduced sulfur compounds where they were abundant. It is relatively easy to remove electrons from these materials to synthesize reduced organic molecules. As local environments became depleted of these sources of electrons, some photosynthetic autotrophs were forced to adapt to weaker reductants. This meant that their reaction centers had to generate comparably stronger oxidants.

A plausible sequence for the evolution of stronger oxidants might have started with reaction centers that had evolved to oxidize $Fe(OH)_2$, or other ferrous iron complexes, with relatively minor modifications, might oxidize similar manganese complexes from the level of Mn(II) to Mn(III). This would require an increase in the oxidizing strength of $P^{.+}$ in the reaction center, because the Mn(II)–Mn(III) redox couple is more oxidizing than Fe(II)–Fe(III) is. An increase in the oxidizing strength of $P^{.+}$ could be achieved by mutations that made the local region of the protein less effective at stabilizing the positive charge of the oxidized chlorophyll. Additional mutations that raised the oxidizing strength still further might have allowed the oxidation of Mn(III) to Mn(IV). A gradual changeover from iron to manganese could have been useful in regions where manganese was accessible, if the greater oxidizing strength of the manganese complex allowed the reaction center to extract electrons from other secondary donors such as amines. In this connection it is noteworthy that the oxygen-evolving center of photosystem II in present-day organisms oxidizes hydroxylamine, as well as water. Further modification presumably led to the formation of still stronger complexes involving two atoms of manganese, and eventually to the complexes of four manganese atoms that are found today. There is still a great deal we do not know about these marvelous tetrameric complexes of manganese that permit photosystems to extract electrons from the oxygen of water and thereby to generate oxygen.

PS-1 Reaction Centers May Have Evolved from the Antenna
Complexes of PS-2 Reaction Centers

There is no firm evidence that reaction centers of the PS-1 and PS-2 groups evolved from a common ancestor. However, the similarity of the photochemical reactions that occur in all reaction centers, as well as the universal dependence on chlorophylls and quinones, suggests that the evolutionary origins of the two groups were linked. How did PS-1 reaction centers come to be associated with PS-2 reaction centers in cyanobacteria and chloroplasts, while the two types of reaction centers occur independently in other types of bacteria? In one scheme these two photosystems diverged from primordial reaction centers at an early stage of evolution, and remained apart in PS-1 containing green sulfur and Gram-positive bacteria and in PS-2 containing lines of green nonsulfur and purple bacteria. Cyanobacteria and progenitors of chloroplasts then arose by a gene fusion that reunited the two types of reaction centers and allowed the development of the z scheme (Fig. 21).

Another possibility is that PS-1 reaction centers evolved from an antenna complex in organisms that contained reaction centers of the PS-2 type. Figure 22 shows a possible evolutionary tree based on this idea. This scenario is supported by the finding that PS-1 polypeptides have a weak homology with a chlorophyll-binding protein that is a part of the antenna system associated with photosystem II.

Because the amount of energy available from an einstein of red light is fixed at about 41 kcal [see Eq. (3)], a reaction center that evolves to generate stronger and stronger oxidants is increasingly limited in its ability to generate strong reductants.

Thus the changes in PS-2 reaction centers that ultimately enabled them to extract electrons from the oxygen of water prevented them from generating strong reductants like NADPH. This situation could have created the evolutionary pressure for the formation of PS-1 reaction centers and for connecting the two types of reaction centers in a series.

The notion that PS-1 reaction centers evolved from an antenna complex that originally served reaction centers of the PS-2 type leaves open the question of why green sulfur bacteria and Gram-positive bacteria only contain reaction centers of the PS-1 type. In the scheme shown in Fig. 22, the green sulfur and Gram-positive bacteria are descendants of organisms that once had both types of reaction centers but lost their PS-2 reaction centers at some point in their evolutionary history. Such losses are not uncommon in evolution. In fact they may account for the fact that many present-day species of eubacteria have no photosynthetic apparatus at all. A loss of PS-2 reaction centers could have occurred as a result of evolutionary adaptation to special environments that contained an abundance of relatively strong reductants. The green sulfur bacteria, for example, generally live in environments that contain hydrogen sulfide. Photosynthetic Gram-positive bacteria also have adapted to live under anaerobic conditions. There are even several species of cyanobacteria that live in sulfide-rich environments and do not evolve oxygen.

Whichever of these scenarios is correct, the ability to oxidize water provided cyanobacteria and chloroplasts with an abundant source of electrons. By connecting

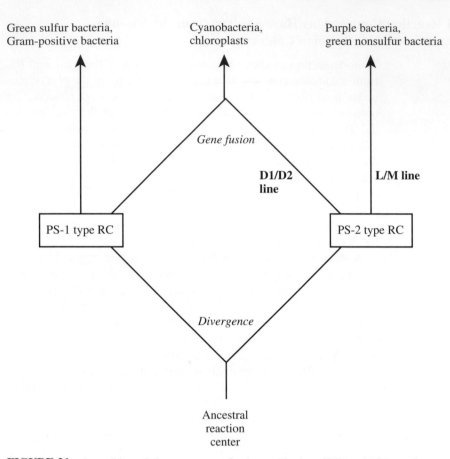

FIGURE 21   A possible evolutionary sequence for the combination of PS-1 and PS-2 reaction centers by gene fusion in a progenitor of cyanobacteria and chloroplasts. (Adapted with kind permission from Kluwer Academic Publishers, from Blankenship R. E. Origin and Early Evolution of Photosynthesis, *Photosynth. Res.* 33:91 1992.)

two reaction centers in series, cyanobacteria and chloroplasts were able to use the energy of two photons to move each electron from water to $NADP^+$. By harnessing the energy of sunlight to drive the formation of carbohydrates and the release of oxygen into the atmosphere, cyanobacteria and chloroplasts made possible the continued evolution of life on Earth.

SUMMARY

In this chapter we have examined photochemical reactions and their evolution.

1. In plants, the photochemical reactions of photosynthesis take place in the chloroplast thylakoid membrane. In bacteria, they take place in the cell membrane.

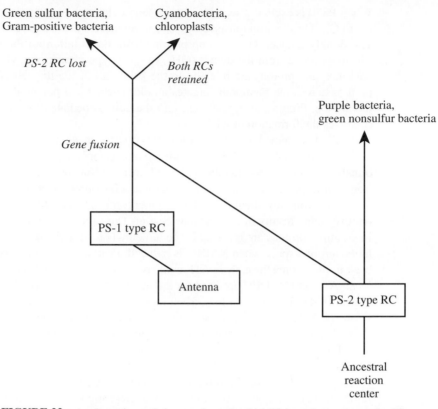

FIGURE 22   An alternative evolutionary scheme, in which PS-1 reaction centers evolved from an antenna complex in organisms that already had PS-2 reaction centers. PS-2 reaction centers subsequently were lost in a branch leading to Gram-positive bacteria and green sulfur bacteria. Antenna complexes continued to evolve independently in all branches of the tree (not shown).

2. Chlorophyll and bacteriochlorophyll become strong reductants when excited by light. Photooxidation of chlorophyll or bacteriochlorophyll occurs in reaction centers. Chloroplasts have two types of reaction centers: photosystem I, which contains a reactive chlorophyll complex called P700, and photosystem II, which contains P680.

3. Antenna systems of small pigment–protein complexes assemble into large arrays that contain hundreds of pigment molecules per reaction center. When the antenna absorbs a photon, the excitation energy moves rapidly from complex to complex by resonance energy transfer until the energy is trapped in a reaction center.

4. The bacterial reaction center contains two molecules of bacteriopheophytin and two of bacteriochlorophyll, in addition to two bacteriochlorophylls of P870. When P870 is excited, it transfers an electron to one of the bacteriopheophytins. The bacteriopheophytin then reduces quinone Q_A, which in turn reduces another quinone Q_B.

When P870 is excited a second time, a second electron is sent through the same carriers to Q_B. The electron released by P870 is replaced by one from a c-type cytochrome. The doubly reduced Q_B picks up protons from the solution on the inside of the cell. Electrons move from the quinol (QH_2) to the c-type cytochrome via a cytochrome bc complex, and protons are released to the solution outside the cell. This movement of protons across the membrane creates an electrochemical potential gradient across the membrane. Proton movement back into the cell is mediated by an ATP-synthase that catalyzes the formation of ATP.

5. Photosystems I and II of chloroplasts operate in series.

6. Photooxidation of P700 in photosystem I appears to reduce a chlorophyll that transfers electrons to membrane-bound iron–sulfur proteins, next to the soluble iron–sulfur protein ferrodoxin, and finally to a flavoproteins that reduces $NADP^+$.

7. Electron movement through the cytochrome b/f complex between the chloroplast photosystems results in proton translocation from the stroma to the thylakoid lumen. In addition, protons are released in the lumen when water is oxidized, and are taken up in the stromal space when $NADP^+$ is reduced. Protons move from the thylakoid lumen back to the stroma through an ATP-synthase driving the formation of ATP.

8. ATP and NADPH are used for the incorporation of carbon dioxide into carbohydrates. In the first reaction for carbon fixation the carbon dioxide reacts with ribulose-1,5-bisphosphate to give two molecules of 3-phosphoglycerate.

9. There is fossil evidence that photosynthetic organisms existed 3.5 billion years ago. Today photosynthetic organisms are found in all three kingdoms. In eubacteria and eukaryotes, the photochemical reactions are based on the photooxidation of chlorophyll or bacteriochlorophyll. By contrast, several archaebacterial species use a retinal-protein complex that undergoes isomerization on excitation with light.

10. Photosynthetic reaction centers of diverse organisms fall into two groups. The PS-2 group includes the reaction centers of photosystem II in chloroplasts and cyanobacteria, and the reaction centers of purple bacteria and green nonsulfur bacteria. The PS-1 group occurs in photosystem I of chloroplasts and cyanobacteria, in Gram-positive bacteria, and in green sulfur bacteria. These are larger and more complex than the PS-2 reaction centers. Various photosynthetic species are related more closely to nonphotosynthetic species of the same phylum than they are to the photosynthetic organisms of other phyla that have similar photosynthetic systems.

11. It has been proposed that the biosynthetic pathway leading to chlorophyll and bacteriochlorophyll in present-day organisms recapitulates the evolution of photosynthesis. This ad hoc hypothesis has been used to suggest that reaction centers started with metal-free, water-soluble systems that evolved into metal-containing, membrane-bound systems.

12. The first photosystems probably used reductants such as ferrous iron, hydrogen, and hydrogen sulfide as a source for electrons. As local environments became depleted of these sources of electrons, photosynthetic organisms were forced to adapt to weaker and weaker reductants. This meant that their reaction centers had to generate stronger and stronger oxidants. Finally, the oxygen of water became the main source of electrons for photosynthesis. It is still unclear how photosynthetic organisms solved

the problem of generating an oxidant that is strong enough to break apart water, without unleashing a reactive species that would destroy the reaction center itself.

13. The basic similarity of the photochemical reactions that occur in all reaction centers, as well as the universal dependence on chlorophylls and quinones, suggests that the evolutionary origins of the two types of photosystems are linked. Another possibility is that PS-1 reaction centers evolved from an antenna complex in organisms that contained reaction centers of the PS-2 type.

Problems

1. Do you think life could have originated in the absence of direct contact with light? Explain your answer.
2. What is the evidence for when oxygen became abundant in the atmosphere?
3. In Chapter 23 we have emphasized the importance of porphyrins in the early evolution of photosynthesis. However, a thorough reading of other chapters in this text indicates important photosynthetic reactions that did not involve porphyrins and that probably preceded the porphyrins. What are they?
4. Granick proposed that the biosynthetic pathway (for bacteriochlorophyll or chlorophyll) recapitulates the evolution of photosynthesis. What major principle of evolutionary change is embodied in this proposal?
5. What is the most convincing piece of evidence that the PS-2 photosystems found in green nonsulfur bacteria and purple bacteria are evolutionarily related to photosystem II found in chloroplasts?
6. Does the distribution of photosystems indicated by the phylogenetic tree in Fig. 17 suggest a pattern of vertical evolution or horizontal evolution?
7. Before oxygen of water became the predominant source of electrons to run most photosystems, other sources of electrons were used. Why do you suppose that it took photosystems a billion years to get around to this transition?
8. The earliest photosystems are believed to have used porphyrins without a Mg ligand to harness light energy. What advantages might this have had in the early evolution of photosystems?

References

Awramik, S. M. The Oldest Records of Photosynthesis, *Photosynth. Res.* 33: 75–89, 1992.

Blankenship, R. E. Origin and Early Evolution of Photosynthesis, *Photosynth. Res.* 33: 91–111, 1992.

Buck, R. The Antiquity of Oxygenic Photosynthesis: Evidence from Stromatolites in Sulphate-Deficient Archaean Lakes, *Science* 255: 74–77, 1992.

Granick, S. The Structural and Functional Relationships between Heme and Chlorophyll, *Harvey Lectures* 44: 226–245, 1987.

Mauzerall, D. Chlorophyll and Photosynthesis. *Philos. Trans. R. Soc. London Ser. B 273*: 287–294, 1976

Nagashima, K. V. P., Hiraishi, A., Shimada, K., and Matsuura, K. Horizontal Transfer of Genes Coding for the Photosynthetic Reaction Centers of Purple Bacteria, *J. Mol. Evol.* 45: 131–136, 1997.

Nisbet, E. G., Cann, J. R., and Dover, C. L. V. Origins of Photosynthesis, *Nature (London)* 373: 479–480, 1995.

CHAPTER 24

Origin and Elaboration
of the Genetic Code

In Chapter 16 we see that translation is the most complex phase of information transfer from the genome to the polypeptide chain. A sequence of deoxyribonucleotides in the DNA is selectively transcribed into a complementary sequence of ribonucleotides in a messenger RNA (mRNA) molecule, which then serves as a template for polypeptide synthesis. The amino acids become linked to transfer RNAs (tRNAs) before they become linked to each other. The charged aminoacyl tRNAs bind to complementary sites on the mRNA. This interaction involves a sequence of three bases on the mRNA, the *codon,* and three bases on the tRNA, the *anticodon.* Our goal in this chapter is to trace the course of evolution of this codon–anticodon interaction.

TRIPLET CODE MAY HAVE EVOLVED FROM A RELAXED SINGLET OR DOUBLET CODE

One of the first questions to arise is whether the code was always a triplet code or was at one time a singlet code or a doublet code. Before attempting to answer this question, let us clarify what is meant by a single code or a doublet code. In a *strict singlet code*, each base in the reading frame of the message would represent an amino acid. Alternatively, one could imagine a *relaxed singlet code*, in which every third base was a coding base. Similarly, a *strict doublet code* would entail a sequence in which every two bases code for a particular amino acid. However, a doublet code also could be present in a sequence of three bases in which two out of every three bases were coding bases. Extrapolating from the evolutionary principle that it must be possible to proceed in simple steps from one stage of evolution to the next, a relaxed code seems much more likely. Consequently, it seems most likely that the code relating base sequences in the RNA to amino acids sequences in proteins was a relaxed singlet or doublet code in which one or two bases in every triplet carried coding information. Then, as greater demands were made on the coding capacity of the translation system, many of the triplets took on essential meaning for all three bases.

MIDDLE BASE IN THE CODON MAY HAVE BEEN THE FIRST ONE TO ACQUIRE MEANING

Another evolutionary problem that must be considered is the significance of each of the three positions in the codon (and anticodon) in the early development of translation systems. Even today it is apparent that the three positions in the codon are used differently. Thus the third (3′) position in the codon is subject to wobble pairing with the first (5′) base in the anticodon (see Chapter 16). The third base in the codon is the least characteristic, while the middle base is the most characteristic, with the first base running close behind. This is apparent from Table 1, in which the triplet nucleotide sequences that specify particular amino acids are indicated. Amino acids are always represented by the same base in the middle position with the exception of serine, which is represented by C in four of its codons and by G in its other two codons. With respect to the 5′ base of the codon, there are three encoded amino acids—leucine, serine, and arginine—that are represented by more than one base. If we look at the 3′ base in the codon, we find much more variability. Only two amino acids, methionine and tryptophan, are represented by a single base in this position, and these two amino acids have only one codon. The remainder are represented by either two or four bases in this position except for isoleucine, which is represented by three bases in this position. Furthermore, in those cases where an amino acid is represented by either of two bases in this position, the two bases are either both pyrimidines or both purines. This final observation is closely related to the wobble rules that either an A or a G in the third position of the codon can pair with a U in the first base of the anticodon, and either a

TABLE 1
The Genetic Code

Group	Amino acid	5'-OH base(s)	Middle base(s)	3'-OH base(s)
I	Phe	U	U	U(1), U(2)
	Leu	U(2), C(4)	U	U(1), C(1), A(2), G(2)
	Ile	A	U	U(1), C(1), A(1)
	Met	A	U	G
	Val	G	U	U(1), C(1), A(1), G(1)
II	Ser	U(4), A(2)	C(4), G(2)	U(2), C(2), A(1), G(1)
	Pro	C	C	U(1), C(1), A(1), G(1)
	Thr	A	C	U(1), C(1), A(1), G(1)
	Ala	G	C	U(2), C(1), A(1), G(1)
III	Tyr	U	A	U(1), C(1)
	His	C	A	U(1), C(1)
	Gln	C	A	A(1), G(1)
	Asn	A	A	U(1), C(1)
	Lys	A	A	A(1), G(1)
	Asp	G	A	U(1), C(1)
	Glu	G	A	A(1), G(1)
IV	Cys	U	G	U(1), C(1)
	Trp	U	G	G
	Arg	C(4), A(2)	G	U(1), C(1), A(1), G(1)
	Gly	G	G	U(1), C(1), A(1), G(1)

U or a C in the third position of the codon can pair with a G in the first base of the anticodon (see Chapter 16, Table 1).

AMINO ACIDS WITH SIMILAR SIDE CHAINS TEND TO BE ASSOCIATED WITH ANTICODONS THAT SHOW CHEMICAL SIMILARITIES

With respect to evolutionary origins it is appropriate to ask whether there are any common chemical properties of amino acids that might lead to a preferential association with their respective codons (or anticodons). Because amino acids become covalently linked to tRNAs, it might be expected that common properties would be found between the anticodon bases and the amino acid instead of between the codon bases and the amino acid. In fact, there is a striking correlation of this sort found in a number of cases (see groups I and III in Table 1). Group I contains most of the strongly hydrophobic amino acids—phenylalanine, leucine, isoleucine, and valine. These amino acids all use the same middle anticodon base A, which is the most hydrophobic of the four bases. Group III contains strongly hydrophilic amino acids; these are all associated with the middle anticodon base U, which is the most hydrophilic of the four bases.

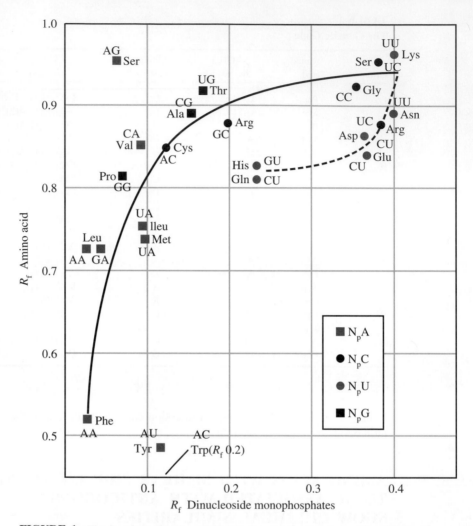

FIGURE 1   The R_f values of amino acids plotted versus the R_f values of the dinucleoside monophosphates. The R_f value is defined as the ratio of the distance traveled by a substance to the total distance traveled by the developing solvent. The solvent was 10 vol% saturated ammonium sulfate, pH 7.0. The dashed lines show the correlations for charged amino acids. (From Weber, A. L. and Lacey, J. C., Jr. Genetic Code Correlations: Amino Acids and Their Anticodon Nucleotides, *J. Mol. Evol.* 11:199, 1978, Reprinted by permission.)

This correlation between hydrophobicity of anticodon and amino acid has been extended by Weber and Lacy to a consideration of both the middle and 3′ positions in the anticodon. They measured the R_f values of amino acids and dinucleoside monophosphates by paper chromatography in a solvent that contained a high salt concentration (Fig. 1). The salt was present to mask the extreme polarity of the charged phosphate group in each dinucleoside monophosphate. A relationship was found between the R_f values for a particular amino acid and the bases found in the middle and 5′ position of its anticodon. For example, one of the most hydrophobic amino acids is phenylalanine, and the most hydrophobic dinucleoside monophosphate is ApA; both codons for phenylalanine start with UU. The most polar amino acid is lysine, and the most polar dinucleoside monophosphate is UpU; both of the codons for lysine start with AA. These correlations in hydrophobicity between amino acids and their anticodons are tantalizing because they may have some bearing on the origin of the genetic code. Possibly the binding between amino acid and anticodon in primitive tRNAs was instrumental in determining the specificity of the linkage between amino acids and the tRNA. In contemporary tRNAs the region of the tRNA that mediates the specificity between amino acids and charging enzymes usually, but not always, includes residues in the anticodon region as well as residues outside of this region.

AMINO ACIDS IN THE SAME BIOSYNTHETIC PATHWAY USE SIMILAR CODONS

Biosynthesis pathways for all 20 amino acids most commonly found in proteins start from carbohydrate skeletons. (Fig. 2). Amino acids are grouped into specific biosynthetic families according to the carbohydrate precursor (Table 2). For example, the pyruvate family comprises the amino acids alanine, valine, and leucine, whose common carbohydrate precursor is pyruvate. It seems likely that within a given amino acid family, the amino acid that is synthesized in the least number of steps was probably the first to be synthesized in primitive living systems. Thus an amino acid like aspartate, which is formed in one step from oxaloacetate, was probably available in living systems before lysine, which requires an additional nine steps.

It occurred to us that amino acids in the same biosynthetic families might have more code letters in common than amino acids found in different biosynthetic families have. Inspection of Table 2 shows that this is the case. For example, all codons of the pyruvate family can be arranged so that codons for alanine, valine, and leucine have two code letters in common. Similarly, members of the glycerate-3-phosphate family have many codons in common with each other. In contrast, there is little similarity between codons of the two families. One explanation for these correlations is that codons for particular families originally served to represent one member of that family, and that as other amino acids were added to that family by an extension of the biochemical reactions associated with a common pathway, some of these codons became assigned to other members of that family.

Another indication that certain amino acids are more primitive than others comes from a comparison of the amino acids synthesized in the Miller spark discharge experiment and those found in a meteorite known as the Murchison meteorite (Table 3). There is a strong correlation between the amino acids originating from these two different sources (starred in Table 3). Five amino acids are found on both lists: alanine,

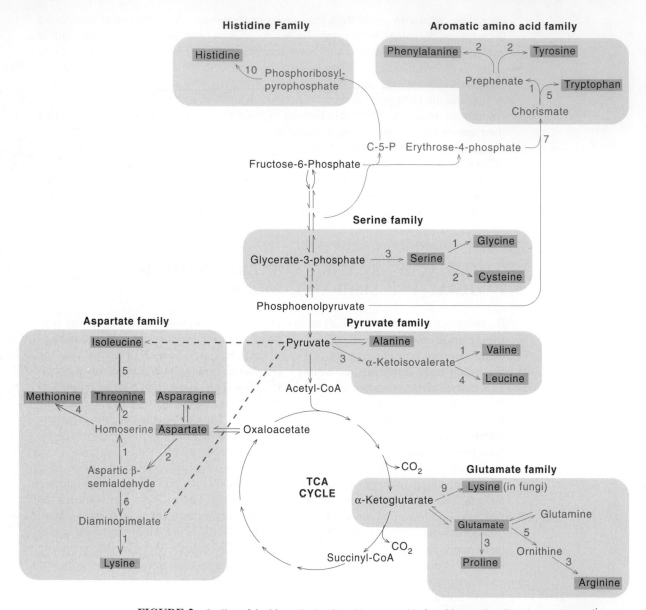

FIGURE 2   Outline of the biosynthesis of the 20 amino acids found in proteins. The *de novo* biosynthesis of amino acids starts with carbon compounds found in the central metabolic pathways. Some key intermediates are illustrated and the number of steps in each pathway is indicated alongside the conversion arrow. All amino acids are emphasized by boxes. Dashed arrows from pyruvate to both diaminopimelate and isoleucine reflect the fact that pyruvate contributes some of the side-chain carbon atoms for each of these amino acids. Note that lysine is unique in that two completely different pathways exist for its biosynthesis.

TABLE 2
Amino Acid Biosynthetic Families[a]

Common intermediate or precursor	Amino acid group[b]	Codons							
Pyruvate	Ala			GCU	GCC	GCA	GCG		
	Val			GUU	GUC	GUA	GUG		
	Leu			CUU	CUC	CUA	CUG		
				UUA	UUG				
Glycerate-3-phosphate	Ser	UCU	UCC	UCA	UCG	AGU	AGC		
	Gly					GGU	GGC	GGA	GGG
	Cys					UGU	UGC		
—	His					CAU	CAC		
Chorismate	Phe				UUU	UUC			
	Tyr				UAU	UAC			
	Trp					UGG			
Oxaloacetate	Met				AUG				
	Ile			AUU	AUC	AUA			
	Thr			ACU	ACC	ACA	ACG		
	Asp			GAU	GAC				
	Asn			AAU	AAC				
α-Ketoglutarate	Lys				AAA	AAG			
	Glu				GAA	GAG			
	Gln				CAA	CAG			
	Pro	CCU	CCC	CCA	CCG				
	Arg	CGU	CGC	CGA	CGG	AGA	AGG		

[a]Bases are written in 5′→3′ direction unless otherwise indicated.
[b]Grouped according to biosynthetic pathways.

aspartate, glutamate, glycine, and valine. Biochemically the first three of these are made in one step from their carbohydrate skeletons. Glycine is made in four steps and valine is made in five steps. The small number of steps involved in the biosynthesis of these five amino acids correlates with their presence from these two nonbiological sources. It could be inferred that amino acids that are easier to make by organochemical procedures are also easier to make by biosynthetic pathways.

AMINO ACIDS THAT CONTAIN CHEMICALLY SIMILAR SIDE CHAINS FREQUENTLY SHARE TWO CODE LETTERS

Another correlation found for codons of different amino acids is related to the amino acid side chains. Amino acids with similar side chains tend to have closely related

TABLE 3

Correlation between Amino Acids Found in Sparking Discharge Experiment and Murchison Meteorite

Amino acid[a]	Steps from central metabolic pathways	Codons
Sparking discharge (CH_4, NH_3, H_2O, H_2)		
*Valine	5	
Leucine	5	
Isoleucine	11	
Serine	3	
*Glycine	4	
*Alanine	1	
*Aspartate	1	
*Glutamate	1	
Murchison meteorite		
*Valine	5	GUN
*Alanine	1	GCN
*Glycine	4	GGN
*Glutamate	1	CAA, CAG
*Aspartate	1	GAU, GAC
*Proline	4	CCN

[a] Star indicates being found both in sparking experiment and in Murchison meteorite.

codons. This can be seen by inspection of the codon chart (see Chapter 9, Table 2). For example, all serine codons are closely related to threonine codons. Both of these amino acids have hydroxyl side chains. The three amino acids with basic side chains also have closely related codons. Thus all histidine codons are closely related to arginine and lysine codons, and two lysine codons are closely related to two arginine codons. Even the three stop codons are closely related. Thus the UAA stop codon is convertible to the UGA or the UAG stop codon by a single base change. It seems likely that this correlation between codons and amino acids is related to the desirability of being able to change from one amino acid to another chemically similar amino acid by a simple mutation involving a single base change. Changing one base in the code word by mutation should be much easier to do than changing two bases is.

THERE IS A THREE-BASE PERIODICITY IN THE READING FRAMES THAT REFLECTS CODON USAGE

In 1981 Shepherd noticed that the coding sequences of most reading frames have a bias for the sequence RNY, where R is a purine, Y is a pyrimidine, and N is either a purine or a pyrimidine. The base preference at different positions of the codon should

TABLE 4

Overall Base Frequencies within Codons[a]

Species	Bases	First position		Middle position		Last Position	
		Number	Percentage	Number	Percentage	Number	Percentage
E. coli	U	47,896	14.83	94,197	29.16	84,896	26.28
	C	76,957	23.82	73,101	22.63	88,314	27.34
	A	82,256	25.46	97,703	30.24	58,421	18.08
	G	115,961	35.89	58,070	17.97	91,437	28.30
Yeast	U	52,458	21.76	67,076	27.82	81,286	33.71
	C	36,388	15.09	54,941	22.79	48,941	20.30
	A	77,761	32.25	84,646	35.11	67,270	27.90
	G	74,500	30.90	34,448	14.29	43,608	18.09
Primates	U	103,396	16.99	161,780	26.58	126,255	20.74
	C	144,695	23.77	138,338	22.73	197,816	32.50
	A	163,967	26.94	193,036	31.71	104,100	17.10
	G	196,649	32.31	115,554	18.98	180,533	29.66

[a]Adapted from Zhang, S. and Zubay, G. The Peculiar Nature of Codon Usage in Primates, In *Genetic Engineering,* Vol. 13. J. K. Setlow, Ed. New York: Plenum, 1991, p. 87.

generate a periodicity that runs through the average reading frames. Shepherd and others have used this device to detect reading frames in previously undesignated sequences by searching for the reading frame that gives rise to periodicities.

Zhang and Zubay's recent analysis of a more extensive database indicates both similarities and differences from the periodicity suggested by Shepherd. Their findings for *E. coli,* yeast, and primates are shown in Table 4. For these three species (and many others) a purine, A or G, is the most frequently occurring base found in the 5′ position. Also, the sum of the purines is greater than the sum of the pyrimdines at that position. At the middle base the hierarchy A > U > C > G is found. In the 3′ position the combination of A and U or G and C dominates in different species. These rules hold when speaking of the average mRNA within a given species. Particular messages of the species, or a subpopulation of the messages within a particular species, shows exceptions to these general rules. In primates, for example, an analysis of 1518 reading frames shows that 931 have A as the most frequently occuring middle base, 376 have U, 147 have C, and 64 have G. Incidentally, almost all proteins that contain U as the most common middle base are membrane-associated proteins, which usually contain large hydrophobic segments suitable for interaction with membrane lipids. This is consistent with our previous observation, that such codons represent hydrophobic amino acids. For the 5′ base, 1321 of the 1518 primate proteins analyzed contain either G or A as the dominant base; C is the most prominent 5′base in 184 cases and U is the most common 5′ base in only 13 cases. In only a very small number of proteins does the sum of the pyrimidines exceed the sum of the purines at the 5′ position. With rare exceptions the sum of G and C or of A and U at the 3′ position exceeds the sum of any other combination of two bases.

SYNONYM CODONS ARE USED TO VARYING EXTENTS IN A SPECIES-SPECIFIC MANNER

Although the code is almost universal, synonym codon usage varies considerably from species to species (Table 5). Furthermore, it seems likely that synonym codon usage is subject to different selection pressures in different species and also serves different functions. It is apparent that different species favor particular synonym codons in some cases and avoid particular synonym codons in other cases. Proteins that are made in large amounts show a strong bias favoring high-usage synonym codons while avoiding low-usage codons. The population of cognate tRNAs for the different codons is generally proportional to the codon usage.

THE CODE IS NOT QUITE UNIVERSAL

It was originally believed that exactly the same genetic code is utilized by all genetic systems. Initial experiments supported this conclusion, because it was found that some mammalian mRNAs could be faithfully translated in cell-free bacterial systems. We now know, however, that significant variations in the meaning of specific code words occur in many genetic systems. The exact scope and nature of these variations are still being discovered, but the current view is that the variations in genetic meaning reflect divergences from the standard or "universal" genetic code described earlier (see Chapter 16), instead of independent origins of the genetic code.

Many of the known variations in the genetic code are found in genes of mitochondria and chloroplasts. It is easy to see why these genetic systems might be more plastic, because they frequently encode only 10 to 20 proteins. The remainder of the organellar proteins are derived by importing nuclear gene products.

The tRNAs used to translate mitochondrial mRNAs are entirely derived from mitochondrial chromosomes. The first clue that something was unusual about the mitochondrial genetic code was that only 24 types of tRNA could be found. According to Crick's rules of wobble, 32 tRNAs minimally are required for the translation of all 61 codons. One possible solution to this conundrum was that mitochondrial genes do not utilize all 61 codons. Another possibility was that the wobble rules might be different for mitochondria. In fact, the latter is the case. Crick's original wobble rules stated that at least two different tRNAs are needed to translate four codon families. In all these cases (with the exception of the codon family specifying arginine—see the footnote to Table 6) single tRNAs have been found responsible for specifying all four code words, and these tRNAs all contain a U in the wobble position of their anticodons. It appears that the mitochondrial ribosome allows these tRNAs to pair with all four members of the codon family. The six mitochondrial tRNAs that pair with the normal two codons contain an altered U in the wobble position, and this modification causes them to conform to the normal wobble rules.

Another peculiarity of the mitochondrial code emerged from a study of yeast mitochondrial codon usage. The mitochondrial code has several differences in code word

TABLE 5
Comparison of Codon Usage in Different Species[a]

Amino acid	Codon	E. coli	Yeast	Primates
Leucine	UUA	11	27	6
	UUG	11	36	11
	CUU	10	11	11
	CUC	10	5	21
	CUA	3	13	7
	CUG	55	9	45
Isoleucine	AUU	47	50	33
	AUC	46	30	53
	AUA	7	20	13
Valine	GUU	29	44	16
	GUC	20	25	25
	GUA	17	16	9
	GUG	34	15	49
Serine	UCU	18	31	18
	UCC	17	18	24
	UCA	12	19	13
	UCG	14	8	6
	AGU	13	15	13
	AGC	26	9	26
Proline	CCU	15	29	27
	CCC	10	13	35
	CCA	19	49	26
	CCG	56	9	11
Threonine	ACU	20	38	23
	ACC	45	24	40
	ACA	12	26	25
	ACG	23	11	12
Alanine	GCU	19	44	28
	GCC	24	24	42
	GCA	22	24	20
	GCG	35	8	11
Arginine	CGU	45	17	9
	CGC	37	4	22
	CGA	5	5	10
	CGG	8	2	20
	AGA	4	54	19
	AGG	2	17	21
Glycine	GGU	38	60	15
	GGC	40	15	36
	GGA	9	15	24
	GGG	13	9	25

[a]Adapted from Zhang, S. and Zubay, G. The Peculiar Nature of Codon Usage in Primates, in *Genetic Engineering*, Vol. 13 . J. K. Setlow, Ed. New York: Plenum, 1991, p. 87.

TABLE 6
The Genetic Code of Yeast Mitochondria[a,b]

		Second position				
		U	C	A	G	
First position (5′ End)	U	UUU UUC } Phe AAG — UUA UUG } Leu AAU[c]	UCU UCC UCA UCG } Ser AGU	UAU UAC } Tyr AUG — UAA UAG } STOP	UGU UGC } Cys ACG — UGA UGG } Trp ACU[c]	U C A G
	C	CUU CUC CUA CUG } Thr GAU	CCU CCC CCA CCG } Pro GGU	CAU CAC } His GUG — CAA CAG } Gin GUU[c]	CGU CGC CGA CGG } Arg GCA[d]	U C A G
	A	AUU AUC } Ile UAG — AUA AUG } Met UAC[e]	ACU ACC ACA ACG } Thr UGU	AAU AAC } Asn UUG — AAA AAG } Lys UUU[c]	AGU AGC } Ser UCG — AGA AGG } Arg UCU[c]	U C A G
	G	GUU GUC GUA GUG } Val CAU	GCU GCC GCA GCG } Ala CGU	GAU GAC } Asp CUG — GAA GAG } Glu CUU[c]	GGU GGC GGA GGG } Gly CCU	U C A G

[a]From Bonitz, S. G. *et al*. Codon Recognition Rules in Yeast mitochondria, *Proc. Natl. Acad. Sci. USA* 77: 3167, 1980

[b]The codons (5′ → 3′) are at the left and the anticodons (3′ → 5′) are at the right in each box.

[c]Designates U in the 5′ position of the anticodon that carries the —$CH_2NH_2CH_2COOH$ grouping on the 5′ position of the pyrimidine.

[d]Although an Arg tRNA has been found in yeast mitochondria, the extent to which the CGN codons are used is not clear.

[e]Two tRNAs for methionine have been found. One is used in initiation and one is used for internal methionines.

meaning. The codons beginning with CU represent threonine instead of leucine, the AUA codon represents methionine instead of isoleucine, and the UGA codon represents tryptophan instead of a stop signal.

RULES REGARDING CODON–ANTICODON PAIRING ARE SPECIES SPECIFIC

The genetic code differs very little between species. By contrast, considerable differences occur between species in the anticodon translation system of tRNA, as evidenced by the mitochondrial tRNA system. In all systems, the bases in the anticodon–codon complex run antiparallel, as in standard double-helix pairing, and in all cases only Watson–Crick-like base pairing occurs between the first two bases in the codon and the opposing bases in the anticodon segment of the tRNA. However, for the 3′base in the codon, the rules for pairing vary with the species and with the base in question.

These rules are as follows. When the 5′ base in the anticodon is a G, it can pair with either a U or a C, and this is true in all organisms. When the 5′ base in the anticodon is an A, it can pair only with a U in the codon. However, it is rare that an A is found in this position of the anticodon. An A in this position in eukaryotes is usually deaminated to an inosine (I) base, which has an expanded capacity for pairing. Base A can pair only with U, but I can pair with U, C, or A. In eubacteria, deamination is limited to the conversion of the ACG sequence to an ICG anticodon. When C is the 5′ base in the anticodon, it pairs with G only. The only known exception to this rule is found in eubacteria, in which the C is covalently modified in the tRNA that recognizes the AUA codon. A U base in the 5′ position of the anticodon shows the greatest variability. An unmodified U can pair with any of the four bases in the 3′ position of the codon. This situation is reflected in the U-family boxes in mitochondria (see Table 6).

U can also be modified in various ways. In mitochondria, one type of modification permits a U to pair with either an A or a G in the two-codon sets. In eukaryotes, another type of U modification limits U to pairing with an A in the codon; and in eubacteria, a third type of U modification permits U pairing with U, A, or G but not with C in family boxes.

CHANGES IN CODON OR ANTICODON USE FOLLOW THE ROUTE WITH THE LEAST NUMBER OF STEPS

The simplest type of codon change involves stop codons, for example, the change of UGA from a stop codon to a codon for tryptophan. In most cellular genomes, tryptophan is represented only by the single codon UGG. In yeast mitochondria, tryptophan also is represented by the codon UGA. For this to happen, all UGA stop codons that are present must first be changed to some other stop codon, either UAA or UAG. The number of changes that have to be made are smaller for smaller genomes. That is why changes in code word meaning are much more frequent in a small genome such as is found in an organelle like the mitochondrion. Because the UGA stop codon can be changed to a UAA codon by one base change, whereas it would take two base changes to convert it to a UAG codon, it is much more likely for any existing UGA stop codons to be changed to UAA stop codons. Once this has happened, the UGA sequence is left open for other uses. The existing codon for tryptophan, AGG, can be converted into a UGA codon by a single base change. Wobble indicates that when the 3′ base of a codon is either an A or a G, it is read by a U in the anticodon. This means that no accompanying change in the tRNA needs to be made as long as U is the 3′ base in the anticodon of the tRNA.

As already indicated, differences in anticodons in different systems are much more common than differences in codons are. The mitochondrial tRNA systems became greatly simplified by a relaxation of the wobble rules. In the nuclear system, Crick's original wobble rules indicate that a U in the 5′ base of the anticodon can normally pair with an A or a G in the 3′ position of the anticodon. In the related situation in mitochondria, a U in the 5′ base of the anticodon can pair with any base in the 3′

position of the codon. As a result, in seven family boxes only a single tRNA, which contains an unmodified U in the 5' position of the anticodon, is needed to recognize all four codons in the family box.

EACH SYNTHASE RECOGNIZES A SPECIFIC AMINO ACID AND SPECIFIC REGIONS ON ITS COGNATE tRNA

Most cells contain only one synthase for each of the 20 amino acids specified by the genetic code. Each enzyme must be capable of recognizing its unique amino acid and one or more cognate tRNAs. Solving the puzzle of how synthases recognize tRNAs has been one of the major challenges in understanding the nature of translation of the genetic code itself. The identity relationship between tRNA and its synthase has, in fact, been called the "second genetic code."

Surprisingly, synthases fall into two categories with respect to the importance of the anticodon as a specificity element that they recognize. Some synthases (17 in E. coli) recognize the anticodon and some (3 in *E. coli*) do not. Thus a few tRNAs can be genetically altered in their anticodon without changing the specificity of their recognition by the cognate synthase. Other synthases apparently require an unaltered anticodon to recognize their cognate tRNAs. Four examples are shown in Fig. 3.

SUMMARY

Our goal in this chapter has been to trace the course of evolution of the genetic code, which entails the interaction between the anticodon and the codon.

1. It seems highly likely that the genetic code began as a relaxed singlet or doublet code in which onc or two bases in every triplet carried coding information. The alternative, a stringent singlet or doublet code in which every base carried coding information, would have been extremely difficult to convert into a triplet code.

2. There is a limited correlation between the hydrophobicity of the anticodon in the tRNA and the hydrophobicity of the attached amino acid. It is possible that the correlation has evolutionary significance, but there is no way of proving this.

3. Amino acids that are related by belonging to the same biosynthetic pathway share more code letters in common that would be expected on a statistical basis. This finding could be interpreted to mean that codon assignments evolved gradually as the number of amino acids encoded increased.

4. Another correlation that has been noted is that amino acids with similar side chains frequently have two code letters in common. This commonality could have the important function of facilitating mutational changes between closely related amino acids.

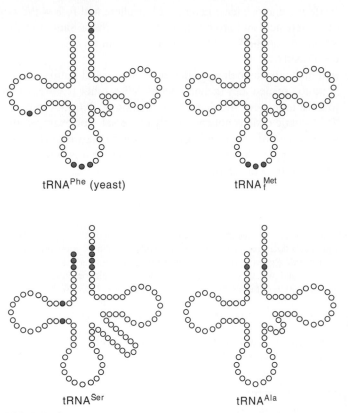

tRNA^{Phe} (yeast)   tRNA$_f^{Met}$

tRNA^{Ser} tRNA^{Ala}

FIGURE 3   Identity elements in four tRNAs. Each circle represents one nucleotide. Filled circles indicate nucleotides that serve as recognition elements to the appropriate aminoacyl-tRNA synthase. It is possible that other identity elements occur in these structures that are still to be discovered. (Reprinted with permission from Schulman, L. H. and Abelson, J. Recent Excitement in Understanding Transfer RNA Identity, *Science* 240:1591. June 17, 1988. Copyright 1988 American Association for the Advancement of Science.)

5. Different species emphasize different synonym codons. In all species, the majority of messages carry a purine in the 5' position. In the middle position, A is more abundant than U, which is more abundant than C, which is in turn more abundant than G. At the 3' position, either A and U dominate or G and C dominate in different species.

6. Organisms show small differences in the meaning of codons and much larger differences in the anticodons they use.

Problems

1. Serine has six codons in two separate locations in the codon chart. Which codons probably were the first to be used? Why?

2. If life were to originate in another biosphere and follow the same general lines of nucleic acid and protein development with the same four bases in DNA and RNA, do you think that the same codon–amino acid relationship would have developed?

3. Amino acids that belong to the same amino acid family are most likely to share common nucleotides in their codons. What other features of amino acids appear to have influenced the sharing of common nucleotides in their codons?

4. The genetic code is not completely universal. For example, different species show more differences in codon meaning in their mitochondria than in their nuclei. What is the most likely explanation for this and how does it reflect on the origin of these differences?

References

Doolittle, R. F. Microbial Genomes Opened Up, *Nature* (London) 392:338, 1998. (The complete genome sequences of a dozen bacteria, and a yeast, provide a wealth of information for tracing evolutionary networks. The article gives a brief guide to what has and can be learned.)

Kyripides, N.C. and Woese, C. R. Archaeal Translation Initiation Revised: The Initiation Factor 2 and Eukaryotic Initiation Factor 2B α-β-γ Subunit Families, *Proc. Natl. Acad. Sci. USA* 95:3726–2730, 1998.

Sueoka, N. Directional Mutation Pressure, Mutator Mutations and Dynamics, *J. Mol. Evol.* 37:137–153, 1993.

Yokobori, S., Ueda, T., and Watanabe, K. Codons AGA and AGG Are Read as Glycine in Ascidian Mitochondria, *J. Mol. Evol.* 36:1–8, 1993.

Prospectus

In science there are two times when it is especially useful to write a textbook. If a field of study has reached a denouement, then a textbook is appropriate if the subject is suitable for classroom use. It provides a way of reviewing a sequence of discoveries and gives a perspective of the discovery process. It is also useful to write a textbook in a field that has reached a point where one can allude to some important discoveries that have been made and to many others that clearly remain to be made. This is the case in the origin of life field. Considerable progress has been made, but much more must be done before we can proclaim a self-consistent scheme for a complete prebiotic pathway.

The function of this prospectus is to highlight some of the major accomplishments that have been made in research on the origins of life and to indicate what remains to be done.

RESEARCH ON THE ORIGIN OF LIFE IS A UNIQUE ENDEAVOUR

Compared with other areas of research we find that origin of life research is virtually unique with respect to its goals and methodology. Most problems in science have rather clear-cut goals such as determining how a certain process works at the molecular level or finding a cure for a disease. In origin of life research the goals are less clear. We hope to recreate the conditions that existed around the time of the origin of life on a microscale. Unlike much research that is in progress today we know that there is a solution to the problem because life is a reality. In the laboratory we may find what looks like a feasible pathway to the origin of life but we never can be totally sure that such is the actual pathway that was taken. We must be content to look for a feasible pathway or pathways to the origin of life. Currently, we are struggling to do just that. We have many observations and many speculations that we hope to lead us to a feasible pathway. Until we have found a feasible pathway it is pointless to worry unduly about whether or not it was the pathway used by Nature. We cannot rule out panspermia either, but as we have indicated (see Chapter 10) Earth's chemical composition and moderate climate, which appear to have persisted for practically 4 billion years, seem like an ideal milieu for the chemical processes necessary for the origin of life.

Our first concern must be to assess the approximate date when life originated and then to seek evidence for what the environment was like around that time. Hand-in-hand with this we must explore the likely chemistry that was essential for the origin of life and ask if it was consistent with Earth's composition and climate at that time. There is a going back and forth here between scientists in different disciplines. The geologists and meterologists must be the ultimate judges of what the climate was like in the distant past, the paleontologists are best trained to estimate the time of the origin of life, and the chemical biologists must determine where and how the reactions leading to the origin of life took place.

EARTH'S ATMOSPHERE HAS CHANGED A GREAT DEAL IN THE COURSE OF TIME

The composition of Earth's atmosphere is a paramount concern because the small-molecule components for the first living organisms are believed to have originated in Earth's atmosphere. Even today the main elements found in Earth's atmosphere—nitrogen, oxygen, carbon, and hydrogen—comprise four of the five most abundant elements found in living organisms (see Chapter 5). The fifth, phosphorus, must have come from the lithosphere. The current atmosphere is a strongly oxidizing one, and as a result of this it would have been totally unfit for the origin of life because most of the basic units in biomolecules exist in a reduced state.

There are excellent reasons for believing that the same elements that populate the atmosphere today were also abundant in the prebiotic atmosphere. However, they must have existed in different more reduced chemical forms. Consider the fact that the nebula from which the Solar System was formed was dominated by molecular hydrogen. Because molecular hydrogen is a strong reducing agent, this means that most of the elements in the solar nebula were in a highly reduced state. Following this line of thought led Urey and Miller in their pioneering experiments in prebiotic chemistry to work with a highly reduced mixture of gases: methane, ammonia, hydrogen, and water. If this mixture was activated by a spark discharge, a variety of organic compounds resulted that included a few amino acids that are found normally in proteins. Although the results of these experiments were stimulating and have been referred to many times, there was a serious flaw in the experimental design. The composition of Earth's prebiotic atmosphere was very unlikely to be as highly reduced. This is because most of the hydrogen found in the solar nebula must have been grossly under-represented in a small planet like Earth due to the low escape velocity of hydrogen and the high temperature of Earth at the time of accretion. Only a small fraction of the hydrogen found in the solar nebula would have been retained by Earth. That would have included the molecular hydrogen that was occluded in the solid mass of Earth and the hydrogen that had formed compounds with other atoms that were retained because of the higher escape velocities of the compounds.

The presence of hydrogen is but one characteristic of a reduced environment. In a strongly reduced mixture of gases most of the atoms tend to adopt a low formal valence state. If we look at Earth as a whole, there are a number of compounds that reflect the state of oxidation of Earth. Not all these influence the surface environment directly, but indirectly they often can have a most significant effect. Iron, which is the most abundant element, is one of these. Earth is about 30% by mass iron. Approximately 85% of this iron is in the zero valence state; this is present in the core. About 8% of the iron is present in the mantle as a mixture of the two commonly oxidized states of iron: ferrous ($+2$) and ferric ($+3$).

There is a controversy over whether the metallic iron presently found in the core and the iron salts in the ferrous and ferric states found in the mantle ever existed as one homogeneous mass. If they did and equilibrium between the different oxidation states of iron was attained, this would have led to the conversion of all the ferric iron into ferrous iron. In the text we state that core formation occurred within about 100 million years. For the reactions in question (see Chapter 5) that would seem like enough time to permit equilibrium to be reached between the various species of iron. If this was the case, then the iron of the mantle shortly after core formation must have been almost exclusively in the reduced ferrous form. This would have encouraged a number of compounds that pass through the mantle to Earth's surface to emerge in a relative reduced state. Conversely, the presence of appreciable ferric salts in the contemporary mantle encourages compounds passing through the mantle to adopt a more oxidized state.

Water was a compound that would have been altered if it passed through a mantle rich in ferrous salts. The result was a reaction of ferrous salts that led to the splitting of the water molecule. The oxygen of the water probably ended up in a ferric oxide while the hydrogen continued the journey to the surface where it was likely to escape into outer space. The continuous loss of hydrogen in this way for a very long time increased the overall redox potential of the planet. The gradual oxidation of the terrestrial ferrous salts also influenced the state of other oxidizable compounds that passed through the mantle. For example, methane was more likely to become partially oxidized as the ratio of ferric to ferrous salts in the mantle increased. The steady loss of hydrogen by this process could have led to a barren planet like Mars if it had not been for another major event. This event was the evolution of oxygenic photosynthesis, which resulted in the production of molecular oxygen (see Chapter 23). In the first few million years of photosynthesis, the oxygen that was produced was mostly consumed by oxidation reactions with ferrous iron. Because ferrous salts are very water soluble and ferric salts are very insoluble, that would have led to a great deal of precipitation from the oceans that resulted in banded iron formations. Once the ferrous iron and other reductants had been depleted by the oxygen, oxygen began to accumulate in the atmosphere. Due to its much higher escape velocity, the oxygen—unlike the hydrogen—was retained. The atmospheric oxygen also put an end to the hydrogen loss. Any hydrogen that reached the atmosphere was consumed by reaction with oxygen to form water. The advent of oxygen synthesis ultimately resulted in the conversion of Earth's atmosphere into a stable oxidizing one. Gradually Earth's atmosphere that had started out as a strongly reducing one became an oxidizing one. What we would like to know most of all is the composition of the atmosphere around the rather short interval of time that included the origin of life. Did it contain a good supply of the reduced molecules like CH_4 and H_2, which are required for the synthesis of the prebiotic molecules essential for RNA synthesis? This information is currently unavailable.

A REDUCING ATMOSPHERE IS NECESSARY FOR THE SYNTHESIS OF HCN AND CH$_2$O

One of the dogmas of prebiotic chemists is that HCN is essential for purine and possible pyrimidine synthesis and that CH_2O is essential for ribose synthesis. These energy-rich precursors can be synthesized in a reducing atmosphere from simpler compounds with the help of solar radiation (see Chapters 12 and 13). The synthesis of the purines from HCN may be controlled so that only adenine and hypoxanthine are produced; these are the two purines postulated to exist in an early all-purine form of RNA (see Chapter 14). In the case of ribose synthesis the main difficulty is the low yields. All the formaldehyde reacts quickly but only a small fraction of it goes into ribose. Even in the best system to date for the synthesis of neutral ribose, only about 5% of the formaldehyde is converted into ribose. Other aldopentoses account for about 15% of the starting material and the remainder of the by-products is still unspecified. Rate studies indicate that of the four straight-chain aldopentoses ribose is synthesized first. Despite this no practical way has been found yet to favor ribose synthesis over the other aldopentoses. In this regard the results of Alfred Eschenmoser should be mentioned. Eschenmoser has found that one equivalent of formaldehyde condenses with two equivalents of monophosphorylated glycolaldehyde to produce ribose $2',4'$-bisphosphate as the primary product. Although the phosphates are in the wrong place for making RNA, a modified pyranosyl RNA can be synthesized in which the phosphates are linked to the $2'$ and $4'$ carbons. The Eschenmoser reaction is not conducted under feasible prebiotic conditions, but despite this the results that he has obtained are fascinating and beg the question of whether primitive RNA had a somewhat different structure from the more familiar RNA. Be that as it may, we prefer to focus our efforts on the synthesis of ribose in a form that can be used to make standard nucleosides.

PHOSPHORYLATION OF NUCLEOSIDES REQUIRES PHOSPHATE ACTIVATION

It is well known that the vast majority of known phosphate on planet Earth is tied up in apatite or hydroxyapatite minerals. This does not take into account the phosphate that has found its way into living creatures. The apatites do not make good derivatives for phosphate activation because it is difficult to release the phosphate from the minerals. There are two possible alternatives to explain how phosphate became available for the essential prebiotic reactions of nucleotide synthesis and possibly nucleotide activation: (1) all the accessible phosphate for conversion into living organisms already has been converted into a special pool that is found mostly in contemporary living systems, recently deceased living things, or commercial phosphate fertilizers produced by specialized industrial processes; (2) the presence of so much hydroxylapatite in Nature may be due to its formation in biosystems. This is a problem for us to consider. Thus far I have not seen much discussion of the turnover in the phosphate pool.

Once phosphate is available in an inorganic or an organophosphate form, it can be used in conjunction with urea for converting nucleosides to nucleotides or it can be converted into trimetaphosphate, which is an excellent reagent for the polyphosphorylation of 5′-mononucleotides (see Chapter 13).

NUCLEOSIDE FORMATION PRESENTS A PROBLEM

The current concoction used for the prebiotic synthesis of nucleosides involves an aqueous mixture of neutral ribose, the purine hypoxanthine, and a divalent cation-containing salt. This suspension is heated to dryness at approximately 75°C. Resuspension and analysis indicates that only about 15% of the hypoxanthine is converted into nucleoside. The same procedure is considerably less effective with adenine and does not work at all with the commonly occurring pyrimidines cytosine or uridine.

A great deal of effort has been exerted to improve the yields of nucleoside without much success. This leads me to believe that there is something fundamentally wrong with the currently used conditions. Perhaps we should try using an activated phosphoriboside as in biochemistry (see Chapter 14). Such studies are in progress.

ACCOMPLISHMENTS AND PROBLEMS ASSOCIATED WITH POLYNUCLEOTIDE SYNTHESIS

It has been possible to link activated mononucleotides into polynucleotide chains using no more than a divalent cation catalyst (see Chapter 14). This can be done most efficiently on a clay surface without a template or it can be done with activated purine mononucleotides using a polypyrimidine template. In the prebiotic world the first polynucleotides to be formed could have been made on a clay surface without a template. Subsequently, these polynucleotides could have been used to make complementary chains. Synthesis does not go beyond the two-strand stage, meaning that true replication is not yet possible. Another limitation of the system is that the template strand in the prebiotic system must be pyrimidine rich.

It is conceivable that a useful ribozyme capable of catalyzing self-replication might be made by chance in an array of random sequence polynucleotides. Alternatively, the first RNAs made in the prebiotic environment could have been structurally modified in some way so that self-replication would be possible only with divalent cation catalysis. It has been suggested that an all-purine RNA might lead to this result (see Chapter 14).

SORTING OUT THE ENANTIOMERS: THE ORIGIN OF CHIRALITY

In ordinary chemistry, enantiomers usually are not distinguishable. In biochemistry it is almost always possible to distinguish between enantiomers because biochemical reactions routinely employ chiral catalysts. The key question that arises is, when and how was chirality introduced into biochemical reactions? Purines and pyrimidines are achiral so this is not an issue in their synthesis. Ribose is a chiral compound existing in two enantiomeric forms, D-ribose and L-ribose; only the former is found in naturally occurring RNAs so the enzymes involved in D-ribose synthesis show an absolute preference for the synthesis of the D-enantiomer.

Because chiral catalysts probably did not exist at the time when the first ribose was synthesized, it seems unlikely that asymmetrical synthesis of D-ribose occurred at this early prebiotic stage. The same should be true for synthesis of the first activated nucleotides. We would expect equal amounts of nucleotides to contain the two ribose enantiomers. Not until the level of polynucleotide synthesis is reached do we have a natural route for discriminating between nucleotides containing the two ribose enantiomers. This results from the fact that a polynucleotide invariably adopts a helical conformation that has one of two chiral forms. Thus a right-handed helix should favor nucleotides containing one sugar enantiomer and a left-handed helix should favor the other. It already is known from the work of Orgel and coworkers that a right-handed helix is more stable if the nucleotides contain D-ribose (see Chapter 14). This leaves little doubt that a left-handed helix would be preferentially adopted from a polynucleotide chain containing L-ribose. We are led to the conclusion that the two chiral forms of ribose could have occurred at the level of the polynucleotide. The prediction we would make is that there was a time when left-handed duplexes and right-handed duplexes were made in approximately equal amounts using L-ribose and D-ribose, respectively. Why then do we only see right-handed duplexes that contain D-ribose? Presumably there was a chance event that led to an altered ribozyme being synthesized in the D-series that was so effective that it completely outpaced the rival enantiomer leading to its extinction. This event marked the birth of chirality in the prebiotic world. From this point on most new RNAs contained nucleotides with D-ribose.

TRANSLATION MUST HAVE FOLLOWED RNA AND RIBOZYME SYNTHESIS

There was a time when it was thought that proteins must have appeared on the scene and evolved in lockstep with nucleic acids. This was prior to the discovery of ribozymes in the early 1980s (see Chapter 14). The notion of the coevolution of the two processes posed one of the greatest dilemmas in prebiotic chemistry. Unlike the ordering of nucleotides on a template by specific hydrogen bonding there was no reasonable scheme of hydrogen bonding that could be found for ordering amino acids on a polynucleotide template or for that matter on a polypeptide template. Furthermore, the scheme used in biochemistry for ordering amino acids in a polypeptide chain (see Chapter 16) requires that each amino acid be recognized not only by a specific transfer RNA (tRNA) but also by a specific enzyme that covalently links that amino acid to its cognate tRNA. For any system that relies exclusively on protein enzyme catalysis this says that you need an enzyme or instead a family of enzymes to make an enzyme or other type of protein. The discovery of ribozymes eliminated this viscious circle. Because ribozymes came first, they could supply all the catalytic functions required for protein synthesis. Eventually most of these functions were taken over by protein enzymes.

A scheme for the evolution of a translation system that assumes the aminoacyl tRNAs initially functioned in other capacities has been proposed (Chapter 17).

EARLY FUNCTIONS OF LIPIDS

Finally, we come to lipids, which probably were not required for the first living systems but probably became important before the advent of protein synthesis. In contemporary biochemical systems, lipids serve a variety of functions: energy storage particles, hormones, and membranes. It seems likely that the membrane function of lipids was the most important one in primitive systems. We speculate that membranes became essential prior to the evolution of protein synthesis as a device for keeping this complex multicomponent system from dispersing.

The prebiotic synthesis of lipids involves compounds and conditions that may be incompatible with those of other prebiotic molecules (see Chapter 19). The lipid membrane precursors could have been synthesized and assembled into stable membrane structures at some distance from the nests of living RNA molecules. Subsequently, they could have found their way by the oceans to the RNAs.

GLOSSARY

Absolute temperature The temperature scale given in units of degrees Kelvin (K). At zero degrees of this scale all atomic movement ceases.

Absolute zero A temperature of $-273°C$ (or 0 K) where all molecular motion stops.

Absorption line spectrum Dark lines superimposed on a continuous spectrum.

Acetal The product formed by the successive condensation of two alcohols with a single aldehyde. It contains two ether-linked oxygen atoms attached to a central carbon atom.

Acid In a chemical reaction, a substance that serves as a source of protons (hydrogen ions, H^+).

Acyl group The univalent group RCO—, where R is any organic group attached to one bond of the carbonyl group.

Acylation Introduction of the acyl group into a compound.

ADP Adenosine diphosphate, a nucleotide diphosphate (NDP) in which the base is adenine.

Adenine A purine base found in DNA and RNA.

Adenosine An adenine-containing nucleoside found in DNA and RNA.

Aerobe An organism that uses oxygen to generate the energy necessary to run its metabolism.

AICA Aminoimidazole carboxamide, an intermediate in the prebiotic synthesis of purines from AICN.

AICN Aminoimidazolecarbonitrile, a branchpoint compound in the prebiotic synthesis of purines from hydrogen cyanide.

Alcohol Any of a class of compounds having the general formula ROH, where R represents an alkyl group and —OH, a hydroxyl group.

Aldehyde Any of a class of organic compounds containing the group —CHO, which yields acids when oxidized and alcohols when reduced.

Algae Any of numerous chlorophyll-containing plant species ranging from microscopic unicellular forms to multicellular forms 100 ft or more long.

Aliphatic Describing organic compounds in which the carbon atoms form open chains.

Alkaline See pH.

Alkaline phosphatase An enzyme that removes terminal phosphates from nucleic acids, whether they are on the 5′ or 3′ ends.

Alkyl group Any of a series of univalent groups of the general formula C_nH_{2n+1}, derived from aliphatic hydrocarbons, for example, the methyl group or the ethyl group.

Allele   One of two or more alternative forms of a gene.

Alpha decay   The decay of a radioactive isotope by the emission of alpha particles.

Alpha particle   The nucleus of a helium atom, consisting of two protons and two neutrons.

Amine   Any of a class of compounds derived from ammonia by replacing one or more hydrogen atoms with organic groups.

Amino acid   The building block of proteins. All amino acids found in proteins contain a central atom to which a carboxyl group, an amino group, a hydrogen atom, and a variable group are attached.

Amino group   The univalent group $-NH_2$.

Aminoacyl-tRNA synthase   An enzyme that activates and attaches a specific amino acid to a specific tRNA molecule.

AMP   Adenosine 5′-monophosphate.

Anabolism   Metabolism concerned with biosynthetic reactions.

Anaerobe   An organism whose metabolism does not depend on the availability of oxygen. Strict anaerobes have little or no tolerance for the presence of oxygen.

Angstrom ($\mathring{A}$)   A unit of length equal to 10^{-10} m.

Angular velocity   The speed with which an object revolves about an axis.

Anion   A negatively charged atom or group of atoms.

Anticodon   A sequence of three bases on the tRNA that pair with three bases in the corresponding codon on the mRNA.

Antielectron   A positron.

Antimatter   Matter consisting of antiparticles such as antiprotons, antielectrons (positrons), and antineutrons.

Apparent brightness   The flux of star's light arriving at Earth.

Apparent magnitude   A measure of the brightness of light from a star or other object as measured from Earth.

Archaebacteria   One of the two kingdoms of bacteria, differing from eubacteria in the chemical composition of their cell walls, in the kinds of lipids in their membranes, and in certain other features.

Aromatic   Describing an organic compound containing an unsaturated ring of carbon atoms; so named because most such compounds have an agreeable odor.

Asteroid belt   A region between the orbits of Mars and Jupiter that encompasses the orbits of many asteroids.

Asthenosphere   The soft outermost layer of Earth located adjacent to the lithosphere.

Astronomical unit (AU)   The semimajor axis of Earth's orbit; the average distance between Earth and the sun.

Atmosphere   The volatiles that are situated above the lithosphere.

Atomic number   The number of protons in the nucleus of the atom of a particular element.

Atomic weight The mass of an atom in atomic mass units.

ATP Adenosine triphosphate, the chief source of chemical energy used to drive biochemical reactions.

Autoradiography The technique of exposing film in the presence of disintegrating radioactive particles. This is used to obtain information on the distribution of radioactivity in a gel of a thin cell section.

Autotroph An organism capable of manufacturing organic nutrients from inorganic raw materials.

Bacteriophage A bacterial virus; also called phage.

Basalt A complex mixture of minerals of which the major components are pyroxene (Fe, Mg) SiO, olivine (Fe, Mg)$_2$ SO$_4$, and a calcium–sodium feldspar (an aluminum silicate of calcium and sodium).

Base In a chemical reaction, a substance capable of accepting protons (hydrogen ions, H$^+$); in a nucleotide, the purine or pyrimidine that is characteristic of the molecule.

Base stacking The close packing of the planes of base pairs, commonly found in DNA and RNA structures.

Basic *See* pH.

Big Bang The unique explosion that triggered the origin of the Universe.

Biochemical pathway A series of enzyme-catalyzed reactions that results in the conversion of a precursor molecule into a product molecule.

Blackbody A perfect radiator that absorbs and reemits all radiation falling on it.

Blueshift A decrease in the wavelength of photons emitted by an approaching source of light.

Bond energy The enthalpy change that results when a chemical bond is broken.

Brønsted acid A molecule with the potential for donating a proton.

Carbohydrate Any of a class of organic compounds composed of carbon, hydrogen, and oxygen in a ratio of about two hydrogen atoms and one oxygen atom for each carbon. Examples include sugars, starches, and cellulose.

Carbonaceous chondrite A chondrite containing carbon.

Carbonyl group The group $\diagdown$C$=$O, occurring in acids, ketones, aldehydes, and their derivatives.

Carboxyl group The univalent group —COOH, present in and characteristic of organic acids.

Carcinogen A chemical that causes cancer.

Catabolism Metabolism concerned with degradation reactions.

Catalyst A substance that accelerates a chemical reaction without itself being permanently changed.

Cation A positively charged atom or group of atoms.

Chemoautotroph An organism that can manufacture organic nutrients from inorganic raw materials, using inorganic molecules instead of light as the source of energy.

Chirality   The quality of left- or right-handedness, which makes two stereoisomers mirror images of each other, due to the presence of an asymmetrical carbon atom in an amino acid molecule, or to the difference in helical twist adopted by a polypeptide.

Chlorophyll   The molecule responsible for the green coloring of leaves and plants; essential to the production of carbohydrates by photosynthesis.

Chloroplast   An organelle in plant cells that contains chlorophyll and that functions in photosynthesis.

Chondrite   A meteorite that contains small spherules known as chondrules.

Chondrule   Small spherules of solid matter found in chondrites that are believed to have formed around the time of the origin of the Solar System.

Chromatography   A procedure for separating chemically similar molecules. Segregation is usually carried out on paper or in glass columns with the help of different solvents.

Chromosome   In the cell nucleus, a refractile body that is the carrier of hereditary information; visible under the microscope just prior to mitosis. A few chromosomes also are found in mitochondria and chloroplasts.

Coding strand   A strand of the duplex DNA with the same nucleotide sequences as the RNA transcribed from that DNA.

Codominance   A situation in which the phenotype is the additive function of two alleles.

Codon   A sequence of three bases in an mRNA molecule that represents a particular amino acid.

Coenzyme   A nonprotein molecule that is required for the catalytic activity of an enzyme to proceed.

Comet   A small body of ice and dust in orbit about the sun.

Condensation reaction   A reaction between two or more organic molecules leading to the formation of a larger molecule with the elimination of a simple molecule such as water or alcohol.

Continental drift   The gradual movement of the continents over the surface of Earth due to plate tectonics.

Continuous spectrum   A spectrum of light over a range of wavelengths without any spectral lines.

Convection   The transfer of energy by moving currents of fluid or gas containing that energy.

Crust   The outermost layer of Earth or other Earthlike planet.

Cyanobacteria   A group of aerobic, photosynthetic eubacteria, formerly thought to be algae and still sometimes referred to as the "blue-green algae."

Cyclic compound   A compound whose structural formula contains a closed chain or ring of atoms.

Cytidine   A pyrimidine nucleoside found in DNA or RNA.

Cytochrome   Any of several oxidoreductase enzymes found in plants and animals, composed of iron, a protein, and a porphyrin, that catalyze intracellular oxidations.

Cytosine   A pyrimidine base found in DNA or RNA.

DAMN Diaminomaleonitrile, a key intermediate in the prebiotic synthesis of purines from hydrogen cyanide.

Deamination Removal of the amino group from a compound.

Decarboxylation Removal of the carboxyl group from a compound.

Degenerate code A code in which a given amino acid may be specified by more than one sequence of bases.

Deletion A mutation in which one or more base pairs are removed from a region of the chromosome.

De novo **pathway** A biochemical pathway that starts from elementary substrates and ends in the synthesis of a biochemical.

Denaturation The disruption of the folded structure of a nucleic acid or a protein molecule.

Density The mass per unit volume; frequently given in units of gram per cubic centimeter (g/cm^3).

Density-gradient centrifugation The separation of molecules by their density in a density gradient by centrifugation.

Deoxyribonuclease (DNase) An enzyme that degrades DNA.

Deoxyribose The five-carbon sugar found in DNA.

Diatom Any of numerous microscopic unicellular marine or freshwater algae having siliceous cell walls.

Differentiation A change in the form and pattern of gene expression of a cell as a result of growth and replication, usually during development.

Dimer A molecule composed of two identical, simpler molecules, a polymer derived from two identical monomers.

Diploid A cell that contains two chromosomes of each type.

Dipole–dipole interaction The favorable electrostatic interaction that results between two dipoles (molecules having uneven distribution of positive and negative charges).

Disulfide bond The single covalent bond formed between two sulfur atoms when two thiol (sulfur-containing) groups undergo oxidation.

DNA Deoxyribonucleic acid, a polynucleotide in which the sugar is deoxyribose; the main repository of an organism's genetic information.

Dominant Describing an allele that is expressed even in the heterozygous state.

Doppler effect The increase or decrease in frequency of a wave motion that is moving toward or away from the observer.

Double helix The normal structure formed by DNA, in which two helically twisted polynucleotide strands are held together by hydrogen bonding and base stacking.

Doublet code The code that would result if each amino acid in a protein polypeptide chain were encoded by a sequence of two bases in a nucleic acid.

Duplex Synonymous term for double helix.

Electron A subatomic particle with a mass of 9.1×10^{-28} g and a charge of -1 unit.

Electron volt   The energy acquired by an electron accelerated through an electric potential of 1 volt (V).

Electrophoresis   A technique for separation of molecules in solution in an electric field.

Elliptical orbit   A planar orbit that can be designated by the major and minor axis of an ellipse.

Emission line spectrum   A spectrum that contains emission lines.

Enantiomers   Two molecules whose structures are mirror images of each other.

Endonuclease   An enzyme that breaks a phosphodiester linkage at some point within the polynucleotide chain.

Enthalpy   The amount of energy in the substances of a system that can be converted to heat energy by a constant-pressure process.

Entropy   The randomness of a system.

Enzyme   A protein that contains a catalytic site for a biochemical reaction.

Equilibrium, chemical   The point at which the rate of forward reaction equals the rate of backward reaction.

Equilibrium, genetic   A term used by population geneticists to indicate that in the population of interest, the various alleles of a gene are present in the proportions calculated for a random distribution as predicted by the initial allele frequencies.

Equilibrium constant   A number that describes the relative concentration of reactants and products for a system in chemical equilibrium.

Erg   A unit of energy; the work done by a force of 1 dyn moving through a distance of 1 cm.

Escape velocity   The speed needed by one object to achieve a parabolic orbit away from a second object and thereby permanently move away from the second object.

Ester   A compound produced by the reaction between an acid and an alcohol with the elimination of a molecule of water.

Esterification   Conversion into an ester.

Ethyl group   The univalent group CH_3CH_2— or —CH_2CH_3, derived from ethane.

Eubacteria   One of the two kingdoms of bacteria, comprising the "true" bacteria, which are spherical or rod-shaped bacteria with rigid cell walls.

Eukaryote   A cell or organism that contains a membrane-bounded nucleus.

Eutectic freezing point   The temperature below which the two components of a mixture make the transition from the liquid to the solid phase in the same relative proportions as those in which they are present in the liquid phase.

Exon   A segment within a gene that carries part of the coding information for a protein.

Exonuclease   An enzyme that breaks a phosphodiester linkage at either one end or the other of a polynucleotide chain.

Facultative aerobe   An aerobic organism that can switch to anaerobic respiration in the absence of oxygen.

FAD   Flavin-adenine dinucleotide.

Fitness   A term used by evolutionists to indicate the capacity to survive and to reproduce.

FMN   Flavin mononucleotide.

Formose reaction   A process in which formaldehyde is used to generate a mixture of sugars in a solution containing calcium hydroxide.

Frameshift   A deviation from normal translation that is traceable to the fact that on the ribosome a message is translated three bases at a time. On rare occasions the chromosome skips over a base, so that subsequent three-base groupings (frames) begin at the wrong base. Another way to produce a frameshift is by an insertion or deletion mutation at some point in the message.

Free energy   The part of the energy of a system that is available for useful work.

Galaxy   A large assemblage of stars, nebulas, and interstellar gas and dust.

Gamma rays   The most energetic form of electromagnetic radiation.

Gauss   A unit of magnetic field strength.

GDP   Guanosine diphosphate.

Gene   A hereditary determinant that specifies a certain trait or traits and that behaves as a unit of inheritance occupying one (usually) contiguous region of a chromosome.

Gene duplication   The duplication of a particular gene within a chromosome.

Generalized acid   A molecule that is capable of donating a proton.

Generalized base   A molecule that is capable of accepting a proton.

Genetic code   The relationship between base sequences found in nucleic acids and amino acid sequences found in proteins.

Genetic equilibrium   *See* Equilibrium, genetic.

Genome   The total complement of DNA in the cells of a given organism; equivalent to the genes carried by a haploid set of its chromosomes.

Genotype   The combination of genes carried by an individual organism, including any that are not expressed by that individual.

Genus   (*pl.* **genera**)   The biological grouping that is just above that of the species. It is customary to print the genus name in italics with an initial capital. If a species name follows, it also appears in italic, but with no capital (e.g., *Escherichia coli*); in such cases, after its first appearance the genus name is often abbreviated to the initial capital letter (e.g., *E. coli*). The names of groupings higher than the genus are printed in roman type.

Gluon   A particle that is exchanged between quarks.

Glycolaldehyde   The molecule that forms between two formaldehyde molecules as a result of an aldol condensation reaction.

Glycolysis   The anaerobic catabolism of carbohydrates to pyruvic acid.

Glycosidic bond   The bond that forms between a sugar molecule and an alcohol or between two sugar molecules on elimination of a single water molecule.

Gram-negative bacteria   Bacteria that do not retain the violet dye when stained by Gram's method.

Gram-positive bacteria   Bacteria that retain the violet dye when stained by Gram's method, because they possess exposed cell walls.

Granite   A mixture of quartz (SiO_2) and a potassium feldspar (an aluminum silicate of potassium).

Greenhouse effect   The trapping of infrared radiation near a planet's surface by the planet's atmosphere.

Greenhouse gas   A gas that reflects low wavelength radiation back to Earth.

Ground state   The lowest electronic energy state of an atom or a molecule.

GTP   Guanosine 5'-triphosphate.

Guanine   A purine base found in DNA or RNA.

Guanosine   A purine nucleoside found in DNA or RNA.

Half-life   The time it takes for loss of half of a specific substance. This term is usually used with reference to an unstable isotope.

HAP   Hydroxylapatite, a mineral with the approximate composition $Ca_{10}(PO_4)_2(OH)_2$.

Haploid   A cell containing one set of chromosomes and correspondingly one allele for each gene.

Helicase   An enzyme that catalyzes the unwinding of a helically coiled molecule, usually a double-helix nucleic acid molecule.

Hematin   An iron-containing pigment that is produced in the decomposition of hemoglobin.

Heme   The deep red pigment found in hemoglobin, consisting of ferrous iron linked to a protoporphyrin.

Hemiacetal   Any of the class of compounds having the general formula RCH(OH)OR, where R is an organic group.

Hemimethylated   Describing DNA that has a methyl group on one strand but not on the comparable site on the other strand.

Hemoglobin   The protein that transports oxygen in the bloodstream.

Heterocyclic compound   A cyclic compound in which at least one of the ring members is not a carbon atom.

Heteroduplex   A duplex structure formed between two nucleic acid molecules that do not show perfect complementarity.

Heterotroph   An organism that cannot manufacture organic nutrients from inorganic raw materials and therefore must take them from the environment.

Heterozygous   Describing an organism that carries two different alleles for one gene.

Holoenzyme   An intact enzyme containing all its subunits and essential cofactors.

Homologous   Describing chromosomes that carry the same pattern of genes but not necessarily the same alleles.

Homozygous   Describing an organism that carries two identical alleles for a given gene.

Hubble constant (H_0)   The constant of proportionality in the relation between the recessional velocities of remote galaxies and their distances.

Hydrocarbon   Any of a class of compounds containing only carbon and hydrogen.

Hydrogen bond   The noncovalent linkage formed between an unshielded proton and a polar compound, the latter usually carrying an oxygen or a nitrogen atom with a fractional negative charge.

Hydrogen burning   The thermonuclear conversion of hydrogen into helium.

Hydrolase   An enzyme that catalyzes bond cleavage by the introduction of water.

Hydrolysis   The breaking apart of a molecule by addition of water.

Hydrophilic   Describing molecules that are polar and therefore readily enter into solution by forming hydrogen bonds with water or other polar molecules.

Hydrophobic   Describing molecules that have a low affinity for water because they are apolar.

Hydroxyacid   An organic acid containing both a carbonyl and a hydroxyl group.

Hydroxyl group   The univalent group —OH.

Hypoxanthine   A white, crystalline, almost water-soluble, alkaloidal purine derivative found in animal and vegetable tissues; used chiefly in biochemical research.

Igneous rock   Rock that is freshly formed from the molten state.

Imidazole   An odorless, crystalline, water-soluble, heterocyclic compound $C_3H_4N_2$, used in organic synthesis.

Imine   A compound containing the =NH group united with a nonacid group.

In vitro   Literally, means "in glass." This describes an experiment that is done in a test tube or other receptacle to simulate what happens in the whole organism *(in vivo)*.

Inosine   The deaminated derivative of adenosine.

Insertion   A mutation in which the normal arrangement of sequences is disturbed by the adding of one or more bases of DNA.

Intervening sequence   *See* Intron.

Intron   A segment of the nascent transcript that is removed by splicing; also refers to the corresponding region in the DNA; synonymous with intervening sequence.

Inversion   A mutation involving the reorientation of chromosomal segment in the opposite direction.

Isomerase   An enzyme that catalyzes reactions involving molecular rearrangements.

Isomers   Compounds that are composed of the same kinds and numbers of atoms but that differ in structural or spatial arrangement and therefore in properties.

Isotope   Any of several forms for the same chemical element whose nuclei all have the same number of protons but different numbers of neutrons.

Jovian planet   Jupiter or planets that resemble Jupiter, namely, Uranus, Neptune, or Saturn. These are the large outer planets that retain a good percentage of their volatiles because of their strong gravitational fields.

Ketone   Any of a class of organic compounds containing a carbonyl group attached to two organic groups.

Lagging-strand synthesis   DNA synthesis that occurs on the template strand that is oriented in the $5' \rightarrow 3'$ direction.

Leader region   The region of an mRNA between the 5′ end and the initiation codon for translation of the first polypeptide chain.

Leading-strand synthesis   DNA synthesis that occurs on the template strand that is oriented in the 3′ → 5′ direction.

Lethal gene   An allele that results in cell death.

Lewis acid   A substance capable of forming a covalent bond with a base by accepting a pair of electrons from it.

Ligase   An enzyme that catalyzes the linking together of two polymeric nucleic acids.

Ligation   Joining together of two nucleic acids by a phosphodiester linkage.

Lithosphere   The thin uppermost layer of Earth.

Luminosity   The rate at which electromagnetic radiation is emitted from a star or other object.

Luminosity class   A classification of a star of a given spectral type according to its luminosity.

Lyase   An enzyme that catalyzes reactions involving removal of a group to form a double bond or addition of a group to a double bond.

Mantle   That portion of Earth's solid mass that is located between Earth's liquid core and the uppermost layer, the lithosphere.

Matter-dominated universe   A universe in which the average density of matter exceeds the mass density of radiation.

Meiosis   The process involving two cell divisions in which diploid cells are converted into haploid cells of opposite mating types.

Messenger RNA (mRNA)   The template RNA carrying the message for protein synthesis.

Metabolism   Collectively, all the processes involved in the maintenance and propagation of a living system.

Meteor   The luminous phenomenon seen when a meteoroid enters Earth's atmosphere.

Meteorite   A small rocky object that travels aimlessly through space and is sometimes captured by the gravitational field of a larger celestial object.

Methanogen   A bacterial species that gets the energy it needs by methane production.

Methyl group   The univalent group —CH_3, derived from methane.

Methylation   Replacement of one or more hydrogen atoms in a compound with the methyl group.

Missense mutation   A single base pair replacement in which a codon that codes for one amino acid is replaced by a codon that codes for another amino acid.

Mitochondrion   A cellular organelle possessing its own chromosomes and whose main role is ATP synthesis by oxidative phosphorylation.

Mitosis   The process whereby replicated chromosomes segregate equally toward opposite regions of the cell prior to cell division.

Mobile genetic element   *See* Transposable genetic element.

Monomer   A molecule of low-molecular weight capable of reacting with identical or different molecules of low molecular weight to form a polymer.

Mutagenesis   The process of producing a mutation.

Mutation   The genetically inheritable alteration of a gene.

Mutation rate   The number of mutations per gene per unit time or per generation time.

NAD   The coenzyme nicotinamide adenine dinucleotide.

NAD$^+$ The oxidized form of NAD.

NADH   The reduced form of NAD.

NADPH   The reduced form of NAD that carries an extra phosphate group.

Nascent RNA   The initial transcripts of RNA, before any modification or processing.

Natural selection   The process that perpetuates those organisms that are best fitted to their environment. *See* Fitness.

NDP   Nucleoside diphosphate, a nucleotide that contains two phosphates linked to one another.

Neutral mutation   A mutation that has no effect on fitness.

Neutron   A subatomic particle with a mass of 1.67×10^{-24} g and zero electrical charge.

Neutron capture   The process whereby an atomic nucleus absorbs a neutron.

Nondegenerate code   A code in which each amino acid is uniquely specified by one nucleic acid sequence.

Nonsense mutation   A point mutation that results in the conversion of an amino acid codon into a nonsense or stop codon.

NTP   Nucleoside triphosphate, a nucleotide that carries three phosphates linked in series to one another.

Nucleic acid   A polymer of ribonucleotides or deoxyribonucleotides.

Nucleoside   An organic molecule containing a purine or pyrimidine base and a five-carbon sugar (ribose or deoxyribose).

Nucleotide   An organic molecule containing a purine or pyrimidine base, a five-carbon sugar (ribose or deoxyribose), and one or more phosphate groups.

Nucleus   In eukaryotic cells, the centrally located organelle that contains the chromosomes.

Oligonucleotide   A short polynucleotide.

Organelle   A subcellular self-contained body found inside the cell, for example, a mitochondrion or a chloroplast.

Orthophosphate   The usual pentavalent form of phosphorus found in nature, in which the phosphorus atom is linked to four oxygens.

Outgassing   Volcanic processes by which gases escape from a planet's crust into its atmosphere.

Oxidation   The removal or loss of electrons.

Oxidoreductase   An enzyme catalyzing the loss of electrons (oxidation) or the gain of electrons (reduction).

Oxyanion   A negatively charged molecule that contains oxygen.

Oxycation   A positively charged molecule that contains oxygen.

Ozone   The triatomic form of oxygen, O_3, produced when an electric spark is passed through air or oxygen.

P_i   The inorganic phosphate ion.

P wave   Longitudinal wave of compression and rarefaction where the seismic waves move in the direction of propagation of the wave motion.

Parsec (pc)   A unit of distance; 3.26 light years.

PEP   Phosphoenolpyruvate.

Peptide   An organic molecule in which a covalent bond is formed between the α-amino group of one amino acid and the α-carbonyl group of another amino acid, with the elimination of a water molecule.

pH   A measure of the concentration of hydrogen ions in an aqueous solution. The scale runs to 14. In a neutral solution the pH is 7; in an acidic solution it is less than 7, and in a basic or alkaline solution it is greater than 7.

Phage   *See* Bacteriophage.

Phenotype   The observable trait or traits resulting from the genotype and possibly other, non-heritable factors; also, an organism displaying the given trait or traits.

Phosphate   An oxide of phosphorus, usually used to refer to orthophosphate.

Phosphodiester bond   The bond that links two alcohols through a single phosphate molecule.

Phosphoryl group   The group $-O-\overset{\displaystyle O}{\underset{\displaystyle OH}{\overset{\|}{P}}}-OH$.

Phosphorylation   Introduction of the phosphoryl group into an organic compound.

Photon   A packet of light energy containing 1 quantum of energy.

Photoreactivation   DNA repair in which the damaged region is repaired with the help of light and an enzyme.

Photosynthesis   Autotrophic synthesis of organic materials that uses light as the source of energy.

Phylogeny   The evolutionary history of a group of species.

Phylum (*pl.* phyla)   A high-level category of biological classification embracing classes, orders, families, genera, and species that are perceived as being related in terms of some overarching criterion.

Planck constant (h)   The proportionality constant between the energy of a photon and its frequency (6.63×10^{-34} J s).

Planet   One of the nine sizable celestial bodies that rotates around the Sun.

Plasmid   A circular DNA duplex that replicates autonomously. Some plasmids can integrate into the host genome. Plasmids differ from viruses in that they never form infectious nucleoprotein particles.

Point mutation   A mutation involving a single base pair and therefore a single amino acid. *See* Deletion, Insertion, Missense mutation, Nonsense mutation.

Poly(A) A polynucleotide in which the base is adenine. Similar notation is used for those containing other bases, for example, poly(G) for guanine or poly(U) for uracil. The notation poly(dA) indicates the deoxy form of the sugar.

Polyamine A hydrocarbon containing more than two amine groups.

Polymer A compound of high-molecular weight derived either by the addition of many smaller molecules or by the condensation of many smaller molecules with the elimination of water, alcohol, or the like.

Polymerase An enzyme that catalyzes polymer formation.

Polynucleotide A chain structure containing nucleotides linked together by phosphodiester bonds. The polynucleotide chain has a directional sense, with a 5′ and a 3′ end.

Polynucleotide kinase An enzyme that catalyzes the addition of a phosphate group onto the 5′-OH terminus of a polynucleotide chain.

Polynucleotide ligase An enzyme that catalyzes polynucleotide synthesis.

Polynucleotide phosphorylase An enzyme that catalyzes the synthesis of polynucleotides from nucleotide diphosphates. The enzyme requires a primer but no template.

Polypeptide A linear polymer of amino acids held together by two peptide linkages. The polypeptide has a directional sense, with an amino and a carboxyl terminal end.

Polysaccharide A carbohydrate that is a polymer of simple sugars.

Polysome A complex between a messenger RNA and several ribosomes.

Porphyrin Any of a group of iron-free or magnesium-free pyrrole derivatives, occurring in all plant and animal protoplasm, formed by the decomposition of hematin or chlorophyll.

PP$_i$ The inorganic pyrophosphate ion.

Prebiotic Refers to something that existed prior to the origin of life.

Precursor A molecule that when reacted with appropriate reagents, gives a desired product; the molecule from which a desired product is immediately derived synthetically.

Primary structure In a polymer, the sequence of monomers and the covalent bonds.

Primase An enzyme that catalyzes the synthesis of certain primer RNAs. RNA polymerase also catalyzes the synthesis of some primer RNAs.

Primer RNA The starting point for DNA synthesis.

Primosome A multiprotein complex that catalyzes synthesis of RNA primer at various points along the DNA template.

Prochiral molecule A nonchiral molecule that may react with an enzyme so that two groups that have a mirror image relationship to each other are treated differently.

Proenzyme Any of various substances that may change into an enzyme as a result of some internal change; also called a zymogen.

Prokaryote A cell or organism that contains a single chromosome and no nucleus.

Promoter That region of the gene that signals RNA polymerase binding and the initiation of transcription.

Prosthetic group   Synonymous with coenzyme except that a prosthetic group is usually more firmly attached to the enzyme it serves.

Protease   Any enzyme that acts on proteins.

Proteins   The most complex macromolecules in the cell, composed of linear polymers called polypeptides.

Protein subunit   One of the polypeptide polymers of a complex multicomponent protein.

Protist   A relatively undifferentiated organism that can survive as a single cell.

Proton   A subatomic particle with a mass of 1.67×10^{-24} g and a charge of 1 unit.

Protonate   To add a proton.

Purine   A heterocyclic ring structure with varying side chains. The purines adenine and guanine are found in both DNA and RNA.

Pyrimidine   A heterocyclic six-membered ring structure with varying side-chain substituents. Cytosine is a pyrimidine found in DNA and RNA. Uracil is found only in RNA, and thymine is found only in DNA.

Pyrophosphate   A molecule formed by two phosphates in anhydride linkage.

Pyruvic acid (pyruvate)   The molecule formed as the end product of glycolysis.

Quark   One of several hypothetical particles presumed to be the internal constituents of certain heavy subatomic particles such as protons and neutrons.

Quasar   A starlike object with a very large redshift.

R process   A process of nuclear transformation initiated when certain isotopes rapidly capture large numbers of neutrons.

Racemic mixture   A mixture containing equal amounts of the two enantiomers of a given compound.

Radioisotope   An unstable isotope that can break up by releasing subatomic particles.

Rate constant   The proportionality constant in the mathematical expression relating rate of reaction to concentration of reactants.

Reading frame   Within a messenger RNA, any of the three possible ways of grouping bases for translation, which depends on which base serves as the starting point.

Recession velocity   The rate at which a star or galaxy is speeding away from another star, galaxy, or Earth.

Recessive   Describing an allele that expresses only when there is no other allele representing the same gene.

Recombination   The occurrence of progeny with different combinations of genes in individual chromosomes than are found in the parent chromosomes.

Red giant   A large, cool star of high luminosity.

Red supergiant   An extremely large, cool star of luminosity class I.

Redox balance   The balance between oxidation (loss of electrons) and reduction (gain of electrons).

Redshift   The shifting to longer wavelengths of the light from remote galaxies and quasars; the Doppler shift of light from a receding source.

Reducing atmosphere An atmosphere that has the chemical potential necessary for donating electrons to other substances.

Reduction The gain of electrons by a substance.

Regulatory enzyme An enzyme in which the catalytic site is subject to regulation by factors other than the enzyme substrate. The enzyme frequently contains a nonoverlapping site for binding the regulatory factor that affects the activity of the catalytic site.

Renaturation The process of returning a denatured structure to its original native structure, as when two single strands of DNA are reunited to form a regular duplex, or an unfolded polypeptide chain is returned to its normal folded three-dimensional structure.

Replica plating A technique in which an impression of a culture is taken from a master plate and transferred to a fresh plate.

Replicase An enzyme that catalyzes the synthesis of RNA from an RNA template.

Residue An atom or group of atoms considered as a group or part of a molecule.

Respiration The release of energy by oxidation of fuel molecules.

Retrovirus An animal RNA virus that replicates through a DNA intermediate.

Reverse transcriptase The enzyme that synthesizes DNA from an RNA template.

Reverse translation The process of reconstructing the mRNA sequence from the amino acid sequence of the polypeptide chain.

Revertant An organism with the wild-type phenotype that arose from a mutant.

Ribonuclease (RNase) An enzyme that cleaves bonds in RNA.

Ribose The five-carbon sugar found in RNA.

Ribosomal RNA (rRNA) The RNA part of the ribosome.

Ribosome The ribonucleoprotein particle on which protein synthesis takes place.

Ribozyme An RNA that functions as an enzyme; constitutes an exception to the rule that enzymes are proteins.

RNA Ribonucleic acid, a polynucleotide in which the sugar is ribose; functions in the synthesis of protein.

RNA primer A template-bound oligoribonucleotide that serves as the initiation point for DNA synthesis.

Rock Solid substance composed of metal oxides, metal silicates, and silicon dioxide in widely varying amounts.

S wave Secondary seismic wave that moves more slowly than a P wave in the direction of wave travel.

Salvage pathway A pathway that leads to the synthesis of nucleotides from partial breakdown products.

Satellite A celestial body that rotates about a planet in a defined orbit.

Secondary structure In a protein or a nucleic acid, any repetitive folded pattern that results from the interaction of the corresponding polymeric chains.

Sedimentary rock Rock that is formed by the slow deposition of solid silicate sediment that becomes compacted.

Seismograph   Instrument for recording waves passing through Earth that are produced by an earthquake.

Semiconservative replication   Duplication of DNA in which the daughter duplexes carry one old strand and one new strand.

Shine–Dalgarno sequence   A sequence of the bacterial mRNA, about nine nucleotides before the initiation site for translation, that is complementary to a sequence on the 3′ end of the 16S rRNA. This sequence simulates ribosomal binding to the mRNA and therefore encourages translation.

Sideral period   The orbital period of one object about another with respect to the stars.

Sigma factor   The protein subunit required for correct initiation by bacterial RNA polymerase. Once transcription has been initiated, the factor is released from the elongating enzyme.

Silicate   A complex in which the anion is silicate (SiO_3^{2-} or the like) and the cation is most frequently Fe^{2+} or Mg^{2+}.

Singlet code   A coding relationship in which a single base in a nucleic acid represents a single amino acid in a protein polypeptide chain.

Speciation   The process whereby organisms diverge into two or more species.

Species   The basic category of biological classification, composed of related individuals able to produce fertile offspring by mating among themselves but not by mating with members of other species.

Spliceosome   An enzyme that catalyzes splicing.

Splicing   Removal of a central segment and rejoining of the remaining segments. Splicing occurs in some DNAs and some RNAs.

SSB   Single-strand binding protein.

Stacking energy   The energy of interaction that favors the face-to-face packing of purine and pyrimidine base pairs.

Star   The most common luminous object found in the Universe. A star consists of hydrogen, helium, and minor amounts of other elements, and glows because of the nuclear reactions taking place in its core.

Start codon   The first codon in a translation reading frame of an mRNA.

Stellar hypothesis   Elements greater in mass than helium that are formed in the center of stars where the necessary temperatures and concentrations can be found.

Stereoisomers   Isomers that are nonsuperimposable mirror images of each other.

Stop codon   One of the three sequences, UAA, UAG, or UGA, that signal the termination of translation.

Stratosphere   Location above the atmosphere and extending to about 50 km above Earth's surface. Most of the protective ozone layer is contained in the stratosphere.

Subduction   Movement of surface layers of rock back to the mantle. This usually occurs when two tectonic plates collide.

Substrate   The molecule on which an enzyme acts.

Supernova   The explosion and fragmentation of a star that is sometimes observed when a star can no longer sustain nuclear reactions.

Survival value   A quantity that is measured in terms of fitness, meaning the capacity to survive and reproduce.

Synthase   An enzyme that catalyzes reactions joining two molecules together.

TCA cycle   The tricarboxylic acid cycle, also called the Krebs cycle of the citric acid cycle.

Template   A polynucleotide chain that serves as a surface for the absorption of monomers of a growing polymer and thereby dictates the sequence of the monomers in the growing chain.

Terminal transferase   The only known DNA polymerase that does not require a template. Terminal transferase adds any deoxynucleotidyl groups in the triphosphate form to the 3′-OH ends of existing polymers.

Terrestrial planet   Any of the planets Mercury, Venus, Earth, or Mars. Sometimes the Galilean satellites and Pluto also are included.

Tertiary structure   In a protein or nucleic acid, the final folded form of the polypeptide chain.

Thymidine   One of the four nucleosides found in DNA.

Thymidine kinase   An enzyme that converts the nucleoside thymidine into the corresponding 5′ nucleotide.

Thymine   A pyrimidine base found in DNA but not in RNA.

Trailer region   That part of an mRNA molecule between the end of the reading frame and the 3′ end of the mRNA.

Transfer RNA (tRNA)   Any of a family of low-molecular-weight RNAs that transfer amino acids from the cytoplasm to the template for protein synthesis on the ribosome.

Transferase   An enzyme that catalyzes the transfer of a molecular group from one molecule to another.

Transition state   The activated state in which a molecule is best suited to undergoing a chemical reaction.

Translation   The process of reading a messenger RNA sequence for the amino acid sequence it contains.

Translocation   A mutation in which a segment of chromosome is moved to another part of the same chromosome or to another chromosome.

Translocation reaction   A reaction in which a molecular grouping is removed from one location and transferred to another location.

Transpeptidation   A translocation reaction involving peptides.

Transposable genetic element   A segment of the genome that can move as a unit from one location on the genome to another, without any requirement for homology; also called a mobile or movable genetic element or "jumping gene."

Triplet code   A code in which three nucleotides in an mRNA code for one amino acid in a polypeptide chain.

Troposphere   The fraction of the atmosphere located in the first 10 to 15 km above the lithosphere that contains about 80% of the total atmosphere.

tRNA-nucleotidyl transferase   An enzyme that adds the CCA 3′-terminal grouping to a tRNA.

Trypsin   An enzyme that cleaves polypeptide chains at peptide linkages in which one of the amino acids is a basic amino acid.

Universal constant of gravitation (*G*)   The constant of proportionality in Newton's law of gravitation.

Universe   All matter, space, and energy of which we are aware.

Uracil   A pyrimidine base found in RNA in place of the thymine found in DNA.

Valence   The number of electron-pair bonds that an atom shares with other atoms. In inorganic chemistry the term often is used to mean oxidation state.

Valence shell   The outermost orbital that is occupied by electrons.

Viroid   A class of infectious RNAs of low molecular weight that is believed to cause plant diseases by interfering with normal splicing.

Virus   A nucleic acid–protein complex that can infect and replicate inside a specific host cell to make more virus particles.

Volatile   Substance that exists in the gaseous state.

Watson–Crick base pairs   The type of hydrogen-bonded base pairs found in DNA, or comparable base pairs found in RNA. The base pairs are A-T (adenine with thymine), G-C (guanine with cytosine), and A-U (adenine with uracil).

White dwarf   A low-mass star that has exhausted all its thermonuclear fuel and has contracted to a size roughly equal to the size of Earth.

Wien's law   A relationship between the temperature of a blackbody and the wavelength at which it emits the greatest intensity of radiation.

Wild type   The predominant allele for a particular gene found in nature.

Wobble hypothesis   A proposed explanation for base pairing that is not of the Watson–Crick type and that often occurs between the 3′ base in the codon and the 5′ base in the anticodon.

Xanthine   A crystalline nitrogenous compound, $C_5H_4N_4O_2$, related to uric acid; found in urine, blood, and certain animal tissues.

Zymogen   *See* Proenzyme.

APPENDIX

Astronomical Quantities

Astronomical unit	$1\text{AU} = 1.496 \times 10^{11}$ m
Light year	$1 \text{ ly} = 9.460 \times 10^{15}$ m $= 63,240$ AU
Parsec	$1 \text{ pc} = 3.086 \times 10^{16}$ m $= 3.262$ ly
Solar luminosity	$1 \text{ L}\odot = 3.90 \times 10^{26}$ w
Solar mass	$1 \text{ M}\odot = 1.989 \times 10^{30}$ kg
Solar radius	$1 \text{ R}\odot = 6.960 \times 10^{8}$ m

Physical Constants

Avogadro's number	$N_A = 6.02 \times 10^{23}$
Boltzmann constant	$k = 1.380 \times 10^{-23}$ J/K $= 8.617 \times 10^{-5}$ eV/K
Gravitation constant	$G = 6.668 \times 10^{-11}$ Nm^2/kg^2
Hubble constant	$H_O = 100$ h km/s/Mpc $= 2.1332$ h $\times 10^{-42}$ Gev
Mass of electron	$m_e = 9.108 \times 10^{-31}$ kg $= 0.5110$ Mev
Mass of H atom	$m_H = 1.673 \times 10^{-27}$ kg $= 938$ Mev
Neutrino mass	$\cong 10^{-35}$ kg
Planck constant	$h = 6.625 \times 10^{-34}$ J s $= 4.136 \times 10^{-15}$ eV
Speed of light	$c = 2.998 \times 10^{8}$ m/s
Stefan–Boltzmann constant	$\sigma = 5.669 \times 10^{-8}$ W/m^{-2}/K^{-4}

Answers to Problems

CHAPTER 1

1. According to the Hubble law

$$v_r = H_0 r,$$

where v_r is the recessional velocity, r is the distance, and H_0 is a constant called the Hubble constant. The exact value of the Hubble constant is a topic of heated debate. However, a commonly used value for this constant is

$$H_0 = 50/km/s/Mpc,$$

where Mpc is megaparsecs. We can calculate the time T_0 it will take for the galaxies to collide if they are run backward by using the simple equation,

$$T_0 = r/v_r.$$

By employing the Hubble law, to replace the velocity v_r in this equation we get

$$T_0 = 1/H_0 = 1/50 \text{ km/s/Mpc} = 20 \text{ billion years.}$$

This calculation assumes that the speed of expansion has not changed since the Big Bang.

2. The assumption that Earth is at the center of the Universe is a very special assumption that has no basis except possibly an ecclesiastical one.

3. A light year is a unit of distance, not time. There are no seconds in a light year.

4. A redshift caused by the expansion of the Universe is properly called a cosmological redshift. A Doppler redshift is caused by an object's motion through space.

5. The answer is no, because the quaser has a very high translational velocity that must be taken into account in any calculation of this sort.

6. A redshift z of 3 may be used to calculate a recessional velocity as long as an equation is used that corrects for the relativistic effect.

$$v_r/C = \frac{(z+1)^2 - 1}{(z+1)^2 + 1} = \frac{15}{17}.$$

7. According to Fig. 3 $r = 3.96 \times 10^9$. By using Hubble's law we calculate a recessional velocity of

$$v = H_0 r = 15 \times 3.96 \times 10^9 = 59 \times 10^9 \text{ LY.}$$

CHAPTER 2

1. The estimate might be off because iron is a heavy metal that might be more concentrated as one goes to the core. If the Sun were to blow up, one might get a better estimate.

2. For all elements more massive than iron the formation is an endothermic reaction. In contrast, all elements less massive than iron including iron are the result of exothermic reactions and therefore much more likely to occur.

3. The atomic number of Al is one greater than Mg. Therefore Al contains one more proton than Mg. In the β-decay process a neutron in the Mg nucleus could break down to a proton and an electron. This would lead to the formation of an Al nucleus. The reverse process could never occur.

4. A nitrogen nucleus ^{14}N could form the β decay of a ^{14}C nucleus.

5. The abundances of nuclei that form in the stars are a function of which types of nuclei can be formed and also of the stability of the nuclei that are formed. Only certain combinations of neutrons and protons form stable units. When two helium nuclei collide, they form a beryllium (^{8}Be) nucleus. The beryllium nucleus does not survive for long because it is very unstable. If another ^{4}He collides with the ^{8}Be nucleus before it decomposes, a carbon nucleus ^{12}C is formed. Similarly, an oxygen nucleus, ^{16}O, is formed when a carbon nucleus reacts with an additional He nucleus.

6. The author is not sure but he thinks so. Extra credit will be given to the student who comes up with a better answer, which would require some sort of documentation.

7. This would require two half-lives or 2000 s.

CHAPTER 3

1. The octet rule is satisfied for all the atoms in a phosphate anion. Neutral phosphate has five electrons in its outer shell. Neutral oxygen atoms have six electrons in their outer shell. If we add three electrons to the phosphate outer shell, then the phosphate has eight electrons in its outer shell. By sharing two of these electrons with each oxygen atom, the oxygen atoms all have eight electrons in their outer shell as well.

2. The electronic structure of the hydrogen molecule has only two electrons in its one an only shell. This is an exceptional case but it leads to a stable configuration.

3. Carbon has four electrons in its outer shell. In methane it shares these four electrons with four hydrogen atoms (see Fig. 5). The electrons constituting the shared pairs with hydrogen are more closely associated with the carbon than with the hydrogen so we may think of the carbon as possessing a formal charge of -4. In CO_2 carbon shared its four valence electrons with two oxygen atoms.

These electrons are more closely associated with the oxygen atoms than with the carbon, which gives the carbon a formal charge of 4.

4. Atoms with a low mass are generally more abundant and more able to form compounds with double bonds.

5. Molecular oxygen resonates between a molecule that obeys the octet rule and a diradical that does not.

$$\ddot{O}::\ddot{O} \longleftrightarrow :\dot{O}\cdot\cdot\dot{O}:$$

6. Ozone resonates between two energetically equivalent forms.

$$\overset{}{\ddot{O}}::\overset{-}{\ddot{O}}\cdot\cdot\overset{+}{\ddot{O}} \longleftrightarrow \overset{+}{\ddot{O}}\cdot\cdot\overset{-}{\ddot{O}}::\ddot{O}$$

CHAPTER 4

1. The immediate effects on the planets would be on the heat effect from solar radiation. At this distance Earth would cool to the point where life as we know it would probably be impossible. Jupiter would get warmer and this should affect the types of reactions seen on Jupiter.

2. Again the primary effect should be on temperature because of a distance change from the Sun. It might have made life on Earth impossible and life on Mars more likely. It is hard to say.

3. Kepler's third law can be derived from Newton's second law of motion and his universal law of gravitation as follows:

$$P^2 = kR^3 \tag{1}$$

where P is the period, k is a constant, and R is the distance between the planet and the Sun (this is Kepler's third law).

Newton's second law of motion states that

$$F = m \cdot a = m \cdot v^2/R. \tag{2}$$

Newton's universal law of gravitation states that

$$F = Gm \cdot M/R^2. \tag{3}$$

In Eqs. (2) and (3), F is for force, G is the universal gravitational constant, m is the mass of the planet, and M is the mass of the Sun.

At equilibrium where P and R are fixed, the forces in Eqs. (2) and (3) must balance. Therefore

$$mv^2/R = GmM/R^2. \tag{4}$$

From

$$4v^2 = GM/R. \tag{5}$$

The period P according to definition is the time it takes the planet to rotate around the Sun. Therefore

$$P = 2\pi R/v \tag{6}$$

and

$$P^2 = 4\pi^2 R^2/v^2. \tag{7}$$

By substituting for v in Eq. (7), we get

$$P = 4\pi^2 R^3/GM = kR^3. \tag{8}$$

4. Escape velocity from Earth is 11.2 km/s for all objects regardless of their mass.
5. The Sun's density (1.4 g/cm^3) is closest to that of Jupiter (1.3 g/cm^3) and Uranus (1.3 g/cm^3). In all three cases this low density can be explained by the large sizes of the planets and of the Sun that prevent much escape of molecules, even the light ones.
6. When we say the Sun is a second-generation star, we mean that the solar nebula that gave rise to the Sun and the bodies that surround it arose partly if not wholly from the supernova explosions of other stars. We are not certain of this but we can be certain of the fact that the Sun is only about 4.6 billion years old.
7. This is a simple calculation because one-eighth remaining indicates that the carbon-containing material has gone through about three half-lives. (Thus the answer is 5,730 $\times$ 3 = 17,190 years.)

CHAPTER 5

1. Mars lacks a substantial liquid iron core if any. This is necessary to power plate tectonics and concomitant volcanoes.
2. In the case of a collision between a basaltic plate and a granitic plate, the basaltic plate subducts because it is made of denser materials. When two granitic plates of approximately equal density collide, the situation is unpredictable. This could lead to a significant mountain range along the line of the collision.
3. This is suggested by the higher concentration of Mg in basalts (2.5%) that come from greater depths than the granites (0.7%) do.
4. This is a complex issue. We mention some of the major issues. The oxygen atmosphere that is mainly a product of photosynthesis would probably be missing. The surface water might be missing as well because there would be no O_2 to trap the H_2 that would otherwise react with the oxygen to reform water. The CO_2 would be very high if water were lacking, as is the case on Venus. The absence of atmospheric oxygen would eliminate the radiation protective ozone shield that would augment the rate of breakdown of water.

5. The answer is most certainly yes because the reduced molecules necessary for the origin of life would have been largely oxidized.

6. Mars has no tectonic plate activity.

7. Earth may not have regained a temperate climate because ice has a greater reflectivity than that of liquid water so that in the frozen state (other factors being equal) Earth would absorb less heat.

8. In prebiotic times iron in the mantle was mostly in the form of ferrous iron. As water was reduced the ferrous iron became oxidized to ferric iron while the hydrogen simultaneously produced gradually bubbled to the surface. In early times when there was little oxygen in the atmosphere, the hydrogen escaped from Earth's gravitational field. After photosynthesis got underway atmospheric oxygen rapidly rose once the ferrous iron had been converted to mostly ferric iron. The ferrous iron caused a significant delay in the rise of atmospheric oxygen but did not prevent it from happening in time.

9. Without oxygen in the atmosphere from photosynthesis, aerobic life would never have developed to any significant extent. Furthermore, without abundant atmospheric oxygen an ozone protective shield would never have developed, making life on the exposed surface of Earth difficult if not impossible. As mentioned earlier, the intense UV may have endangered the water supply. At best one can imagine a subterranean anaerobic form of life.

10. They are the only ones that have a liquid iron core *and* that rotate at a significant rate on their own axis.

CHAPTER 6

1. The emphasis on size as a criterion for distinguishing between prokaryotes and eukaryotes really is not adequate. This is underscored by a find of living giant sulfur bacteria (see *Science* 284: 493, 1999). Fossils of the ancestors of these organisms would probably have been mistakenly identified as eukaryotes.

2. Fossil-like structures that are too small to accommodate ribosomes should not be discounted because it is possible that living, membrane-enclosed cells existed prior to the advent of translation.

3. It is very unlikely that organisms differing only by asymmetry would have coexisted for long. Sooner or later one of these organisms would have taken the upper hand through a beneficial mutation that would have given it a selective advantage.

4. An organism composed of nothing but carbohydrates seems unlikely. The weaknesses of such an organism is that carbohydrates seem unlikely to be capable of self-replication.

5. A protein with a hydrophobic surface would most likely associate itself with other structures that bear hydrophobic surfaces. Most likely such proteins would be membrane-bound.

6. The simplest type of life imaginable would be an RNA molecule. A living system must be capable of replication and gradual change. Only a nucleic acid has these properties. Arguments raised in the text favor the evolution of RNA before DNA.

7. This has always puzzled me. Macromolecule is defined as a large molecule and lipids that are large molecules are quite common.

CHAPTER 7

1. It facilitates an arrangement in which processes dealing with one function can be localized.

2. A committed step is the first step directed at the synthesis of a particular compound. It is the most effective step to regulate so as to minimize waste and the accumulation of intermediates for which the need already has been satisfied. In times of sufficiency the committed step should be inhibited. In times of need the committed step should be allowed to function freely or even to accelerate.

3. If the conversion of chorismate to tryptophan takes place in five steps (see Fig. 5), there must be four intermediates between chorismate and tryptophan. If the third step is regulated, this would lead to an awkward situation in which an intermediate might accumulate. This would occur under conditions where the end product tryptophan was present in adequate amounts to satisfy metabolic needs. This would be a wasteful process and might even lead to complications if the metabolic intermediate at high concentrations were toxic.

4. Catabolism results in the breakdown of compounds that are no longer needed. Frequently, this breakdown is coupled to the production of starting materials, reducing power, and energy for the synthetic processes that are associated with anabolism.

5. Biosynthesis usually requires more ATP than catabolism requires. If a biosynthetic process requires a single ATP, the reverse conversion of catabolism should require less ATP, which in this case would eliminate the need for ATP altogether. The two processes, biosynthesis and catabolism, use different enzymes.

6. Cells maintain a favorable ratio of ATP to ADP meaning that this ratio is greater than would exist at equilibrium. Only under these conditions can ATP be useful in supplying free energy, when it breaks down.

7. A reaction at equilibrium such as the reaction to break down ATP to ADP and P_i and the reverse reaction of ADP and P_i reacting to form ADP still takes place at equilibrium but it supplies no free energy.

CHAPTER 8

1. A catalyst increases the rates of the forward and backward reactions equally. As a result a catalyst does not shift the equilibrium for a reaction but instead lessens the time it takes to go from a nonequilibrium situation to equilibrium.

2. In the trypsin, chymotrypsin, and elastase enzymes the N—H groups of Ser195 and Gly193 must be oriented so that they can form H-bonds simultaneously with the carbonyl group of the peptide that is being hydrolyzed.

3. A simple way to attack this problem is to notice that 87% for a first-order reaction is equivalent to three half-lives. The math follows (let P = product and R = reactant):

For a first-order reaction $dR/dt = -kR$ initial rate, where R is the amount of starting material at any time $dP/dt = k(R - P)$, where P is the amount of product at any given time.

By integrating from $t = 0$ we get $\ln(R/R - P) = kt$.

Let $R = 100$ at the start of the reaction. We know that when $P = 87$, then $t = 7$ min.

From this we can calculate k as $\ln(100/13) = k(7)$ so that $k = \ln 7.6/7 = 2.3 \times 0.88/7 = 28$ min^{-1}.

By using this value we can calculate that 50% of the reaction occurs at $t = 2/3 \log 2/0.28 = 2.4$ min.

4. The transition-state intermediate is the highest energy conformation attained by the enzyme–substrate complex.

5. (a) All three hydrolytic enzymes use the same mechanism.

(b) The main difference between these three enzymes is that they are specific for different peptides. This is because of the specificity pocket on the enzyme that determines which amino acid side chain binds at the active site.

CHAPTER 9

1. Proteins are informational molecules because they carry the information from the mRNA in the form of a related sequence of amino acids.

2. (a) . . . CUCUCU . . . Such a sequence should make one type of polypeptide. Because CUC is a leucine codon and UCU is a serine codon, it should be expected that an mRNA bearing this sequence would stimulate the synthesis of a copolymer carrying an alternating sequence of leucine and serine.

(b) . . . GAUGAU . . . This sequence should behave very differently from the previous one. Depending on the reading frame, it carries three different trinucleotides: GAU–asp; AUG–Met; and UGA, which is a stop codon. In the GAU reading frame a message with this sequence should stimulate the synthesis of a polymer of aspartic acid residues. In the AUG reading frame it should produce a polypeptide containing only methionine and in the third reading frame it would probably not produce anything.

3. Polypeptide chains are more flexible than nucleic acids and they contain a greater variety of side chains that should be able to function as catalysts.

4. Every nucleic acid is a potential catalyst because it is capable of serving as a template for the synthesis of a complementary chain.

5. It seems very unlikely that another biosphere that evolved without being influenced by our biosphere would have the same codon–amino acid relationship. We have raised the issue of hydrophobicity and hydrophilicity between anticodons and amino acids. There may be something to this but there is still a great deal of flexibility in making specific assignments. Beyond this it is not even possible to say that the same bases or the same nucleotides would be used in another biosphere.

CHAPTER 10

1. Nucleic acids probably came before proteins because of the synthesis problem. To synthesize a protein you need enzymes to link specific amino acids to specific tRNAs. You also need enzymes for the polymerization process but this is less serious. It seems highly likely that the first enzymes that were involved in protein synthesis were ribozymes. Ribozymes have the advantage that they can serve as a template for their own synthesis.

2. For Problem 2 see Problem 5 in Chapter 5.

3. Silicon probably was not used much for several reasons: (1) it does not adopt as many oxidation states as carbon does; (2) it does not exist in a volatile form where it could have interacted with the other atoms and molecules found in living systems; and (3) it cannot form double bonds with other atoms like carbon does.

4. It surely is possible to imagine living RNA molecules without much assistance coming from the other classes of molecules. This is discussed extensively in the text.

5. The oligomerization of most biopolymers could have been driven by a wet–dry cycle. In the dry state the water is a product of the reaction and therefore the removal of water by evaporation to the dry state would have favored oligomerization.

6. See Problem 3 in Chapter 6.

7. On the one hand, it is possible that compounds that were not precisely what was needed to make the first living organisms could have accumulated to a great extent before the correct compounds were made. In this case, catabolism could have played a crucial role in supplying the system with the most appropriate substrates for the origin of life. At the same time, it is possible to construct a scenario in which RNA precursors were synthesized from atomic components available in the prebiotic atmosphere. Ultimately, it seems likely that so much junk piled up that catabolism was very useful and probably essential to constructive evolution.

8. Life prior to cells seems possible. For example, simple RNA molecules loosely bound to clays in a sequestered region where precursors to more RNA could be made and accumulated seem possible.

9. See Problem 8 in Chapter 5.

10. No reaction can proceed if the thermodynamics is unfavorable.

11. Nucleic acids can store information, serve as a template for their own synthesis, function as enzymes, and undergo gradual change. Proteins make superb enzymes but they lack most of the potential for replication.

12. DNA is more stable than RNA is to degradation. Thus as the size of the genome increased, the premium on stability rose. Exactly at what stage DNA became the custodian of the transmissible genetic information is hard to say. However, I suspect it was before the system learned how to make protein.

13. See Problem 4 in Chapter 6.

14. See Problem 4 in Chapter 5.

15. See Problem 5 in Chapter 5.

16. Volcanoes are a major route from inner Earth to the surface for those gaseous and liquid compounds that in many cases are vital to the origin and maintenance of life.

CHAPTER 11

1. Carbohydrates constitute a major component of plant and bacterial cell walls. The sugar ribose is a vital component of nucleic acids. All metabolism is focused on the so-called metabolic pathways that involve the glycolytic pathway, the Krebs cycle, and the pentose phosphate pathway.

2. Close inspection of phosphoenolpyruvate (PEP) reveals that the central carbon is locked in an enol configuration. Removal of the phosphate from PEP allows the central carbon to return to the keto form with the release of 10 to 15 kcal of energy.

3. In the glyoxylate cycle the intention is to build multicarbon molecules from C-2s. The strategy that is used involves the TCA cycle for many of the reactions. Only two new enzymes are introduced and those steps in the TCA cycle that result in the loss of two carbons are bypassed.

4. (a) If a system were unable to synthesize malate synthase, it would lose the glyoxylate bypass.

   (b) If a system had a pyruvate carboxylase that was not activated by acetyl-CoA, it would overproduce acetyl-CoA at the expense of the Krebs cycle.

   (c) If the system possessed a pyruvate dehydrogenase that is inhibited by acetyl-CoA more strongly than the wild-type enzyme is, this would result in the underproduction of acetyl-CoA.

CHAPTER 12

1. It is unlikely that sugar synthesis occurred in the oceans because the formaldehyde that is needed to get such a reaction underway is unlikely to have reached the necessary concentrations in such a large body of water.

2. The strategy of taking a reaction mixture to dryness shifts the equilibrium for the reaction to the right only if water is a reaction product. In sugar synthesis water is not a product of the reaction.

3. Glycolaldehyde is a catalyst for the formose reaction even though it is also a participant. This is possible because glycolaldehyde is generated in larger quantities than it is consumed. In this reaction it is essentially impossible for two formaldehyde molecules to interact to form glycolaldehyde because this would involve an aldol condensation in which both donor and acceptor carbon atoms have a partial positive charge. Glycolaldehyde makes a good acceptor because the interacting carbon atom can accommodate a partial negative charge following the dissociation of a proton.

4. The second reaction in Problem 4 proceeds more readily for the reason already given in answering Problem 3.

5. One way of solving this is to follow the hint given in the problem and then to find the answer in Chapter 18.

6. (a) $2CH_2OHCHO \longrightarrow CH_2OHCHOHCHOHCHO$
         $CH_2OHCHO + CH_2O \longrightarrow CH_2OHCHOHCHO$
         $CH_2OHCHO + CH_2OHCHOHCHO \longrightarrow CH_2OHCO(CHOH)_3CH_2OH.$

(b) Dihydroxyacetone is the acceptor. Therefore glyceraldehyde is the donor.

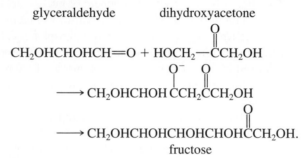

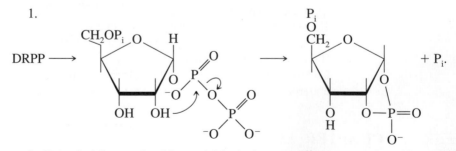
fructose

<hr>

CHAPTER 13

1.

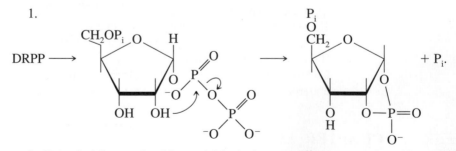

2. In pyrimidine nucleotide synthesis the heterocyclic ring compound is made first; this is followed by formation of a nucleotide from a phosphorylated sugar and a pyrimidine ring. In purine nucleotide synthesis the heterocyclic ring is built atom-by-atom on the 5'-phosphorylated sugar.

3. The prebiotic route to purine nucleotide synthesis most likely starts with purine ring synthesis from HCN. Because of their stability, the purines do not need to react immediately like the sugars do. The most likely sugar for condensation with the purine would be something like the one used in biochemical reactions with pRpp. How this sugar could be made under prebiotic conditions is not known.

4. The most likely source of ammonium formate is HCN, which can be converted into ammonium formate in two hydrolytic steps.

5. Three reactions in which water removal is essential include the AICA conversion to hypoxanthine, the AICN conversion to adenine, and the formation of phosphorimidazolide from nucleotide and imidazole.

6. The most sensitive site on neutral ribose is the C-1 carbon, which contains an aldehyde in the open chain form. In a nucleoside the C-1 group is involved in a reaction with a purine base that locks the sugar in a pentose conformation.

7. The amount of a compound present at any given time is a function of its rates of synthesis and of degradation. It is difficult to estimate the synthesis rate for purines and for sugars. However, due to its greater stability we predict that the purines are more likely to be present in excess.

8. No answer is given here because this problem should be solvable by consulting the text.

CHAPTER 14

1. All protein synthesis uses the same rRNA and tRNA. Therefore there is no reason to have a rapid turnover of these two species of RNA. By contrast, mRNAs encode specific proteins. When the need for a specific protein has been met, it makes sense to degrade its mRNA so that the nucleotides can be used to make other mRNAs.

2. Nucleoside triphosphates instead of nucleoside diphosphates are used in the synthesis of RNA to ensure the irreversibility of the reaction. After the internucleotide linkage is made, the pyrophosphate side product is hydrolyzed to P_i to ensure irreversibility.

3. If the rate of RNA synthesis is 3000 nucleotides per minute, then it should take $2,000,000/3,000 = 667$ min to synthesize an RNA with 2000 kbp.

4. See Problem 12, Chapter 10.

5. Clearly, it is easier to get RNA synthesis underway if the first RNAs contained only two bases instead of the canonical four. The purines adenine and hypoxanthine can form a stable double helix with a pair of hydrogen bonds between the adenine and the hypoxanthine. The synthesis of these two purines is simpler and more efficient than the synthesis of any other purines of the pyrimidines.

6. There is no one best answer to this question. However, the enzymes that one picks should be enzymes for which there was clearly a great need in primitive times. I would suggest that three of the most important enzymes that were

among the first ribozymes to be made are RNA polymerase, enzymes for synthesis of pRpp or a reasonable facsimile, and an enzyme for connecting pRpp and purine.

7. The activity is a hydrolytic activity embedded in the RNA that is cleaved. Because of the nature of this activity and its action, it does not pass the rigorous definition for an enzyme because an enzyme must be able to carry out the same reaction many times without itself undergoing any permanent change.

8. See Problem 4 in Chapter 6.

9. Let us say that in the array of oligonucleotides produced in the Ferris system there were occasional polymerase ribozymes produced. For these to be useful in the evolutionary process they must be able to catalyze their own synthesis.

10. The affinity of the protein for cytosine is bolstered by the nonspecific electrostatic affinity between the positively charged proton and the negatively charged nucleic acid polymer.

11. Folding is crucial because it brings functional groups together and creates sites for binding specific groups. With only the two purine bases in a random sequence half of the bases from two chains would be expected to make H-bonded base pairs. For a two purine–two pyrimidine sequence only 1 in 4 would be expected to make H-bonded base pairs.

12. There is no one correct answer to this problem. In each example picked it should be explained why the particular enzyme chosen would have been especially useful in the prebiotic world.

CHAPTER 15

1. The glutamate used in transaminations is regenerated in one of two ways: by a transamination reaction using ammonia or by a transamination reaction involving the amide nitrogen of glutamine.

2. It is hard to imagine a simpler one-step reaction for this process. Furthermore, the supply of the D-amino acid may be controlled by simple feedback inhibition of the racemase.

3. Tryptophan is synthesized by a chain of five enzymes. Feedback inhibition of the first enzyme in the tryptophan pathway is used to regulate the synthesis of tryptophan.

4. In the porphyrin pathway there is a tRNA that forms a complex with glutamate that is an activated intermediate in porphyrin synthesis. Similarly, there is a tRNA that forms a complex with glycine that serves as an active intermediate in bacterial cell wall synthesis.

5. No. It usually is the other way around.

6. In the biochemical world glycerate-3-phosphate is converted into serine in three enzymatically catalyzed steps. In the prebiotic world glycolaldehyde is converted into serine with the help of the Strecker synthesis reaction.

7. Imidazole or imidazole derivatives have been implicated at many points in prebiotic synthesis: the synthesis of adenine, of hypoxanthine, of histidine, and of imidazole-activated mononucleotides.

8. When polypeptide chains fold, they usually fold so that the hydrophobic regions are buried in the interior of the protein where they are not exposed to water. If all or most of the amino acid side chains were hydroxylated, there would be no regions remaining that could be described as hydrophobic.

9. Based on their use as precursors in nucleic acid synthesis, the amino acids aspartic acid, glutamine, and glycine would be most useful. Serine and glycine are the easiest amino acids to make from carbohydrate precursors and on this basis might have been used a great deal in prebiotic and early biotic systems.

CHAPTER 16

1. Four equivalents of ATP are required for each peptide bond that is synthesized in translation. This is much greater (about four times) the bond energy of the peptide linkage. A good deal of this energy is needed to ensure accurate translation.

2. The following amino acids and ratios would be incorporated:
   lys phe asn met stop leu tyr ile
   AAA UUU AAU AUA UAA UUA UAU AUU.

3. The anticodon of this tRNA is AGU.

4. The energy for actually making the peptide bond in translation comes from the energy stored in the aminoacyl linkage to tRNA. This energy originated from ATP.

5. GAA is a codon for Glu. The effects of the altered amino acids would be
   AGG lys, gly, glu
   CCC gln, ala, val
   UUU stop, val asp.

CHAPTER 17

1. You should be able to do this by referring to the text.

2. See Problem 3 in Chapter 9.

3. There is no best answer for this problem.

4. One might suspect that polypeptides with repeating amino acid sequences that served structural functions were important prior to the development of the translation system.

5. We still have structural polypeptides that have repetitious amino acid sequences for the most part. These are found in collagen: glycine, proline, and hydroxyproline; and in silk proteins in which three amino acids dominate: glycine, ala-

nine, and serine. In silk proteins every other residue is a glycine. Although there are none known now, it would be possible to make a repeating unit that would favor the formation of long-chain α helices (see text).

CHAPTER 18

1. The reactions numbered four, five, and six are very similar to the reactions in Chapter 11, Fig. 4 involving the conversion of succinate to fumarate, fumarate to malate, and oxaloacetate in reverse. In both cases we start from a keto compound that gets reduced to an alcohol, which gets dehydrated to an olefin that gets reduced to a carbon carrying two hydrogens. Both reductions in the lipid example (see Fig. 5) involve the coenzyme NADPH while only the first reduction in the Krebs (TCA) cycle reaction run in reverse uses NADH. The second reduction in the Krebs cycle uses a flavin coenzyme ($FADH_2$). The chemistry of these reactions is very similar but the remaining molecules have quite different structures.

2. β Oxidation is best described as a spiral process because the reaction in every turn of the spiral is virtually identical, but the substrate in each turn involves a substrate that has two fewer carbons.

3. A thioesterase activity hydrolyzes palmitate from the fatty acid synthesis enzyme and thereby terminates the growth process.

4. The order is methane $\rightarrow$ methanol $\rightarrow$ formaldehyde $\rightarrow$ formic acid $\rightarrow$ carbon dioxide.

5. The process of extracting chemical energy from a fatty acid involves a complex series of catalyzed reactions.

6. The most likely reactions of fatty acids or lipids to become important in primitive times probably involved the formation of membranes and of course the prebiotic reactions leading to these compounds.

7. The energy comes mainly from the Sun, and secondarily from the heat from inner Earth and the minerals that are found in a reduced form.

8. ΔG is about -5 kcal (see Fig. 11 and also see Problem 5 in Chapter 12).

CHAPTER 19

1. Phospholipids are amphipathic. The fatty acid portion is hydrophobic while the end containing the negatively charged phosphate is hydrophilic. Triacylglycerols are hydrophobic throughout.

2. In most globular proteins, nonpolar amino acid side chains are buried on the interior of the protein where they are isolated from solvent water. Polar amino acids are located on the surface where they can interact with solvent water. For intergal membrane proteins, the situation is the opposite because the

exterior of the protein is usually in direct contact with the hydrophobic surface of lipids.

3. Peripheral membrane proteins have an affinity for the membrane surface because of the electrostatic forces between charged portions of the proteins and the lipids. These proteins may be selectively dissociated from the membrane by salts and chelating agents. Integral membrane proteins are firmly bound by the hydrophobic forces of the membrane lipids. Their dissociation requires that the membrane structure be disrupted, for example, by detergents.

4. The revolving door comparison applies best to passive transport where no energy is required for the transport. In the case of active transport against a concentration gradient, energy is required and the membrane transport components are asymmetrically oriented. Signals also may be passed across membranes without the passage of any material substance.

5. Yes. Semipermeability is essential. See Problem 7.

6. It seems likely that membranes were needed prior to the evolution of translation systems because of the necessity to sequester so many components.

7. Membranes are composed primarily of lipids and proteins. The membrane must be designed so that it retains vital cellular components, permits the entry of needed commodities, and encourages the exit of undersirable reaction products. All this is believed to have required a good deal of evolution even in primitive membranes.

CHAPTER 20

1. A 1:1 sheet clay has two distinct layers. Usually 1:1 clays make tight complexes in which each surface is strongly bonded by an adjacent surface. A 2:1 sheet clay is arranged like a sandwich in which the buried sheet is tightly complexed with identical sheets on either side. The outer sheets are only loosely attracted to like surfaces so that the individual sandwiches can be easily separated by water or other solvents. These exposed surfaces on the sandwich have a potential for binding a wide range of small molecules.

2. The clay montmorillonite binds RNA on its positively charged edges. Although this binding is quite strong, the RNA can move freely as long as it stays on the edges.

3. Three possible roles of clays in the origin of life are providing (1) a binding surface for RNA molecules before the advent of membranes, (2) a place where the RNA molecules and possibly other polymers could be formed, and (3) catalytic sites for prebiotic synthesis.

4. Because clays are believed to have formed from aqueous suspensions, such a finding would suggest that the meteorite was a fragment from a much larger celestial body where water played a role in its formation.

5. It might not be practical to form a clay in the laboratory. In nature, clay formation is believed to have been a very slow process with the necessary monomers

filtering from rain-soaked dry lands into the oceans where very slow reactions of clay monomers with the growing clay polymers took place.

6. (a) Kaolinite is not a good binder because it is a 1:1 clay (see Problem 1 for an explanation as to why 1:1 clays do not make good binders).

    (b) Affinities for HAP are dominated by ionic interactions between negatively charged small molecules and positively charged sites in the mineral.

    (c) Binding to montmorillonite presents a more complex picture. These clay particles have a high affinity for organic ring structures that is augmented if they are positively charged. The binding probably takes place on the negatively charged faces of these sheetlike clay particles. Additional binding sites on the edges of these sheets have a moderate affinity for negatively charged molecules (see answer to Problem 2).

CHAPTER 21

1. Vertical evolution is the classical type of evolution in which species gradually change and different variants of that species are favored mating types. Horizontal evolution involves the very rare exchange of genetic material between organisms that do not exchange genes by the normal mating process.

2. Connecting two segments of land by an isthmus would lead to a fierce competition in which only the fittest would survive. Thus, making such a land connection would result in a decrease in the number of species.

3. Approximately 2 billion years ago the oxygen content of the atmosphere began to rise rapidly. The presence of atmospheric oxygen was documented by its reaction with ferrous salts, which were very abundant in the early oceans. These salts that were very soluble became oxidized by exposure to atmospheric oxygen into ferrous salts that are very insoluble. The insoluble ferric salts formed sedimentary rock layers that could be accurately dated by isotopic methods.

4. This is a difficult distinction to make. Currently, it is believed that prokaryotes came more than a billion years before eukaryotes. In general, prokaryotes are smaller and show less branched structure than eukaryotes.

5. Selection processes that stem from human-made directed mutagenesis should result in species that are of immediate use to man, while the Darwinian selection processes were not as focused. The long-range consequences of this shift in selection processes are most difficult to predict.

CHAPTER 22

1. The TCA cycle operates in the forward direction at least down to α-ketoglutarate (which is the starting point for the four amino acids, glu, gln, arg, and pro). Apparently it does not continue on to oxaloacetate or a host of additional [^{14}C]amino acids would be found.

2. *Chlorobium limicola* uses the entire TCA cycle in the reductive direction. The function of the cycle is to fix carbon and other substrates. Because *Chlorobium* is a photosynthetic organism, it most likely gets its energy from sunlight.

3. α-Ketoglutarate is the starting point for the synthesis of several amino acids. Oxaloacetate is the starting point for the synthesis of aspartate and a host of other amino acids. Succinoyl-CoA is the starting point for heme synthesis.

4. Pyruvate is not an intermediate in the TCA cycle but it normally supplies two carbons per pyruvate to the TCA cycle.

5. It is useful when the latter half of the TCA cycle operates in the counterclockwise direction because it makes it thermodynamically feasible to do so.

6. He considered the reaction wasteful because pyruvate should be most useful for biosynthesis if it is not entirely consumed by the TCA cycle.

CHAPTER 23

1. Yes. Light could have been used to warm the planet and to supply the energy necessary for the origin and evolution of life but direct contact with light is not necessary.

2. See Problem 3, Chapter 21.

3. Photosynthetic reactions were probably extremely important in the atmospheric formation of HCN and CH_2O. These two molecules were the starting point in the prebiotic world for the synthesis of purines and ribose, respectively.

4. Because a pathway such as that to chlorophyll probably evolved slowly in stages, it stands to reason that the early intermediates would have been synthesized first. If they could have functioned in a useful capacity, then it seems likely that their use would have preceded that of later intermediates in this pathway.

5. In all PS-2 reaction centers, the initial electron acceptor is a pheophytin or bacteriopheophytin and the second and third acceptors are quinones. This group includes the reaction centers of photosystem II in chloroplasts and cyanobacteria, and the reaction centers of purple bacteria and green nonsulfur bacteria.

6. The distribution of photosystems indicated by the phylogenetic tree in Fig. 17 suggests a pattern of horizontal evolution. Thus the phylogenetic tree based on the 16S-rRNA sequences shows a very different pattern of evolution.

7. The use of water as the predominant source of electrons presents enormous technical problems, therefore, as long as more convenient sources of electrons were readily available, there was no need to evolve this more complex system.

8. The earliest photosystems without Mg had a substantial negative charge (see Fig. 20) making them much more water soluble. More advanced porphyrin complexes with Mg divalent cations had very little net charge, making them more amenable to forming complexes with hydrophobic membranes.

CHAPTER 24

1. The six serine codons are UCU, UCC, UCA, UCG, AGU, and AGC. It seems highly likely that the first four codons evolved independently of the last two codons but it is most difficult to decide which came first.

2. No. There might be some similarity but codon assignment looks like an event that could be equally well resolved in more than one way.

3. Amino acids with similar side chains might undergo more conversions than amino acids with very different side chains. The sharing of two out of three codons would facilitate the process of interconversion.

4. Mitochondria have far fewer codons because of the much smaller amount of mRNA. Therefore it is easier for them to make small changes in their use of codons. If a mitochondrial protein were to be present in the cytosol, it could cause problems. A mitochondrial message in the cytosol would probably not be translated into a functional protein for just such reasons. The use of small differences in the nuclear and mitochondrial codes would help in this type of control.

INDEX

Page numbers followed by italic f refer to illustrations, and page numbers followed by italic t refer to tables.